D Smitley
W9-CGL-231

Biology of Leaf Beetles

Biology of Leaf Beetles

P. JOLIVET AND K.K. VERMA

Intercept

Andover

British Library Cataloguing in Publication Data
Biology of Leaf Beetles.—

A CIP catalogue record for this book is available from the British Library
ISBN 1–898298–86–6

Published in October 2002 by Intercept Limited,
PO Box 716, Andover, Hampshire SP10 1YG, UK
Email: intercept@andover.co.uk
Website: www.intercept.co.uk

Cover photograph: *Timarcha tenebricosa* Fabricius on host plant *Galium mollugo* Linne (by Pierre Jolivet).

Typeset in Times by
Ann Buchan (Typesetters), Shepperton, Middlesex.
Printed by Athenaeum Printers Limited

Contents

Dedication

Dedicated to all chrysomelidologists, past and present, whose work and contributions have led to the current knowledge of leaf beetles.

About the Authors

Pierre Jolivet, DSc

Dr. Jolivet has written many research papers and books on Chrysomelidae. His areas of special interest include: the biology of *Timarcha*, food plants of chrysomelids, and ants and plants. He has co-edited three important books on chrysomelidology: *Biology of Chrysomelidae* (1988); *Novel Aspects of the Biology of Chrysomelidae* (1994); and *Chrysomelidae Biology* in three volumes (1996).

In recognition of Dr. Jolivet's valuable contribution to the subject, the Fifth International Symposium of Chrysomelidae, held in August 2000 in Brazil, was named after him.

K.K. Verma, MSc, PhD

Dr. Krishna Kumar Verma taught zoology and entomology for over 35 years, both at undergraduate and postgraduate levels in M.P. Government colleges, India. He has worked for a long time in the field of the functional morphology and physiology of Chrysomelidae. He has to his credit a number of outstanding papers published in both Indian and international journals. He has contributed to *Novel Aspects of the Biology of Chrysomelidae* (1994) (edited by P. Jolivet, M.L. Cox and E. Petitpierre) and *Chrysomelidae Biology* Volume 1 (1996) (edited by P. Jolivet and M.L. Cox).

Preface

The study of phytophagous beetles is a vast subject. Here, two specialists with similar concepts, though different experiences, have joined forces to produce a very successful piece of work. The authors complement each other so well that their writing seems to be the work of one person.

When we consider a book dealing with a group of organisms, a number of questions arise. These concern taxonomy, anatomy, palaeontology, evolution, reproduction, ecology, food and so on. When a subject is approached with the curiosity of discovery, the results can be especially interesting and, here, the authors, both of whom have had long experience of outstanding research, have taken such an approach. The work promotes further curiosity.

Pierre Jolivet and Krishna Kumar Verma take us on a journey through the world of leaf beetles and their host plants, frequently pointing to areas which need deeper exploration. Basic evolutionary processes, including the interdependence between the insects and their food plants, are made clear to the reader. Even the archaic cycads, which are now nearly extinct, attract specialized insect feeders such as *Aulacoscelis melanocera* and related forms. Entomologists have described specialized buccal organs in these forms for rasping cycad tissues. The great diversity presented by Angiosperms coincides with the diversity of the phyllophagous beetles, which are generally mono- or oligophagous. Studies of such beetles suggest that the choice of a particular plant species is determined by the chemical and physical properties of the plant. Colour, texture, chemical composition, succulence of the leaves or stem, etc., are all characteristics which determine the choice of a plant food species by a particular leaf beetle. The presence of toxic chemicals, latex, wax bloom or hardness of tissues all deter a plant-feeding insect from attacking a plant. It has been reasonably suggested that, in order to prevent damage by a plant-feeding insect, a plant species may evolve such deterrents, and, in due course, an insect population may develop the morphological or physiological properties necessary to overcome the unfavourable features of the plant. Thus, there has been a co-evolution of plants and their insect feeders.

The book provides pleasant and interesting reading and the authors' enthusiasm for detail is apparent. For instance, *Timarcha* spp. in Oregon lead a nocturnal life. The body colour of one species (*T. intricata*) is black and another species (*T. cerdo*) is dark red. But species of the same genus in the Balearic Islands are bright blue like the Mediterranean sea.

While the book is primarily meant to introduce the reader to the biology of chrysomelid beetles, the phytophagy of these insects means that they are also of interest to agricultural scientists as well as front line entomologists. Among chrysomelids, *Diabrotica*, *Aulacophora*, *Galerucella*, *Leptinotarsa*, *Crioceris*, *Hispa*,

Dactylispa, and many others are crop pests. Perhaps in view of this, Jean Henri Fabre, more than a century ago, pointed out the necessity to include details of insect diet in his account of insect fauna.

Interesting deviations from phytophagy among leaf beetles have been described in the book. Such deviations include carnivorous diet, oophagy, and cannibalism. Chrysomelids associated with ants are polyphagous and saprophagous.

Another engrossing part of the book deals with the various defence mechanisms developed by leaf beetles to protect themselves against parasites and predators.

The book is essentially biological. It successfully brings out 'naturalness' in the organization of the Chrysomelidae, in spite of the great size of the family and the large diversity it presents. Such a work could only have been realized in the final stages of a lifetime's study.

Yves Delange, Scientist Emeritus,
Department of Botany,
National Museum of Natural History,
Paris, France

Foreword

It is almost impossible to cover every aspect of the biology of leaf beetles in a book of this size. The family of leaf beetles is vast (37 000 described species and perhaps twice the number alive) and the papers published on various aspects of leaf beetle biology are many. There are probably 10 000 papers dealing with the Colorado Potato Beetle, *Diabrotica* and other galerucines, alticines and chrysomelines. Thanks to some timely and useful reviews of the many works which have been published on leaf beetles, a wealth of good ideas and discoveries are now available.

This small book is intended to highlight some such observations and theories. We have tried to familiarize ourselves with all that has been published to date, but may have missed some interesting detail or interpretation and some of our own theories may not be acceptable to all. We ask to be excused for any lapses and hope to be informed of any shortcomings so that our work in the future will benefit.

There is great biodiversity among leaf beetles. The genetic differentiation of populations of several species has been measured by the use of allozymes in order to study the gene flow (Simonsen *et al.*, 1999). Genetic exchange between populations does occur, but no more for apterous species in a fragmented habitat. Gene diversity at allozyme *loci* has also been investigated in some leaf beetles (Krafsur *et al.*, 1993; Krafsur, 1995, 1999; Krafsur and Nariboli, 1995). Some of the species studied are economically important crop pests, and display adaptive traits that may be correlated with genetic diversity. Certain species of *Timarcha* demonstrate a great diversity in food plant selection, but are externally little differentiated. Some groups of *Timarcha*, such as those found in North America, need to be thoroughly restudied. Genetic diversity is particularly obvious as polymorphism in the case of leaf beetles with a wide range of distribution, for example the cassidine *Conchyloctenia punctata* Fabricius (Heron, 1999), another cassidine *Aspidomorpha miliaris* Fabricius (Maulik, 1919) and *Lema coromandeliana* Fabricius and *L. semifulva* Jacoby among Criocerinae (Kalaichelvan *et al.*, in print).

Chrysomelidae appear to have attracted the attention of naturalists to a much greater extent than other beetle families. Recently, several books have appeared on the biology of leaf beetles, including volumes edited by P. Jolivet and his associates (1988, 1994, 1996), Furth and his associates (1985, 1988, 1992, 1998), Scherer (1982) and Cox (1999). '*Biologie des Coléoptères Chrysomélides*' by P. Jolivet (1997) covers a number of main topics on leaf beetle biology. There are a number of other coleopterous families as large, or even larger than Chrysomelidae and at least some of them are as important economically. These include Curculionidae (53 000), Cerambycidae (35 000), Staphylinidae (30 000) and Scarabaeidae (25 000) (the values within parentheses indicate the approximate number of described species). For reasons that are not clear, none of these large families has received as much attention from insect biologists as Chrysomelidae. Hence, there is need for frequent reviewing and summarizing in Chrysomelidology.

This book is intended to be a 'guided tour' through the massive literature on Chrysomelidae, and not an extensive treatise. If the reader is especially interested in a certain area in this large field, he may turn to detailed reviews and treatises concerning that area. If any author finds that his or her study has not been cited in this book, we hope he/she will excuse us.

Though chrysomelids have been studied from various viewpoints, including toxicity and cytotaxonomy, some aspects have not received enough attention. For example, the relationship with ants has been inadequately studied, especially in the southern hemisphere. The biology of leaf beetles is little known in Madagascar, New Caledonia and New Zealand, and forest canopy-dwelling leaf beetles also remain to be studied more thoroughly.

The taxonomy of Chrysomelidae has been approached cladistically too, but, as pointed out by John Turney (2000), the results emerging from such studies have been often difficult to understand and accept. However, a number of recent studies involving cladistic analyses using DNA sequences (Hsiao, 1994a,b; Farrell, 1998; Gomez-Zurita *et al.*, 1999, 2000a; Hsiao and Pasteels, 1999; Hsiao and Windsor, 1999, etc.) have led to some interesting inferences. Such studies should help in giving chrysomelid taxonomy a more satisfying profile.

Several questions concerning leaf beetles are unanswered, and are likely to remain so for a long time. For instance, why are there no donaciines in New Guinea outside the Fly River area, in Australia outside Queensland, in Latin America outside Cuba, and in a few places in Mesoamerica? Or why is there lack of parity in the direction of *'retournement'* of the chrysomelid aedeagus, clockwise turning being the rule and anticlockwise rotation a rarity, although the hormonal and muscular mechanisms of the rotation suggest equal probability for the turning of the organ in either direction?

P. Jolivet and K.K. Verma

1

Introduction

Leaf beetles, as their name indicates, are beetles which feed on the leaves of a variety of plants. Under the insect order Coleoptera, the order of beetles, there is a large and natural superfamily, which has been variously referred to as Phytophaga/Cerambycoidea/Chrysomeloidea. It includes three families. These are the Cerambycidae, larvae of which bore into the woody stems of bushes and trees, the Bruchidae, which develop within seeds, mainly leguminous, and the large number of remaining members of the superfamily which constitute the Chrysomelidae, the leaf beetles.

The family Chrysomelidae is one of the largest insect families and includes some 37 000–40 000 described species. Curculionidae (weevils) and Staphylinidae are other such large families. Every year, a hundred or more species come to light. It is thought that tropical forest canopies still lodge a large number of species awaiting discovery, but this is perhaps more true for curculionids than for chrysomelids. At the same time, a number of species of leaf beetles is being forced into extinction by the reduction of forest cover, the extension of agriculture, indiscriminate use of pesticides, and urbanization. However, Jolivet (1997) believes that before long the number of known species will be 50 000.

Leaf beetles present a great diversity. The diversity is so great that chrysomelid taxonomists tend to identify these beetles by directly referring to subfamily characters, instead of first looking for family characteristics. The subfamilies are very well defined by their characteristic features and are obviously very different amongst themselves. We have recently begun to realize the reason for this diversity. The evolution of leaf-beetles has been closely associated with the evolution of flowering plants or Angiosperms. The Angiosperms are a huge, successful group of about 250 000–300 000 species and have become greatly diversified. Chrysomelids, in evolving with them, have developed a similar diversity. However, not all the Angiosperms have thus far been colonized. Farrell *et al.* (1992) and Farrell (1998) have convincingly demonstrated a coevolution of plants and phytophagous beetles. Coevolution is sometimes weak or nil at the genus level, if we consider that selection can be done sometimes at random or linked with the availability or chemistry of plants (Becerra and Venable, 1999). Some species or genera are polyphagous. Cladogenesis does not always coincide between plants and their insects.

Such great diversity among chrysomelids has often led to the view that the group has been polyphyletic in origin. But there are notable similarities among different looking subfamilies (Chapter 12). Moreover, complex and well-integrated systems such as wing venation and the genital system (including the aedeagus and aedeagal musculature), though much varied in chrysomelids, are derivable from a common basic plan. This situation suggests close polyphyly or parallelophyly, i.e. parallel evolution from a common ancestral stock. In fact, a critical study of leaf beetles from the viewpoint of phylogeny is likely to reveal some basic problems of evolution and systematics.

In this monograph on leaf beetles, the group has been taken as a family, the family Chrysomelidae, which is divided into twenty extant subfamilies, following roughly the classification pattern of Seeno and Wilcox (1982). Further, the subfamilies have been placed into five sections, viz. Eupoda, Camptosomata, Cyclica, Trichostomata and Cryptostomata. The classification of Chrysomelidae is discussed in Chapter 2. Bruchidae are excluded from the group (Kingsolver, 1995; Verma and Saxena, 1996; Schmitt, 1998).

Being plant-feeders, chrysomelids include some important agricultural pests. Many species, such as those belonging to the genera *Diabrotica*, *Aulacophora*, *Galerucella*, *Entomoscelis*, *Plagiodera*, *Leptinotarsa*, *Lema*, *Crioceris*, *Hispa*, etc., are active pests. In addition, many species are potential pests. As leaf beetles feed on various plants, and, as predator and parasitoid populations depend for their nourishment on phytophagous forms, the occurrence and abundance of chrysomelids is an indicator of existing biodiversity. It has been pointed out earlier that many species of leaf beetle are facing extinction. This is particularly true with apterous or brachypterous species which cannot recolonize, like flying species, when their habitats are ruined. It is also the case with canopy-frequenting species when forests are cut. It is important, therefore, to familiarize ourselves with the biology of these species so that conservation strategies can be planned.

Besides their economic and biological importance, many leaf beetles are of aesthetic value. Beauty is to be found in the metallic brilliance of *Sagra buqueti* Lesson, the body shapes and iridescent colours of many South American tortoise beetles and the spiny armour of hispine beetles. All species of *Promechus* (*Aesernia*) of New Guinea, *Sagra* species of south-east Asia, and *Polychalca* (*Desmonota*) of Brazil have been described as 'living jewels' by Jolivet (1997). *Polychalca variolosa* (Weber), from the state of Bahia, was actually sold by jewellers, but now it has become so rare in Brazil that the trade has been stopped.

It is important to study the association of chrysomelids and their host plants as it sheds some light on the phylogeny of the former. This is covered in Chapter 4. Chapter 6 deals with the ecological distribution of these beetles.

Though chrysomelids feed mostly on leaves and other plant tissues, entomophagy and cannibalism are also known among them. This is also dealt with in Chapter 4.

Leaf beetles mostly live well exposed on the plant body, and, in general, they are not efficient flyers. Hence, they run a high risk of predation and parasitic attack. Quite naturally, they have taken to a variety of remarkable defence devices. Chapter 8 considers defence mechanisms in leaf beetles.

The great structural diversity presented by chrysomelids encompasses some interesting morphological features, including those of the internal anatomy. But, because a large number of fresh specimens are needed for the study of such anatomical details and because lengthy techniques are required, our knowledge in this area is scant. Such morphological features of special interest have been described in Chapter 9 and we hope that this section of the book will be useful to those who are looking for unexplored, or less well-explored, areas for investigation.

In recent years, due to the pioneering efforts of some chrysomelidologists, many different and widely-scattered studies have been reviewed and brought together, giving a boost to the study of chrysomelids. These include the organization of a series of international symposia on the Chrysomelidae by D.G. Furth *et al.*, the publication of the newsletter *Chrysomela* by Terry N. Seeno, the editing of a series of volumes on the *Biology of Chrysomelidae* by P. Jolivet, M.L. Cox, E. Petitpierre, T.H. Hsiao and others, and, most recently, the editing of the first volume of *Advances in Chrysomelidae Biology* by M.L. Cox in 1999. All these developments have led to a feeling that a summary of present knowledge about the Chrysomelidae was overdue; hence this book.

2

Classification

The Chrysomelidae is a very large and much diversified group. Classification is therefore complex and demonstrates a history of 'lumping' and 'splitting'. That the classification in its classical shape is still generally accepted goes to support the notion of the 'natural state' of the group.

Chapuis (1874) described a system which provided a foundation for the development of the modern, and generally accepted, classification of the family. The history of chrysomelid classification has been well traced by Seeno and Wilcox (1982), Schmitt (1996) and Suzuki (1996).

Distinguishing features of the family

It is not possible to pinpoint a character or characters which would define the family Chrysomelidae. In this context, it will be useful to keep in view the principle of polythetic taxa.

Often higher taxa, which are seemingly natural, are not characterized by a single or a few characters. They are delimited by a considerable set of characters. None of these characters is present in all members of the taxon. Such a taxon has been referred to as a polythetic taxon (Sneath, 1962; Mayr and Ashlock, 1991). Perhaps it would be correct to add that some of the characters in the set may be shared by other taxa, and to re-emphasize that it is the set of characters as a whole which sets aside the taxon in question from related groups.

The set of characters that delimit the family Chrysomelidae include the following:

(a) Adults have the tarsal count 5, 5, 5, but are pseudotetramerous or subpentamerous, as the fourth or the penultimate tarsomere is very small, or even minute, and hidden in an emargination in the distal face of the enlarged third tarsomere. The first three tarsomeres are provided on their underface with a dense arrangement of hairs, forming a flat brush.
(b) Body oval or elongated, length 1 to 27 mm.
(c) Head not produced into a beak; it shows two gular sutures.
(d) Antennae usually small, simple and filiform, 11 jointed. No frontal tubercules on the head for insertion of the antennae, but the feelers are variable. Number of joints in them may be a little more (12) or less than 11 (4 to 10). Instead of being simple and filiform, they may be serrate or slightly clubbed in some cases. They present atypical forms, especially among the subfamily Galerucinae (*Figure 2.1 A–F; Figure 2.2 A–E*).
(e) Hindwing venation presents the following special features. In Chrysomeloidea the basal part of the vein M1+2 is always lacking. The distal and surviving part

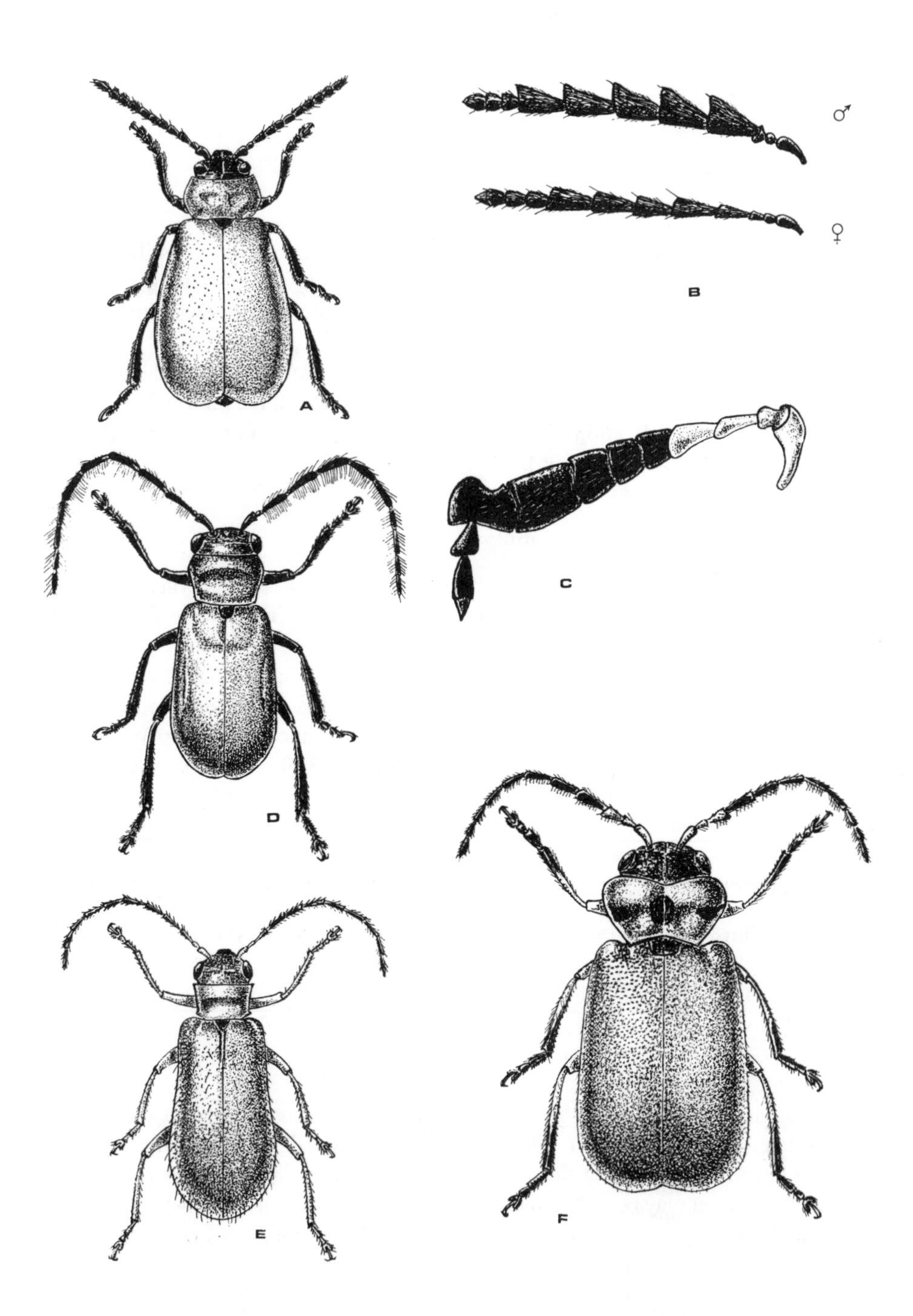

Figure 2.1. Some examples of antennae among the Galerucinae: (A) *Galeruca (Haptoscelis) melanocephala* Ponza from France; (B) *Arthrotus cyanipennis* (Laboissière): antennae, from Vietnam; (C) *Nirina imitans* Jacoby: male antenna from Congo; (D) *Hyphaenia cyanescens* Laboissière from Vietnam, × 8; (E) *Trichomimastra pectoralis* Laboissière from Tonkin, × 15; (F) *Parapophylia cordicollis* Laboissière from South Africa (Laboissière 1921, 1924, 1929, 1934, 1936).

of this vein is connected through the cross-vein m-m with M3+4 to form the characteristic recurrent vein (characteristic of the cantharid type of wing venation). The basal part of the vein Cu1a has disappeared in all Chrysomeloidea. Hindwing venation in them shows considerable variation, but all variants are derivable from a common basic plan (see Chapter 9: Hindwings and hindwing venation, and *Figure 9.2*).

(f) Abdomen with only five complete sterna. The first two abdominal sterna are reduced and more or less incorporated in the following sternum.

(g) There are six Malpighian tubules, which are distally associated with the rectal wall (Cryptonephridic system). The six tubules present a characteristic arrangement (see Chapter 9: Cryptonephridic arrangement of Malpighian tubules). In Donaciinae, due to their aquatic habits, the cryptonephridic association of the excretory tubules with the rectum has been lost, but remnants of the characteristic arrangement of the tubules may still be made out (see Chapter 6: Aquatic chrysomelids, and *Figure 6.1*).

(h) There are two testis follicles on each side. The follicles may be simple, sac-like and non-septate (in Galerucinae and Alticinae) or with its interior divided into a number of loculi by radial septa or septate (in the remaining chrysomelid subfamilies). This occurrence of two types of testis follicles, septate and non-septate, suggests a great divide (or early divergence?) between Alticinae–Galerucinae on one hand and the remaining Chrysomelidae on the other. This situation lends support to those who advocate splitting the family. But the observations of Wieman (1910a,b) on the development of septate testis follicles in *Leptinotarsa* clearly suggest derivation of the septate condition from the non-septate condition. Besides, Scarabaeidae is a well defined, compact and natural family, and some members of this family have septate testis follicles, while others have non-septate testicular organization (Virkki, 1957), the two types of testicular organization being closely comparable to similarly named testicular structure in Chrysomelidae.

(i) Aedeagal structure is much varied among Chrysomeloidea, but the basic structure is *'mode en cavalier'* of Jeannel (1955) with a ring-like tegmen provided with a parameral cap on the dorsal side. Different types of aedeagi found in the large group are all derivable from this basic aedeagal organization (see Chapter 9: External genitalia, and *Figures 9.41* and *9.42*).

(j) *'Retournement'* of the aedeagus seems to be universal among Chrysomeloidea. *'Retournement'* means a rotation of the aedeagus about its longitudinal axis, through about 180°, during development (see Chapter 9: External genitalia, and *Figures 9.44*, *9.46* and *9.47*).

(k) Metasternum with a transverse suture.

(l) Chrysomelidae are mostly leaf-feeders, feeding on leaf surface or mining into the leaves (and almost all are phytophagous). But there are deviations from this feeding habit. Some feed on pollen, some on stem and some (the larvae) are root feeders.

Crowson (1994) has pointed out some evolutionary tendencies among Chrysomelidae. They include the following:

(a) Loss of dorsal part of the tegmen of the aedeagus.

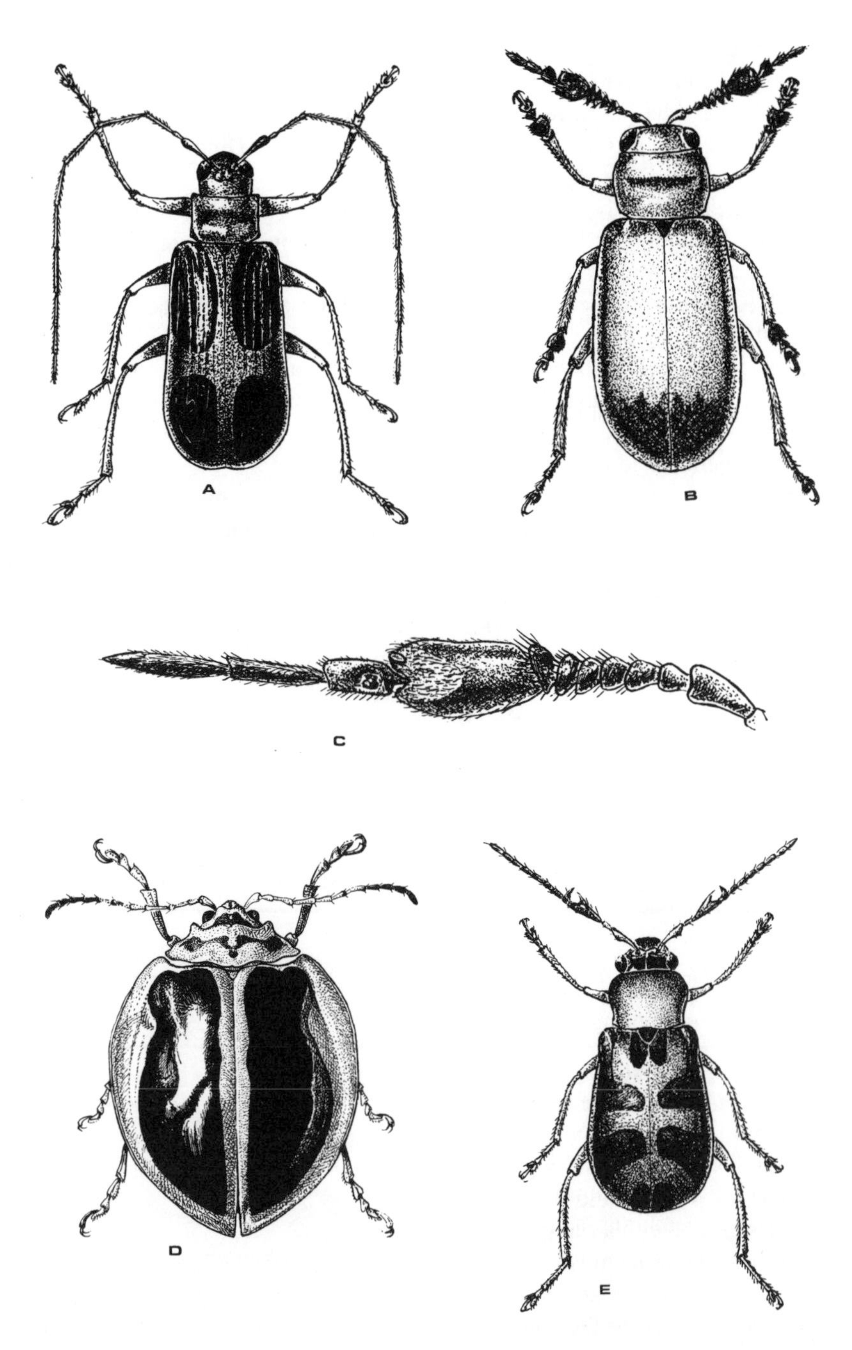

Figure 2.2. Examples of antennae among Galerucinae: (A) *Mimastra pygidialis* Laboissière, 1929 (Galerucinae) from Vietnam, × 6; (B) *Cerophysa pulchella* Laboissière, 1930 (Galerucinae) from Tonkin, × 7.5; (C) Antenna of *Agetocera chapana* Laboissière, 1929 (Galerucinae) male from Burma; (D) *Oides laticlava* (Fairmaire, 1889) (Galerucinae), × 7; (E) *Cerotoma arcuata* (Olivier, 1791) (=*Cerotoma adami* Laboissière 1939) (Galerucinae) from Brazil, × 8. Antennae are often different between males and females and produce pheromones (Laboissière, 1929, 1930, 1939).

(b) Loss of certain veins in the hindwing venation.
(c) Loss of lateral plates of the metendosternite.
(d) Connation of the first two abdominal sternites with the following one.
(e) Number of tibial spurs becoming reduced from 2 to 1 or 0.

To this list of evolutionary tendencies may be added a reduction of the tegminal apodeme and a tendency to the formation of a basal muscular bulb for the aedeagus.

Crowson has also pointed out that these tendencies have appeared polyphyletically among Chrysomelidae; hence a cladistic approach, based on derived characters, is not likely to yield reliable results.

The distinguishing features described above clearly suggest that Chrysomeloidea are a natural assemblage. Common evolutionary tendencies may have appeared along several different lines of descent, but the nature of the changes suggests inheritance of the tendencies from a common ancestral stock. It seems reasonable, therefore, to assume that in the origin of the assemblage there has been close polyphyly or parallelophyly. The validity of this notion is discussed at some length in Chapter 12. Simpson (1965b) and Mayr and Ashlock (1991) are clear that close polyphyly or parallelophyly in practice amount to monophyly.

If Phytophaga (or Chrysomeloidea) were taken as one family, it would be too large. Moreover, it includes three fairly well defined ecological groups, namely wood-boring cerambycids, leaf-feeding chrysomelids and seed-feeding bruchids. Ecological criteria may help in defining families. Mayr and Ashlock (1991) said, 'Like the genus, but perhaps to an even greater degree, the family tends to be distinguished by certain adaptive characters that fit it for a particular adaptive zone, e.g. the woodpeckers or the family Picidae, the leaf beetles or the family Chrysomelidae'. Hence, keeping in line with the taxonomic practices, Phytophaga or Chrysomeloidea are divided into three families: Cerambycidae, Chrysomelidae and Bruchidae.

When the descent of a large assemblage is from a common ancestral stock through close polyphyly or parallelophyly, the subgroups into which the assemblage may be divided would not be 'a series of nicely nested boxes into which all its (evolution's) products fit' (quoted expression from Simpson, 1965a). Intermediate forms, like bruchids *Rhaebus* and *Eubaptus*, which are very much like the chrysomelid subfamily Sagrinae, and stem-boring Megalopodinae, would not be unexpected.

Chrysomelid subfamilies

Because of the large size and great diversity of the Chrysomelidae, it has been found convenient to divide them into several subfamilies. We have taken as standard the division of leaf beetle subfamilies adopted by Seeno and Wilcox (1982) (except for the misplacing of Megascelinae, despite three papers on the topic published before), but some additions and modifications have been made. The system in Seeno and Wilcox includes 19 subfamilies. To this list another subfamily, Palophaginae, has been added. If one fossil subfamily, Protoscelinae, which links Cerambycidae with Chrysomelidae, were also added, the total number of subfamilies would be 21. Protoscelinae are very close to Aulacoscelinae, which are actual living fossils.

As stated earlier, Chapuis (1874) drew up a classification of leaf beetles which provided the foundation for the present one. He divided the family into four sections.

Jacoby (1908) modified Chapuis' scheme, making five sections with 14 subfamilies under them. Jacoby's scheme is presented here:

(1) Mouth placed anteriorly,	
(i) Antennae widely separated at base, elytra of hard texture.	
(a) Intermediate ventral segments not medially constricted, pygidium not exposed.	
(a') Thorax without distinct lateral margins, head produced, eyes prominent, prosternum exceedingly narrow	Section EUPODES (includes Sagrinae, Donaciinae, Criocerinae).
(b') Thorax with distinct lateral margins (rarely without), head not produced, eyes not prominent, prosternum broad	Section CYCLICA (includes Lamprosominae, Eumolpinae, Chrysomelinae).
(b) Intermediate ventral segments constricted, pygidium usually exposed	Section CAMPTOSOMES (includes Megalopinae, Clytrinae, Cryptocephalinae, Chlamydinae).
(ii) Antennae not widely separated at base, generally closely approximate, elytra more or less soft in texture	Section TRICHOSTOMES (includes Galerucinae, Halticinae).
(2) Mouth not normal, small, hidden or nearly so	Section CRYPTOSTOMES (includes Cassidinae, Hispinae).

After that, some new subfamilies were added to the list, and various changes for their section placement were suggested (Seeno and Wilcox, 1982).

We have adopted the arrangement given below. In this arrangement perhaps Synetinae would be more naturally placed close to Chrysomelinae, Galerucinae and Alticinae or under a section of their own.

Section 1. Eupoda

1. Sagrinae
2. Aulacoscelinae
3. Orsodacninae
4. Palophaginae
5. Zeugophorinae
6. Donaciinae
7. Megalopodinae
8. Criocerinae
9. Synetinae

Section 2. Camptosomata

10. Clytrinae
11. Cryptocephalinae
12. Chlamisinae

Section 3. Cyclica

13. Lamprosomatinae
14. Megascelinae
15. Eumolpinae

Section 4. Trichostomata
16. Chrysomelinae
17. Galerucinae
18. Alticinae

Section 5. Cryptostomata
19. Hispinae
20. Cassidinae

The distinguishing features of chrysomelid subfamilies can be found in books on the fauna of different regions, such as the old volumes of the *Fauna of British India* series (Jacoby, 1908; Maulik, 1919, 1926, 1936). Borror *et al.* (1976) also provided a useful key for the identification of leaf beetle subfamilies. See also many recent books, such as Jolivet (1997), Arnett (1968a, 1993), Lawrence and Britton (1994), etc. for the classification of leaf beetles. Recent opinions have been discussed in various CD-ROMs (Lawrence *et al.*, 1999a,b) and in new editions of Arnett's books (2000), but divergence still exists amongst the different authors. Occasionally, drastic changes have been made, but not always judiciously.

3

Palaeontology: Food Plants and Evolution

Records of leaf beetles as fossils are incomplete, and, as little is known about plant–insect associations in the past, we can only surmise on this aspect. One thing is certain, that leaf beetles evolved with the green plants, probably the early cycads and Benettitales in the first place, and then with the Angiosperms during the Mesozoic period. But how far the other plants: algae, mosses, ferns, horse-tails and the Coniferae were involved, we can only guess. Only one thing is definite, and that is that the evolution of leaf beetles during the Cretaceous was parallel with the Angiosperms. Ranunculales, with Magnoliales, Amborellaceae and Winterales, are primitive orders and families of plants and it is not surprising that several living *Chrysolina*, for instance, have maintained an atavistic taste for *Ranunculus*. It is also not surprising that, at the water's edge of the Santana formation, from the Brazilian Lower Cretaceous, *Ranunculus* remains are abundant as well as a rich phytophagous beetle fauna (Grimaldi, 1990). It would be interesting (certainly for botanists and molecular biologists} to know which chrysomelid fed on *Amborella* (Amborellaceae), the most primitive flowering plant, Unfortunately, the food web of this plant is totally unknown in New Caledonia.

We do not know much of the chrysomelids in the Trias, during the split with the cerambycids, and nothing at all, if there was a beginning of the split, at the end of the Permian. Only in the Jurassic did leaf beetles become abundant and they were numerous during the Cretaceous and the Tertiary. When the fossils from the Cretaceous amber become better understood, we will know more. Amber fossils during the Cenozoic are close to the living, even if many genera from that period are now extinct. The explosion of the chrysomelid complex followed the diversification of the Angiosperms during the Mesozoic, although several species or genera reverted to Ferns and Gymnosperms during late evolution.

Palaeontology

Many books have been written on fossil insects. The best up-to-date revision of fossil chrysomelids, written by Santiago-Blay (1994), appeared in *Novel Aspects of the Biology of Chrysomelidae* (Jolivet *et al.*, 1994). Before and since, several papers and books have been published, some on leaf beetles, but mostly on amber insects (Larsson, 1978; Krzeminska *et al.*, 1992; Poinar, 1992; Poinar and Poinar, 1994, 1999, etc.). It is well accepted that beetles appeared at the end of the Palaeozoic (early Permian) and probably the leaf beetles separated from the cerambycids during the Permo–Trias. The parallel evolution between chrysomelids and plants is outlined in Jolivet's book (1997).

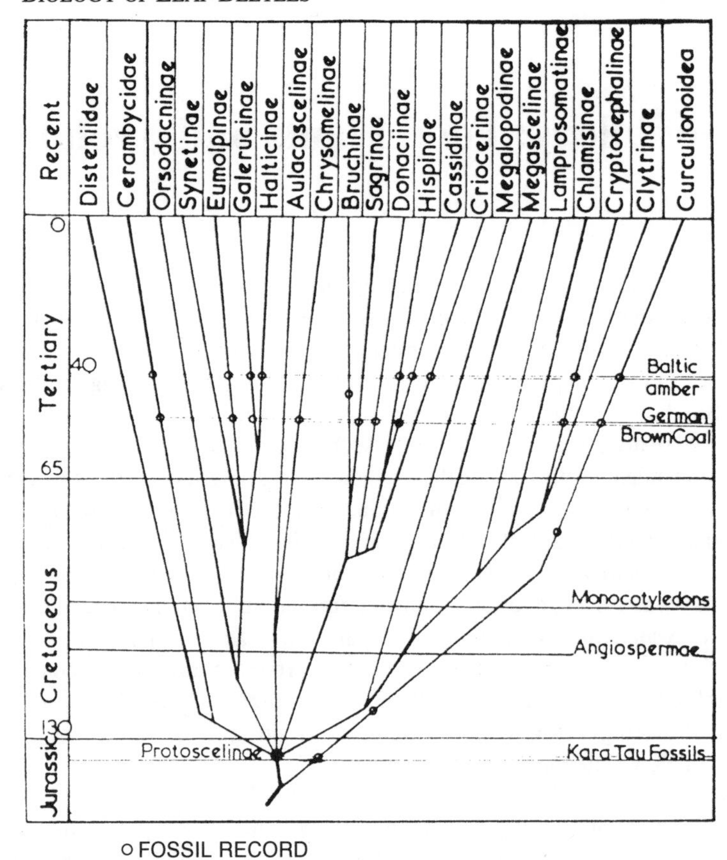

Figure 3.1. A diagram showing the phylogenic relationships amongst the major groups of the Chrysomeloidea presented by Mann and Crowson (1981). Small circles indicate the authentic fossils (after Schmitt, 1988).

Figure 3.2. *Pannaulika triassica* Cornet, tracing the venation with reconstructed margins and apex. One of the first Angiosperms (after Cornet, 1994).

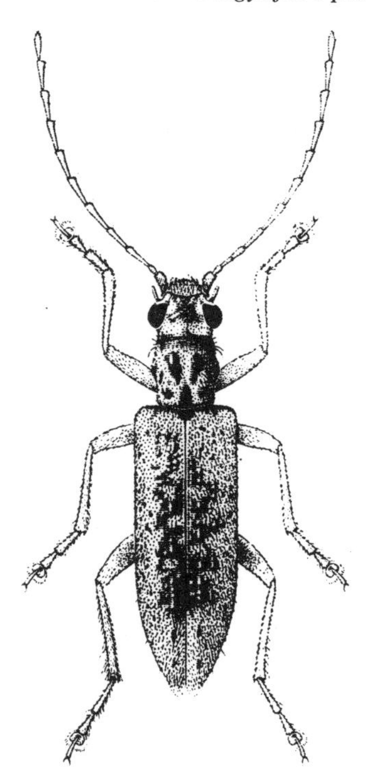

Figure 3.3. A living fossil on *Araucaria* cones: *Palophagus bunyae* Kuschel and May from Bunya Mountains, Queensland (after Kuschel and May, 1996a).

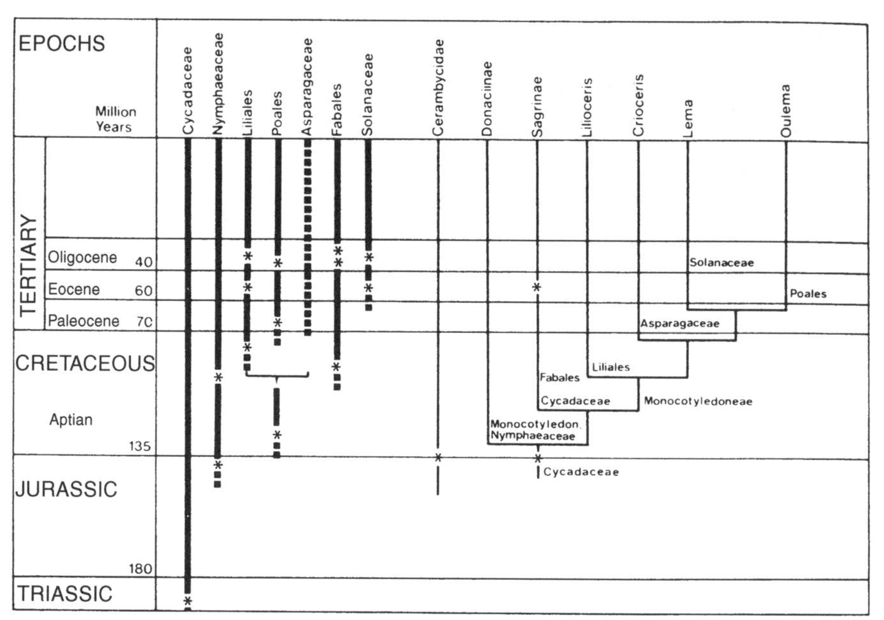

Figure 3.4. The crioceriforms and their food plants. On the right of each stem lineage the probable food plants are given at the time when they were presumably being used. Asterisks indicate fossils (after Schmitt, 1988).

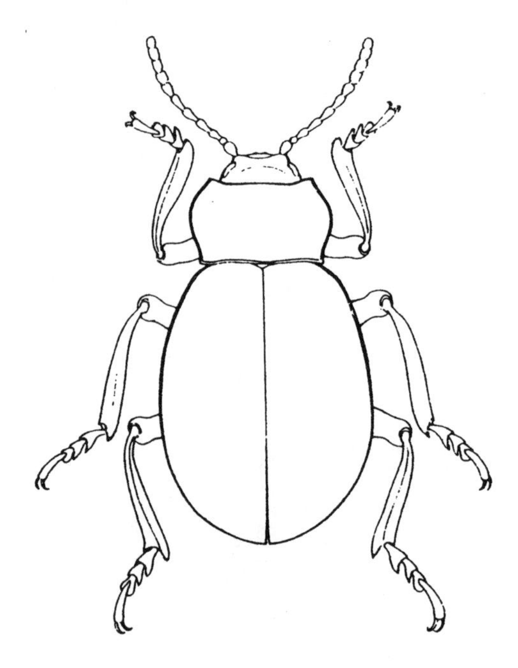

Figure 3.5. A living fossil from the Jurassic, *Timarcha pimelioides* H. Schaeffer, female, from Egades Islands, Sicily (after Jolivet, 1996b).

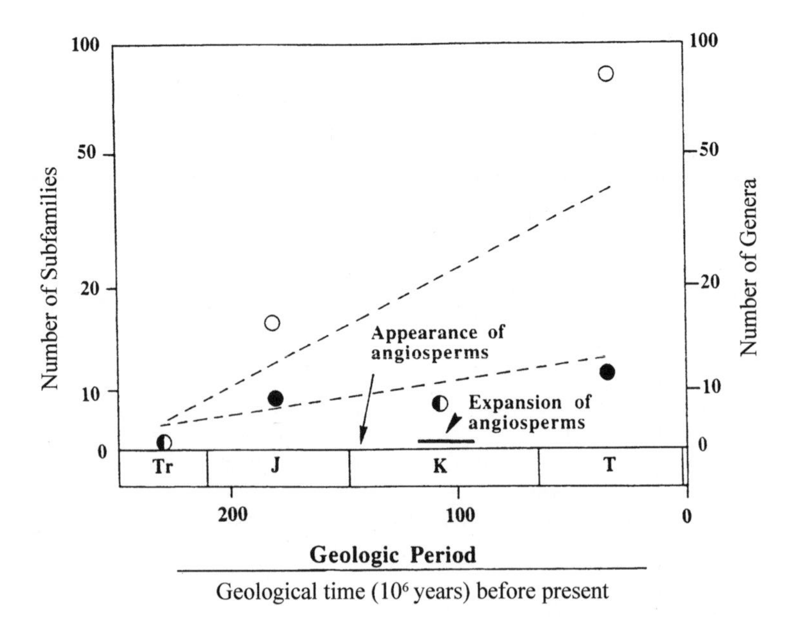

Figure 3.6. Diversity of subfamilies (●) and genera (○) for Chrysomelidae from the Triassic (Mesozoic) to the Tertiary (Cenozoic) on semilogarithmic scale. Dashed lines are possible interpretations of exponential diversification of chrysomelids with Angiosperms. There are about 1700 extant genera of chrysomelids (after Santiago-Blay and Craig, 1999).

PALAEOZOIC

Beetles appeared, undoubtedly, during the Lower Permian (Peyerimhoff, 1934; Ponomarenko, 1969; Shear and Kukalova-Peck, 1990) but, according to Carpenter (1992), the leaf beetles were present from the Trias to the Holocene. Any trace of chrysomelid before the Lias is doubtful. The Phytophaga seem to have originated at the Permian–Triassic line (Santiago-Blay, 1994). Beetles evolved by acquiring strong anterior and protective wings (elytra) and invaginated genitalia. According to Shear and Kukalova-Peck (1990), the selective force was protection against predators by the turtle strategy. Look at the living Cassidinae: "they are difficult to pierce, to crush, hold, lift up, or otherwise harm". Many of them adhere strongly to the plant with oily trichomes and some South American species even cover their eggs with their flattened body. As pointed out by earlier authors, the bodies of beetles are sealed against desiccation by the close-fitting elytra; this is especially true for the apterous forms (*Pimelia*, *Timarcha*) with fused elytra. This tank-like exoskeleton probably helped to make the beetles the most successful and most numerous of animals (Haldane, taken from many authors; Crowson, 1960; Farrell, 1998. See also Evans and Bellamy, 1996).

MESOZOIC

Little is known about Triassic chrysomelids, but the Protoscelinae with cerambycid affinities (long antennae), appeared in the late Jurassic (Medvedev, 1968). These were probably the ancestors of the Aulacoscelinae, still almost exclusively cycad-feeders today. After a timid start during the Trias, cycads and Benettitales exploded during the Jurassic and the Cretaceous to become extinct (the Benettitales) after the Mesozoic. Living cycads are the remains of an abundant flora of the past. It seems almost certain that the Protoscelinae were also cycad or cycad-like feeders (Medvedev, 1968; Labandeira,1999). From the Canadian Mesozoic amber, one sagrine has been recovered (79 mya) (Poinar and Poinar, 1999). Hypotheses have been put forward about the association of the leaf beetles with cycads (Jolivet, 1998a) and cycadoidea (Windsor *et al.*, 1999). It seems that Hispine beetles evolved at least 20 million years earlier than suggested by fossils (Wilf *et al.*, 2000). Ginger leaves from latest Cretaceous to early Eocene show feeding damage attributable solely to rolled-leaf hispines. It seems that the present diversity of leaf beetles occurred during the Cretaceous and probably earlier, contemporaneously with the evolution of the Angiosperms.

Nine existing subfamilies of chrysomelids were represented during the Mesozoic: Sagrinae, Clytrinae, Cryptocephalinae, Chrysomelinae, Eumolpinae, Galerucinae, Alticinae, Cassidinae, and as we have seen above, Hispinae. Most of them appeared during the Jurassic, and some genera of that period already belonged to existing genera like *Cassida*, *Gonioctena*, *Oreina*, *Plagiodera*, *Chrysolina*, *Chrysomela*, etc. *Timarchopsis czckanowskii* Brauer, Redtenbacker and Ganglbauer, from the early Jurassic (Lias) of Siberia superficially resembles *Timarcha*, but no-one can be sure about its real affinities (Brauer *et al.*, 1889; Rohdendorf, 1957), since only the head and prothorax have been preserved. It seems to have the typical tarsi of a female chrysomeline, but the print is not clear enough to tell. The plesiomorphic characters retained by *Timarcha* are certainly a sign of their great antiquity, but an apomorphic character, like apterism, must be also very old indeed. It must have developed as a mutation, when the original population was small, since, all over their holarctic distribution, the whole genus is apterous. Scales in place of wings can very rarely appear atavistically (Ruschkamp, 1927), but most of the time the vestiges, if present, escape the observer because the elytra have to be removed in order to see this condition.

CENOZOIC

Twelve chrysomelid subfamilies have been reported from the Tertiary and the early Quaternary. Donaciinae, Criocerinae and Hispinae are known from the Cenozoic, but they must also have existed during the Cretaceous. Most of the existing families were present in the Eocene and many specimens are known from the Miocene amber. Orsodacninae, Megascelinae, Synetinae, Megalopodinae, Chlamisinae, not to mention the rare Palophaginae, are not represented in the fossil strata or the ambers of this period. Most of the representatives of the above subfamilies are rare, localized in time and space (Megascelinae, Orsodacninae, Aulacoscelinae) and the chances of them being fossilized in a tropical environment are few. In the tropics fungi, predators and bacteria immediately invade any dead insect and reduce it to nothing. There is also little chance that the above subfamilies would mummify in amber because of their habits and the food plants they frequented. However, 325 species, noted by Santiago-Blay (1994), are known already from the Eocene and before. These include the ancestors of the *Sagra* (*Eosagra*), a very primitive group. Donaciinae were quite common and diversified during the Cenozoic and some specimens were very well preserved in the Pleistocene. Many existing genera can be recognized among them.

Timarcha seems to have existed during Eocene and it, more than likely, appeared in the Jurassic–Cretaceous periods. Curiously, this beetle has retained, as previously mentioned, plesiomorphic characters (male aedeagus, nervous system, etc.) until now, together with apomorphic ones (apterism, fused elytra). It is certainly the most primitive tribe of the Chrysomelinae which, according to several authors, deserves being treated as a separate subfamily, the Timarchinae (Sharp and Muir, 1912; Powell, 1941; Verma, 1996c; Jolivet, 1999a). The beetle probably originated in western or central Asia and crossed to America during the Jurassic or later. A cross-Pacific migration through Beringia remains a possibility (Elias, 2000; Poinar *et al.*, 2001).

In his study of amber chrysomelids, Poinar (1999) mentions 26 genera known from Baltic amber, 11 genera from Dominican amber and 2 genera from Mexican amber. It is evident that these findings are just the tip of the iceberg. Santiago-Blay (1994) states that 24 genera from that period are now extinct. It is also evident that species and genera found in the various ambers (Hispaniola, Baltic) are still represented in warmer climates and in localities nearby, but have been eradicated from insular or cold areas. Poinar and Poinar (1999) pointed out that the reason for this (in the Dominican Republic) was the Pliocene–Pleistocene glaciations, the following dryness of the climate, and inbreeding.

CLIMATIC CHANGES AND BIODIVERSITY

It is clear that biodiversity diminished after the Mesozoic and early Eocene explosion. However, this extinction rate cannot be compared with that of the present day, which has been largely brought about by man's intervention (Labandeira and Sepkoski, 1993; Santiago-Blay, 1994). According to Paul Whalley (1987), no evidence of worldwide catastrophic events over the Cretaceous–Tertiary boundaries exists for the insects in general, and the beetles in particular. Leaf beetles also continued to evolve after the supposed 'fall of the comet', without much apparent disturbance, the main types having probably appeared during the Cretaceous. It is also likely that the beetles, being more mobile than the dinosaurs, moved into more hospitable areas (the tropical highlands of Cornet, 1994) when there were abrupt climatic changes. According to Labandeira and Sepkoski (1993), diversity increased steadily after the Permian–Triassic, which is thought of rather as a 'bottleneck', and rose sharply in the Middle Mesozoic, i.e. during the Jurassic. Insects, like flowering plants, multiplied sharply after the mass extinction at the end of the Cretaceous, 65 mya ago. As far as phytophagous insects are concerned, generalists are more likely to survive mass extinctions than specialists (Erwin, 1998). Climatic changes during the Tertiary must have had little effect on the leaf beetles and only during the Plio–Pleistocene glaciations in Siberia and northern Europe were several genera and species eradicated. It is probably around this time that we must place the disappearance of *Timarcha* in Siberia, the genus being essentially thermophilic and fundamentally adapted to steppic living conditions.

How many species of chrysomelids are actually living? The answer is around 50 000, of which 37 000–38 000 have been described. The canopies of tropical forests probably harbour the biggest number of undescribed species, and other surprises are probably to be found in Australia, New Guinea and its surroundings, parts of Africa, Madagascar and South America. We can reasonably estimate the total number of fossil and living chrysomelids to be around 100 000 species. How many of them will be described before extinction and how many fossils will be found? These are difficult questions to answer. We cannot expect to carry out any reasonable cladistic analysis of the fossils, partly because of their poor state of preservation, and also because of the inadequacy of the actual data about existing subfamilies and genera. There are too many missing links from Madagascar and from the rare genera and subfamilies. Even *Timarcha*, a living fossil in itself, is missing in most available cladograms!

Phytophagy, palaeoecology and biodiversity

Leaf beetles are fundamentally plant-eaters and, with the advent of the Angiosperms mostly during the Cretaceous and the Tertiary, plant choice was extended .The basic trophic machinery of insects was in place nearly 100 million years before the Angiosperms appeared (Labandeira and Sepkoski, 1993). Of course, timidly flowering plants appeared early in the Trias, but they were so rare that we can assume that Protoscelinae were cycadoid-eaters only. The existing and primitive Aulacoscelinae are still cycad-eaters and probably come from the Protoscelinae. It is likely that *Carpophagus banksiae* MacLeay, an archaic Australian sagrine, feeds on *Macrozamia* (Cycads), but the fact needs confirmation. Actually, in Thailand, Vietnam, northern Australia and New Guinea, several species of *Lilioceris* feed on *Cycas* spp. (Szent-Ivany *et al.*, 1956; Hawkeswood, 1992; Wilson, 1993; Shepard, 1997; Jolivet, 1998b). Without the plants, it is very difficult to guess what the extinct chrysomelids actually did eat. Santiago-Blay (1994) suggests that *Donacia* fed on water-lilies,

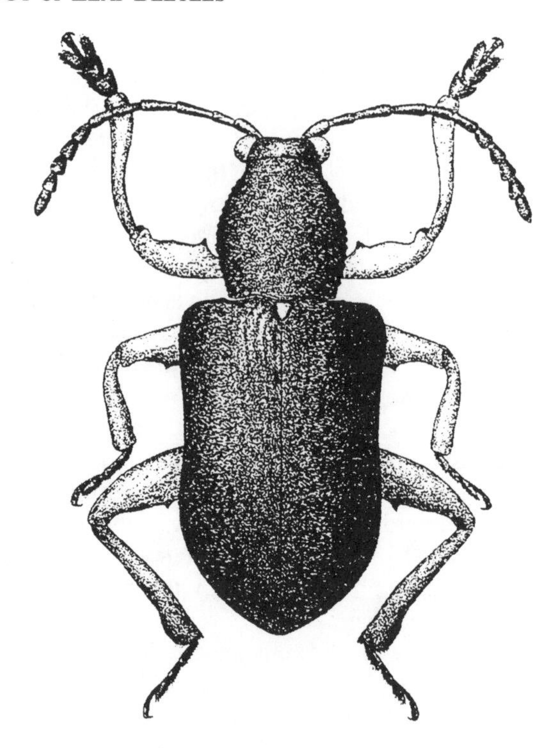

Figure 3.7. *Profidia nitida* Gressit (Eumolpinae). Oligocene–Miocene. Chiapas Amber, Mexico (after Santiago-Blay, 1994).

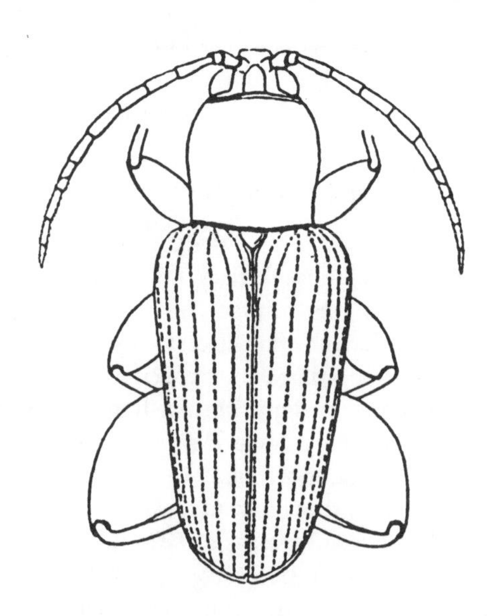

Figure 3.8. *Eosagra subparallela* Haupt (Sagrinae) (after Santiago-Blay, 1994).

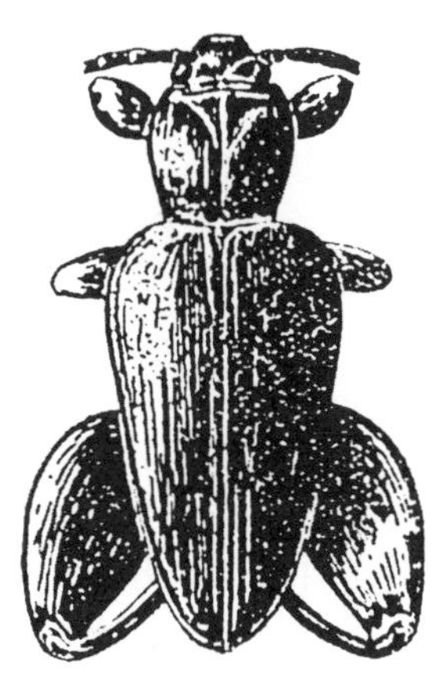

Figure 3.9. *Eosagra obliquata* Haupt (Sagrinae). Middle Eocene (after Santiago-Blay, 1994).

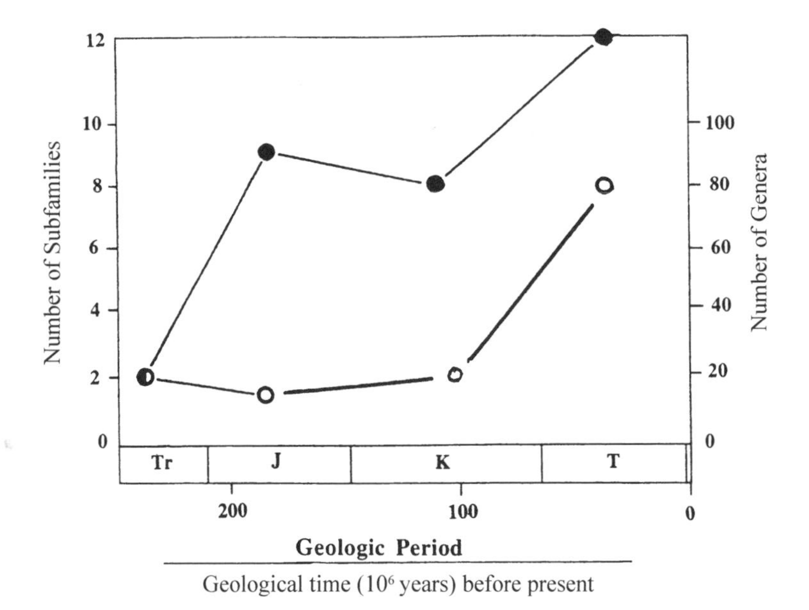

Figure 3.10. Diversity of subfamilies (●) and genera (○) for Chrysomelidae, from the Triassic to the Tertiary (after Santiago-Blay and Craig, 1999).

Gonioctena on poplars, *Lema* on liliaceous plants, *Plateumaris* on *Iris*, etc. We can reason by analogy, but it is very difficult with the Donaciinae which were basically Nymphaeaceae feeders but polyphagous on subaquatic plants, to assign a given genus of plant. Only with *Altica* sp., can we be sure that the insect fed on *Alnus* sp. (Betulaceae) during the middle Eocene, a relationship which exists to this day (Lewis and Caroll, 1991).

As we will see later, it seems almost impossible for the living species to make insect and host plant cladogenesis coincide (Futuyma and McCafferty, 1990), in spite of the fact that some believe to have done so (Farrell *et al.*, 1992; Farrell, 1998; Farrell and

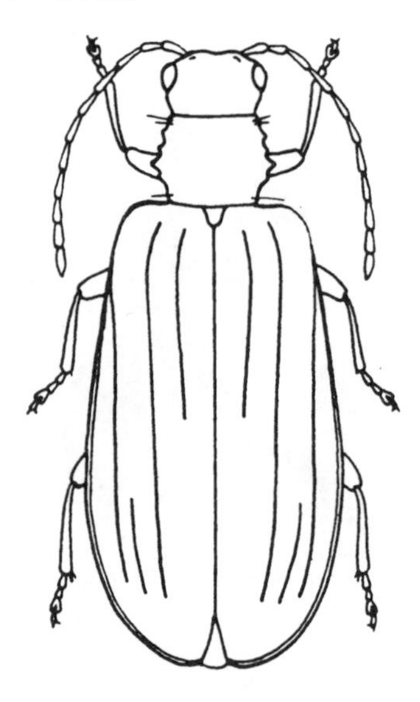

Figure 3.11. *Syneta adamsi* Baly (Synetinae), adult. An enigmatic leaf beetle, the genus position of which is still questionable (after Yu, Peiyu *et al.*, 1996).

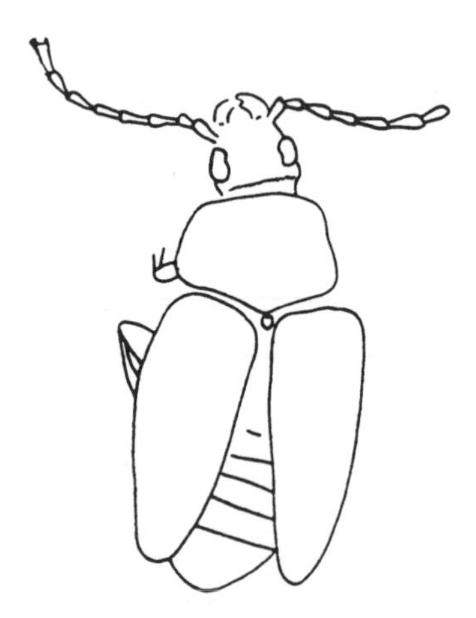

Figure 3.12. *Protoscelis jurassica* Medvedev (Protoscelinae). Jurassic (after Santiago-Blay, 1994).

Mitter, 1998). However, concordance does seem to exist between *Phyllobrotica* and its host plants, reflecting a parallel cladogenesis. In a recent paper, Becerra and Venable (1999), studying the New World food plants of the alticine *Blepharida*, found that host chemistry is the main factor explaining the patterns of host use, but the influence of host availability seems practically unexplored. As far as we can tell, food selection must have started for a given genus from a polyphagous group having selected a group of plants in a given area. Completely different groups and selections would have occurred in another isolated area separated by mountains or water. We believe that food choice is random at the beginning, according to the dominant and

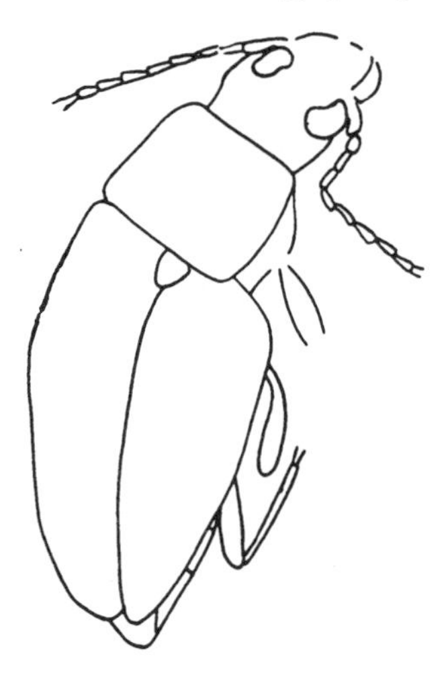

Figure 3.13. *Protosceloides parvula* Medvedev (Protoscelinae). Jurassic (after Santiago-Blay, 1994).

available plants of a given area. The main criteria would be availability and chemical similarity. According to Becerra and Venable (1999), there may be correlations between host phylogeny, host chemistry and host geography. The two *Timarcha* complexes from the Old World and the New World, separated since the Pliocene at least, feed on completely different orders of plants: the Rosales in America and the Dipsacales–Gentianales (occasionally Asterales and Lamiales) in Europe, Asia and Africa. Selection by related species of the same genus of leaf beetle is often for a particular plant group (*Chrysolina*), but it is difficult to link the plant evolution and the insect food choice. Probably, the earliest *Chrysolina* fed on *Ranunculus*, a very primitive plant, and many subgenera have retained this choice. Several authors have tried to link host plants with genera or subgenera of leaf beetles, such as *Chrysolina* or *Timarcha* (Garin *et al.*, 1999; Gomez–Zurita *et al.*, 1999; Hsiao and Pasteels, 1999), but often failed to take into account secondary and facultative plant associations. *Chrysolina*, *Timarcha* and many other chrysomelids have a much wider trophic spectrum than the one initially imagined. In our opinion, any reasonable cladogram cannot make plant and leaf beetle phylogeny even loosely coincide.

Larvae of *Cryptocephalus* with scatoshell (protective case) are known from amber (Santiago-Blay *et al.*, 1996). Probably, as now, the larvae, being polyphagous, fed on bark and plant debris, but they were also loosely associated with ants. The scatoshell protects mainly against ants and, for the Clytrinae at least, the ant–insect association certainly existed since the Eocene, even since the Cretaceous, when ants started to diversify. The palaeoecology of leaf beetles is reviewed by Poinar (1999). The larvae of cassidines of the Cenozoic are very similar to those of the present day and must have had the same biology. Cretaceous amber fossils, if they ever become available, would reveal much more biological data about the evolution of the specialized groups. The association with *Araucaria* and primitive coniferous trees, as is the case with the Palophaginae, must certainly go back a long way, but many leaf beetles feeding on Gymnosperms or ferns, horse-tails or mosses are recent adaptations.

According to Santiago-Blay and Craig (1999), there exists a kind of parallel between the increasing diversity of the leaf beetles and the appearance and development of the Angiosperms. However, this parallelism is evident for the Insecta as a whole (Labandeira and Sepkoski, 1993). In view of this, and with information from available chrysomelid fossils, the idea of parallel cladogenesis in plant and leaf beetles cannot be accepted outright. More data are needed to substantiate this view. There are more amber specimens waiting to be found and described but many are lost forever, since a large element of chance comes into resin-trapping and much depends on the biology of the insect.

Unfortunately, we know nothing of the biology of the Gondwanian sagrines from Australia, Madagascar and Brazil, except *Atalasis* from Argentina (Monros, 1955a; Jolivet, 1988; Jolivet and Hawkeswood, 1995; Jolivet, 1997). Adult sagrine beetles are abundant in Australia and Madagascar, but their host plants and galling habits are practically unknown. Any data on this would greatly help in the understanding of the trophic evolution of the Chrysomelidae. Many leaf beetles feed normally (Palophaginae) or occasionally (Galerucinae, Eumolpinae, Orsodacninae, etc.) on pollen, but generally it is only an additional intake of proteins. The fact that some Australian sagrines have been caught feeding on pollen does not imply a close relationship with the plant. Probably, the early chrysomelids were cycad-eating, pollen-feeders and dug into the microsporophylls of the cycadoids.

Evolution

It is extremely difficult to trace the evolution of the leaf beetles. We do not know much about the proto-chrysomelids from the Permian, nor the forms which probably began to split from the cerambycids, the bruchids and others. The proto-Coleoptera were cupedid-like and had by no means a phytophagous shape. The separation with the cerambycids probably started during early Trias when the Coleoptera had a shape which was more familiar to us. According to Crowson (1994), cycadales and the male cones of *Araucaria* and related species were the first plants available to and selected by the chrysomelids. Actually, some primitive chrysomelid subfamilies, such as Palophaginae and Aulacoscelinae, do demonstrate that type of selection. However, Orsodacninae feed on Angiosperms, but their life cycle is still a mystery. Sagrinae, somewhat polyphagous at the larval stage, also feed on Angiosperms, although there are some doubts about the most primitive ones. According to Labandeira (1998a,b), sporivory preceded pollinivory, and probably polyphagy preceded oligophagy, and then came monophagy. By the end of the Middle Cretaceous, most of the Gymnosperm feeders had been transferred onto Angiosperms, either by descent from Gymnospermous ancestors or by direct host transfer. Actually, authentic cycad or Gymnosperm feeders are really the exception among the chrysomelids. Most of the chrysomelids which actually feed on ferns, mosses, horse-tails, coniferous trees, are secondary adapted.

In recent years, a number of cladograms, based on various parameters, have been produced in order to show the relationships and evolution of the genera and the subfamilies. Every cladogram is a hypothesis and depends on the algorithm used to generate the hypotheses (Niklas *et al.*, 1999). The cladistic hypotheses will change as new data or taxa are added to the matrix. This is very true for the chrysomelids. The

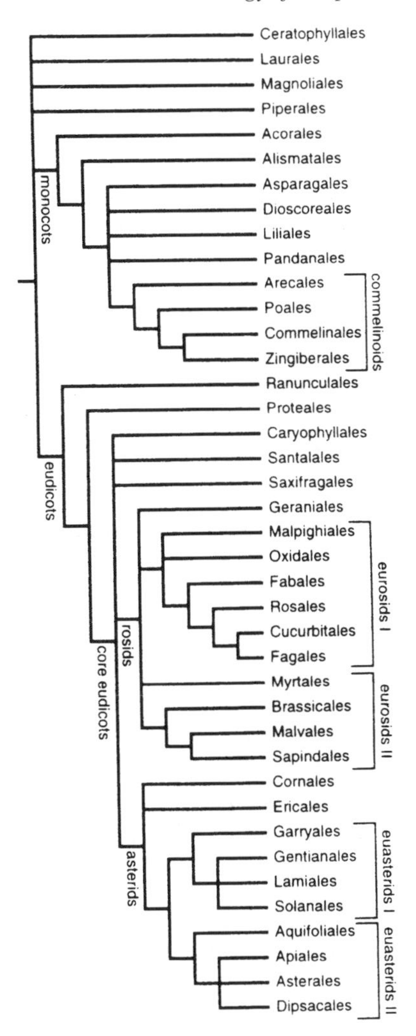

Figure 3.14. Phylogenetic relationships amongst the orders of flowering plants (after Nyffeler, 1999).

inferences made by some of the cladograms were so difficult to accept that we must question the selection of the data on which they were based.

Plant phylogeny has been the subject of a great number of resource books and papers (Takhtajan, 1969, 1980, 1991, 1997; Thorne, 1976; Cronquist, 1981, 1988; Swain, 1963, etc.). More recently, several important papers, using a cladistic analysis, have tried to re-evaluate the evolutionary history of plants. The results have not been new or revolutionary when compared with the ideas of Cronquist or Takhtajan (Thorne, 1992; the Angiosperm Phylogeny Group, 1998; Nyffeler, 1999, etc.). These authors sometimes criticize Takhtajan's system (1997) for being extremely elaborate and for its propensity for splitting families and orders. The new systems are mostly based on molecular studies but are incomplete in that they keep many primitive, or not so primitive, plant families outside the ordinal classification.

The fact that leaf beetles feed on particular plants is not always a case of

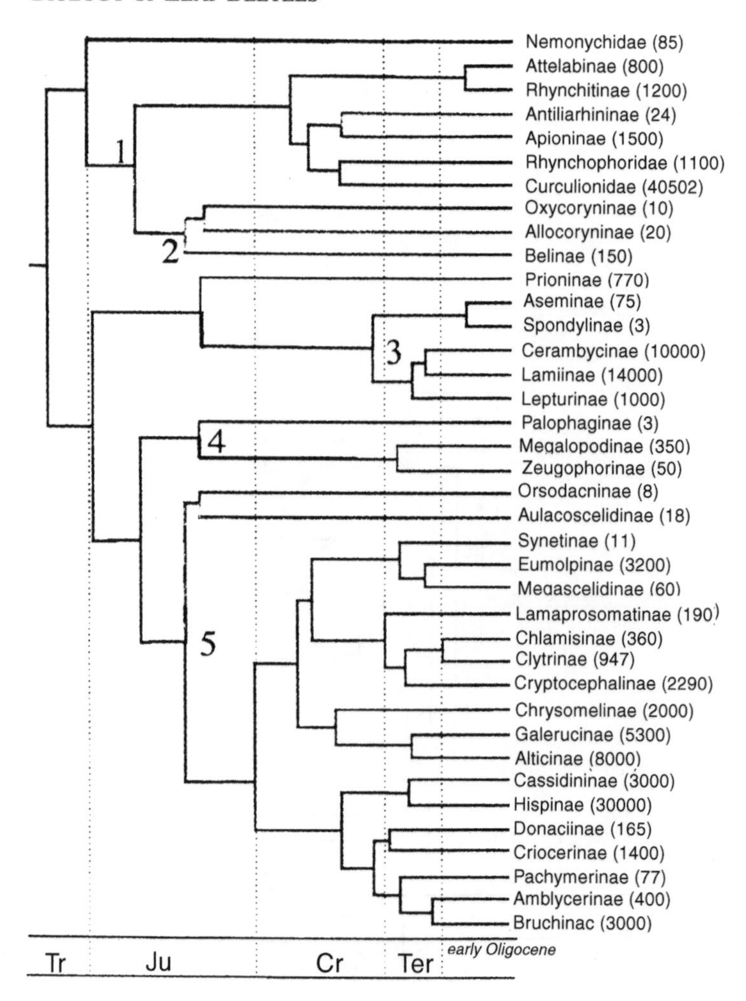

Figure 3.15. The phylogeny of the families of Phytophaga, according to Farrell (1998).

coevolution. The concept of Ehrlich and Raven (1964) has been slightly overused over the years. Many polyphagous plant-eaters do not show a special adaptation to the host chemicals or to the plant morphology.

When did the flowering plants originate? The answer is probably during the Trias, some 220 million years ago, with *Pannaulika triassica* Cornet, a dicot-like leaf. However, the evidence of the leaf is still controversial in spite of the fact that Triassic fossil pollen was found (Cornet, 1994). The Angiosperms are believed to have appeared 130 million years ago at the beginning of the Cretaceous period, and they went through their great phase of diversification 30 million years later. What happened during the 'cryptic evolution' in the Jurassic period (duration 30 million years) during which there is a strange lack of fossil evidence? Were the flowering plants confined to the tropical upland regions where insects could have evolved with them without leaving fossils? This is Cornet's hypothesis, which is still to be confirmed. It seems that most of the primitive chrysomelids, like Sagrinae,

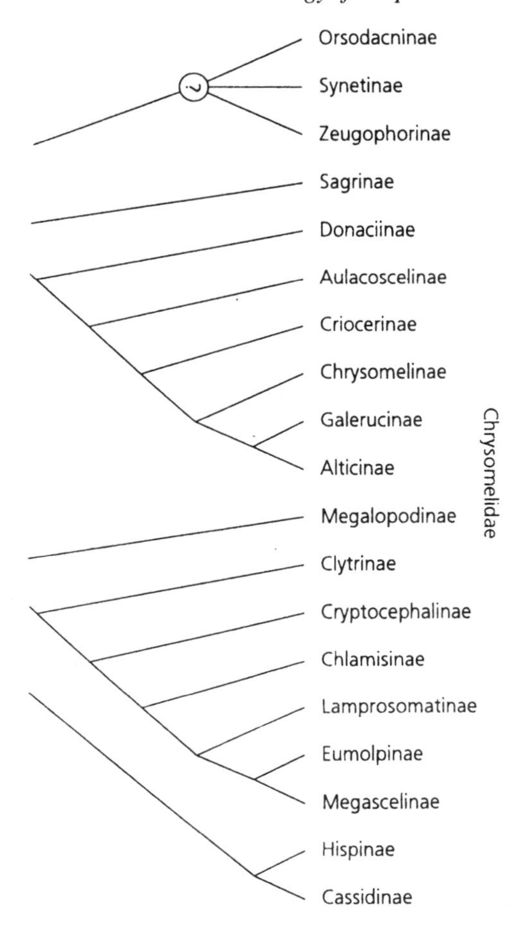

Figure 3.16. A classical view of the phylogeny of the subfamilies of Chrysomelidae (Jolivet, 1997; redrawn by Schmitt, 1996).

Aulacoscelinae, Palophaginae and related, as well as primitive weevils, like the Nemonychids, were associated with the cycadoids or the coniferous plants during the Mesozoic, as they are now. Orsodacninae, being mostly pollen-feeders as adults, are also very primitive, but are not associated with Gymnosperms as quoted by Farrell (1998). Several other chrysomelids feed also on cycads, like many *Lilioceris* in Asia (Jolivet, 1998b) or perhaps *Carpophagus*, a sagrine. Criocerinae can be also considered as primitive despite the structure of the tegmen of the male aedeagus.

The latest classification of flowering plants gives 462 families in 40 putatively monophyletic orders (Nyffeler, 1999). This classification dismisses the classical primary division of the flowering plants into dicotyledons and monocotyledons (hereafter dicots and monocots). Neither of them is more primitive than the other and Hispinae and Criocerinae, which feed often on monocots, are not more or less evolved for that reason. Donaciinae, and also several Galerucinae, which feed on Nymphaeaceae, are linked to a primitive plant family. Nymphaeaceae and Amborellaceae are found at the base of the system, with Winteraceae, and others like Laurales, Magnoliales, Piperales, etc. Mathews and Donoghue (1999) recently placed

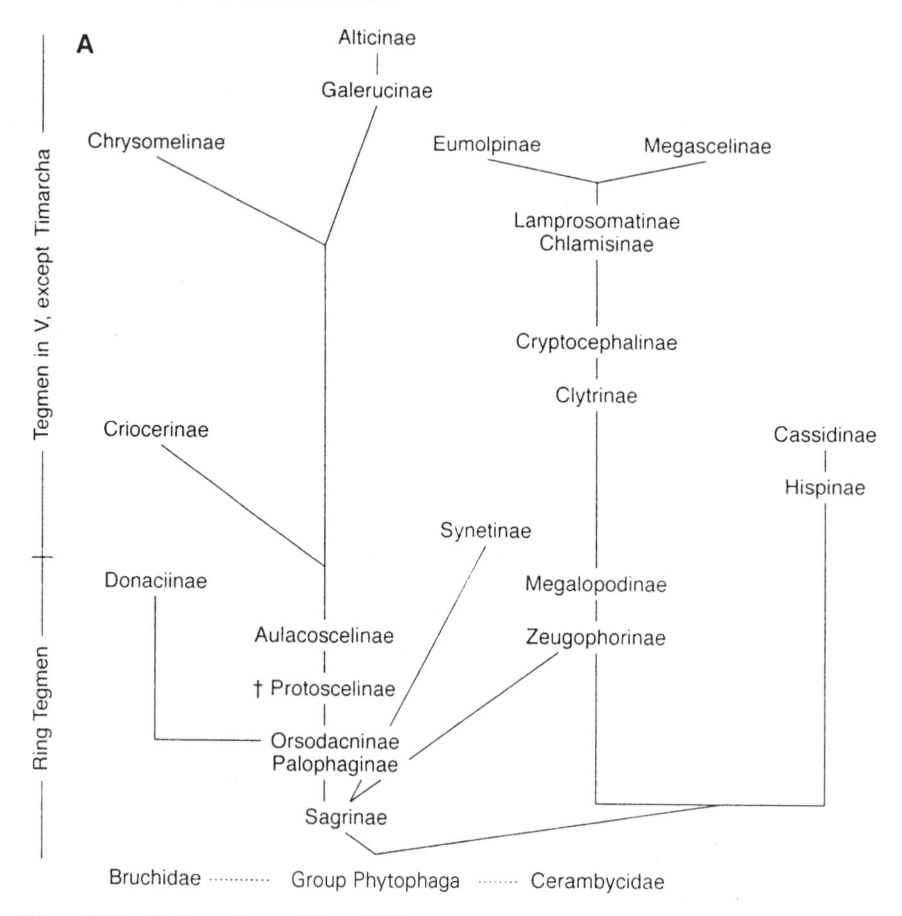

Figure 3.17. Phylogenetic tree (Jolivet 1997).

the 'roots' of the Angiosperms (70 species) near *Amborella* from New Caledonia (Amborellaceae) and water-lilies (Nymphaeales), with *Austrobaileya* from Australia (Austrobaileyaceae) being early 'branches'. Ranunculales are thus confirmed at the beginning of the 'tree', exactly as in the old classifications. We do not know, unfortunately, which leaf beetles feed on the Neo-Caledonian *Amborella*, the most primitive flowering plant, but we know of the tendency for many of them, *Neocrepidodera* (*Asiorestia*), *Longitarsus* (pars), *Argopus*, *Chrysolina*, *Phaedon*, *Hydrothassa*, *Galeruca*, etc. to feed on *Ranunculus*, *Aconitum*, *Clematis*, *Caltha*, etc. It is an ancestral selection, sometimes retained in several *Chrysolina*, and has not been taken into account in various cladograms. In the tropics, Ranunculaceae are also frequently selected by various beetles, but the data are scarce.

It is impossible to link Farrell's cladistic 'tree' with the botanical ones, at least in comparing the genera and families. It is evident that any cladogram is affected by the addition of new genera, and so many are missing in Farrell's hypothesis (1998). A transitory hypothesis is replaced rapidly by another different hypothesis, and the missing links, such as *Timarcha* and many important and aberrant genera from Australia, New Guinea, Madagascar, Africa and South America, could completely change the whole picture. As noted by Brown (1999), the emerging phylogenetic framework of the plants is preliminary and will surely change with time. In this cladogram, many primitive families are 'orphan' families as they are not even assigned to orders, unlike in the system adopted by Cronquist (1988) and Takhtajan (1997). Our knowledge of the host plants of chrysomelids is still in its infancy. Around 15 000 species of leaf beetles can be correctly associated with a host plant, out of more than 37 000 described species. Even so, this does not cover the potential trophic spectrum which is surely much richer than described so far (Jolivet and Hawkeswood, 1995). No cladogram has been based simply on the choice of food plants but this could reveal the ancestry of the group.

Despite its imperfections, Farrell's cladogram (1998), based partly on ribosomal DNA sequences, is probably the most accurate, even if it is very incomplete. Placing Megascelinae and Eumolpinae close to Synetinae is probably a mistake (Verma and Jolivet, 2000). Many *Syneta* characters seem convergent with eumolpines, and this can be due to the similar larval history. However, Palophaginae, Megalopodinae, Zeugophorinae, Orsodacninae, Aulacoscelinae are well placed, even if, curiously, Sagrinae are missing and Donaciinae and Criocerinae are found well apart. Donaciinae are the only truly aquatic group among the chrysomelids and only a few species are known as aquatic among the Hispinae (*Cephaloleia*). The subfamily is placed close to Donaciinae in Farrell's 'tree' (Jolivet, 2001). Both Hispinae and Criocerinae are partly monocot eaters, a discarded group for the molecular botanists. Are they loosely related? It is possible, and were once so considered by Crowson. However, Criocerinae have some primitive morphological features and sometimes demonstrate archaic food selection (cycads). In Farrell's cladogram, Galerucinae and Alticinae, Cassidinae and Hispinae, Clytrinae and Cryptocephalinae are grouped but kept separate. We still hold that Bruchinae, certainly related to Sagrinae, should be kept as Bruchidae, as in a natural classification by Kingsolver and Pfaffenberger (1980), Kingsolver (1995), Verma and Saxena (1996), Schmitt (1998). However, if more genera were studied, the system would begin to become more coherent. The number of species in Farrell's cladogram is also questionable (nearly 3000 for Chrysomelinae).

In Chapter 12 the phylogeny of the subfamilies will be considered in detail, chiefly from the angle of morphology. *Figures 3.14* and *3.15* show phylogenetic trees of the chrysomelids and the plant orders. Many different phylogenetic solutions have been proposed and some are summarized in Jolivet (1997); Suzuki (1988, 1994, 1996); Lee (1993); Hsiao (1994a,b); Reid (1995); Schmitt (1996). Finally, the cladograms based on molecular biology by Hsiao (1994a) and Farrell (1998) are certainly providing a more accurate picture of evolution, despite the missing data in both of them.

The classification adopted here remains classical, following Seeno and Wilcox (1982), modified, and Jolivet (1997). We keep Synetinae after and outside Eupoda, not inside Eumolpinae. The affinities of the subfamily have been discussed in Verma and Jolivet (2000).

The probable evolution of Chrysomelidae was superficially discussed in Jolivet (1988, 1997). Many of the living genera seem stable and are remnants of a more abundant biodiversity existing during the Tertiary. The genus *Timarcha,* primitive as it is, is a strange paradox. It is confined to isolated areas and mountains, being well adapted but still morphologically variable. It is frequently difficult to separate its species, the morphological and anatomical characters being often extremely variable.

Food plants vary, but generally the cytology and morphology amongst species is fairly uniform. Recently, Gomez-Zurita *et al.* (1999) studied the molecular phylogeny of *Chrysolina* and *Timarcha* in Spain, but it is extremely difficult to establish the relative ancestry of the different clades in parallel with their food plants. By contrast, the Colorado Potato Beetle, *Leptinotarsa decemlineata* Say, seems to remain very stable morphologically and biologically in Mexico, its place of origin, and elsewhere. Besides several colour mutations, very few differences at the morphological, anatomical or molecular level have been detected.

The origin of the chrysomelids, although certainly polyphyletic or parallelophyletic, still remains a mystery. The biggest diversity occurred during the Tertiary, but at that time the present fauna, a relict of a much richer diversity, was already totally represented even with the present ant and plant associations. We need more data, more molecular analysis and more material from the whole planet, before we can expect to come to an accurate assessment of the branching of the subfamilies. The present classifications are satisfactory. The recently proposed phylogenies are not particularly insightful, nor do they differ significantly from the old system

4

Food and Feeding

When describing the food and feeding patterns of different chrysomelid groups, it is useful to look at some details of habits and biology.

Almost all leaf beetles are phytophagous or plant-eaters (leaves, bark, stems, flowers, buds, roots) or at least potentially phytophagous. Some species of Alticinae will feed on fungal spores, algae and detritus. Some larvae (Camptosomata) often feed on bark or are detriticolous, feeding on decaying vegetal matter, or can even be coprophagous, carnivorous or oophagous inside or around ant nests (Clytrinae, some rare Cryptocephalinae and Eumolpinae). In reality, these larvae, which also feed on plant debris, are practically omnivorous. Other exceptions to phytophagy are coprophagy, nematophagy, entomophagy and cannibalism. All except the last are rare and localized. Oophagy does not exist among the miners, the stem-borers and in cases where the mother covers or protects the eggs. Otherwise, it is quite a universal phenomenon.

Over 37 000 species of leaf beetles are actually known, and a few hundred new species are described every year. We know something about the food plants of about a third of them, but the data are very scarce in Central Africa and totally non-existent in certain other countries, for example, Madagascar. This great island contains one of the richest original flora of the planet. It also has a fascinating original fauna, with many endemics among the chrysomelids. This is probably the only white spot on the trophic map of the leaf beetles. We do not know much about the host plants of poor but original faunas, like those of New Caledonia or New Zealand and have to depend on lists dating from the beginning of the 20th century. Even where the food plants have been well studied (Europe, Asia, North America), the total food spectrum of a given species is very rarely known. In the canopies of tropical forests, we can really only guess at the food preferences of the beetles (often regarded as morphospecies). This has been researched to a small extent (Novotny *et al.*, 1999) but questions still remain as to the specificity of a beetle for a given tree or a plant. With that knowledge, we can only speculate about the biodiversity of the canopy itself. It has been estimated at 30 million (Erwin, 1983a,b), 10–80 million, including the soil and leaf litter (Stork, 1988; Stork and Gaston, 1990), 5 million plus the mites (May, 1988) and 'a few' million (Wilson, 1992). These suggested numbers do seem to imply our total ignorance in the matter. Much depends on the specificity of the beetles found on the trees and climbers. The curculionids captured recently in the canopies of the eastern and western forests of Panama are new. This does not mean that all the canopy chrysomelids are new, but that many remain undescribed.

Many 'canopyists' have researched the specificity of the tree-frequenting chrysomelids (Farrell and Erwin, 1988; Basset, 1992; Farrell *et al.*, 1992; Basset and

Samuelson, 1996; Novotny *et al.*, 1999; Wagner, 1999, etc.). Most of the work was carried out in the neotropics, but New Guinea, Australia, Borneo, south-east Asia and, more recently, tropical Africa have also been investigated. Wagner (1999) collected 309 species and 9909 specimens of chrysomelids over 130 trees fogged in east Africa. Alticinae remain the most abundant, even during the dry season in the canopy. Wagner thinks they aggregate along a humidity gradient, but we know that leaf beetles will feed to some extent on non-host plants, and even tree leaves, during the absence of their host plants. The author believes in a low specificity of chrysomelids to tree species. As everywhere, Alticinae, Galerucinae and Eumolpinae, which are good flyers, remain the most common species on tree crowns. In America, Megascelinae also frequent the canopy. Other subfamilies, heavier and not such good flyers, are not so well represented, and Chrysomelinae and Cassidinae are exceptional on the tree tops. Hispinae are not usually found and Eumolpinae are less specific than Galerucinae (Novotny *et al.*, 1999). This can be linked to the polyphagous larval habits among Eumolpinae.

Sometimes, what appears to be a choice, like *Ranunculus* spp. for *Chrysolina*, can be a survival of an antique selection of a primitive plant. The food plant choice made by a leaf beetle can be investigated by observation and testing. The testing of food preferences has only been carried out among a few economically important species, like the Colorado Potato Beetle (Hsiao, 1974, 1981, 1986, 1988) and a few others. *Diabrotica, Aulacophora* and *Systena*, for instance (Jolivet, 1999c) are primary cucurbit feeders and secondarily polyphagous. Other species of ornamental or biological importance have also been well tested (*Timarcha*, *Chrysolina*, *Oreina*, *Chrysomela*) but for them, except for *Oreina*, we are still a long way from knowing all the host plants. This is why we should continue to be sceptical about the cladistic concordances between host plants and their specific insects.

The food selection habits of many European species also remain unknown. These include *Cyrtonastes* (Greece, the eastern Mediterranean islands and Syria) and *Cecchiniola* (Crimea), etc. Of course, we are dealing here with rare, localized species, leading hidden and nocturnal lives and slowly vanishing from the surface of the planet. Here also the information given is sometimes incorrect; *Promechus* (Chrysomelinae) in New Guinea seem to accept plants other than Araliaceae, and *Iscadida* (Chrysomelinae) in South Africa and Zimbabwe feed on plants other than Vitaceae, although these plants would be their first choice. The chosen plants are often not related. The reliability of lists of insects and plants is often questionable and simple spatial coincidences are frequently reported by lazy observers as food relationships. Their data are vague and inconclusive. It is a question of educating the collectors and stressing that observations should be verified by laboratory tests. Most of the observations in the tropics are done on agricultural pests, but in Brazil, Argentina, Chile, Panama, Costa Rica, Nicaragua and Mexico, data on many other leaf beetles have been collected by dedicated specialists. Unfortunately, lists which were drawn up some time ago, such as those collected from Zaire, reveal the unreliability of the information transmitted to the taxonomist by the collectors (Selman, 1972). It is also very difficult during the summer or the winter diapause to revive the beetles and to make them feed. In conclusion, too many of the existing lists are unreliable and some places, such as Madagascar and New Caledonia, lack the specialists to carry out the research.

Plants and phytophagy

In the geological past, leaf beetles were probably polyphagous, but, before the appearance of the Angiosperms during the Mesozoic, the choice of food material was limited. Cycadoidea were the most likely plants of choice, along with coniferous trees, ferns and related forms. It is almost certain that vascular cryptogams, like ferns, Pteridospermae, horse-tails, lycopods and cellular cryptogams (to use an old classification), like mosses, algae, lichens and even bacteria, were also eaten, as a supplement, during the Trias and early Jurassic, but the diversity of leaf beetles coincided with the explosion of the Angiosperms during the late Jurassic and the Cretaceous. Later on, some species of Alticinae and others reverted to ferns, horse-tails, coniferous trees and mosses. Some primitive groups (Aulacoscelinae and Palophaginae) remained faithful to cycads and *Araucaria*, but there are exceptions. Other primitive, or relatively primitive, subfamilies (Donaciinae, Sagrinae, Megalopodinae, Chrysomelinae, Timarchinae) have adapted to other developed Angiosperms, even though some of them have maintained their relationships with Ranunculaceae and Nymphaeaceae. The number of Angiosperms increased over time (Burges, 1981). From around 500 species in the late Carboniferous (Hughes, 1976), the Angiosperms increased to 3000 species at the beginning of the Cretaceous, to around 22 500 species by the end of this period. That was followed during the Tertiary by an explosive diversification which gave rise to a contemporary land flora of about 300 000 vascular plant species. By comparing the relative size of the wings of Mayflies (Ephemeroptera) with the 'weight' of the ancient atmospheres, John L. Cisne (Cornell News Releases, 19.12, 1999) infers that the atmosphere was practically the same from the Permian to the present. Coevolution, if there was coevolution between plants and insects, existed under conditions similar to the present. However, pre-Cretaceous flora was not as strongly differentiated as it is today. Note that we do not have real fossil Angiosperms before the Trias and so we are forced to make assumptions. The diversity of chrysomelids does not match the diversity of the Angiosperms and so there will be many available niches which are not occupied by the insects. That is also the conclusion of Spencer (1990) dealing with Agromyzidae. Now, more than a tenth of the world's plant species are heading towards extinction. These are mostly cycads, like *Encephalartos woodi,* which is now already extinct. Insects living on these plants will also disappear fast.

As far as the evolution of phytophagous insects is concerned, the possibility of 'bridge species' (Hering, 1950, 1951; Jolivet, 1998c) should also be taken into account. According to this theory, two species that live on two different family groups of plants, not closely related, are able to transfer from one to another group through other plant species, the bridge species. This common denominator needs only to possess some common chemical substances. Pilson (1999) believes that plant hybrids bridge the genetic gap between actual and potential host species. So, herbivorous insects are more likely to evolve an expanded host range in the presence of hybrids. The theory was already presented by Hering (1950).

Southwood (1985), who also thought that insects did not use all the available plant diversity, recognized two evolutionary routes to herbivory: via saprophagy, i.e. eating dead and dying plant material, and via pollen and fruiting bodies, strobili of Pteridophyta or early Gymnosperms. A third route could have been provided via algae

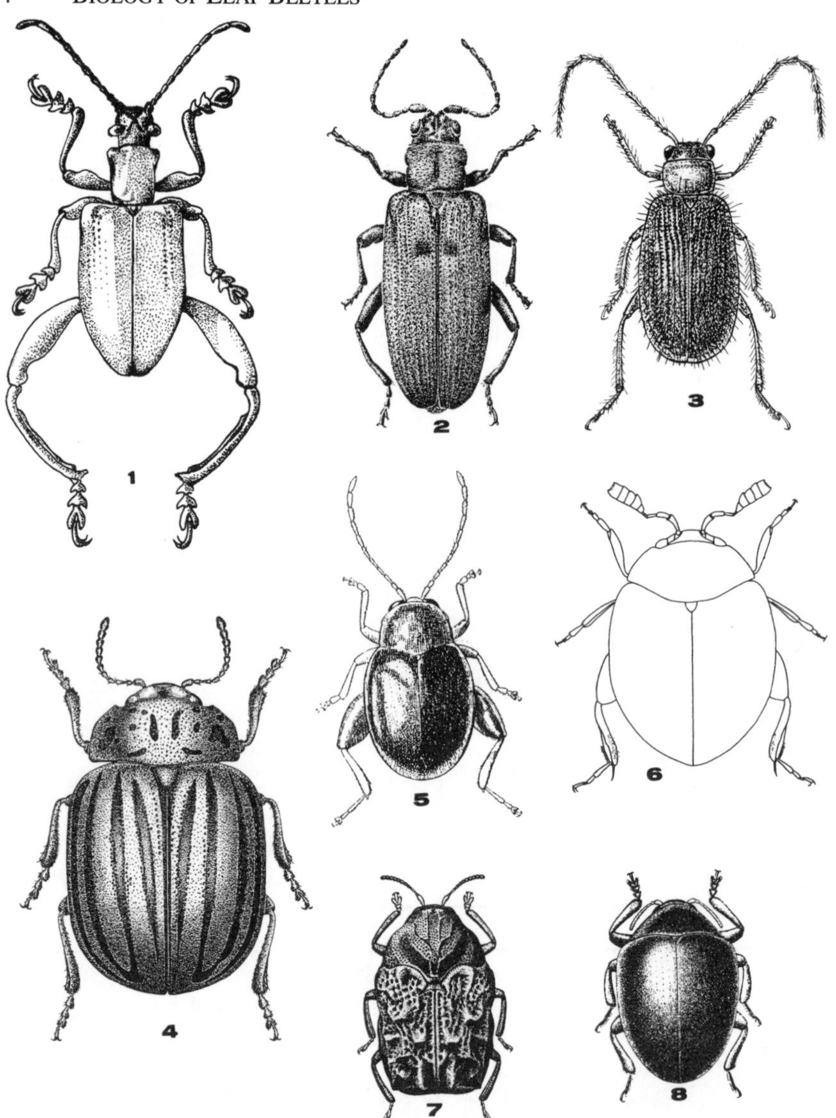

Figure 4.1. (1) *Sagra femorata* Drury, 1773 (Sagrinae) from Thailand. Feeds on the stems of various legumes (mostly Fabaceae) and is gallicolous at the larval stage (after Jolivet, 1995). (2) *Donaciasta goeckei* Monros (=*D. garambana* Jolivet) (Donaciinae) from Eastern Zaire and Burundi. Polyphagous but also on Nymphaeaceae (after Jolivet, 1972). (3) *Strobiderus (Cerophysella) tonkinensis* Laboissière, 1930 (Galerucinae), North Vietnam. On *Ipomoea batatas* L. and wild species (Convolvulaceae) (from Laboissière, 1930). (4) *Leptinotarsa decemlineata* (Say) 1824 (Chrysomelinae). Specimen from western Europe. On *Solanum* spp., including *S. tuberosum* L. and *Lycopersicum esculentum* (L.). It has invaded the USA, southern Canada, Europe and western Asia from its Mexican origin It will probably reach Korea within a few years (Jolivet, 1995). (5) *Longitarsus nigripennis* (Motschulsky) 1866 (Alticinae) from India. The larva hollows out the pepper berries and the adult shaves the pepper leaves (*Piper nigrum* L., Piperaceae) (after Maulik, 1926). (6) *Clavicornaltica besucheti* Scherer, 1974 (Alticinae) from Sri Lanka. The larva feeds on the roots and is subterranean. The adult is partly endogenous. Some species are found in the canopy, probably among the epiphytes (after Scherer, 1974). (7) *Chlamisus superciliosus* Gressitt (Chlamisinae) from China. Host plant unknown, but probably a member of the Rosaceae (from Gressitt and Kimoto, 1961). (8) *Oomorphoides pallidicornis* Gressitt and Kimoto, 1961 (Lamprosominae) from China. Feeds on Araliaceae (from Gressitt and Kimoto, 1961).

(see Crowson, 1998). According to Southwood, interactions between plants and animals started 250 myr ago, and plants evolved 70 myr before insect herbivory. The earliest damages to fossil leaves started in early Permian (Rohdendorf and Raznitsin, 1980), when probably cerambycids and chrysomelids were splitting off. See also Smart and Hughes (1973).

Probably, during the Cretaceous and the Tertiary, leaf beetles became oligophagous, or even monophagous, and only some of them remained (or became secondarily) polyphagous (Donaciinae, *Systena* among the Alticinae, numerous Eumolpinae having lost the Asclepiadales basic selection and many others). Some highly polyphagous genera, like *Diabrotica* and *Aulacophora*, kept their ancestral taste for the pollen and flowers, and eventually the leaves of Cucurbitaceae. Jolivet calls this 'ecological selection' (Jolivet, 1988a): the selection of Donaciinae limited to aquatic and semiaquatic plants, for adults and larvae, and the selection of species of the tribe Paropsini often restricted to *Acacia* (Mimosaceae) and *Eucalyptus* (Myrtaceae) trees, the dominant trees in Australia. Selection by *Papilio* spp. of Apiaceae and Rutaceae in Israel and elsewhere depends on the chemistry, the families being unrelated. That is also ecological selection, the two plants not being available together. Ecological selection is then the acceptance by phytophagous insects of plants not necessarily related taxonomically but present in a given biotope. In this case, and in many others, the supporters of the comparative cladogenesis between insects and host plants find their notion thwarted. Very often, plant selection occurs at random, even if, in certain cases (Donaciinae, *Galerucella*), the primitive selection (Nymphaeales) reappears. Becerra and Venable (1999) outlined the need to consider the correlation between host phylogeny, host chemistry and host geography in the insect plant selection. Host availability and chemical similarity both play a role in that field.

Certainly, green algae like *Pleurococcus*, growing on the trunks of trees, were a possible source of food in the earlier stages of chrysomelid evolution (Crowson, 1998). Jolivet (1991) mentioned the actual use of fungi, *Venturia* conidia (some alticine adults), algae and lichens (Camptosomata larvae), bacteria (feeding of some galerucine larvae on nodules of the roots of Leguminosae) etc. by living chrysomelids. It is true that many case-bearing Camptosomata larvae, like Lamprosomatinae, Chlamisinae, Cryptocephalinae, feeding on trunks or stems of plants, also commonly ingest bits of algae, mosses and lichens (Erber, 1988).

Uriarte (2000) examined how the interactions between nitrogen availability, leaf tissue quality and insect herbivore pressure vary over time. She took the relationships between the galerucine *Trirhabda virgata* Leconte and the goldenrod (*Solidago altissima* L.), its host plant, as an example. It is evident that the attractiveness of a given plant can vary over time for a chrysomelid, but this topic cannot be discussed here.

Given below are details of the main selections, as far as it is known, of the chrysomelid subfamilies, following the classification by Seeno and Wilcox (1982), Verma (1996c) and Jolivet (1957–59, 1988a, 1997). Other classifications are discussed in Chapters 2 and 11, but these do not differ much, even those which have been modified by young specialists! (*Nil novi sub sole:* 'nothing new under the sun'). I think that cladistics has benefited botany more than chrysomelidology, even if molecular biology can improve the understanding of generic and specific classification. For more details on host plant/chrysomelid relationships see Jolivet and

Hawkeswood (1995) and the papers of Jolivet (1977, 1978, 1987a,b, 1988a,b, 1989d, 1991, 1995).

SAGRINAE

This subfamily is certainly primitive, and its roots are more connected with Bruchidae (*Rhaebus*, *Eubaptus*) than with Cerambycidae. The galling habits of the larvae, unique among the leaf beetles, are certainly not comparable with the seed-boring habits of the Bruchidae. Sagrinae are found chiefly in south-east Asia and Australia, but are also present in all the southern tropics. They exist in Madagascar, tropical Africa and southern America. They were present during the Cretaceous and the Cenozoic in the Northern Hemisphere, but disappeared probably with the glaciations of the end of the Tertiary. The stem-boring habits of the Megalopodinae and probably Aulacoscelinae larvae are typically a cerambycid character, but Sagrinae and Megalopodinae have adopted a great variety of plants among the Angiosperms without much selectivity.

Fossil sagrines are known during the Eocene and even earlier during the Cretaceous (Canadian amber) (Poinar, 1999). There is even a possibility that *Mesosagrites multipunctatus* Martynov from the Jurassic is a real sagrine (Martynov, 1935). The group is primitive, old and very probably moved from the Cycadoids to the Angiosperms during the Cenozoic. Early coniferous feeding, as for Palophaginae, is probably a parallel selection to that of Cycadoidea. It is evident that the Proto-coleoptera from the Permian were feeding on more primitive plants and probably on spores and strobili.

Sagrines do not jump like Alticinae, but their strong, inflated hind femora are used to grasp firmly the stems of the food plant. This is a similar system to that found in Megalopodinae. Paulian (1942) studied the femoral endoskeleton in the sagrines and described a sclerite, corresponding to the Lever organ, an apomorphic structure, rather enigmatic in its function, but very probably reinforcing the muscular connections of the tibio–femoral articulations. A similar structure, according to Paulian, exists in the mesofemora. According to Furth and Suzuki (1990a, 1992), the Lever's organ is the flexor tendon and is very prominent in *Sagra*. However, the extensor tendon (= metafemoral spring or Maulik's organ) is considerably less developed in *Sagra* than in Alticinae. Therefore, it is clear that earlier authors were correct in their view that sagrines use their large legs and associated muscles for flexion and not extension, i.e. for grasping and not for jumping. According to Furth (pers. comm.), male metafemora are often somewhat larger and are probably also used to hold the female during copulation.

As mentioned above, Sagrinae disappeared from the boreal regions at the end of the Pliocene with the glaciations, but remained common in Africa, Madagascar, New Guinea, and south-east Asia (*Sagra* spp.). Archaic Gondwanian genera survived in Australia, New Guinea, Madagascar, Argentina and Brazil. A small species of the genus *Megamerus*, *M. alvarengui* Monros was described from Brazil (Monros, 1956) in the Noroeste region, where, at the beginning of the rainy season, one can see the eclosion of the adult. The genus is known to occur also in Australia and Madagascar. It is the same story in the south of the Sahel area (11.12 N) in Africa where *Sagra* appears in mass just one week after the first rains. Blue species of *Sagra* are relatively common in tropical Africa, while brightly coloured ones are represented in south-east Asia. Australia hangs on to the greatest diversity of the archaic genera, generally uniformly light brown in colour. The same general colour is found in South America and Madagascar. The biology of the genus *Atalasis* is well known in Argentina (Monros, 1955a) and the larva well described. It is a stem-borer with a pupa in a gall as in other known Sagrinae. *Sagra* larvae have been well described by Maulik (1941) and several Indian entomologists. So far, outside *Atalasis* and *Sagra*, no other sagrine larvae have been found or described.

The sagriniform Bruchidae live inside the seeds or drupes of different plants. *Eubaptus* in Argentina feeds on *Ruellia* (Acanthaceae) and the bright green *Rhaebus*, in Oriental Asia, on *Nitraria* (Zygophyllaceae). The spermatophagy of these insects remind us of the Bruchids, but the external morphology superficially resembles sagrines, which are not spermatophagous but borers and gallicolous into the semirigid stems of various Angiosperms. Feeding externally, and not internally, on seeds is exceptional among chrysomelids and can be found only among several Criocerinae on Commelinaceae. Kingsolver and Pfaffenberger (1980) keep *Rhaebus* among Bruchidae, creating a special subfamily, the Rhaebinae. Previous authors studied five species of the genus and the larvae of one species. Some characters are certainly primitive and connect *Rhaebus* to *Megamerus* and to *Carpophagus*. In any case, the wing venation does not belong to the sagrine pattern.

Unfortunately, we do not know much about the archaic sagrine genera from the Southern Hemisphere, except for the Argentinian *Atalasis*. These genera, viz. *Coolgardica*, *Ametalla*, *Megamerus*, *Mecynodera*, *Neodiaphanops*, *Pseudotoxotus*, *Polyoptilus*, *Diaphanops*, *Duboulaya* and *Carpophagus* are, in most cases, attracted during the night by light traps. Lionel Stange (pers. comm.) recently collected a *Megamerus* with a light trap in south Western Australia, near Perth. The beetle is kept in the DPI collection in Gainesville, FL. *Megamerus* is known as gallicolous from Madagascar. One species is known in NE Brazil, on the Natal coast, on bushes. It is a dry area with intermittent violent rains, and the pupa very probably waits inside a gall for eclosion. I (P.J.) have seen *Megamerus alvarengai* Monros in Sâo Paulo. It is much smaller than the Madagascan or Australian species but has the same colour and pattern.

The data we actually get on Australian species are vague, with little information on the true host plants of the larvae and the adults. However, Australian Sagrinae can be polyphagous like *Sagra* themselves, but we must note also that *Atalasis sagroides* Lacordaire in Argentina remains oligophagous on various genera of Malvaceae. According to Reid (1995), one species of *Diaphanops* has been observed feeding on the pollen of *Melaleuca* sp. (Myrtaceae), and one *Mecynodera* on the pollen of *Acacia* sp. (Mimosaceae) (Poinar, 1999). In New South Wales, Australia, *Mecynodera coxalgica* (Boisduval) has been collected (Webb, 1986) on flowers of *Kunzea* sp. (Myrtaceae), feeding on nectar and pollen. Feeding on pollen does not indicate any host plant relationship and that seems to be a kind of ecological selection, Mimosaceae and Myrtaceae being the most abundant trees there. More interestingly, *Carpophagus banksiae* MacLeay, a bruchiform genus, has been noted on *Banksia* sp., an xerophytic bush of the Proteaceae family (Rosales), but it has also been described on *Macrozamia* sp. (Zamiaceae, Cycadales) in temperate Australia (Jolivet, 1988a). Probably, all those sagrines are more or less polyphagous, at the larval and also the adult stage.

They need only an arbustive tender stem, sufficiently rigid to hold a gall with a pupa, to develop. These species are probably all borers and gallicolous and are attracted by flower nectar and pollen as adults. But we must be careful when putting forward these hypotheses. Some species can pupate into the ground, like Megalopodinae, and any scenario is possible.

The genus *Sagra* is distributed in the tropical areas of the Southern Hemisphere, except southern Australia and South America. The genus has been divided into three subgenera (Monros, 1958): *Prosagra* from SE Asia, *Sagra* from SE Asia, Indonesia, New Guinea, and *Tinosagra* from tropical Africa and Madagascar. Its biology is cerambycid-like not bruchid-like. *Sagra* species, and mostly the males, have enormous inflated hind femora to grasp the stems of the plants on which they walk after eclosion, and on which they feed. Normally, sagrines have very well-developed wings and, despite their weight, they occasionally fly. *Sagra* spp. do not jump and, when their function was not well understood, the size of their femora was considered as a case of hypertely. Colours vary between the beautifully multicoloured Asian species and the smaller blue, green, reddish or violet African species. Size and colour vary also considerably inside one species. Such variation can be due to genetics (colour) or the quality and abundance of larval food (size).

All *Sagra* larvae are borers into stem or, rarely, root. The larva digs inside semilignous bushes or plants rigid enough to contain it and the gall produced later. According to Maulik (1941), up to 20 larvae can be found inside the gall formed around the stem. The larva feeds on internal tissue and, when big enough, builds a cocoon. Pupation takes place inside the inflated stem. Later on, the adult will pierce a hole before leaving, but already the gall wall is suberified and weak enough to make escaping easy. So, the larva spends its entire life inside an edible tissue, the stem, as do Megalopodinae but, as distinct from them, this larva is gallicolous. Pupation takes place there and not in the ground, as far as our present knowledge goes.

Sagra larvae live inside stems of a great number of Fabaceae (*Dolichos*, *Faba*, *Mucuna*, *Phaseolus*, *Indigofera*, *Pueraria*, *Pachyrhizus*, *Vicia*, *Canavalia*, *Vigna*, *Eriosema*), of Caesalpinaceae (*Cassia*, *Delonix*, *Moghania*), of Cucurbitaceae (*Cucurbita*), of Menispermaceae (*Cocculus*), of Combretaceae (*Terminalia*), of arborescent Convolvulaceae (*Ipomoea*), of Verbenaceae (*Tectona*), of Dioscoreaceae (*Dioscorea*), of Rutaceae (*Coffea*), of Rhizophoraceae (*Rhizophora*) in the mangrove, of Acanthaceae (*Thunbergia*) and many other semiarbustive plants. Mass eclosion of the adults occurs at the beginning of the rainy season in Africa and can be more erratic in the humid tropics like Vietnam or Indonesia.

Larvae and adults are polyphagous, and imported *Eucalyptus* trees, *Solanum melongena* L., the egg-plant, or *Vanilla* climbers can also be attacked by *Sagra* in tropical Africa.

Sagra eggs are laid on the bark and cocoons are sometimes situated in the aerial roots of *Rhizophora* trees in the mangrove in Java or in the roots of *Coffea* in Africa.

Atalasis sagroides, the Argentinian species, a Megamerini, not a Sagrini, as pointed out by Clavareau (1913), feed on various Malvaceae (*Sphaeralcea*, *Malvastrum*, *Gossypium*, *Sida*, *Abutilon*). The stems must be rigid enough to contain the larvae. Monros (1955a, 1959) has described in detail the larvae and biology of the genus. Adults feed on the leaves, the flowers, the pollen and the stamens. As for *Sagra*, the larva, a borer, produces a gall and very probably nymphosis takes place there. *Atalasis sagroides* is localized in the dry areas of NE Argentina. The larvae are gallicolous and not only stem-borers as are the megalopodines. It is possible, though very improbable, that, in certain sagrines, pupation takes place in the ground.

As we can see here, we do not know anything of the biology of the archaic genera in Australia, Madagascar and NE Brazil. However, the life histories of *Atalasis* and *Sagra* are actually well known. The general biology of the group is probably similar.

AULACOSCELINAE

Crowson (1946), in revising the chrysomelid group Sagrinae, placed *Aulacoscelis* within a special tribe. Then Monros, in 1949, excluded the Aulacoscelini from the Sagrinae. Finally, the new subfamily Aulacoscelinae was described by Monros in 1953 and the group was revised in 1954 by Monros himself, with few data on its biology. Two genera were described: *Aulacoscelis* from North and Central America and *Janbechynea* from Mexico, USA, Peru, Bolivia and Brazil.

The systematic position of Aulacoscelinae is questionable. The group is very primitive and probably related to the Protoscelinae (Medvedev, 1968). *Aulacoscelis* spp. have certainly some affinities with Chrysomelinae and Orsodacninae (wings), with *Timarcha* (aedeagus and tibial spurs), and they were placed by Suzuki (1988) in the chrysomeline subfamily. In 1992, Suzuki finally placed the aulacoscelines between Megalopodinae and Sagrinae, which seems reasonable.

Both genera, *Aulacoscelis* and *Janbechynea*, seem to feed exclusively on cycad leaves but, while we have many records from Mexico and Central America for *Aulacoscelis* (Jolivet, 1998b; Cox and Windsor, 1999a,b; Windsor *et al.*, 1999), data on *Janbechynea* are scarce. We know only that one species, *Janbechynea elongata* (Jacoby), was found in Laredo, Texas, on imported cycads (Monros, 1954b). It is evident that native *Zamia* are found in the United States only in Florida, where no aulacoscelines exist, and that in Texas, New Mexico and Arizona, aulacoscelines were found only on imported *Cycas revoluta* Thunberg or other species of cycads. Only in Mexico are native and imported cycads the food plants. *Aulacoscelis* spp. seem also to feed on flowers, probably the pollen, of various Asteraceae and they have been reported on *Hechtia texensis* S. Watson, a Bromeliaceae. Crowson (1991), dissecting the gut of *Aulacoscelis melanocera* Duponchel and Chevrolat, found abundant pollen grains inside. However, *Aulacoscelis* spp. feed normally in Central America and Mexico on *Zamia* fronds in the wild. *Zamia* seems to be the preferred host. Jolivet has seen in Xalapa, Mexico, that *Aulacoscelis melanocera* occurs commonly in spring on the local species of *Zamia* but also on many other genera or species of cycads grown in the botanical garden and the local greenhouses of the Institute of Ecology. According to local gardeners, even the Cuban *Microcycas calocoma* Miquel is eaten. When I (P.J.) was there, the leaves had been partly eaten. Windsor and Jolivet in Panama, and Ness and Gomez in Costa Rica (Windsor *et al.*, 1999) observed *Aulacoscelis appendiculata* Cox and Windsor and *A. costaricensis* Bechyne feeding on *Zamia fairchildiana* Gomez and other species of *Zamia*. The mandibles of *Aulacoscelis* spp. are used to pierce the epidermis of the fronds, thereby releasing liquids which alone are ingested. *Aulacoscelis* thus feed on young and tender fronds appearing just before the spring rains. These fronds are a tender green in contrast with the dark green leaves of the previous year. Often, *Aulacoscelis* spp. feed on fronds already damaged by adult

Languridae (*Nomotus* spp.) and larvae of the lycaenid *Eumaeus goddarti* (Hübner) or other species of *Eumaeus*. *Aulacoscelis* also feeds on *Eumaeus* pupal exudates. Feeding occurs in Central America only at the beginning of the wet season. In Florida, one species of *Eumaeus* feeds on the local *Zamia* fronds, and one langurid and several curculionids feed on the male cones, but no aulacosceline is present. Aulacoscelines do not feed normally on the cones and so do not intervene in the pollination. See in Crowson (1989, 1991), Jones (1993), Norstog and Nicholls (1997) and Jolivet (1998b) a review of the relationships between the beetles and cycads.

So *Aulacoscelis* spp. and probably *Janbechynea* spp., occur in semiarid areas after the rains, as in Xalapa, Mexico, for instance, but also in the wet understory forest as in Panama, Costa Rica and Nicaragua. *Aulacoscelis appendiculata* Cox and Windsor sometimes migrates with the wind in Panama near the continental divide, where the beetle may rest on non-host plants, before moving again. The motive behind this migration is still obscure, but probably it is in connection with the colonization of new territories by cycads.

In Panama, Windsor and Jolivet used to rear *A. appendiculata* on the local species of *Zamia* in the laboratory. Windsor obtained the neonate larvae which were carefully described later on (Cox and Windsor, 1999a,b). *Aulacoscelis* and *Nomotus* adults often occur in spring in aggregations on new fronds of *Zamia*. These relationships with *Zamia* and other cycads seem to be very old and probably go back to Jurassic. Cycasin and other compound glucosides present in *Zamia* and other cycad tissues are very toxic to mammals and birds. Aposematism on the tender green fronds is probably useful to the insect but not yet learned by chickens in the New World, which die quickly after feeding on them. In the laboratory, *Aulacoscelis* and *Nomotus* adults drink eagerly on sugary liquids of mango fruit, a south-east Asian imported plant common in Panama. Very probably, Protoscelinae were cycadophiles since cycad foliage is also known from the Kara Tau series (Medvedev, 1968).

Aulacoscelis is not unique among chrysomelids in ingesting leaf juices. The female Megalopodinae cuts the tip of the stems of plants and drinks the juices before laying eggs inside. Megalopodinae are also a very primitive family, related to cerambycids, and the species has retained old habits.

As mentioned previously, aulacoscelines probably feed occasionally on pollen. So far, no real observations have been made in the field of adults eating pollen grains. It seems that *Aulacoscelis* spp. oviposit at the base of *Zamia* plants. Do the beetle larva bore into the plant rhizomes? It seems probable, since adults have been seen emerging from the soil around a cycad. Pupation must occur in the ground. Another possibility is that *Aulacoscelis* larva is a stem-borer, but it seems doubtful.

Aulacoscelis species can be found at sea level in Central America to about 1500 m. The large species of the genus *Janbechynea* can be found at very high altitude in Bolivia.

According to Cox and Windsor (1999a,b), who described larvae of *Aulacoscelis appendiculata* from Panama, this larva shares eight synapomorphies with Orsodacninae and four with Chrysomelinae. The aedeagus closely resembles that of *Timarcha* and this confirms the special position of the last genus.

ORSODACNINAE

This subfamily is mostly holarctic, but it seems that *Cucujopsis* Crowson, 1946, from

Queensland, is by its wing venation more orsodacnine than palophagine. First stage larvae of two species of *Orsodacne* have been described by Cox (1981) and Mann and Crowson (1981) and have been obtained from eggs or from bird gizzards (Van Emden, pers. comm.). It is very probable that those larvae are borers inside the buds of various plants but, as for the Aulacoscelinae and even the Megascelinae (very probably root-eaters on Fabaceae), nothing definite is known about their biology. The adult mouth parts are certainly adapted to pollen eating and the adults frequent mostly white flowers of many plants like Rosaceae. They are also strongly attracted by the racemes of *Acer* trees (Aceraceae), where pollen is abundant and honey is freely exposed on the disk. Adults, and very probably larvae, seem to be fully polyphagous. According to Cox (1981), the larva is closely related to Galerucinae and Alticinae, and he believes that it could be expected to feed extensively on the roots of plants, like *Syneta*.The fact that *Orsodacne* larvae seem to be common in tits' gizzards supports the theory that they are either stem- or leaf-mining, but a bud-borer habit seems to be the most likely.

The adults are floricolous (Jolivet and Hawkeswood, 1995) and consume the pollen or anthers of mostly white flowers of Rosaceae (*Crataegus*, *Cerasus*, *Prunus*, *Malus*, *Mespilus*, *Rosa*, *Sorbus*, *Spiraea*, *Filipendula*), Caprifoliaceae (*Viburnum*), Orchidaceae (*Listera*), Dioscoreaceae (*Tamus*), Oleaceae (*Ligustrum*), Aquifoliaceae (*Ilex*), Fagaceae (*Quercus*), Apiaceae (*Heracleum*, *Bupleurum*), Euphorbiaceae (*Mercurialis*), and Valerianaceae (*Valeriana*). In the USA, *Orsodacne atra* Ahrens has been captured on *Galium* (Rubiaceae), Caesalpinaceae (*Cercis*), Anacardiaceae (*Rhus*), Hamamelidaceae (*Hamamelis*), Lauraceae (*Lindera*), Rosaceae (*Rubus*, *Spiraea*, *Prunus*, *Malus*, *Amelanchier*), Aceraceae (*Acer*), Cornaceae (*Cornus*), Carpinaceae (*Carpinus*), Betulaceae (*Alnus*), Corylaceae (*Corylus*), Salicaceae (*Salix*), Scrophulariaceae (*Verbascum*) and Iridaceae (*Iris*). Adults do not seem to feed on leaves and petioles but only on pollen. According to Mann and Crowson (1981), the mouth parts of adult *Orsodacne* and *Aulacoscelis* are pollen-eating adaptations. That is true for *Orsodacne* but not so for *Aulacoscelis*, which pierces and rasps cycad fronds to extract the juices. These beetles are juice-addicted and suck mango fruits with the same avidity. According to Cox (1981), feeding on pollen, which is rich in proteins, enables the female beetles to lay many eggs (20–35). Cox places the subfamily Orsodacninae close to Zeugophorinae and Donaciinae (basis: adult characters), but also close to Galerucinae and Alticinae (larval characters). In reality, it seems that both adults and larvae are primitive and must be placed at the base of the genealogical tree.

According to Kuschel and May (1990) and Crowson (pers. comm.), *Cucujopsis* develops in the male cones of *Araucaria* or *Agathis* (Araucariaceae) in Australia and probably on *Wollemia nobilis*, the new member of the group. The larva is very similar to that of the Cerambycidae and if this relationship is confirmed (the notion is still hypothetical), this would be further confirmation of the antiquity of the group. Coniferous trees or cycads were surely ancestral choices among the chrysomelids.

PALOPHAGINAE

This new subfamily created by Kuschel and May (1990) is found at present in Australia and Chile–Argentina (Kuschel and May, 1990). It is highly probable that it exists also in New Guinea, Vietnam, Malaysia, New Zealand, The Philippines, Norfolk Island, S. Brazil and New Caledonia where it has not yet been searched for. *Araucaria* and *Agathis*, the host plants, have not survived in South Africa where probably the group is now extinct. The beetles must also be present on the new species of araucarian tree found recently near extinction in the Blue Mountains, near Sydney (*Wollemia nobilis*) (da Silva, 1997). Recently, also, Palophaginae, *Palophagus vargasorum* Kuschel and May, have been found in Chile and Argentina (Kuschel and May, 1996a,b). It seems that the Araucariaceae family first appeared in the Triassic, some 250 million years ago, then started a dramatic retreat at the end of the Cretaceous and at that time the plant family disappeared from the Northern Hemisphere and South Africa altogether.

Adults and larvae of *Palophagus* live on the male strobili (cones) of *Araucaria* trees and feed on pollen. The larva usually inhabits the central pith of the male strobili, chewing the stalks of the spent sporophylls. In the younger cones, there is never more than one present at a time feeding on the pollen in the company of the larvae of Nemonychidae weevils. The ambulatory ampullae on both dorsal and ventral surfaces enable the larvae to move with ease among the sporophylls (Jolivet and Hawkeswood, 1995). The larva pupates in the soil where it passes through an additional but inactive instar (hypermetamorphosis) (Kuschel and May, 1990). The fully-grown larva, having produced a pupal cell, moults to an inactive pupal stage which is morphologically similar to the previous instar but lacks pigmentation and has simplified mouth parts. Moulting, according to Kuschel and May (1990), to the true pupal stage occurs after 5–6 months and the final ecdysis to adult after 30–50 days of pupal period.

Adults and larvae of *Palophagus bunyae* Kuschel and May feed on *Araucaria bidwillii* Hooker and *P. australiensis* feed probably also on another species of *Araucaria*, *A. cunninghamii* D. Don. *Palophagoides vargasorum* Kuschel and May, from Chile, feeds on *Araucaria araucana* (Molina) K. Koch, a tree with leaf duration of 10–15 years. Larvae of *P. vargasorum* develop also in the male strobili and they are fully grown by the time the cones burst open and release the pollen. As for the Australian species, the mature larva drops to the ground and enters the soil to pupate. It also shows a prepupal stage. Only a few adults have been reared and, as is the case for *Orsodacne*, *Aulacoscelis*, *Megascelis*, rearing artificially is very difficult, due to fungal infection and predators.

Palophaginae are related to Orsodacninae and Megalopodinae. Their morphology and feeding habits are very primitive and the choice of Araucariaceae a very ancestral characteristic. Many new discoveries are expected when the *Araucaria* and related species are thoroughly investigated.

ZEUGOPHORINAE

This small holarctic family is also present in the Old World tropics. *Zeugophora* larvae are miners in the young leaves of various trees and shrubs and the adults feed on the leaves of the same plants. *Zeugophora* spp. are common on leaves of Salicaceae (*Salix*, *Populus*), Juglandaceae (*Juglans*), Corylaceae (*Corylus*) and Betulaceae (*Betula*). Until now, *Zeugophora* species have not been collected, like *Donacia*, in the Neotropical region, but they may have been overlooked. *Donacia* is present in Cuba.

Pedrilla is tropical, but is also present in Japan, where several other tropical or subtropical species are breeding. It feeds on *Santalum* (Santalaceae), *Populus* (Salicaceae), *Evonymus* and *Tripterygium* (Celastraceae), etc. *Pedrilla* seems to be polyphagous but perhaps it can be also monophagous, at least at the species level. The presence of an adult of *Pedrilla annulata* Baly in China on *Isachne* (Poaceae) may be accidental, but it may also be for pollen feeding.

Zeugophorinae have primitive genitalia with a circular tegmen and the base deeply divided. Leaf-mining habits are not specially primitive, but can be considered as a means of protection against predators.

DONACIINAE

Adults and larvae of Donaciinae are polyphagous on aquatic and semiaquatic plants (ecological selection). Nymphaeaceae can be considered as a primitive selection and actually this family is placed at the very beginning of the Angiosperms. Donaciinae are primitive insects, related to cerambycids, and their genitalia and wing venation have ancestral characters. They are also the only purely aquatic chrysomelids. Their biology will be discussed in another chapter. Only semiaquatic chrysomelids are known among some other families. Hispinae, regarded by some as related to Donaciinae, contain a few species of the neotropical genus *Cephaloleia* living inside the inflorescences of the *Heliconia*, a true phytotelmata. They are specially adapted to a real aquatic life (Jolivet, in print).

The distribution of Donaciinae is holarctic, but they have made southern invasions in the West Indies, Central America, Malaysia, Indonesia, India, Thailand, southern New Guinea, Queensland, tropical and southern Africa, and Madagascar. They are found in lakes, ponds and slow rivers. Why should they be localized in New Guinea only along the Fly River and not in the north or the highlands? Perhaps an excess of predators is responsible. *Donacia* must be a recent colonization in the south. Why southern Australia or South America have not been invaded is another mystery. *Donaciasta* from Africa is more dull in colour than its holarctic counterparts, exactly like *Haemonia* or the rare, entirely aquatic species of *Donacia*. This probably suggests a more aquatic way of life. A few species of holarctic *Galerucella*, like *G. nymphaeae* (L.), feed on Nymphaeales, but this could be a secondary adaptation, like Alticinae reverting to mosses, ferns, horse-tails, and chrysomelid lines recolonizing coniferous trees.

Donaciinae fly well and fast. They fly out of water and from reed to reed, the males looking for females. Most of them are aquatic only in the larval stage (*Plateumaris*, *Donacia*), others remain in water (*Neohaemonia*, *Macroplea,* some rare *Donacia*), except for disseminating flights and mating.

A detailed list of host plants, none really specific, is given in *Host Plants of the Chrysomelidae of the World* (Jolivet and Hawkeswood, 1995) for the Holarctic and the tropics. Everywhere, mostly in the tropics, Nymphaeaceae seem to be the preferred food of the adult. Larvae feed mostly on *Myriophyllum* and related species. They take in air through specially adapted crochets. One species found in the Baltic and probably in other seas (North, Caspian) with low salinity, is the only marine leaf beetle, *Macroplea mutica* (Fabricius). It lives in brackish water on *Ruppia*, *Zostera*, *Potamogeton* and *Myriophyllum*, plants belonging to four different families. It is most remarkable to find a donaciine on *Zostera*, since the big marine ribbons do not harbour any other beetle.

Plateumaris prefer lentic habitats (lakes, ponds, swamps), *Donacia* is more common on slowly running water (rivers, canals), but both can be found in small lakes. *Macroplea* and *Neohaemonia* live at the bottom of small ponds among *Potamogeton* and other aquatic plants, where you find also the cocoon. It is certain that monocots dominate among the chosen hosts but that could be due to the fact that, statistically, these plants are more common in aquatic habitats. Donaciinae come to light mostly in the tropics. The adults often feed on the pollen of flowers.

MEGALOPODINAE

Megalopodinae is a small pantropical group (350 spp.), probably extinct in the north except in Japan, Korea, Tibet and China. One species (*Temnaspis nigropunctata*) has been described by Pic from Syria (1896). There are more than 15 genera known, including an endemic one in Madagascar. Their very peculiar biology was studied by various entomologists (Tella, 1952; Monros, 1954a; Yu Peiyu, 1977; Santos, 1980, 1981a, b; Schulze, 1996) and a resemblance with cerambycid biology has been noted. Now, we have data on South America, China, and South Africa, but few genera and very few species have been studied. We have no biological data, but for a few host plants, from Korea, Japan, Vietnam, the Philippines, India, Sri Lanka, Indonesia and Madagascar. The same is true for tropical Africa. Data are better for South and Central America. In Korea, the main host plant of eastern Asia, *Fraxinus*, in the Oleaceae, has been confirmed as the valid host.

While amongst chrysomelids the females are generally bigger than the males, amongst Clytrinae and Megalopodinae the males are bigger. The powerful hind legs, with enlarged femora, are used for copulation, and also serve to grasp the stems like they do in Sagrinae. Megalopodinae do not jump.

Host plants are very well known (except in Madagascar). In South America, the beetles seem to prefer Solanaceae and Asteraceae, in Asia Oleaceae and in Africa Anacardiaceae. Selection of food by the beetle is not very specific, adults and larvae preferring the tender stems of herbaceous or shrubby plants. Only in Asia are the flexible stems of *Fraxinus* chosen. It is common in Brazil, Argentina and Central America, to see megalopodines active on young plants, but few detailed observations have been made. The complex behaviour of the female, digging and laying eggs in the stem, seems to be the same everywhere, with small behavioural differences. Chosen plants are really diverse (see for more details on the biology: Jolivet and Hawkeswood, 1995) and, in Asia, outside Oleaceae, the beetles have been seen on Rosaceae, Leguminosae *sensu lato*, etc.

Schulze (1996) has studied *Sphondyla tomentosa* (Lacordaire) on *Rhus zeiheri* (Anacardiaceae), a shrub in her garden in Africa. Behaviour is rather similar in all the species and genera and it seems constant all over the tropics and subtropics, including the oriental temperate zone. Adults of *Temnaspis nankinea* (Pic), in China, hibernate in a cell, in the soil, and emerge during spring time. The female cuts the extremity of young shoots of *Fraxinus chinensis* (Oleaceae). She then digs small holes, 2–3 mm below the tip, and lays one egg in each hole. She seals the hole after egg-laying or the sap soon congeals. Then she cuts the leaves, on each side of the stem, away from the holes. The female feeds on the sap coming out from the cuts and this causes the wilting and narrowing of the tip of the stem around the eggs. South American species chew a little bit of the stem, 10 cm below the eggs. The incision, spiralling or not, produces the same result: to decrease the production of sap around the eggs. Incubation lasts 10–15 days, then the larva bores inside the stem and feeds inside, moving slowly down. Excreta are rejected towards the oviposit hole or through newly dug holes. The larva at its maximum size digs a hole at the base of the stem and pupates in the soil in a built-up cell. Adults emerge from the pupae before winter but remain in the nymphal chamber until spring in China (Yu, Peiyu and Yang, 1994). In Central America and Brazil, eclosion and adult activity coincide with the spring time (May in mid Brazil and April–May in Panama).

Megalopodinae fly well in the sun and produce a striking noise, already mentioned by Lacordaire in Brazil, by striking the pronotum against a striate part of the mesonotum. There are generally, at least in America, four larval stages and a prenymphal stage, a beginning of hypermetamorphosis. The males, with their great legs, fight for the females and, like many chrysomelids, guard their female when found. Females can eventually reject the male, with their hind legs, when not ready for or after mating.

Megalopodinae are also a primitive group, closer to the cerambycids than many others. Their morphology (wings), anatomy (aedeagus) and biology place them at the beginning of the tree with other Eupoda.

CRIOCERINAE

Criocerinae also constitute a primitive group but on the way to Chrysomelinae. Schmitt (1988), in his biological review of the subfamily, insists on the phylogenetic relationships amongst what he names the crioceriformes (Criocerinae, Donaciinae and Sagrinae) and also amongst the plants they eat or they have eaten. He believes that cycad relationships, existent with several SE Asian *Lilioceris* feeding on *Cycas* (Jolivet, 1998b), are very old in the group and started during the Jurassic. Nymphaeales relationships must have started during the Cretaceous with several loosely related archaic aquatic monocots. However, only Donaciinae among Eupoda have actually kept Nymphaeaceae and related species in their food spectrum. Male aedeagus structure and wing nervation in Criocerinae are slowly becoming more 'modern', closely approaching Chrysomelinae, and the V-aedeagus of the subfamily is leaving the ring schema of the primitive groups. Only *Timarcha,* among Chrysomelinae, retains the ring-like primitive structure of the tegmen in the male genitalia.

Criocerinae are rather homogeneous and are common all over the world, even in oceanic islands, but they are missing in New Zealand, while they are abundant in Madagascar and not too rare in Australia. They exist also on New Caledonia. Australo-Papuan species of the genus *Stethopachys* prefer orchids and these beetles seem to be the only criocerines in New Caledonia. Criocerines constitute a very important subfamily from the economic point of view and they feed on both monocots and dicots, though more commonly the former, as for Hispinae. Monocots are not more primitive than dicots and only some families are more archaic than others. Various studies on the host plants have been published (Jolivet, 1977; Schmitt, 1988; Jolivet and Hawkeswood, 1995). Despite important gaps in the host plant records (Madagascar for instance), one can easily deduct the food plants of the different genera from published reports.

Around 1400 species of Criocerinae are known and divided into 13 or 14 genera. Some genera like *Lema* are enormous (800–900 species). Some small islands like La Réunion have their own species. They fly moderately, live mostly on small herbaceous plants and are generally absent from the canopy of the tropical forests. At least one larva, *Ortholema abnormis* Heinze, is gallicolous; other ones, *Neolema* spp., are spermophagous on ovules of Commelinaceae. Several feed on the red fruit of Liliaceae and their red colour can be homochromic or aposematic against the green background. Several larvae are miners but most of them are free on the leaves. Some larvae protect themselves with excreta, as *Lilioceris* or *Plectonycha*, a different method from the one adopted by Camptosomata, but in some ways similar to many Cassidinae. Pupation takes place in a cocoon, either into the soil at the foot of the plant or on the leaves. *Crioceris,* which feeds on monocots, has a tendency to eat the tissues in a straight line between the parallel veins, leaving one side of the epidermis intact. Species feeding on dicots are more voracious and destroy the whole leaf lamina with veins, with the exception of large ones.

Lilioceris feed mostly on Liliaceae, but several species have been captured on *Cycas* spp. in New Guinea, Vietnam, Thailand, and Australia. The recorded history of *Lilioceris* on *Cycas* started with Szent-Ivany *et al.* (1956), who mentioned in New Guinea *Lilioceris clarki* (Baly) on the new fronds of *Cycas circinnalis* L. Then Hawkeswood (1992) has quoted *Lilioceris nigripes* (Fabricius) in Queensland on *Bowenia spectabilis* Hooker, a Zamiaceae. Similar observations have been made in Vietnam and in Thailand. Shepard (1997) has captured one species of *Lilioceris* on the young fronds of *Cycas siamensis* Miquel in Thailand. *Lilioceris nigripes* was also mentioned on the fronds of *Cycas ophialitica* K. Hill, a Cycadaceae, in Central Queensland by Gary W. Wilson (1993). The emerging fronds of the *Cycas* were browsed by the larvae and adult forms of *L. nigripes.* They eat the epidermal tissue of the pinnae, starting at the tips and working towards the rachis of the frond. According to Wilson, the larvae are gross feeders and have a distended and unprotected abdomen. The metabolic waste material is accumulated around the larvae as frass. There seems to be an association in Australia between a butterfly larva and an ant. Similar associations with other insects are seen among *Aulacoscelis* beetles on *Zamia* fronds in America. In Thailand, on *Cycas siamensis*, the larvae are below the fronds and rasp the abaxial epidermis and part of the mesophyll. In all these cases, larvae and adults are orange-red and very visible over the green foliage. They are protected by their toxicity and the cycasterones absorbed from the plant tissues.

Crioceris is attracted by Liliaceae, mostly the Asparaginae. Other genera live on Liliaceae, Commelinaceae, Smilacaceae, Orchidaceae, but, as for *Lema* and *Oulema*, they are sometimes attracted by Poaceae and their pollen. The subgenus *Quasilema* is specialized on Solanaceae. Several *Lema* and related species feed on various wild and cultivated Poaceae, like corn or rice. Roughly speaking, monocots outnumber dicots among the chosen plants, but, curiously, Solanaceae remain a hidden tendency of certain genera (Tempère, 1935, 1946, 1967). We can call that atavism. Solanophagy can reappear suddenly among Liliaceae-feeders like *Lilioceris lilii* (Scopoli), when generations of the insect had fed on *Lilium.* For food plants, see Jolivet and Hawkeswood (1995). It may be added that we have no information about the food habits of some genera.

SYNETINAE

This subfamily is difficult to classify due to its peculiar morphology and its larval habits. Crowson (1953), Kurcheva (1967), Reid (1995) placed the beetles near or among the Eumolpinae. A similar wrong assertion was made by many others. Recently, Yu (1988), Yu *et al.* (1996) have reviewed the question. Many specialists have followed the eumolpine relationships, probably due to a convergence in larval habits (radicicolous like eumolpine larvae) but the adult morphology is very different: wing venation and aedeagus are close to Criocerinae and Chrysomelinae. Other characteristics point to a basic position at the end of Eupoda and the beginning of the Trichostoma. These problems have been debated in a recent paper (Verma and Jolivet, 2000).

Larvae of several species are actually well known and well studied. There are only two genera, *Syneta* and *Tricholema*, very close to each other (Edwards, 1953), so the biology is uniform. The female drops its eggs onto the ground where they hatch, without any special protection, after 2–3 weeks. Another leaf beetle, a chrysomeline, *Iscadida*, does the same, but it is a very rare exception. When they hatch, the larvae dig into the soil and find the roots.

Syneta is polyphagous and lives mainly in the northern cold forests, penetrating south a little into the US. In Eurasia, it occurs in northern parts in Siberia, China and Japan. It feeds on Gymnosperms and various dicots. Several species are pests of fruit trees (*Pyrus*, *Cydonia*, *Cerasus*, *Prunus*) or feed on Betulaceae, Salicaceae and Aceraceae. *Abies*, *Tsuga*, *Picea*, *Larix*, *Pinus* are also hosts in the northern parts of the distribution. A list of food plants is given in Jolivet and Hawkeswood (1995). As well as in the Kurcheva (1967) paper, some larvae have been described, such as *Syneta albida* LeConte, in the USA, (Wilson and Moznette, 1913; Wilson and Lovett, 1911; Moznette, 1916).

CLYTRINAE

The larvae of Clytrinae, like the larvae of Camptosoma or Camptosomata and the Lamprosomatinae, are case-bearers, i.e. they live inside scatoshells or cases built by themselves with faecal material.

Apart from some rare cases among the Cryptocephalinae and one genus of Eumolpinae, Clytrinae constitute the only subfamily which is myrmecophilic at the larval, and sometimes also even the adult stage. *Hockingia curiosa* Selman adults live with the ants inside the big stipular thorns of acacia trees in east Africa. Some clytrine larvae are only submyrmecophilic, i.e. live like several tenebrionids or cetoniids, around, not inside the ant nest. Clytrine larvae become, by necessity, polyphagous or better, omnivorous, i.e. they feed on detritus of any kind, animal or vegetable, ant eggs, excreta, etc. inside the ant nest. These larvae are well protected from the ants by the scatoshell and, in case of danger, they retract their heads inside. Adults also, when they escape from the nest, try to avoid the ants and probably use reflex bleeding or secretions for protection. Clytrines penetrate into the ant nest at the egg stage, the egg being covered with excreta artistically distributed to resemble a seed, or at the larval stage, carried by the ants themselves. It seems strange that the ants bring inside their nest future predators, even if they do help to clean the nest. Clytrine larvae do not have trichomes and do not produce sweet secretions like the adults of *Hockingia* do, for instance.

Recently, Schöller (1996a,b) studied the food selection by the larvae of a clytrine, *Macrolenes dentipes* Olivier, and by that of a cryptocephaline, *Pachybrachis anoguttatus* Suffrian. He found that the clytrine fed on dried and rehydrated leaves of *Quercus ilex* L. and dead larvae and adults of a lepidoptera, *Anagasta* (*Ephestia*) *kuehniella* (Zeller), whereas, the larvae of the cryptocephaline fed only on dry leaves. This confirms that myrmecophilous and submyrmecophilous leaf beetles are totally polyphagous and zoosaprophagous, while the non-myrmecophilous case-bearer larvae are detriticolous on bark and plant debris only. Case-bearer larvae of ant-frequenting clytrines and cryptocephalines must feed also on the eggs and larvae of ants. Before Schöller's experiments, only sporadic observations were available.

As for clytrines, the eggs of many tropical phasmids in Australia, America, Africa, are carried inside the nest by the ants, but the eggs are provided with a caruncle which plays the role of an elaiosome and is eaten by the ants. Females of certain Clytrines and cryptocephalines, *Iscadida*, *Syneta*, some curculionids, like *Otiorrhynchus*, and many tropical phasmids drop their eggs on the ground. In *Syneta* and weevils, larvae hatch on the ground and penetrate into the soil. Amongst others, including tropical phasmids and Clytrinae, the eggs are carried by the ants themselves into the ant nest. The reason in the case of the clytrines is not clear, the egg not having any edible part. Perhaps there is a pheromone or a still undescribed secretion which attracts the ants, since the excreta-covered egg is not really convincing as a seed. Some clytrine eggs are fixed by the female on the leaf by a peduncule (*Labidostomis*, Jolivet, 1952; *Coptocephala*, Pietrykowska, 2000) and the ants cut the stem and carry the eggs to the nest.

Adult clytrines are polyphagous on shrubs, monocots or dicots, but in Europe they seem to prefer Fagaceae and Betulaceae and in Africa, legumes and Malvaceae. The choice of plants depends mostly on their availability, their succulence, the age of the tree and tenderness of the leaves but it may also depend on the proximity of the ant nests, since, once they are dropped, the eggs are immediately taken away by the ants. Once, ants were observed carrying a larva which means that there must be something attractive about larvae, too (Samsinak, 1965). In Africa, clytrine larvae with their scatoshells follow the ant columns when they move. The attraction seems to be mutual.

Legumes, in a wide sense, Rosaceae, Polygonaceae, are often eaten by the adults. They eat indiscriminately the flowers, buds, young leaves and pollen. Males appear first, followed a few days later by females. After copulation, males and females lead a hidden life among the shrubs. At least 37 plant families are eaten by this polyphagous subfamily.

Clytrines are extremely rare in the islands and very few species have been recorded from New Guinea, Australia or Madagascar. They are missing in the Mascarenes, in the Galapagos, and in most of the oceanic or volcanic islands, and even in small continental islands, however rich or poor their ant fauna (Cabo Verde islands, most Pacific islands, etc.). The native ant fauna is often probably too poor to support the clytrines and to maintain eventually the imported species. However, there are notable exceptions. For instance, *Lachnaea paradoxa* (Olivier) has been mentioned from Pantellaria, a small volcanic island near Sicily. It feeds on *Ferula* sp., a member of the Apiaceae (Ratti, 1986; Jolivet, 1996a).

Many host ants are known amongst North American and European species. There is really no specificity about the ant nests, though some clytrines seem sometimes to prefer some ant genera to others. Little is known about the relationships between ants and clytrines in the tropics of all the continents. The case-bearing larvae are very common in Brazil, India, Japan, China, but so far practically nothing is reported about their biology. The biology must be similar but it would be interesting to have data from Australia and from America about local ants. Clytrines are rather different in America and there are quite primitive ants in Australia. Further details on clytrine biology are given by Erber (1988) and Jolivet and Hawkeswood (1995).

CRYPTOCEPHALINAE

Very few larvae of cryptocephalines are myrmecophilic or submyrmecophilic. Most of them are free on the stems, leaves or at the base of herbaceous plants and lead hidden lives. At least one species, *Isnus petasus* Selman, lives, at the larval and adult stage, inside stipular thorns (the basal appendages) of the leaves of the east African *Acacia* trees. Probably there are more species with similar habits which are still unknown. The clytrine, *Hockingia curiosa*, and several eumolpines of the genus *Syagrus* share that kind of life with the ants. *Isnus* adult shows yellow secretory trichomes appreciated by the local ants, species of *Crematogaster* and *Camponotus*.

Adult cryptocephalines are common on the flowers and leaves of various trees, shrubs or herbaceous plants, feeding on plant tissue or pollen. In the holarctic zone, most of the Cryptocephalinae live on Pinaceae, Betulaceae, Fagaceae, Asteraceae and Cistaceae. Adults resemble clytrines in their quiet behaviour, but are lighter and can be carried by the wind over long distances. In this way, they populated the oceanic islands (Cabo Verde archipelago, Mauritius, etc.) when Clytrinae could not do so (Jolivet, 1979). As for clytrines, cryptocephaline adults prefer the tender leaves of young plants. Normally, the adults feed on dicots, but Poaceae are often selected for their pollen.

Cryptocephalinae are polyphagous and frequent at least 64 different families of plants, but the beginning of a specialization is becoming apparent. A few species are captured on coniferous trees, evidently a recent adaptation. Schöller (1995) found that some species of *Cryptocephalus* prefer petals over leaves (*Adenocarpus viscosus*, Fabaceae) in the Canary Islands. Petalophagy is very common also among galerucines. Cryptocephalines fly well, walk slowly and, in case of danger, drop dead easily. Erber (1988), LeSage (1984) and Schöller (1999) studied the biology of several species of Cryptocephalinae.

CHLAMISINAE

It is not clear whether the Chlamisinae are all specific about their host plants (Brown, 1943, 1952, 1959) or if some of them are oligophagous. It is a fact that several SE Asian species seem to be specialists of *Rubus* spp., for instance (Jolivet, 1954b; Jolivet and Hawkeswood, 1995). Others seem rather generalists, at least at the larval stage. The species of the subfamily feed on various plants: trees, shrubs, and herbaceous species. Some species like *Chlamisus cribripennis* (LeConte) feed on the leaves, stems and fruits of blueberries at the larval and probably at the adult stage.

Adults, like the larvae, browse on the bark. Other species, like *Exema pennsylvanica* Pierce, on their hosts (*Gutierrezia* sp., *Artemisia* sp., *Helianthus* sp., *Solidago* sp.), have a striking resemblance to caterpillar faeces (Karren, 1964; Verma and Vyas, 1987). Close examination will cause them to fly or to drop to the ground, exactly as do the mimetic weevils of the genus *Peridinetus* in tropical America on *Piper* trees.

Among the 33 selected plant families, Rosaceae, Betulaceae, Fagaceae, Ericaceae, Asteraceae are often selected in the temperate zone and Malpighiaceae, Myrtaceae, Sapindaceae, Melastomataceae, in the tropical zone. Legumes *sensu lato*, including Mimosaceae, are also commonly selected in South America.

LAMPROSOMATINAE

This group can be placed at the beginning of the Cyclica (wings) with Megascelinae and Eumolpinae. However, the larva is also a case-bearer, being protected against predators by a scatoshell, like other Camptosomata. The biology has been studied mostly in South America, but we also know details about the European and Japanese species. In tropical America, some larvae inside their scatoshells are the shape of a Phrygian cap and can mimic spines on the stem of trees or shrubs. In Brazil, it is the case with *Lamprosoma bicolor* Kirby common on the stems of *Terminalia catappa* L., an imported Combretaceae there (Costa Lima, 1955). Lamprosomine adults walk on the leaves or stems and fly very well. They show reflex immobilization and drop down when disturbed (Lacordaire, 1848).

Monros (1949), Moreira (1913) and others such as Fiebrig (1910) have summarized the biology of some species in South America. The scatoshell is made of small wooden pieces intermingled with excreta and is well camouflaged against the stem. Inside the case, the larva is bent and nymphosis takes place inside against the trunk. Monros (1949) has described carefully the metamorphosis and morphology of *Lamprosoma chorisiae* Monros on *Chorisia speciosa* St. Hill or *Chorisia insignis* H.B.K. (Bombacaceae).

Kasap and Crowson (1976) studied the biology of the European species, *Oomorphus concolor* Sturn. So far, no relationships with ants have been described among Lamprosomatinae but they may exist in certain cases.

The group is somewhat polyphagous in the tropics. In the Holarctic, Araliaceae, with the closely related Apiaceae, seem to be the favourite food. In the neotropics, association with Myrtaceae, Combretaceae, Bombacaceae, and Mimosaceae are often described. Adults and larvae chew the bark and the larva uses this method to get the pieces of wood used to build its case. The larvae also chew the spines along the stem. They also feed on fresh green tissue. In the Florida Keys, *Oomorphus floridanus* Horn, very probably an Araliaceae feeder, has been observed feeding on rodent excreta (Woodruff, pers. comm.). According to Peck and Thomas (1998), the adults feed also on *Pinus* and *Ficus*, probably on the trunk. Coprophagy is extremely rare among chrysomelids but is common with clytrine larvae in ant nests.

MEGASCELINAE

Megascelinae larvae have been reared from eggs in Panama by Don Windsor and described by Cox (1998a) and Cox and Windsor (1999a). The adults are abundant on various Fabaceae all over Tropical America. In Panama, adults can be found abundantly on *Stizolobium* (*Mucuna*) *mutisiana* (Fabaceae) and the larvae probably feed on the roots of the same plant. Adults nibble the leaves. In Nicaragua, Jolivet captured adults on Fabaceae, but also on many other plant species which were unrelated to the Fabaceae, probably, accidentally. The fact that megasceline adults are constantly being found on legumes in tropical America supports the theory of the oligophagy of the adults even if the larvae are possibly polyphagous on roots. The eggs are laid in clusters on leaf-folds of the host plant and then covered by a thin membrane secreted by the female. The larvae probably drop to the ground after hatching.

The wings of Megascelinae, the aedeagus and their tarsal morphology place the subfamily close to Eumolpinae (Jolivet, 1954a, 1957; Bechyne and Bechyne, 1971b). Seeno and Wilcox (1982) ignored those observations and placed them wrongly near the Megalopodinae at the beginning of the classification. Mann and Crowson (1981) placed them between Megalopodinae and Lamprosomatinae, a more reasonable, but not convincing, placement. The adult morphology speaks in favour of a separate subfamily (Jolivet, 1997; Cox, 1998a).

Megascelis contains more than one hundred species from southern USA to the extreme south of the Americas. Adults fly well and often frequent the forest canopies. They are quick to escape and fly away and can only successfully be caught with a sweeping net. However, they are mostly to be found on herbaceous legumes near the ground, often in good numbers. *Megascelis* do not stridulate, do not jump and rarely practice reflex immobilization. Detailed food preferences are listed in Jolivet and Hawkeswood (1995).

Mariamela, from southern Argentina, lives and feeds on *Nothofagus*, a Gondwanian Fagaceae. Very little is known of its biology. Even if certain leaf beetles have a panantarctic distribution, like many plants, there is no Megasceline in Australia and outside America. All other data seem to be due to labelling mistakes. Also, no real amphipolar distribution of leaf beetles is actually known (Crowson, 1980). Megascelines have penetrated in the southern part of North America, but not in Florida separated as it is from Mexico by sea and land.

EUMOLPINAE

Generally compact and rounded, Eumolpinae are often brightly coloured. They are rarely black when alive but they sometimes turn black after collection (*Stenomela*). They are naturally grass green when alive on leaves. Many keep their metallic colours after death, like many cassidines. More than 500 genera and about 3400 species have been described. Many more remain unknown to science, mostly in tropical mountains where the adults frequent flowers and in the canopy of the forests. They are generally good flyers and a few of them are brachypterous or apterous. Deduced from DNA frequences and morphological characters, the phylogeny according to Farrell (1998) places them near the Megascelinae, which is correct, but close to Synetinae, which is much more questionable. It all depends on the significance given to the characteristics chosen.

Eumolpine tarsi have the third segment deeply divided. The cotyloid cavities are rounded. Wing venation with its cells is very specific to the group and the primitive genera, as well as the Megascelinae, do not have the medio-cubital binding patch (subcubital fleck) on the wings, which otherwise seems very characteristic of the subfamily (Jolivet, 1957).

Several species are crop pests, mostly the frugivorous and polyphagous ones. *Colaspis* spp., often abundant on the wild and bushy solanaceous plants, are considered as garden pests in the neotropics. The larvae also feed on the roots of many plants, including cultivated and wild Poaceae. Several eumolpine adults feed on the leaves, flowers and fruits of various crops including cotton, cocoa, alfalfa, peanuts, potatoes, eggplants, sweet potatoes and bananas. Many genera are polyphagous, with certain species living on coniferous trees. Eumolpines are abundant in tropical America, feeding on 112 monocots or dicots and also by recent adaptation on Gymnosperms. Several polyphagous or oligophagous genera (43 spp.) feed on cocoa leaves and their biology has been studied by Edna Ferronato (1986, 1988, 1999) in Bahia state, in Brazil. *Taimbezinhia theobromae* (Bryant), *Percolaspis ornata* (Germar) and *Plaumanita* sp. are considered as dominant on the cocoa trees. The larvae feed on the roots of the *Theobroma* spp. and on many other plants, including Poaceae.

A detailed review of the food plants of the Eumolpinae, genus by genus, is given by Jolivet (1982, 1987c) and Jolivet and Hawkeswood (1995). Many recent and old references about the biology of eumolpines are quoted by Cox (1998b). Recent papers by Jerez (1996) summarize many unknown data among Chilean archaic and recent species (*Hornius*, *Stenomela*, and *Dictyneis*). *Hornius* spp. are associated in Tierra del Fuego with various species of *Nothofagus* (Fagaceae), the antarctic beech (Monros, 1952). The brownish adults feed on the bark and the leaves of the tree and the larvae live in the smaller branches. First instar larvae live and feed on leaf bud and later on new leaves. A primitive feature of *Hornius*, and unique among eumolpines, is the tegmen in the male which has a complete ring around the median lobe, like primitive Chrysomelidae (Crowson and Crowson, 1996). *Timarcha,* among Chrysomelinae, shows a very primitive tegmen and only very few other chrysomelines have a ring tegmen, but not so primitive, without a cap. They also do not have a split at the base of the aedeagus. *Stenomela pallida* Erichson feeds in Chile on various Myrtaceae: *Blepharocalyx crukshanksii*, *Luma chequen* and *L. apiculata*. The big adults (12–14 mm), which are entirely green on the trees, feed on the stems and leaves, whilst the larvae feed on foliar lamina only. Eggs of *Stenomela* and *Hornius* are covered by faeces and the larvae are not root-feeders like most of the eumolpines. Another very primitive eumolpine is *Spilopyra sumptuosa* Baly, from Australia. It feeds on rainforest bushes (*Cupaniopsis anacardioides* (A. Rich.), a Sapindaceae) (Hawkeswood, 1987, 1991). Unfortunately, nothing is known about its biology, but it is probably peculiar. The species of the genus *Dictyneis* lives in dry xerophytic forests. *Dictyneis asperatus* Blanchard is polyphagous and feeds as well on *Aristotelia chilensis* (Elaeocarpaceae), *Lomatia dentata* and *L. hirsuta* (Proteaceae) and *Pinus radiata* (Pinaceae). Another species (*D. parvus*) is associated with *Mulinum spinosum* (Umbelliferae), a high altitude plant (2100 m). All *Dictyneis* are generalists and the larvae are root-feeders. Unfortunately, nothing is known of the host plant of the so-called 'neo-caledonian eumolpine', *Stenomela*-like, *Bohulmijania antiqua* Monros. It probably feeds also on Myrtaceae, abundant in the island. The exact status of this beetle is to be re-studied. Records in Chile of *Stenomela pallida* Erichson on *Gunnera chilensis* (Gunneraceae) by Monros (1958) were totally wrong, even if *Gunnera* plants are parts of the association. To be noted '*Stenomela pallida*' (pale) is badly chosen for a specific name, because if *Stenomela* spp. are light black in the collections, they are tender green in nature, matching the colour of the host plant. A synonym, *Stenomela herbacea* (Blanchard) was a much better choice.

Fundamentally, it seems that the basic food of the Eumolpines is Asclepiadales (Asclepiadaceae, Apocynaceae), eventually also the Convolvulaceae (Jolivet, 1982). From this fundamental choice, many species became oligophagous or even polyphagous. They rapidly attacked the succulent crops and many wild plants. In Africa, several genera feed on Asclepiadaceae, but some also feed on other latex-bearing plants like Euphorbiaceae (*Platycorinus* spp.). *Platycorinus* spp. and related (*Chrysochus*, *Chrysochares*) feed on *Calotropis gigantea* (Asclepiadaceae) in Asia and on *Vincetoxicum*, *Asclepias* (Asclepiadaceae) and on *Apocynum* (Apocynaceae) in Europe and N. America. *Eumolpus* in South America feeds also Asclepiadaceae. Several species of eumolpines like to feed on flowers like those of *Lavoisiera* spp. in the Brazilian mountains (Serra do Cipo and Diamantina).

Finally, in brief, the host plants for eumolpines are: 4 Gymnosperm families, 15 monocots and 97 dicots. In most of the species, the larvae are polyphagous on roots and the adults oligophagous on leaves. Only Euryopites and Corynodites are specialists on Gentianales but several genera like *Eumolpus* have similar food relations. Larvae and adults of *Platycorinus* on *Calotropis* (Asclepiadaceae) in Asia and on Euphorbiaceae in Africa cut the veins of the leaves to slow down the latex flow. It contains cardiac glucosides for *Calotropis* and diterpenes for Euphorbiaceae. These eumolpines are aposematic and carry bright metallic colours. Vein-cutting and petiole-chewing seem to be the rule for many eumolpines. The same rule applies for the chrysomeline *Labidomera clivicollis* (Kirby) which reduces the flow of latex from the petiole in Asclepiadaceae plants by cutting the leaf veins (Hartmann, 1977; Dussourd and Eisner, 1987; Dussourd and Denno, 1991; Becerra, 1994a,b). So do *Chrysochus auratus* (Fabricius), feeding on *Apocynum cannabinum* (Apocynaceae) (Williams, 1991) and *Platycorinus* spp. on *Calotropis gigantea* Ait. (Asclepiadaceae). The biology and host plants of many eumolpines in North America and Russia (Kurcheva, 1958, 1967, 1975) has been reviewed in Jolivet and Hawkeswood (1995). Several can be serious pests of crops, including vines (*Bromius obscurus* (L.) and others). *Cleptor inermis* Lefèvre from Queensland may be also a pollen-feeder (Samuelson, 1994). According to Cox (1998b), the larva of *Odontionopa sericea* (Gyllenhal) from South Africa feeds within galls of the stems of *Leucodendron salignum* (Proteaceae). It may also be pointed out that eumolpine species can be frugivorous (*Colaspis* spp.), granivorous, and florivorous. The larvae also feed outside or inside roots according to the species. They are polyphagous and *Theobroma* adapted species have larvae which feed on cocoa roots or on root of any other plant, including Poaceae. Since eumolpine species can detoxify many chemicals and are adapted to latex-bearing plants, they generally have a tendency to a large oligophagy or even to polyphagy.

The fact that primitive *Hornius* or *Psathyrocerus* species feed on the subantarctic *Nothofagus* is probably an indication of a primitive selection of Fagaceae shared with many Gondwanian species of Chrysomelidae in Australia and Tasmania. However, Fagaceae are not specially primitive but this could be what I have named an ecological selection dating from the Upper Cretaceous, selection based on the most common and locally available trees or plants.

CHRYSOMELINAE

This subfamily is fundamentally divided into two groups: the primitive and enigmatic Timarchini (for some the Timarchinae) and the Chrysomelini. Only the genus *Timarcha* is part of the Timarchini and other genera placed by Weise in this group belong to the Entomoscelina (Daccordi, 1994, 1996; Jolivet, 1998a). Chrysomelinae constitute an important subfamily (3000 species and 132 genera) and the host plants are rather well known, except in Madagascar, New Caledonia (where there are a few species), New Zealand and tropical Africa. Also, the data obtained from central Africa (Zaire) are not always reliable. Several European genera, on the verge of extinction, like *Cecchiniola* and *Cyrtonastes*, have no host plant record, like most of the small ground-living species in the southern tropics (Patagonia, Chile, Cape region, Australia). About 45% of the genera have their host plant entirely or partly known. Genera like *Chrysolina* and *Phaedon* have species in various subgenera which accept in nature or in the laboratory primitive plants like Ranunculaceae, probably an atavistic selection. Such hidden selection is often ignored by the cladists. Even in well-studied genera like *Chrysolina*, *Gonioctena*, *Phaedon*, many Doryphorina, lacunae are big, but often the main selections belong to a common spectrum. This is the case for *Chrysomela* species, for instance, specialized mostly on Salicaceae. Basic selections of the group have been recorded in Jolivet and Hawkeswood (1995). Finally, the food spectrum known for *Chrysolina* is rather restricted and relatively specialized. (Garin *et al.*, 1999; Gomez-Zurita *et al.*, 1999; Hsiao and Pasteels, 1999). Arctic species, probably by necessity, demonstrate a wider selection, eating locally available plants (Kincaid, 1900; Jolivet, 1992; Khruleva, 1996).

Timarcha, one of the most studied genus and the most primitive of the subfamily, feeds on Rubiaceae and Plantaginaceae in Europe, Asia and North Africa, and also, but rarely, on Dipsacaceae, Scrophulariaceae, Asteraceae and Brassicaceae. Chemical and taxonomical relationships exist between Plantaginaceae and Rubiaceae, Dipsacaceae and Scrophulariaceae, but not with Brassicaceae and Asteraceae. However, in the southern mediterranean, *Timarcha* becomes generalist on Rubiaceae and Plantaginaceae, while it seems more restricted to Rubiaceae in central Europe. In North America, *Timarcha* feeds on Rosaceae only, *Rubus* and *Fragaria*, at least on several species with tender leaves. Fifty percent of the food plants of *Timarcha* are known and we can expect few surprises from the remaining stock. However, Poinar (pers. comm.) has recently found in Oregon the larva of *Timarcha intricata* Haldeman feeding on *Gaultheria shallon* Pursh (Ericaceae), when normally it is a *Rubus* feeder. *Timarcha metallica* Laicharting also shifts from *Galium* and *Plantago* to *Vaccinium* in the north of Europe (Kuntzen, 1919; Grafteaux, pers. comm.) Perhaps the *Timarcha* complex in the USA need more research and the species are more numerous (Jolivet, 1994b). A recent paper by Gomez-Zurita *et al.* (2000b) is on a re-study of the biology and evolution of the genus *Timarcha* in the Old World. The relation between phylogeny and host plant use seems to indicate a widening of trophic range as a derived character in *Timarcha*. If we consider *Timarcha* as being a steppic species, *Plantago* must be the original selection from which all others have been derived. *Vaccinium* can also link *Metallotimarcha* and *Americanotimarcha*.

Precursor work on the biology of *Plagiodera versicolora* (Laicharting) and other

European species was carried out by Balcells (1946, 1955). The larva has pre-social behaviour and *P. versicolora*, when introduced into North America, became a pest of *Salix* (Wade, 1994).

Theoretically, Chrysomelinae feed exclusively on dicots. We do not know anything about the food of the Madagascan genera (*Hispostoma*, *Barymela*, and others). Very little is known about the food habits of genera from the southern tip of the world. However, the *Eucalyptus* and *Acacia* feeding Paropsini are well studied in Australia and Tasmania. Food plants of the Entomoscelina, *sensu Daccordi,* have been reviewed recently (Jolivet, 1998d) but, as we can see, few data are available. Host plant relationships confirm the closeness of *Chalcolampra*, *Allocharis* and *Phola*.

Some plant families have been frequently cited as chrysomeline food: Asteraceae, Pittosporaceae, Brassicaceae, Solanaceae, Rosaceae, Myrtaceae, Mimosaceae, Asclepiadaceae, Apocynaceae, Verbenaceae, Lamiaceae, Plantaginaceae, Rubiaceae, Fabaceae, Urticaceae, and many others. Several genera can select different plant families according to the subgenera (*Timarcha*, *Chrysolina*, *Phaedon*) or the species (*Leptinotarsa*). For instance, the genus *Leptinotarsa*, one of the Doryphorina, can select Asteraceae, Solanaceae or Zygophyllaceae, according to the species. Many genera among the Doryphorina select Solanaceae, the dominant plant in South America, or Asteraceae, like *Elytrosphaera* (Macedo *et al*., 1998; Vasconcellos-Neto and Jolivet, 1998).

Many genera, such as *Timarcha*, *Cyrtonus*, *Cyrtonastes*, *Iscadida*, *Algoala*, *Elytrosphaera*, *Brachyhelops*, are apterous. Others are brachypterous or unable to fly because of loss of wing muscles, as is the case for many females of *Chrysolina*. Most of the non-flying species are restricted to herbaceous plants or to small bushes, such as *Iscadida* on *Rhoicissus*, a Vitaceae. However, several species of *Chrysolina* fly well with the sun (*C. americana* (Linné), *C. hyperici* (Forster), *C. quadrigemina* (Suffrian), *C. aurichalcea* (Mannerheim)), but most of the species of the genus are bad or generally non-flyers. Many genera, common during the day under stones or at the foot of plants, are apterous (*Xenomela*, *Oreomela*, *Cyrtonus*, *Cyrtonastes*, *Cystocnemis*). They often feed on Asteraceae or Lamiaceae, but it is not an absolute rule. Species of genera like *Jolivetia, Pataya* and *Brachyhelops*, are small, dark species, apterous and seem to lead a hidden life on the ground in Chile. *Brachyhelops* survives in Patagonia in cold surroundings and must diapause in winter.

One species of Chrysomelinae, *Chalcolampra* (*Phola*) *octodecimpunctata* (Fabricius), has very peculiar life habits as a larva. This larva feeds in China on *Vitex cannabifolia* and builds tubiform nests, made of excrements, on or between the stems of the plant. The larva stays hidden in its case with half of the anterior part of the body inside and the rest of the abdomen outside (Chen, 1964). Up to now, this is a unique instance of a case-bearing larva among the Chrysomelinae. We do not know if any other species of the genus shares these habits.

Finally, most of the known Chrysomelinae are monophagous or oligophagous on herbaceous plants (*Timarcha*, *Chrysolina*) or on small bushes (*Calligrapha*, *Promechus*, *Mesoplatys*, *Platyphora*) or on trees (*Paropsis*). Many species like young shoots or leaves (*Chrysomela* or *Paropsis*). No real polyphagous species are known among them.

Two hundred genera of plants belonging to 40 families have been mentioned as hosts of Chrysomelinae. *Paropsis* and Paropsina feed mainly on *Eucalyptus*

(Myrtaceae) and *Acacia* (Mimosaceae), which constitute a large part of the Australian forests. *Paropsis* feed also on *Eucalyptus* in southern New Guinea, while *Paropsides* feed on Rosaceae in the Far East, Philippines and India. There are twelve paropsine genera and over 400 species (Simmul and de Little, 1999). As for *Chrysomela* spp., paropsines feed preferentially on new growth.

The main host plants are Salicaceae, Betulaceae, Polygonaceae, Brassicaceae, Fabaceae, Mimosaceae, Malvaceae, Rubiaceae, Araliaceae, Myrtaceae, Apiaceae, Solanaceae, Plantaginaceae, Lamiaceae, Verbenaceae, Asteraceae and many others. Tree-feeders are good flyers, but very few frequent the canopy of the forest because, generally, chrysomelines are heavy. Solanaceae, originally from and dominant in the New World, are a frequent choice there. It is normal that solanophagy is so common among the Doryphorina in tropical America. Few Chrysomelinae frequent the Araliaceae, a common choice among Lamprosomatinae. Along the rivers of New Guinea, on *Schefflera* and *Boerlagiodendron* leaves, abound the beautiful *Promechus* (*Aesernia*), the jewels of the leaf beetles. Even in the tropics there are seasons, and *Promechus* larvae and adults appear only during some months of the year. *Promechus* have also been found by Gressitt and Hart (1974) on various Pittosporaceae and Cunoniaceae, related families, but not close to Araliaceae. Remote links between plants can be chemical, but can also reveal hidden relationships or be an atavistic selection, as for Ranunculaceae for *Chrysolina*. We do not know much about the host plants of the diverse *Labidomera* (Daccordi and LeSage, 1999), but *Labidomera clivicollis* (Kirby) feeds exclusively on *Asclepias* spp. Another species from Costa Rica (*L. suturella* Guérin-Méneville) feeds on *Asclepias curasavica* L. (Asclepiadaceae) and on *Lycianthes* (= *Witheringia*) *heteroclita* (Sendtner) (Solanaceae). Many Doryphorina select Solanaceae, Asclepiadaceae and Asteraceae and this seems to be an ancestral selection. In laboratory studies, *Leptinotarsa decemlineata* accepts not only some Solanaceae but also *Asclepias* and *Lactuca*.

GALERUCINAE

According to Wilcox (1973), Galerucinae includes 5802 described species but their number keeps growing, mostly in the tropics (Silfverberg, 1998). From time to time, new genera are described. According to Silfverberg (1998), the European fauna is derived mainly from an eastern Eurasian one. India seems to have acquired its present fauna from the north. We can reach similar conclusions for many groups of beetles, including the Chrysomelinae, even if the central Asian fauna was empoverished by the Plio-pleistocene glaciations. Many black, low-running *Galeruca* are apterous in the Asian or European mountains (Wang, 1990) and others are even unable to fly in lowland areas. *Arima marginata* (Fabricius) in Europe is brachelytrous and apterous in both the sexes and, in Mexico, only the females of *Metacycla* spp. are apterous and brachelytrous, the male remaining macroelytrous. Such apterous species are always endangered and in Holland (Beenen, 1998) the boundaries of *Galeruca* spp. are shrinking, even if they are macropterous. Often, those species cannot fly due to degeneration of the wing muscles.

Most of the galerucines feed on dicots but there are certain exceptions. Many adults feed on Poaceae and Cucurbitaceae pollen, like the *Aulacophora* and *Diabrotica*. Both the genera, as well as many other Luperini, are originally cucurbit feeders which became secondarily polyphagous on crops and wild plants (Jolivet, 1999d). *Coelomera* spp. feed mostly on *Cecropia* (Cecropiaceae) leaves, pollen and, like the symbiotic *Azteca* ants, on Müller trophosomes, special bodies growing on a silky cushion, the trichilium, at the base of the foliar petioles. In several species of *Coelomera*, like *C. cajennensis* (F.), the female digs into the prostomata (preformed openings) of *Cecropia* internodes and lays eggs inside. Then she seals the openings, which are later reopened by the larvae after hatching. Other species (*C. lanio* (Dalman)) lay eggs at the end or under the leaflets. It is an area normally not patrolled by ants.

Many galerucines, mostly those from the canopy of tropical forests, remain to be described. They generally fly well, and even big species like *Coelomera* can be found at the canopy level. The complete cycle takes place on the leaves and trunk.

Several larvae, such as eumolpines, megascelines, synetines and many alticines are subterraneous and radicicolous. Among them are *Phyllobrotica*, *Platyxantha*, *Luperus*, *Megalognatha*, *Aulacophora*, *Diabrotica*, *Cerotoma* and many others mostly among Luperini. Despite the fact that those larvae are polyphagous on roots, they can also feed on the same plant as the adult. Alticinae, adults and larvae, on the contrary remain generally on the same plant all their life. Some larvae (*Exosoma*) live inside the bulbs of Liliaceae and Amaryllidaceae in a semiaquatic medium (Jolivet, 2002, in print), while the adults feed on various bushes. *Exosoma* larvae have several special anatomical features, like the bicameral spiracles, perhaps an adaptation to their special biology (Böving and Craighead, 1930; Laboissière, 1934; Crowson and Crowson, 1996). Some larvae, like *Monoxia*, are miners inside the leaves. *Galeruca*, *Agelastica*, *Arima*, *Galerucella*, *Monocesta*, *Lochmaea* larvae are free on herbs, bushes and tree leaves eaten also by the adults.

Pupation takes place in the soil inside a cavity (*Aulacophora*) or an earthen cell (*Exosoma*), in the bark crevices at the foot of the tree, inside buds or directly over the leaves according to the species. Eggs are laid on the leaves, or in the soil crevices, or stuck by a sticky secretion or a faecal mess to the plant, according to genera or species.

Galerucine larvae are often gregarious on the leaves and *Coelomera* larvae show the ring defence behaviour named cycloalexy (Jolivet *et al.*, 1990; Vasconcellos-Neto and Jolivet, 1994) as a means of protection against aggressive ants on *Cecropia* trees. This defensive system, mainly directed against ants and bugs, is found also among certain criocerines, chrysomelines and cassidines. *Cerotoma* larvae and others feed specially on the nitrogen-fixing nodules on Fabaceae roots. *Diabrotica* and *Cerotoma* and many Luperini are specially attracted by cucurbitaceous roots by the cucurbitacins (Nishida and Fukami, 1990; Nishida *et al.*, 1992).

Hibernation can happen in all instars according to the climate and the species concerned. Reflex bleeding is relatively rare but is mostly frequent among larvae, as in *Diabrotica* (Wallace and Blum, 1971). However, some adults show also prebuccal haemorrhage: *Galeruca* spp., *Luperus* spp., *Agelastica alni* (L.), *Sermylassa halensis* (L.), etc. Secretory glands of larvae seem efficient in protecting the insect against predators. Repugnatory glands are well known in *Agelastica* larvae. Defence in the group is mostly represented by reflex immobilization, reflex bleeding, repugnatory glands, buccal or enteric discharges, haemolymph toxicity, aposematic coloration, mimetism or homochromy, cycloalexy, flight, etc. Galerucinae do not jump and normally their femora are not enlarged as for Alticinae. There are few known predators since they are toxic, but many have pathogens and parasitoids.

Diabrotica and Diabroticines constitute an enormous group of nearly 1000 species. Larvae and adults often feed on different plants but Cucurbitaceae are the dominant choice. The adults have a tendency to feed on the flowers and pollen of plants of the pumpkin family, but will also frequent hundreds of different plants belonging to many families. Many genera, like *Diabrotica* and *Aulacophora*, fundamentally cucurbitaceous feeders, may become polyphagous by secondary adaptation (Jolivet, 1999d). Eben (1999) outlines the preference of *Diabrotica* species for bitter cucurbits and cucurbit flowers. Cucurbitacin attraction could be a relict of an ancestral host plant relationship. The biology of *Diabrotica* has been recently reviewed by Krysan (1999). Diabroticines are also attracted by various pollens like sweetcorn, winter squash and goldenrod. Phagostimulant activity comes from various free amino acids (Hollister and Mullin, 1999). Amino acids and cucurbitacin are probably directly involved in the attractiveness of Cucurbitaceae flowers.

The food plants are known for about 30% of the galerucines. Eighty-three genera are mono- or oligophagous (61.4%) and 52 are primarily or secondarily polyphagous (38.5%). Basically, galerucines feed on Cucurbitaceae, Fabaceae and Verbenaceae but they may have adapted to many other families. Adults often feed on the pollen of Poaceae and the larvae feed on their roots. More than 100 plant families, mostly dicots, have been listed as hosts. However, 17 monocot families have been adopted by certain genera, including Amaryllidaceae, Zingiberaceae and Liliaceae.

Galerucine larvae are either elongated or incurved. *Coelomera* larvae and others have a big anal shield, often ciliated, which protects the larvae against aggressors. The larvae use it to repel predators. This shield is more or less developed according to the genera. The tenth anal segment, below the ninth, is a pygopod used for clinging to the substrate or for progressing. Among larvae, ocelli may be absent (root-feeders) or present. Antennae are mono- or biarticulate.

ALTICINAE

Alticinae and Galerucinae are sister groups and it is not surprising that some people have tried to merge them. Transitional genera exist. However, Alticinae have jumping abilities due to strong muscles inside the enlarged hind femora. This character separates them from Galerucinae.

Normally, Alticinae are specialists and only a few genera (*Systena*) are really and totally polyphagous. The apparent polyphagy of a given genus is often the result of the observation of a complex and artificial genus resulting from the amalgam of several mono- or oligophagous species. A phenomenon, not peculiar to alticines, is of the adult becoming attracted to plants not related to the normal host plants. This can be seen in temperate (*Quercus*, *Populus*, *Pistacia*) and also in tropical (*Persea*, *Carica*, *Psidium*) countries. It happens in cases of stress, sporadic or exceptional dryness, or seasonal disappearance of the normal hosts, and it always involves easily edible and succulent plants. Crops can be involved, and even imported trees like guava outside America. I have often noticed the phenomenon in the Cabo Verde islands in the Atlantic when the irrigation channels dry out. The islands are very dry naturally but the mountains accumulate enough water during the only annual day of torrential rains. Perhaps we could name that kind of abnormal trophism: 'alienotrophism'. It has nothing in common with allotrophism, which is the sudden adoption of a different plant under the effect of a probable mutation but not under a physiological necessity. Larvae do not feed on these new plants and adults only nibble on the leaves to get water or juices. Petitpierre (1985) regarded these secondary plants as a refuge. The situation has been described by several authors (Peyerimhoff, 1915, in Algeria; Furth, 1983, in Israel; Doguet, 1984, in Algeria; Jolivet, 1985, in the Cabo Verde archipelago; etc.). The succulence of cultivated crops has a strong attraction for alticines in the Cabo Verde islands during the driest part of the year and these crops have no connection with the normal host plants.

Most alticines are terrestrial but several, like *Altica* and others, are attracted by subaquatic plants along streams. Other species are subaquatic themselves, like *Pseudolampsis*, *Agasicles*, *Disonycha*, *Phrenica*, *Lysathia*, without much modification of their morphology. Several species have a setose venter which is not a real plastron. *Longitarsus nigerrimus* (Gyllenhal) feeds in Europe on a carnivorous hydrophyte, *Utricularia vulgaris* L. and other species of *Utricularia* (Jolivet, 2002, in print). It is not specially adapted to aquatic life but it manages very well in water.

Like Eumolpinae and Galerucinae, Alticinae can be serious crop pests but, generally, these species remain very susceptible to insecticides. Few cases of resistance have been found so far.

Alticine larvae are miners inside leaves (*Sphaeroderma*, *Dibolia*, *Febra*, *Mantura*, *Mniophila*, *Argopistes*, several *Phyllotreta*, *Argopus*), fruit-eaters (*Longitarsus nigripennis* Motschulsky), stem, petiole, flower or bud-eaters (*Chaetocnema*, *Psylliodes*, *Clitea*), or are subterraneous and feed on roots (*Phyllotreta*, *Longitarsus*, *Podagrica*, *Nisotra*, *Aphthona*, several *Chaetocnema*, *Xenidea*, *Epitrix*, *Asiorestia*, *Crepidodera*, *Batophila*, *Clavicornaltica*, *Psylliodes attenuata* (Koch), *P. affinis* (Paykull), *Erystus* and several *Altica*). Other larvae are free-living (*Podontia*, *Blepharida*, and *Altica*) on the leaves of herbaceous plants or bushes. Several larvae, such as those of *Podagrica* and *Nisotra*, *Phyllotreta vitula* Redtenbacher, most of the *Aphthona*, *Epitrix*, several *Longitarsus*, feed only on the external part of the roots, others penetrate inside them among the vascular bundles. Free larvae usually, but not always, remain under the leaves.

The Alticinae subfamily contain 520 genera and more than 7000 species. Probably more than 10 000 species actually exist in the whole world. Roughly, the trophic spectrum of 160 genera is known. However, this sampling is enough to deduce the tendencies of many other genera.

As usual, we do not know anything about Madagascar and many blanks remain in Australia, Africa and in the South American mountains, including the Andes. Several aspects of alticine–plant relationships have been discussed in Jolivet (1986b) and in Jolivet and Hawkeswood (1995). A number of papers by Matsuda, Nielsen, Begossi and Benson have reviewed several problems (in Jolivet, Petitpierre and Hsiao, 1988). Many genera and species of Alticinae are studied in the various books of the series *Biology of Chrysomelidae*, including recently the biology of the Chilean genus *Procalus* (Jerez, 1999), the Canadian *Altica corni* (Lesage and Denis, 1999), the South African *Longitarsus* (Biondi, 1999), etc.

Alticinae, Hispinae and several Galerucinae, including *Coelomera*, are common on *Cecropia* trees, myrmecophilous or not (*C. hololeuca* in Brazil), and some of them, mostly the galerucines, can cause serious damage. Only Galerucinae have free-living larvae on *Cecropia*, protected by cycloalexy, reflex bleeding or repugnatorial glands.

Alticine and hispine larvae are exclusively leaf miners. As happens with curculionids and other beetles, adult galerucines (*Coelomera, Monocesta, Austrochorina, Gynandrobrotica, Dircema, Syphaxia*) are not attacked by ants (*Azteca*) with which they seem to live harmoniously. Some of these galerucines (*Gynandrobrotica*) sometimes live above the ant limit (3000 m) on *Cecropia* trees with many other folivorous insects. Alticines are very common above the ant limit.

Several Alticinae live on epiphytic plants and are common in the canopy of tropical forests. *Clavicornaltica*, for instance, are abundant in epiphytic soils and leaves in south-east Asia, and it would be interesting to compare the flying abilities between the ground- and epiphytic-dwelling species. Probably, the canopy-frequenting species have more efficient wings. There remain many undescribed species in the canopy and, even if they are described one day, their trophic relations with the plants will remain unknown. Knock-down with pyrethroids makes any trophic observation impossible. Basset, working on the canopy of various tropical forests in New Guinea and Panama (Basset and Samuelson, 1996; Novotny *et al.*, 1999), tried to evaluate the specificity of the leaf beetles on several trees. On these data, and only on them, depends the evaluation of the biodiversity of the tropical forests. No system for studying the canopy fauna is perfect: the platform, the tower, the climbing gear, the raft. Only giant cranes can permit study of the fauna without too much disturbance, but the system is inadequate in one way. The areas covered have to be very limited as it is difficult to move the crane once it is fixed to a given spot.

Alticinae are protected by their toxicity, reflex bleeding, aposematism, jumping abilities, prompt flight, and reflex immobilization. Alticine rarely relies on mimicry or homochromy. Several species (*Diamphidia, Polyclada*) are very toxic and aposematic, like the poisonous species of the Namib desert which are used as arrow poison by the Bushmen (Jolivet, 1967a; Jolivet and Hawkeswood, 1995). *Podontia* larvae, also very toxic, use an immobilization reflex and drop to the ground when disturbed. *Procalus* spp. in Chile feed on Anacardiaceae and seem to be related to *Blepharida,* which are also toxic, and feed on the same species in the New World (Furth and Young, 1988; Jerez, 1999).

While most of the species can fly, some are brachy-, micro- or apterous (Jolivet, 1959; Furth, 1979a,b, 1980; Shute, 1980). Wing polymorphism can be seen in some species and the phenomenon is well known among Chrysomelinae. In *Chrysolina banksi* (F.), for instance, in the Mediterranean region, three quarters of the specimens are brachypterous (Jolivet, 1959). According to Peyerimhoff (1911, 1915), wing reduction can be associated to a different food plant. However, there is no proof of that kind of speciation among alticines. Apterous genera include *Trachytetra, Batophila, Orestia, Minotula* (insular), *Minota* (alpine), *Podagrica* (some species only), *Apteropeda, Mniophila, Ivalia, Kamala* and *Clavicornaltica* (partly subterranean and micropterous) (Scherer, 1979) and there are others. *Clavicornaltica* do not fly but can jump well.

Alticinae feed on Brassicaceae, Malvaceae, Fabaceae, Asteraceae, Solanaceae, Onagraceae, Lythraceae, Polygonaceae, Amaranthaceae, Lamiaceae, Verbenaceae, Boraginaceae, Chenopodiaceae, Euphorbiaceae, Linaceae and many others. Most species frequent dicots but some choose monocots such as Liliaceae, Amaryllidaceae, Poaceae, etc. Floral feeding is not unknown among them. Pollen is also collected by many species and is digested after it has started to germinate inside the gut (Samuelson,

1988, 1994). Arnett (1968b) was the first to record this internal germination of pollen in insects. Mosses, horse-tails, ferns (5 families), tree ferns in the tropics and coniferous trees are also selected, but as secondary adaptations. Many species are miners on fern leaves (Kato, 1991). However, in contrast to eumolpines, radicicolous larvae are not really polyphagous and usually feed on the same crops as the adults. Several species of alticines are strictly monophagous (*Phyllotreta armoriciae* (Koch) on *Armoricia*), most of them are oligophagous and very few, for example, *Systena*, are polyphagous, *Altica ampelophaga* Guérin-Méneville feed on grapes, *Vitis vinifera* L. (Vitaceae) and the related families Onagraceae and Lythraceae. It also probably feeds on many other members of the Vitaceae since it has been noted on *Parthenocissus quinquefolia* (L.), an introduced garden plant (Balcells, 1953, 1955; Doguet, 1994). Biondi (1999) notes that many *Longitarsus* species feed on Boraginaceae in central Asia and the Mediterranean and Afrotropical regions.

HISPINAE

Hispinae and Cassidinae are also sister groups. They form the Cryptostoma and represent ultra-specialized lines. Intermediate forms are known and are often placed in one or the other subfamily. There are no true generalists among them, but some Hispinae do tend towards members of the Poaceae.

There are actually more than 3000 species of Hispinae and as many in Cassidinae. However, many remain to be described, mostly in tropical forests. Cassidinae seem to have reached their *numerus clausus*. According to Vasconcellos-Neto (1988), many of cassidine 'species' in the tropics are really coloured mutations of the same taxon. Flat hispines and their larvae are often found protected between the young rolled-up leaves of palm trees or Zingiberales in America. Spiny hispines are generally free as adults on Poaceae and dicots. Larvae can be free, but they are usually leaf-miners. Since many of them feed on Poaceae such as rice, corn, maize, sugar cane, bamboos, and also palm trees (Arecaceae), hispines have an agricultural importance. *Coelaenomenodera* species are serious pests of oil palm (*Elaeis guineensis* Jacq.) in Africa.

Larvae of certain species sometimes modify their behaviour during their cycle; from miners they can become free-living on leaves. Generally speaking, Hispinae are relatively well protected from predators by their hiding and mining habits at the larval stage, and by spines as adults. Adult mimicry is also well known between Hispinae and other insects. Group mimicry (more *müllerian* than *batesian*) is common in Asia and Africa and is clearly an efficient system. Several Hispinae from Central and South America (*Cephaloleia* spp.) are often red in colour on the red bract of the host plant, *Heliconia*. This is surely a case of homochromy since other species living on other parts of the plant do not have this colour. Mimicry among *Cephaloleia* species was studied by Staines (1999).

The larval biology of Hispinae was studied by Maulik (1919) and, since then, many observations have been made. These include those made during the mass-rearing of certain species in the laboratory (Mariau, 1988). Teixera *et al.* (1999) studied the biology of a leaf-mining species: *Octuroplata octopustulata* (Baly) in Brazil. It mines in different plants belonging to different families (Leguminosae, Myrtaceae, Ochnaceae). Mining hispines are often not very selective about their host plants

(Hering, 1951; Cox, 1997). However, *Apteropeda orbiculata* (Marsham) choose the remotely related families of the Plantaginaceae, Labiateae, Asteraceae and Scrophulariaceae for their mining activities.

The differences between Hispinae and Cassidinae larvae are rather subtle. Hispine larvae can be divided into four main groups:

(1) Those which live between the folded leaves of monocots (Poaceae) as miners, between folioles of new fronds or at the base of palm tree petioles (Arecaceae): Cephaloleini, Botryonopini, Callispini, Leptispini, and others.
(2) Those which bore into the stems of various herbaceous or semiligneous plants: Anisoderini.
(3) Leaf-miners: Coelaenomenoderini, Hispini, etc.
(4) Those living freely on or below the leaves.

In general terms, many species of hispines are miners, a great number live between the appressed young leaves of various monocots and only a minority are stem-borers. *Oediopalpa* larvae (Oediopalpini) behave like Cassidinae and build excremential and exuvial shields at the end of their abdomens. In India, the larvae of *Leptispa pygmaea* Baly live on the leaves with the adults. The larvae of *Brontispa froggatti* Sharp of Vuanatu (New Hebrides) live with the adults in the opened buds of coconut trees (*Cocos nucifera* L., Arecaceae) (Maulik, 1919). For a classical study of the biology and parasitic characteristics of the coconut leaf-mining beetle pest, *Promecotheca reichei* Baly in Fiji, see Taylor (1937).

Adult hispines feed on leaves and eat the mesophyll, leaving white longitudinal traces. Many of these adults are nocturnal but some, like *Dicladispa* spp., are diurnal. Budhraja *et al.* (1979) state that a larva of *Dicladispa armigera* (Olivier) eats an average of 132.4 sq mm of rice leaf during its life. This can be mean a huge loss of crop potential in Asia or Africa as these insect invasions can be of massive proportions. Adults start mostly with the leaf apex and then attack the parenchym. An adult can eat 24.4 sq mm of tissue per day.

Generally, hispine tribes have a uniform larval biology. One exception is Gonophorini which have larvae which can be leaf-miners or free-living. Several tribes demonstrate a high specificity in the choice of plant families. For instance, in Indonesia, Kalshoven (1951, 1957) speaks of six tribes, each of which use only one plant family: Botryonopini on Arecaceae, Leptispini on Poaceae, Eurispini on Cyperaceae, Coelaenomenoderini on Arecaceae, Oncocephalini on Commelinaceae, and Pharangispini on Zingiberaceae. These selections are, in reality, even more subtle as the Pharangispini are actually synonymized with the Coelaenomenoderini. In Asia, the Anisoderini live on 4 plant families, Callispini on 5 families, Cryptonychini on 3 families, Gonophorini on 7 families, and Hispini on 9 families. In America, the selected plant families are related but more numerous. Among *Dactylispa* in the Old World, the association is mostly with the Poaceae, but other species live on Rubiaceae, Sterculiaceae and Bombacaceae, all dicots.

There seems to be some correlation between taxonomy and trophic selection among many hispines. Kalshoven (1957) grouped *Callispa* species, according to their diet: bamboos, palm trees or Orchidaceae.

Among Hispinae on *Cecropia* trees in South America (*Metaxycera*, *Sceloenopla*), any contact with *Cecropia* ants (*Azteca*) is reduced to a minimum since the females, like *S. maculata* (Olivier), lay 8 eggs in a cylindrical ootheca (2.5 × 1.5 mm) and insert it into the mesophyll of a young leaf on the abaxial side. The biology of those larvae has been carefully described by Andrade (1981) from observations made in Brazil in the Rio area. See also Jolivet (1986a, 1996a).

Strong (1977–1982) and Strong and Wang (1977) studied the neotropical Arescini and Cephaloleini on Zingiberales. These are flat and smooth species living in a semiaquatic and viscuous medium between the appressed immature leaves of the host plants, *Heliconia*. These species can be mono- or oligophagous. Sometimes, not less than five hispine species can live on different parts of a simple *Heliconia* plant. *Cephaloleia consanguinea* Baly is strictly monophagous on *Heliconia imbricata* (Kuntze) Baker. Two species of *Cephaloleia* (*C. puncticollis* Baly and *C. neglecta* Weise) live inside the water-filled bracts of several *Heliconia*. They strip and mine the internal epidermis of the bracts (Seifert and Seifert, 1976a,b; Strong, 1977a,b; Jolivet, 2002, in print). The bract acts as a true phytotelm. Sometimes, several genera or species of Hispinae coexist in the same mine (*Arescus*, *Chelobasis*) in *Heliconia* and *Calathea* (Maulik, 1937).

Hespenheide and V. Dang (1999) studied the biology of the Hispinae of Costa Rica. Among 135 species collected, 77 were leaf-miners and they were found to use 5 monocot and 17 dicot families as hosts. Despite the protection of the mines, parasitism seems to be high.

No genus or tribe of Hispinae is common to the New and the Old World. This is the reason why Uhmann (1957–1958) divided the subfamily between Hispinae Americanae and Hispinae Africanae, Eurasiaticae and Australicae. Palm trees are also different between the two worlds, except for three genera, including the coconut tree. Thirteen species of Hispinae feed on coconut leaves in the Old World. Different species (Alurnini, Hispoleptini) attack the same plant in the New World and so it is difficult to deduce from the hispine the origin and distribution of the coconut. Several American Hispinae are adapting slowly to the oil palm (*Elaeis guineensis* Jacq., Arecaceae) which was imported from Africa to Brazil.

Hispinae are essentially a tropical group. Only a few species live in temperate areas. There are no Hispinae in New Zealand and hispines do not live above 2000 m in the tropics, even if their host plant grows there.

Monocots are preferred in Asia and America: Zingiberaceae, Arecaceae, Poaceae, Cyperaceae, Musaceae, Orchidaceae, Pandanaceae, etc. The association of a certain number of hispines with dicots is considered by Kalshoven (1957) as a recent adaptation comparable to the case of Criocerinae and dating from the early Eocene. At the time of Kalshoven's publication, monocots were considered to be more primitive than dicots. The new botanical classification does not make the distinction and the notion of mono- and dicots has completely changed.

Transfer from monocots to dicots or vice versa still takes place. Tempère (1946, 1967) observed in his garden in France *Lilioceris lilii* (Scopoli) (Criocerinae), from *Lilium* originally, suddenly feeding on *Solanum*. Kalshoven (1957) observed a similar trophic transfer in *Dicladispa armigera* (Olivier) in Indonesia. It normally fed on *Saccharum officinarum* L. or on other Poaceae but was seen to be mining *Tectona grandis* L. (Verbenaceae), a dicot. The latter case can be attributed to the phenomenon of proximity (xenophagy).

Hispine larvae often leave their original mine and penetrate into another leaf. This happens after the destruction of the chlorophyllian tissue. Nymphosis can also occur in a different mine. There are a few differences between the palm-frequenting species living between the appressed leaves and the real leaf-miners, but, in both cases, the protection against predators is excellent and the food abundant and always available.

Hispinae have been used in the biological control of *Lantana camara* L. (Verbenaceae), but with little success. Hispines, like cassidines, do not acclimatize easily to a new environment and are difficult to rear in a laboratory.

Finally, 110 genera of hispines (74.4%) have their host plant known, 63 plant families are recorded, 19 mocots and 44 dicots, but monocots are more commonly eaten. Zamiaceae (Cycadales) and Ophioglossaceae among the ferns are hosts to a few species, probably as a secondary adaptation.

From the 110 genera of hispines investigated, 86 (78%) feed on monocots and 24 (22%) on dicots. Most of the known hispines are mono- or oligophagous (86.1%) and the rest are polyphagous (13.9%), at least at genus level.

Poaceae, Arecaceae, Zingiberales are mostly eaten, but among dicots Fabaceae, Caesalpinaceae, Malvaceae and Asteraceae are often selected. No hispine feeds on Brassicaceae; sinigrin seems to be a repellent for the whole group. A few species attack Solanaceae and Convolvulaceae. No hispine feeds on Chenopodiaceae, the host plants of many cassidines.

CASSIDINAE

The last and most specialized family of Chrysomelidae comprises about 160 genera and 3000 species. The recent catalogue compiled by Borowiec (1999) quotes 2760 valid names after the synonyms. After the revision of the South American species (Vasconcellos-Neto, 1988), the number will probably decrease still further, while hispine numbers will always increase. Host plants for more than half the genera are well known. More and more data are becoming available on cassidine biology, mostly from tropical America where it is the most varied. Cassidine species are easy to observe on leaves since they feed openly, adults cling to their plants and they fly very little. When disturbed, cassidine produce sounds by rubbing their heads against the anterior side of the prothorax. They then adhere very strongly to the substratum, retracting their head and legs (the tortoise strategy), and may finally drop to the ground. The tarsal pad, with its special lubricated hairs acts as a sucker on the plant.

Cassidines fly little, but well enough to invade tropical bushes and trees. In the temperate zone, most of the species feed only on small herbaceous plants. Whereas Hispinae are always macropterous, cassidines can be totally apterous (*Delocrania*, *Elytromena*) or micropterous (*Stoiba*, *Fornicocassis*, *Pilemostoma* and *Mionycha* among *Cassida*). These species, being unable to fly, choose host plants which are easily accessible by climbing (Jolivet, 1959).

Neotropical species have been studied by Windsor (1987); Buzzi (1988); Vasconcellos-Neto (1988); Windsor *et al.* (1992); Buzzi and Djunko Miyazakf (1999); Hsiao and Windsor (1999), and many others. The main host plants were published by Jolivet (1988) and Jolivet and Hawkeswood (1995). Since then, many others have been recorded by various authors. The recent work of Vasconcellos-Neto (1988) has shown that a species like *Chelymorpha cribraria* (F.) is really a complex of many 'described' species which are mutations. The criteria are only based on colours or elytral markings. Many other species of *Botanochara, Stolas,* etc., are also certainly composite, and the actual trend is to reduce the number of species. Roughly speaking, 3000 species would be a maximum for Cassidinae *sens strictu.*

Contrary to the opinions of some authors, there are fundamental differences between Cassidinae and Hispinae in general biology and in adult and larval morphology. Intermediate forms have occasionally been transferred from one group to the other (see the recent paper by Hsiao and Windsor, 1999). Cassidine larvae are generally free and clearly visible on or under the leaves. They carry their exuviae and faecal matter in the form of a shield on the furca which is itself a prolongation of the tergum of the ninth abdominal segment. Several larvae have been described as miners.

Most of the Cassidinae (tortoise beetles) are characterized by their oval, rounded bodies which have extended sides, flattened around the body, including the area above the head, in front of the prothorax. *Delocrania* and *Notosacantha* are more or less intermediate between hispines and cassidines. They have elongated bodies with a small elytral margin and the frontal side of the prothorax flattened with a notch in the middle, exposing the dorsal part of the head. An incomplete tortoise shell, so to speak.

Cassidine eggs are deposited isolated (*Coptocycla*) or covered with excreta and secretions (*Cassida, Laccoptera*). They are normally deposited in packages, called ootheca, made of lamellae (*Aspidomorpha* and *Basipta*). These ootheca, which are cemented on leaves or stems, have a preformed exit above for the larvae.

Cassidine larvae have often a supra-anal extension (furca) above the ninth abdominal segment (*Aspidomorpha, Laccoptera*), but there is only a simple bifid extension in *Oocassida.* The structure of these extensions varies in the Neotropical region and certain species and genera do not have any furca (*Eurypepla*).

Certain genera and species of Cassidinae can change colour, either gradually or rapidly. A gradual change during development is associated with the age and the nature of food, or with a change of season (*Cassida*). A rapid change is due to the hydration and dehydration of the chitin lamellae which make up the elytra (Jolivet, 1994c). This rapid change can take one minute at the most. This phenomenon is limited to some species or genera and is linked with general humidity but also, and mainly, to stress factors. It can be related to sexual attraction, homochromy, or aposematism. During normal conditions of humidity, temperature and resting on the host plant, there exists a 'normal' coloration, golden for some species. The colour changes suddenly when the insect is disturbed by a potential predator. Some American species go from red to brownish, but others pass through a large coloration spectrum.

The protection of the larvae by the dorsal shield seems to be efficient. The structure of the shield varies and can be very complex. Neotropical cassidines, such as *Hemisphaeronota* and *Dorynota*, carry a long ball of excremential filaments over their abdomen. Nymphosis occurs under this faecal umbrella which protects well against ants, dryness or rain. Many cassidine larvae manipulate their shield against predators (Eisner *et al.*, 1967; Olmstead and Denno, 1993; Olmstead, 1994, 1996; Vencl and Morton, 1999).

The adults of many American Cassidinae cling to their host plants using the sucker-like ends of the tarsal hairs which are covered by an oily secretion (Eisner, 1971). It is very difficult to dislodge them. Cassidines have various means of defence and certain species are very toxic.

Normally, cassidines live on herbaceous plants in temperate areas, while tropical species can be found on bushes or even trees (*Cordia* spp., *Ipomoea arborescens*, *Tabebuia* spp.). Windsor *et al.* (1992), who have made a special study of the Panamean species, write that half of the species there use lianas or climbers while the rest live on low plants, bushes or trees. According to these authors, climbers or forest lianas (Convolvulaceae) are the only host plants of subsocial species. This can also be verified in Brazil. *Echoma* species in Brazil and Central America seem to feed only on the white flowers of *Mikania* (Asteraceae) (Windsor *et al.*, 1995). Eggs are laid inside the flowers and the larvae grow there eating the buds. The larvae are pale and covered with petal fragments and are not easily visible.

Cassidine larvae normally feed under the leaves of plants, while the adults on the upper leaf surface are potential prey for predators. However, apart from parasitoids, there are very few predators. The tortoise strategy protects the adults very well against ants and bugs. The larvae chew the leaf parenchyma, with the exception of the upper epidermis and the veins, while the adults feed on the leaf lamina or the petiole, making irregular holes. The few species of the palaearctic region (70) feed mostly on Asteraceae, Lamiaceae, Caryophyllaceae, Chenopodiaceae and Convolvulaceae. Everywhere, and mostly in the tropics, the selection of Convolvulaceae is dominant. If we can estimate the total existing Cassidinae to about 3000, most of the species live in the neotropical region and offer an enormous diversity of shapes, colours, patterns and biology. The only known cases of parental care among the cassidines are found with some neotropical tribes. No case has been found in the Old World. Cycloalexy is well known among cassidine larvae in tropical America and was only sporadically found in the Old World tropics with *Aspidomorpha* and *Conchyloctenia* spp. (South Africa: Heron, 1992, 1999; India: Verma, 1992). Generally, parental care can be seen in America among the larvae of subsocial species only, but cycloalexy can happen among species devoid of parental care (Jolivet *et al.*, 1990). Heron (1999) made a special study of the biology of *Conchyloctenia punctata* (Fabricius) in South Africa, while Buzzi and Miyazaki (1999) studied *Stolas lacordairei* (Boheman) in Brazil. Maternal care of the eggs, larvae and pupae was never observed in the last species.

Fiebrig (1910), working in Paraguay, noted 8 cassidines on Bignoniaceae, 4 on Boraginaceae (Ehretiaceae), 4 on Convolvulaceae, 3 on Asteraceae, 2 on Lamiaceae and 1 on Arecaceae. It was a good sampling for the time. Windsor *et al.* (1992) obtained the following results in Panama: Arecaceae 5, Zingiberaceae 3, Lamiaceae 0, Boraginaceae 10, Convolvulaceae 28, Solanaceae 7, Bignoniaceae 11, Asteraceae 9, Lecythidaceae 0, Sterculiaceae 1. The percentages can slightly differ from one place to another but remain significantly close.

Tempère (1935) emphasized the preference of European cassidines for Centrospermae, a natural order of dicots, examples being Amaranthaceae, Caryophyllaceae, and Chenopodiaceae. All plants in these European families are herbaceous and small. The choice of Convolvulaceae, with its sister group the Cuscutaceae, is a natural one, probably the most primitive, but rather distant from the families mentioned earlier.

For the general selections, we refer to Jolivet and Hawkeswood (1995). They have mentioned 83 genera of cassidines. All are oligophagous or monophagous and none is really polyphagous. When Cassidinae attack a different plant, such as those in the Moraceae, *Citrus* spp. or Fabaceae, this is thought of as a survival reflex similar to the behaviour of Alticinae in periods of dryness. Twenty percent of the world's Cassidinae feed on Convolvulaceae; others on Verbenaceae, Lamiaceae, Bignoniaceae, Cuscutaceae, Solanaceae, Boraginaceae, Caryophyllaceae, Chenopodiaceae, Amaranthaceae, Asteraceae, Sterculiaceae, Oleaceae, Combretaceae, Nelumbonaceae and Lecythidaceae. The other choices are only exceptional (12 families in all).

The cassidine genera feeding on Convolvulaceae are sometimes oligophagous and often accept Bignoniaceae. Palm tree-frequenting species are very different and close to the hispines. Monocot feeding is exceptional among Cassidinae.

The choice of host plants

Despite being recently questioned, the suggestion by Crowson (1989, 1991), that the Protoscelinae were feeding on benettital or cycad strobili or leaves at the end of Jurassic seems very likely. Cycad fossils are known from the same formations. Living Aulacoscelinae still frequent *Zamia* leaves in tropical America. Perhaps some Australian sagrines, and certainly some *Lilioceris* in south-east Asia, feed on cycads. They can be all considered as primitive beetles. *Araucaria* feeding by Palophaginae is also primitive. Alticinae which have adapted, in certain cases, to ferns, mosses, horse-tails and coniferous trees have done so only recently.

Many chrysomelids, adults or larvae, feed on lichens on the bark, on soil algae, on root or leaf fungi. Many subterraneous larvae of Chrysomelidae (*Cerotoma*) feed on bacterial nodules of Leguminosae. *Crepidodera aurea* Geoffroy feed on the fungal conidia of *Venturia* spp. on *Populus* leaves. Many chrysomelids eat the pollen of various flowers, namely galerucines and alticines. Many beetles of other subfamilies (Orsodacninae, Megascelinae, Eumolpinae, Hispinae, etc.) do the same. A pseudo-germination takes place with the effect of sugars inside the insect gut and that allows a rapid digestion of the pollen grains.

Adult chrysomelids are often captured and eaten by carnivorous plants, namely *Sarracenia* species in America, *Drosera* everywhere but, until now, no chrysomelid has been observed feeding on their leaves. However, many chrysomelid species, mostly alticines, feed in forests on epiphytic or myrmecophilous plants. One notable exception among carnivorous plants is quoted by Ihssen (1936) and Mohr (1962): *Longitarsus nigerrimus* (Gyllenhall) (Alticinae) lives among *Sphagnum* mosses in European swamps and feeds on *Utricularia intermedia* Hayne (Lentibulariaceae), a small subaquatic plant. So far, no chrysomelid has been noted feeding on *Pinguicula*, another ubiquitous and hydrophytic carnivorous plant. It is possible that the sticky fluid secreted by its glands acts as a repellent for the insects.

Chrysomelids need fresh leaves for food and are relatively rare in high mountains, except for some specialized genera. Galerucinae reach at least 2000 m in Africa (*Xenarthracella* on *Clerodendrum*), 3000 m in the Ruwenzori (*Ruwenzoria viridis* Laboissière on *Erica arborea* L.), 4000–4400 m in the Yunnan (*Xingeina* and *Shaira* on *Potentilla*) and at the same altitude, or even more, in the Andes. Alticinae reach the snow limit in the Andes and can be found on the summits with various carabids. Several *Elytrosphaera*, such as *E. nivalis* Kirsch (3700–4150 m) and *E. cartographica* Bechyne (3550 m) live on the high volcanoes of Ecuador, They probably feed on Asteraceae there but *Espeletia* is missing in this area (P. Moret, pers. comm.). It is notable that many of the Andean chrysomelids are black, probably as a protection against ultra-violet. *E. nivalis* is blackened, but the metallic colours persist a little on the prothorax side. *E. melas* Jolivet is present in Bolivia at 3000–3500 m and is totally black. Blackening tendencies are also for other leaf beetles in the Andes like cassidines (Jolivet and Vasconcellos-Neto, 1993; Vasconcellos-Neto and Jolivet, 1998). Naturally, mountain species are often apterous (*Elytrosphaera*) or micropterous (*Galeruca*) and their activity is slowed down. Often, they are found under stones at high altitude and they can be nocturnal, like *Metallotimarcha*. High altitude chrysomelids will be discussed at some length in Chapter 6.

Only good flyers (Megascelinae, Galerucinae, Alticinae, Eumolpinae) frequent the canopy of tropical forests, from where they, particularly the Megascelinae, can fly down to lower-growing plants. Several Chrysomelinae may frequent the canopy, but only the good flyers. Certain species probably remain in the trees all the time and lianas, like *Coussapoa* or *Pourouma* (Cecropiaceae), are home to certain number of *Coelomera* (Galerucinae).

There are records of leaf beetles feeding on mosses (*Mniophila*, *Minota*), but precise observations are missing. *Minota* spp. has been found by Hering (1951) mining *Plantago* leaves (Plantaginaceae) and *Digitalis* (Scrophulariaceae) but, according to Heikertinger (1950), they feed also on *Vaccinium myrtillus* L. (Ericaceae) and ferns (Kato, 1991). The larvae and adults of the alticine genera *Ivalia*, *Kamala*, *Mniophilosoma*, *Clavicornaltica* probably also feed occasionally on mosses. Larvae of *Apteropeda orbiculata* (Marsham) were found by Cox (1997) mining the leaves of *Plantago lanceolata* L., *Teucrium scorodonia* L. and *Centaurea nigra* L. in England. *A. orbiculata* has been confused with *Mniophila muscorum* (Koch), but, if the latter is not a leaf-miner, it occurs in and feeds on moss all the year round. Adult alticines in the Far East (*Clavicornaltica*) seem to be root-, stem- or detritus-feeders at certain stages of their life, but aerial species of the same genus have been captured, probably living in the suspended soils of certain epiphytes, or possibly free-living. We do not know anything of the biology of these alticines, but Crowson found a galerucine larva in Queensland with moss debris in its gut.

Horse-tails (*Equisetum*) seem to feed *Hippuriphila* and reports of alticines on ferns are many (Kimoto, 1964–1966; Hendrix, 1980; Kato, 1991). They mine in the leaves but are less common on the herbaceous ferns of temperate areas than on arborescent species of the tropics.

Kimoto (1984) noted miners among the chrysomelids in Japan on Ranunculaceae, Lardizabalanaceae, Polygonaceae, Oxalidaceae, Oleaceae, Asteraceae, Liliaceae, and Poaceae. According to Kimoto, most of the species quoted in Japan on Pteridophytes (*Manobia*, *Minota*) would not all be miners, but chewers of leaves. Many mining species are noted from Okinawa. A few galerucines live on ferns and, for instance, *Galerucella birmanica* (Jacoby), which normally feed on the water nut (*Trapa* sp.) and *Nymphaea*, was seen on *Salvinia*, a floating fern. However, this does not prove that it was feeding on it. Very few chrysomelids feed on *Ginkgo biloba* L. (Ginkgoaceae), on cycads (*Aulacoscelis*, *Lilioceris*, *Carpophagus*). Palophaginae feed on *Araucaria* cones in Australia, Chile, Argentina and probably elsewhere where *Araucaria* grow naturally. Other species feeding on coniferous trees (Galerucinae, Clytrinae, Eumolpinae, Synetinae, Alticinae) are generally recent adaptations, except perhaps for Synetinae.

The selection of Angiospermae by leaf beetles has been studied subfamily by subfamily in Jolivet and Hawkeswood (1995) and is outlined below. Criocerinae and Hispinae seem to be monocot specialists and have adopted dicots only recently. This situation should not be taken as indicative of the greater antiquity of the subfamilies, however, as multiple adaptations to monocots happen in various subfamilies, sometimes rather unexpectedly as, for example, with the case of *Exosoma* on Liliaceae and Amaryllidaceae.

Part of the Eumolpinae adapted exclusively to Asclepiadaceae and Apocynaceae in the Old and New World. They all found a way to get rid of the latex flow. Selection of the Asclepiadales could be the fundamental selection of the group which later became oligophagous or, in some cases, polyphagous. *Timarcha* species in the Old World feed on Rubiaceae and Plantaginaceae, the families being chemically and botanically related. Species attracted by sinigrin or mustard oil feed on Brassicaceae and related, Capparidaceae, Resedaceae, Tropaeolaceae, Cleomaceae, etc. *Timarcha* species will very occasionally feed on the Brassicaceae in the Mediterranean region but have never been found on the related families mentioned above. Relationships with *Ranunculus* spp. among *Chrysolina,* which feed normally on Asteraceae or Lamiaceae, is difficult to explain except by reference to some ancestral trophism.

The Doryphorina in America are especially attracted to *Solanum* but will also accept members of the Asteraceae, Apocynaceae and several other families. Doryphorina are a neotropical tribe and the Solanaceae are numerous and most diversified in America, where they probably originated. There are very few solanophagous species of leaf beetles in the Old World, and certainly fewer than in the New World (Hsiao, 1986).

All these selections, which are regarded as botanical, are fundamentally linked to the attraction of certain chemicals, the absence of repulsive or toxic substances and probably secondary compounds. Certain chemical substances are common among certain families of plants and this is the reason why they are selected by some chrysomelid subfamilies. An insect is attracted by a colour, an odour, a texture, but without its palps and antennae it loses its selective power. With the use of sophisticated methods (antennograms, palpograms, walk compensator, etc.), the reactions of the Colorado Potato Beetle and others to odours and flavours have been studied in detail by the Wageningen laboratories in the Netherlands.

In nature, an insect can sometimes be attracted to a plant, like certain species of *Solanum*, which will in fact kill or sterilize it. *Leptinotarsa decemlineata* (Say) normally feed on *Solanum*, but, in the laboratory, will accept lettuces (Asteraceae), a neutral plant, or *Asclepias* (Asclepiadaceae), a highly toxic plant. Asclepiadaceae is a common selection of Doryphorina with Solanaceae and Asteraceae. Perhaps the choice of the Colorado beetle is dictated by its genome.

Selection factors for phytophagous insects, and particularly leaf beetles, have been studied and discussed in various books (Jolivet and Hawkeswood, 1995; Jolivet, 1997). Insects such as *Chrysolina geminata* (Paykull), are capable of circumventing the phototoxic effects of hypericin by appropriate behavioural and biochemical strategies (Guillet *et al.*, 2000). They often use the toxins for their own defence. By contrast, reduced feeding on transgenic potato plants by *Epitrix cucumeris* (Harris), an alticine, is partly due to the toxicity of the plants to immature growth stages and the preference for adults to feed on nontransgenic material (Stewart *et al.*, 1999). It has been confirmed experimentally with *Longitarsus jacobaeae* (Waterhouse) that leaf beetles orient themselves to host plants which are upwind (Zhang and McEvoy, 1995).

The choice of leaves

It is evident that the chemical components of plants vary during the day–night cycle. New leaves are very different in this way from old leaves. In the tropics, young leaves or bracts are often red or white while others appear a tender green. Kursar and Coley (1992) claim that if the young leaves are red or white it is for reasons of economy of energy. Insects feed on leaves which are poor in nitrogen and chlorophyll instead of the green, chlorophyllian, though older leaves. It is not always true since *Coelomera* spp. (Galerucinae) will feed on the old or new, tender or tough green leaves of *Cecropia* trees in America. They avoid red leaves.

Oedionychus sp. in Costa Rica (Rockwood, 1974) will go straight for the new leaves of the host plant, *Crescentia alata* (Bignoniaceae), and do not normally eat the older leaves. According to Rockwood, seasonal changes in the chemistry of the leaves have important effects on herbivores. *C. alata* leaves are eaten by *Oedionychus* only for a short period, when they are young or senescent. In Kenya, and probably elsewhere in Africa, the male of *Gabonia gabriela* Scherer (Alticinae) is attracted by the wilted leaves of *Heliotropium* spp. (Boraginaceae). This strange behaviour could be the beginning of a lek-behaviour, i.e. the tendency for males to group together to attract the females using certain chemicals, but there may be another reason, which still remains to be discovered (Boppré and Scherer, 1981). This attraction of a chrysomelid for withered leaves is not unique and becomes the normal behaviour of many phytophagous insects in the autumn. Scherer and Boppré (1997) found that *Gabonia* and *Nzerekorena* spp. (Alticinae) visit the withered plants of *Heliotropium* to gather pyrrolizidine alkaloids (PAS). It is possible that PAS are used by the males as precursors of male pheromones or as defensive chemicals. Pyrrolizidine alkaloids can be used in different ways. For instance, in *Utetheisa ornatrix* (Lep. Arctiidae), the female receives pyrrolizidine alkaloids from the male, with the sperm package, and passes them on to the eggs, which become, because of their alkaloid content, less vulnerable to predation (Iyengar and Eisner, 1999).

Thus, insects have a choice of new or older leaves which is based not only on the fact that the chemistry varies but also on the toughness, the suberization, the concentration of tannins etc. of the leaves. Generally, leaf beetles prefer young shoots (Megalopodinae) or tender leaves, which explains why crops are usually attacked when the plants are immature. Tender crops are sometimes chewed by Alticinae when nothing else is available and the area is totally dry (Cabo Verde islands), even if the plants have nothing in common with the normal food (alienophagy).

According to Hayashi *et al.* (1994), the larvae and adults of *Chrysolina aurichalcea* Mannerheim prefer the young leaves of *Artemisia princeps* when mature leaves would seem to be more beneficial in terms of promoting long life and good fertility. When the larvae were fed with senescent leaves, their growth is rapidly decreased and their mortality increased. However, neonate larvae do not show any preference about leaf quality, despite the fact that young leaves are necessary for their development. Authors conclude that a synchronization has evolved to promote, after the autumnal reproduction, hatching at spring time when the host plant germinates. Raupp and Sadof, 1991, in their studies on *Plagiodera versicolora* (Laicharting) noted that leaf injury reduces the acceptability of the host plant. Leaf toughness acts as a potent defence affecting morphology, feeding behaviour, and ultimately the spatial and temporal patterns of herbivores, according to research on *Plagiodera versicolor* (Raupp, 1985). *Timarcha tenebricosa* (Fabricius) accepts young leaves of *Galium aparine* L. but rejects them when they become too hairy. It would probably have accepted shaved leaves. It is also the case with *Chrysomela lapponica* L. larvae which clearly prefer shaved discs to unshaved ones (Zvereva *et al.*, 1998). Larvae feeding on highly pubescent plants move about up to three times as much as larvae feeding on less pubescent plants. Leaf hairiness disturbs the behaviour of many leaf beetles. A recent paper by Obermaier and Zwölfer (1999), studying three chrysomelids, *Galeruca tanaceti* (L.), *Cassida rubiginosa* (Mueller) and *Oreina luctuosa* (Suffrian), all Asteraceae feeders, shows that the three investigated chrysomelids adapt differently to overcome the dilemma of variable food quality (concentration of protein nitrogen).

The shape of the *Passiflora* leaves (Passifloraceae) is very variable among the 350 known species, mostly American. Some have entire lobed or curious crescentic or bilobed leaves, with extra floral nectaries on the petiole. The shape of the leaves influences the choice, for oviposition, of the female of the butterfly *Heliconius* (Heliconiidae), The shape of the leaf is less important among chrysomelids, although of more significance to good flyers. It is probably as important as colour and odour, or special allelochemical factors (kairomones). The type of plant, whether a climber or not, is certainly important for the adult female and perhaps even for the larvae. Some Cassidinae in the neotropics are specialized for climbers when non-climbers belonging to the same family are available.

Leaf beetles have been fed on synthetic or semisynthetic materials in research carried out mostly on agricultural pests: *Diabrotica* spp., *Leptinotarsa decemlineata*, etc. For the time being, no mass rearing for biological control has been attempted. To rear the Colorado Potato Beetle, various chemicals have to be added, such as enzymes (Waligora *et al.*, 1993) and hexanes from the potato plant (Krzymanska *et al.*, 1993). The addition of enzymes improves assimilation and hexanes provide the necessary phagostimulants.

The gustative sensilla of the Colorado beetle galea are responsible for the acceptance or rejection of a given plant. A single cell (as demonstrated with the electropalpogram) responds actively to the potato sap and is even sensitive to extreme dilutions (10 times the normal). The separation by the Colorado beetle of host and non-hosts among the Solanaceae (*Datura*, *Atropa*, *Lycopersicum*, *Solanum*, etc.) is certainly due to the response of certain specialized cells. The codage of complex stimuli exists in nature and certainly intervenes when using synthetic media (Mitchell *et al.*, 1990).

Addressing another question, the famous hypothesis of talking trees seems to have been recently confirmed. According to Dolch and Tscharntke (2000), after defoliation, herbivory by *Agelastica alni* L. (Galerucinae) seems to increase with distance from the defoliated tree. Resistance seems therefore to be induced not only in defoliated alders, but also in their undamaged neighbours. The conclusion of the previous authors is ‘defoliation in alders may trigger interplant resistance transfer, and therefore reduce herbivory in whole alder stands’.

Cannibalism

Cannibalism can be considered as deviant behaviour for phytophagous insects, but it can help the development of the larvae in certain cases (Mafra-Neto and Jolivet, 1996). It is widespread among vertebrates, herbivorous or not, and it is not always accidental. Filial cannibalism exists in certain fishes amongst which filial cannibals or heterocannibals are very common. Fitzgerald (1992) regards this behaviour as adaptative. Baby hyenas almost always eat their twin (Frank, 1994). Aggressiveness or instinct for aggression is probably at the base of this behaviour. Intraspecific struggle is for the selection of the better, and insects are no exception to this. Cannibalism is not confined to insects among invertebrates; it is known also in spiders.

Cannibalism may be defined as the act of eating conspecific individuals. They can be brothers or half-brothers or offspring. Though cannibalism is known among leaf beetles, a female never eats her mating partner, a phenomenon which occurs with mantids and some spiders. Stevens (1992) has pointed out that cannibalism in beetles includes mobile instars (i.e. adults and larvae) eating up immobile stages (i.e. eggs or pupae), or the mobile instars eating larvae.

It seems that an attack on an egg batch or isolated eggs is more likely to be a demonstration of cannabilism among herbivorous insects than an attack by a predator which would not need any supplementary proteins.

Dickinson (1992a) states that cannibalism towards an insect's own eggs is a kind of infanticide. However, adult chrysomelids which guard their eggs and larvae, or larvae which show some kind of collective defence (cycloalexy), do not eat each other. *Coelomera* larvae (Galerucinae), feeding on the leaves of *Cecropia*, a parasol tree of America, do not eat each other, neither do they eat their own eggs. The same thing can be said of Cassidinae females which guard their eggs and larvae (Vasconcellos-Neto and Jolivet, 1988, 1994; Jolivet *et al.*, 1990). These authors have, however, observed cases of cannibalism in *Plagiodera versicolora* (Laicharting), but the insect does not demonstrate complete cycloalexy, only a loose aggregation, at the larval stage.

Multiple copulations are frequent among the chrysomelids and there are cases of multiple paternity. Cannibalism may occur between brothers and sisters and half-brothers and half-sisters.

CANNIBALISM AMONG BEETLES

Stevens (1992) mentioned cannibalism among Coccinellidae, Chrysomelidae, Bostrychidae, Cucujidae, and Gyrinidae. It is evident that this list is incomplete.

Herbivores, granivores and bark beetles may eat their own eggs and larvae. True carnivorous beetles, such as Carabidae, may take to a mixed diet, which includes cannibalism in addition to predation. The same is true for omnivorous beetles such as Tenebrionidae. Familial cannibalism has been recorded mostly among Coccinellidae, Chrysomelidae and Silphidae.

Stevens recognized familial and non-familial cannibalism and filial cannibalism. All types of cannibalism have been observed among chrysomelids, but normally neonate larvae or adults eat eggs or larvae from their own clutch. According to the same author, cannibalism is favoured by natural selection since it is supposed to increase the fitness of an individual. We shall later consider this assertion.

Cannibalism is frequent among Coccinellidae such as *Adalia bipunctata* (Linné) in the laboratory and in the field. According to Agarwala (1991) and Agarwala and Dixon (1993), female beetles and young larvae are capable of familial recognition, and these females show a certain aversion to attacking their own eggs, while males do not share this hesitation. No such observations have been recorded among chrysomelids and they are very difficult to make.

PROTECTION OF THE EGG CLUTCH. LACK OF CANNIBALISM

It is evident that chrysomelines, such as *Timarcha*, cassidines and others, which build a primitive ootheca from secretions, regurgitated food or excreta or protect their eggs individually, as do the Camptosoma, do not eat their own eggs. Besides, *Timarcha* is a K-strategist which lays big eggs in small numbers and has no interest in eating them. Other species, like *Iscadida* in southern Africa, feed on the leaves of a small bushy plant (*Rhoicissus* spp.), a Vitaceae. Females drop their eggs, which are rounded and naked, to the ground. Several insects do the same: *Syneta* spp., Clytrinae, most of the Cryptocephalinae, several curculionids like *Otiorrhynchus*, many phasmids, etc. In certain cases, the eggs are protected individually. This way the insects cannot eat their own clutch, the eggs being dispersed and the larvae free and separated. Phasmid and clytrine eggs are carried by the ants into their nest. They both mimic seeds, but, in the case of certain phasmids, the eggs have a capitulum, mimicking the elaiosome of real seeds which is also eaten by the ants (Compton and Ware, 1991; Hughes and Westoby, 1992; Windsor *et al.*, 1996). It seems very probable that clytrine eggs possess an attractive chemical for the ants such as exists in phasmids and certain seeds in ant gardens.

Cannibalism is also impossible, at least very difficult, in galerucines like *Coelomera* which hide their eggs inside the internodes of the host plant, a *Cecropia* tree, or agglutine them in an ootheca at the extremity of the leaf. The eggs are protected from the siblings by the secretions and from the parents by their hidden position. The adults and larvae of *Coelomera* spp., like ants, eat the trophosomes (Müller bodies) of the plant which are rich in glycogen and so do not need a special supply of proteins such as provided by living prey.

Roughly, chrysomelids which build their ootheca or those Hispinae, Zeugophorinae, Megalopodinae and Sagrinae which bury their eggs inside vegetal tissues, cannot eat their own eggs. They are also mostly K-strategists; their eggs are relatively big and not numerous.

Many Alticinae, Galerucinae, and most of the Eumolpinae lay their eggs into soil crevices, often at the foot of their host plant, and the larvae penetrate into the soil to feed on the roots. Cannibalism must be absent or, if it exists, could only happen when the eggs are hatching. Hispinae insert their eggs into vegetable tissues or lay them between the appressed leaves of monocots, into a semiaquatic medium. Certain species of *Platyphora* (Chrysomelinae) are viviparous in Brazil and the females drop their small larvae one by one when walking over leaves. These larvae aggregate themselves over *Solanum* leaves. Other species of *Platyphora* lay their larvae in clusters and the larvae immediately assemble themselves in a defensive ring (cycloalexy). There is no report of larvae eating each other in a cycloalexic ring.

CANNIBALISM AMONG CHRYSOMELIDS

Doryphora punctissima (Olivier) lays its eggs in a clutch below the leaves of its host plant, *Prestonia isthmica* (Apocynaceae), and the eggs are covered with a gelatinous and light substance (Eberhard, 1981). However, adult *Doryphora* occasionally eat part of their eggs and their larvae when they come out of the mass. In this case, there is no protection with a faecal or secretory substance.

Colorado Potato Beetle, *Leptinotarsa decemlineata* (Say), lays its eggs freely under *Solanum* leaves. It eats them willingly. It is probably the first case of oophagy mentioned among leaf beetles (Durchon, 1946; Trouvelot and Grison, 1946). It is likely that all species of this genus behave in the same way, but some species are much less fertile.

Durchon (1946) reports that Colorado beetle females eat one third of their own eggs. Males also eat eggs and, if we consider that a female can lay up to 3000 eggs alone, predation does not reduce much of the population through oophagy. However, most of these observations have been made inside cages and it can be different in the field. For instance, Dickinson (1992a) mentions that adult females of *Labidomera clivicollis* (Kirby) cannibalize the eggs in the laboratory, but only females of a Texas subspecies do it in the field.

Stevens (1992) mentions cannibalism among chrysomelids but only in four species. In reality, many other species are oophagous or practice siblicide, i.e. cannibalism between larvae.

Eumolpines lay their eggs covered at the apex with endosymbionts which are rapidly ingested by the larva with the eggshell when hatching (Ferronato, 1988). There is no report in the literature of adult eumolpines eating their own eggs because adults live free on the leaves, the eggs are hidden and the larvae radicicolous. Cannibalism remains, however, likely among free primitive eumolpines. In Argentina, *Hornius* eggs are protected by excreta and the primary larvae dig inside the foliar buds of *Nothofagus* (Jerez and Ibarra-Vidal, 1992; Jerez, 1996).

Selman (1994b) reports that in Australia the eggs of *Chrysophthartra* spp. are laid together and eaten partly by the first larvae to hatch. Many of these eggs are probably infertile (trophic eggs) and are a source of proteins for the cannibals. However, oophagy exists among larvae, gregarious or not, among Paropsini. Carne (1966) mentions that neonate larvae of *Paropsis atomaria* Olivier in Australia eat the shells of their own eggs before attacking leaves. It is also probable that the larvae also eat the eggs of their own clutch.

Among *Gastrophysa viridula* (De Geer), studied by Helen Kirk (1988), cannibalism occurs when larvae eat the eggs of the same laying. Viable or trophic eggs are eaten the same way. In the case of *G. viridula*, the eggs are laid unprotected. What is interesting is that larvae, fed upon only the eggs, are unable to complete their first larval instar. It is notable that larvae fed with *Rumex* whilst also eating eggs during the first instar grew better than the larvae fed on *Rumex* only. For the larvae, a certain cannibalism seems beneficial.

Eggs are protected either by scattering or by the chemicals contained in them. These are oleic acid, and probably isoxazolinone glucoside. According to Pasteels, those compounds are repulsive to the ants, but not to the larvae themselves. The orange colour of the eggs on the green background of the leaf is certainly aposematic.

Table 4.1. Data on cannibalism among chrysomelids

Species	References
Plagiodera versicolora (Laicharting)	Wade and Breden, 1986
Labidomera clivicollis (Kirby)	Eickwort, 1971, 1973, 1977; Dickinson, 1992a
Leptinotarsa decemlineata (Say)	Durchon, 1946; Trouvelot and Grison, 1946
Leptinotarsa juncta (Germar)	Stevens, 1992
Chrysophthartra sp.	Selman, 1994b
Doryphora apud *punctatissima* (Olivier)	Eberhard, 1981
Chrysomela aeneicollis Schaeffer	Rank, 1991
Gastrophysa viridula (De Geer)	Kirk, 1988

The same happens with the Colorado Potato Beetle. Egg colour can vary between the genera and species. Often, it is linked with the colour of the haemolymph.

Dickinson (1992a), studying *Labidomera clivicollis* (Kirby) in Texas and in New York state, distinguishes three cases of cannibalism in the field: oophagy, or egg cannibalism, or siblicide between larvae; cannibalism by older larvae against younger ones of the same clutch or of different origin and cannibalism by adult females or, rarely, by males.

Such distinctions are rather subtle, and we must remember that different types of cannibalism are known among chrysomelids. Actually, most of the observations concern laboratory beetles and little research has been done in the field.

According to Dickinson (1994), females prefer to eat eggs from other females rather than their own eggs, but the selective effect of cannibalism is extremely difficult to demonstrate.

Palmer (1985) says that the adult *Labidomera clivicollis* is unable to survive for a long period of starvation using the food accumulated by the larva. It is probable that cannibalism allows the female to survive during difficult periods. The phenomenon can be adaptative.

Among *Plagiodera versicolor* (Laicharting), hatching larvae are fiercely cannibalistic, at least in the laboratory (Breden and Wade, 1985, 1987, 1989; Wade and Breden, 1986; Stevens and McCauley, 1989) and up to half of the clutch can be eaten (Wade, 1994). The same behaviour exists in *Chrysomela aeneicollis* Schaeffer in California (Rank, 1991). According to Wade and Breden, eggs eaten by *Plagiodera versicolor* are not trophic eggs but eggs with embryos inside.

Breden and Wade hold that cannibalistic larvae grow more quickly than non-cannibalistic ones. The size of the group enhances plant-feeding activity.

Cannibalism could also exist in some subsocial groups but Wade (1994) emphasized that cannibalism acts rather negatively on group survival.

Parental relationships in the group (sisters or half-sisters), or non-parental relationships in certain cases of cycloalexy, do not seem to be important. Among *Paropsis* or tenthredinid larvae in Australia, interspecific or intergeneric groups may be formed, so strong is the attraction between the individuals. It would be interesting to see if predation is worse in multispecific cases of cycloalexy.

According to Breden and Wade (1985), the variation of cannibalism between races has a genetic component.

CONCLUSIONS ABOUT CANNIBALISM

Cannibalism seems more common among herbivorous insects than among predators.

Is it only a speculation but do plant-eating invertebrates actually need to eat meat sometimes? If phytophagous individuals complete their diet with cannibalism, it could be advantageous for the plant in limiting any damage. For the cannibal, the advantages are nutritional and in the reduction in the interspecific competition. An intra- and interspecific competition is well known among *Diabrotica* species (Woodson, 1994).

Lack of food can bring about a reabsorption of the ovaries inside the body in the Colorado beetle (Selman, 1994a), representing internal cannibalism. This remarkable adaptation is well known among tropical rodents.

According to Dickinson (1992a,b), oophagy, generally attributed to laboratory overcrowding in cages, really exists in the field. This could be the result of an intraspecific competition, but the phenomenon is difficult to observe *in vivo*.

Generally speaking, only chrysomelids which do not protect their eggs, cannibalize their eggs, or their eggs are subjected to oophagy by an alien adult. When the eggs are protected by vegetal tissues or faecal or glandular masses, they are outside the reach of adults or larvae. Cannibalism seems to be widespread among chrysomelids, but has only been recorded in a few species, mostly among beetles of agricultural importance. Probably, cannibalism has been selected by evolution for its nutritional advantages. If it can really be proved that a female discriminates between her own eggs and alien eggs, this could represent a form of protection of own progeny (Dickinson, 1992a). According to Hinton (1981), viviparity among *Oreina* and others is also a kind of parental care. It is true that no case of cannibalism has been observed among viviparous chrysomelids, which does not mean that it does not exist.

According to Fox (1975), cannibalism is not an aberrant laboratory behaviour, but a normal answer to environmental pressure. Cannibalism is largely spread among herbivorous animals and Kirkpatrick (1957) points out that several phytophagous insects are inveterate cannibals even in the presence of abundant food. According to the same author, cannibalism is less frequent among terrestrial predatory insects, perhaps because they already have a source of protein. Cannibalism seems also to occur in response to environmental factors. It can vary among diverse populations of the same species, but data in this context are poor for chrysomelids. Cannibalism can also be a regulatory mechanism but, in the case of leaf beetles, food is abundant and the increase of the group is often important in a subsocial species. Cannibalism can thus have a negative impact, at least for subsocial species, and it seems exceptional among them. More important for them is the group, which is helpful in the defence against predators.

Also, according to Breden and Wade (1985), a cannibal individual in a group receives nutritional benefit but reduces the survival capacity of the group itself. It is true, however, that the comparison between subsocial or solitary larvae with regard to cannibalism remains to be studied carefully.

Entomophagy, coprophagy, nematophagy

Entomophagy is eating insects which are not conspecific. Surely, it is a way of obtaining proteins but it can be also a deviance from a classical phytophagy under the influence of similar attractants. It is really exceptional among phytophagous insects and only one case is actually known among chrysomelids. Mafra-Neto and Jolivet

(1994) described the case of *Aristobrotica angulicollis* (Erichson), a diabroticine, feeding on live adult meloids in Brazil. Really, this strange behaviour must be a recent adaptation, since *Diabrotica* spp. are the neotropical equivalent of other Luperini, such as *Aulacophora* from the Old World, which are predominently cucurbitophagous (Jolivet, 1999d). *Diabrotica*, like *Aulacophora*, is also attracted by the flowers, pollen and leaves of Cucurbitaceae, but it became more polyphagous, which accounts for its extraordinary behaviour, i.e. entomophagy. *Aristobrotica angulicollis* induces reflex immoblization among its prey, the meloids. They fall from the *Solanum* leaves and the beetle eats them alive (in a state of thanatosis). These meloids (*Epicauta aterrima* (Klug)) live in the larval stage as parasites of Acridian eggs in the ground and are relatively common in Brazil on various plants and crops. They are highly toxic, rich in cantharidins, and show an abundant reflex bleeding in the leg joints. This does not prevent the galerucine from attacking the meloid, being toxic itself and naturally immune. It is probable that feeding on a beetle like *Epicauta* increases its personal toxicity and so its defensive means. Recently, Hemp *et al.* (1999) mention clerid males being attracted to cantharidin traps. Cantharidin possibly acts as a kairomone. Two galerucines are also attracted by the cantharidin traps: *Barombiella vicina* Laboissière and *Barombiella* sp.

Another exceptional case is the galerucine *Aplosonyx nigripennis* (Jacoby). Normally herbivorous on Araceae and Vitaceae, it was found in Indonesia sucking the wound of a snake, a local coluber, a *Ptyas* sp. This is not, of course, entomophagy, but a rather abnormal behaviour.

Coprophagy is still rarer among chrysomelids, and we can hardly consider as coprophagous the clytrine larvae which live inside ant nests. They are really detriticolous and feed on ant eggs and larvae, vegetal debris, and ant excreta.

However, there is a so far unique observation of a case of coprophagy in the Florida Keys: *Oomorphus floridanus* Horn feeding on rodent excreta (*Neotoma floridana*). Normally, holarctic Lamprosomatinae feed on Araliaceae and Peck and Thomas (1998) mention *Oomorphus* on *Pinus* sp.and *Ficus* sp., which seems only a spatial coincidence.

Really, entomophagy and coprophagy are exceptional among chrysomelids. We can even wonder if *Diabrotica* feeding on the meloid is not an isolated case of a mutating population of a normal phytophagous species. *Diabrotica sensu lato* is polyphagous and we can imagine that *A. angulicollis* has evolved towards a canthariphagous diet when coevolving with plants containing molecules similar to cantharidin. We could also think that meloids accumulate products acting as powerful phagostimulants for *Diabrotica*. These two hypotheses are acceptable, but this strange behaviour facilitated by the abundance of potential preys must be isolated in time and space.

According to Kulik (1991), the Colorado Potato Beetle, *Leptinotarsa decemlineata* (Say), often feeds on abandoned potato tubers in Russian fields. Along with these potatoes, the beetle also eats abundantly specific and saprophagous nematodes, a carnivorous diet for a phytophagous insect.

Parallel diversification of plants and leaf beetles

Is there always coevolution between plants and beetles? It seems that the term has been over-used and that real coevolution can be considered only in specific cases such as myrmecophilous plants or carnivorous plants, for instance, where both sides are adapted to each other. It is difficult to see real coevolution between a plant and an insect, even if there is developing detoxification of the chemicals in the latter.

Coevolution, as defined by Ehrlich and Raven (1964), is an 'endless evolutionary arms race'. To Ehrlich in Daly *et al.* (1978), coevolution exists when two different kinds of organisms are so intimate ecologically that any evolutionary changes in one are likely to induce changes in the other. Matthews and Matthews (1978) give a similar definition: when two or more populations interact so closely that each serves as a strong selective force on the evolution of the other, resulting in reciprocal stepwise adjustments.

The adaptation of an insect to a plant toxin can be seen as an evolution, but the plant is supposed to react and produce more toxins to repel the insect. It is, however, almost impossible to trace a parallel evolution of the insects with their food plants. Many attempts to ascertain such changes have failed (see Chapter 3). However, Farrell and Mitter (1998) see a possible coevolutionary diversification between the North American cerambycid, *Tetraopes*, and its main food plant, *Asclepias* spp. Cladogram correspondence seems in this case to reflect synchronous diversification of the two clades. Becerra (1997) reconsidered the question and tried to determine the importance of plant chemistry on herbivore host shifts comparing the molecular phylogenies of *Blepharida* (Alticinae) with the toxic plant *Bursera* (Burseraceae), rich in terpenoids. Historical patterns of host shifts correspond to the patterns of host chemical similarity. Becerra concludes that plant chemistry had played an important role in the evolution of host shifts by phytophagous insects. This is confirmed later on by Becerra and Venable (1999) with the same genus of beetles.

Trials to reconstruct the phylogeny of Phytophaga (Curculionidae, Cerambycidae, Chrysomelidae, Bruchidae) from DNA sequences and morphological characters have been carried out. They are briefly reviewed in Chapter 2. As mentioned previously, several fundamental genera and subfamilies are missing in these analyses and this decreases the value of the results. However, Farrell (1998) admits that basal lineages for the phytophaga were associated in the Jurassic with conifers and cycads as they are today with certain groups (Aulacoscelinae, Palophaginae, Nemonychinae, Allocoryninae, etc.). Other groups diversified over Angiosperms in general. According to Farrell *et al.* (1992), the evolution of affiliated insect and plant lineages has occurred over similar geological time intervals, although cases of strictly parallel diversification are rare. The authors also remark that some leaf beetles did not evolve at all, like the palm-feeder hispid *Delocrania*, which did not vary during the last 35 million years since the Dominican amber sample. It seems that the Cretaceous fauna was much different from the present, but there is also some allochrony in the evolution of the insects, some remaining the same all over the Tertiary. However, Angiosperms were still in full evolution during this period.

Garin *et al.* (1999) and Gomez-Zurita *et al.* (1999), discussing the classical research of Petitpierre on the genetics of leaf beetles, found that there is a specialization of small groups (subgenera or species) of *Chrysolina* and *Timarcha* on several genera or families of plants. The host plants for *Chrysolina*, as well as for *Timarcha*, suggest an absence of a parallel evolution of these beetles with their host plants. Probably, the colonization happens according to available plants having similar compounds.

It is a fact that it is almost impossible to compare the relative evolution of plants and phytophagous insects using parallel cladistic analysis. According to Futuyma and McCafferty (1990), studying species of *Ophraella* (Galerucinae) and their Asteraceae hosts, the phylogeny of *Ophraella* appears not to agree with that of its hosts. Species of *Ophraella* are North American galerucines, which feed on eight genera in four tribes of Asteraceae. The genetics were reviewed by Futuyma (1992). Finally, host shifts in *Ophraella* seem to be facilitated by commonalities in plant chemistry. The authors do not see here any parallel evolution.

Hsiao and Pasteels (1999) studied the evolution of host plant affiliation and chemical defence in a natural group (*Chrysolina-Oreina*) using mtDNA phylogenies. *Chrysolina-Oreina* are evidently highly specialized herbivores. The authors suggest several modifications in the forming of subgenera, lumping some (*Hypericia* + *Sphaeromela*), separating others. Attempts have been made to reconstruct ancestral host plant associations, but the importance of *Ranunculus* selection among several species has not been taken into account. Is it an ancestral selection reappearing as a major or potential selection? That remains to be seen. At least, there is no relationship between that primitive family (Ranunculaceae) and the ones normally selected.

Grouping together Cassidinae and Hispinae, Hsiao and Windsor (1999) tried to establish a preliminary phylogeny of the subfamilies. Monocot feeders were separated from dicot feeders. The parental guarding of offspring appears to have a dual origin, evolving first among Convolvulaceae feeders and later among Asteraceae feeders. Evidently, this study concerns only the New World.

It remains, however, that insect–plant interactions in general are responsible for much terrestrial diversity (Ehrlich and Raven, 1964; Farrell *et al.*, 1992).

Much of the diversity of leaf beetles and of phytophaga in general occurs actually in the tropics. Bechyne and Bechyne (1971a,b) showed that biodiversity in Peru is enormous: over an area of 1 000 000 sq km there are about 4000 species of chrysomelids, whereas there are only 400 species over 10 000 000 sq km in Europe. There are, thus, 60 times more chrysomelids in Peru than in Europe. Sampling is almost complete in Europe and in its initial stages in Peru. Farrell *et al.* (1992), studying leaf beetles, confirm the difference in biodiversity between Peru and Indiana, a state of the USA. However, their statistics seem a little biased, apparently based on canopy studies and not much on soil sampling with a sweeping net. However, they conclude that the differences in latitudinal diversity gradient are enormous. We know also that in several valleys in Columbia, Venezuela or Peru, the flora diversity is the greatest on the planet. Permanent heat and no real seasonal differences allowing many generations per year, diversity of the flora, conservation of the cenozoic fauna, absence of glaciations and catastrophic events, are certainly the reasons for the abundance of the fauna in the tropics. There is also a great originality and endemicity of the flora in Madagascar and New Caledonia, for instance. However, Madagascar separated early from the continent (Cretaceous) and New Caledonia, being small and a long time isolated, have an impoverished original leaf beetle fauna. Temperate insect–plant communities seem to be only relicts of a formerly richer past due to the cooling of climate and not really being compensated by reinvasions after catastrophic events.

Futuyma and Keese (1992) think that insect divergence postdated plant divergence and that there is little evidence that parallel cladogenesis of plants and herbivores is common (Mitter and Farrell, 1991).

Recent findings in New Jersey, USA, of fossil flowers 90 million years old, i.e. 30 million years earlier than previously suspected, suggest that insects were diversifying alongside plants during middle Mesozoic. This could explain why chrysomelids were starting to diversify during the Jurassic.

5

Developmental Stages

Chrysomelidae may lay large eggs, which are few in number and well protected (K-strategists), or they may deposit numerous, small, rounded or oval eggs which may be simply scattered on the ground or laid on the food plant and provided with some sort of protection. (About 3000 eggs are laid by the female Colorado Potato Beetle and maybe more for other species of *Leptinotarsa*). In either case, the eggs may be laid singly or in batches. Eggs may be protected by a covering of sticky secretion, an ootheca, or by a deposit of stomach regurgitation or excreta. For extra protection, eggs may be deposited in crevices in plant tissues or in gaps made through incisions in the plant body. In brief, by producing a large number of eggs or by providing an efficient protection to the eggs, or by both, the survival and continuity of race are ensured.

Several chrysomelids are viviparous (Bontems, 1985, 1988). The females deposit small larvae which aggregate immediately or become dispersed over the leaves. Bontems tends to confuse viviparity and ovoviviparity. Ovoviviparity is the rather common situation in which females deposit big eggs which already contain advanced embryo and which hatch soon after they are laid. Some chrysomelid species have the tendency to be viviparous, ovoviviparous or oviparous according to the latitude or altitude. Generally speaking, viviparous or ovoviviparous females tend to lose their spermatheca. This generally happens in the colder parts of their range. Some tropical chrysomelids are also viviparous, like many species of *Platyphora* in Brazil.

The number of eggs laid by a chrysomelid varies much from one genus to another. The case of 3000 or more eggs deposited by the CPB (Colorado Potato Beetle) or others can be contrasted with that of the hispine *Promecotheca reichei* Baly which lays no more than 16–25 eggs (Taylor, 1937). Some ground-living species, like *Clavicornaltica* in south-east Asia, also lay eggs in very small numbers. So do Antarctic species, like the ovoviviparous *Brachyhelops*, which, like *Timarcha*, is a K-strategist. According to Selman (1994a), eggs laid in groups contain non-developed embryos, while big eggs laid in small numbers are already much developed.

Larvae and pupae have been well studied morphologically and biologically by a relatively large number of researchers. These studies were summarized and reviewed by Cox (1982, 1988, 1996b) Recently, larvae of *Megascelis* and *Aulacoscelis* have been described in detail (Cox, 1998a; Cox and Windsor, 1999a,b). Of course, we must also mention the classical works of Böving and Craighead (1930), Paterson (1930, 1931, 1941), Marshall (1979) and so many others dealing with chrysomelid larvae.

The eggs

Apart from Hinton's classical work (1981), there are relatively few papers on chrysomelid eggs. They have been reviewed and summarized by Selman (1994a), Petitpierre and Juan (1994) and Nordell-Paavola *et al.* (1999).

Many species, some entire subfamilies/tribes such as the Cassidinae, the Stenomelini, secrete more or less complex oothecae. Clytrinae and most of the Camptosomata produce scatoshells around the eggs. These are mostly of faecal origin and differ according to species and genera. This form of protection is discussed in another chapter. Here we are dealing only with the structure of the eggs and oviposition.

THE STRUCTURE OF THE EGG

Chrysomelid eggs are relatively simple. They consist of an embryo inside a shell or chorion which is variously ornamented according to the species. The chorion is often enclosed by a secretion of the mother's colleterial glands, which are modified accessory glands, opening into the common oviduct (Selman, 1994a). When it hardens, this secretion forms outside structures, like polygons, papillae or longitudinal ridges. This is the extrachorion. According to Petitpierre and Juan (1994), four ultrastructural elements are found in egg-chorion of chrysomelids: polygons, scales or warts, dot-like reliefs and fenestra. Hexagons and pentagons are common among species of the genus *Timarcha*, but eggs are not always distinguishable between species. Eggs are often glued together and to the substrate by a layer soluble in alcohol (Hinton, 1981).

Sometimes, there are pockets of air inside the extrachorion which have a respiratory function in aquatic (*Donacia* spp.) and subaquatic species (*Galerucella nymphaeae*) (Selman, 1994a; Nordell-Paavola *et al.*, 1999). The eggs of *Diabrotica* spp., which are laid in soil subject to flooding, also have this plastron-like structure (Jolivet, 2002, in print). Hinton (1981) described vertical columns containing a film of air in the chorion of the egg of *Galeruca tanaceti* (L.). Acropyles and micropyles allow the air to move inside (Scherf, 1956, 1966). This respiratory device is missing in purely terrestrial groups like the Clytrinae, which lay eggs on leaves or on trees, but many subfamilies like the galerucines or the donaciines have a well-developed chorionic respiratory system. There are often large differences between species of chrysomelids to be found in drawings of the extrachorion. For instance, Rowley and Peters (1972) found differences in chorion structures among four species of *Diabrotica* and that is also true for other galerucine species. There are also noticeable differences between the chorion structures of the species of *Galerucella* (Nordell-Paavola *et al.*, 1999) and in various species of Chrysomelinae, though it is not always clear among the red eggs of *Timarcha* (Petitpierre and Juan, 1994). Atyeo *et al.* (1964) identify species of *Diabrotica* by chorion structure in the same way as is done with mosquitoes, e.g. in the *Anopheles maculipennis* complex. The egg morphology of European species of chrysomelids has been described by Klausnitzer and Förster (1971).

The micropyles of chrysomelid eggs are always situated at the anterior pole. The micropylar structure is not very visible in *Gastrophysa viridula* DeGeer (Hinton, 1981), a purely terrestrial species. Eumolpinae which lay their eggs in soil have an extrachorion which contains packages of symbionts with which to infect the larvae (Ferronato, 1988).

Timarcha eggs are red when fresh and then they become orange and yellow after some time. Generally speaking, chrysomelid eggs are pale yellow to pale orange or brown. Selman (1994a) describes how variations among the colours and sculptures of Australian Paropsini can provide good taxonomic characters. Of course, he is dealing with fresh, not desiccated eggs. Red, orange and brown are the most common colours, but grey, pink and blue eggs are also found. Green seems rare, but it is sometimes also the colour of the haemocoel fluid. According to Doguet (1994), alticine eggs are elongated, subcylindrical, generally whitish in coloration, or yellow or orange. The chorion is as usual, reticulated or subgranular.

THE COVERING OF THE EGGS

Many chrysomelids use faecal matter to cover and protect their eggs. Besides being a mechanical protection, this system acts also as a repellent against predators. Several subfamilies (Clytrinae, Cryptocephalinae, Chlamisinae, Lamprosomatinae) use excreta for egg camouflage. This scatoshell acts as a base for the faecal cone, which will grow with the larva with the accumulation of excreta (Hinton, 1981; Erber, 1988; Hilker, 1994; Selman, 1994a). Pupation occurs inside the scatoshell. Lecaillon (1898a,b) studied the mechanism and this is reviewed in Chapter 8.

In the Clytrinae, the egg chorion is thin and lacks any respiratory device. It is covered by faecal material and by the secretion of an anal gland, which functions like the colleterial glands but is not homologous.

The female turns her egg with her posterior tarsi when the egg is still partially inside the body. Scatoshells formed around the egg have variable structures characteristic for each species of clytrines or cryptocephalines. In *Labidostomis* spp., scatoshells are attached by threads to branches or leaves and each scatoshell has an opercle at its end. These shells, surrounded by plates, look very much like plant seeds.

In *Timarcha* (Chrysomelinae) around twenty eggs and in *Coelomera* (Galerucinae) sixty eggs are glued together in a kind of primitive ootheca and then stuck to the substratum (the soil or plant matter) either by stomach regurgitation or secretions, or both. This gives relatively good protection, though among *Coelomera* some of the eggs can be prey to bugs or ants (Jolivet, 1987a). *Bohumiljania caledonica* (Jolivet) (Eumolpinae) in New Caledonia lays four eggs covered by ootheca.

Cassidinae make complex ootheca using their colleterial glands. These were studied by Muir and Sharp (1904) and others, as quoted by Hinton (1981). Less well known are the short articles of Rabaud (1915, 1919) on the biology of the cassids in France and those of Fiebrig (1910) in Paraguay. There is generally one egg per chamber and the number of chambers is very variable. *Chelymorpha cribraria* (F.) keeps around 80 eggs in each ootheca attached below the leaf by a strong pedicel. The African species *Basipta solida* Boheman builds an ootheca around a stem. Many cassidines make simple ootheca covered with faecal material and enclosing only 1–7 eggs. One ootheca is made each day, which can mean a great number of eggs for certain species (around 400 for *Gratiana spadicea* (Klug)). There is an enormous amount of literature on South American cassidines (Windsor, 1987; Buzzi, 1988; Vasconcellos-Neto and Jolivet, 1988; Windsor *et al.*, 1992; Rodriguez, 1994a,b, 1995a,b, etc.) and a complete bibliography can be found in Borowiec (1999).

An ootheca containing 80 eggs could be built in about 20 minutes. When the larvae hatch, they can exit easily only at the distal extremity of the ootheca. Muir and Sharp (1904) established a classification of the oothecae based only on South African

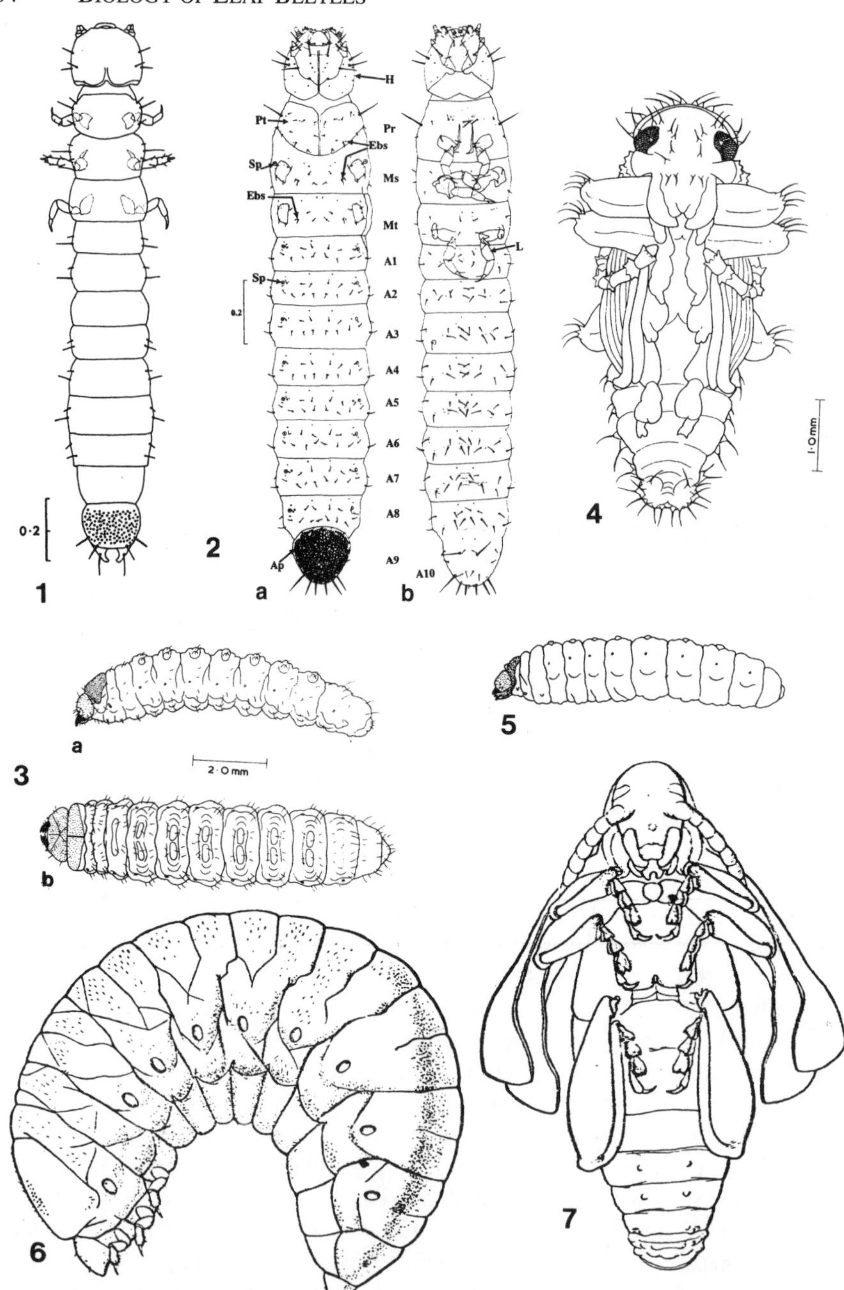

Figure 5.1. (1) *Orsodacne cerasi* (L.) (Orsodacninae). First instar larva (after Cox, 1981). (2) *Aulacoscelis appendiculata* Cox and Windsor, 1999 (Aulacoscelinae). a: dorsal and b: ventral view (after Cox and Windsor, 1999b). (3) a and b: *Palophagus bunyae* Kuschel and May, 1990 (Palophaginae) (after Kuschel and May, 1990). (4) *Palophagus bunyae* Kuschel and May, 1990 (Palophaginae). Pupa (after Kuschel and May, 1990). (5) *Palophagoides vargasorum* Kuschel and May, 1996 (Palophaginae). Larva (after Kuschel and May, 1996b). (6) Larva of *Sagra femorata* (Drury, 1773) in its natural position of the bent body (Sagrinae) (after Maulik, 1941). (7) Ventral aspect of pupa of *Sagra femorata* (Drury, 1773) (Sagrinae) (after Maulik, 1941).

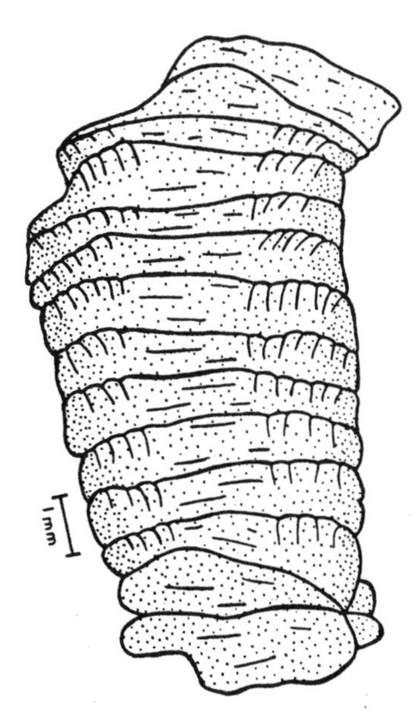

Figure 5.2. Elaborate ootheca of *Aspidomorpha miliaris* (Cassidinae). It is made up of an overlapping arrangement of leaflets of a parchment-like material. Each leaflet includes 4 elongated chambers on each side of the median line. Out of these, the two inner chambers usually contain an egg each, and the outer ones act as air chambers (after Verma, unpublished).

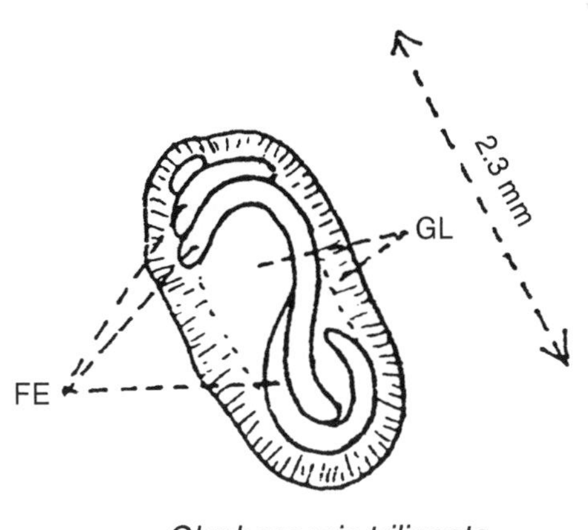

Figure 5.3. Egg cover for an egg of *Glyphocassis trilineata* Hope (from Kalaichelvan and Verma, 2000). FE = faecal deposit, GL = hardened glandular secretion.

species. Group 1 comprises small and imperfect oothecae with faecal matter (*Laccoptera*, *Cassida*). Group 2 comprises oothecae without faecal matter attached to a leaf or surrounding a stem or a calyx. There is even more variation among South American species.

The females of certain cassidines (Stolaini and others) in South America guard their eggs, covering them with their entire body. Then they watch over the larvae and the pupae (Windsor, 1982; Jolivet, 1988b; Windsor and Choe, 1994; Hsiao and Windsor, 1999). On the whole, it seems that the males do not watch over the larvae, but it may be the case with certain species. It is well known that, in many species of chrysomelines, males actually watch over their females to prevent other individuals from inseminating them. Cycloalexy is seen in the tropics, but restricted to some species. Curiously, parental care is restricted to the New World.

OVIPOSITION

The number of eggs per female is very variable among the chrysomelids, even between species (Hinton, 1981). It varies from 16 (*Promecotheca reichei* Baly) to more than 2800 among the *Leptinotarsa* species, especially *L. decemlineata* (Say). Some arctic *Chrysolina*, when they are not viviparous, do not lay more than 12 eggs. *Ambrostoma quadriimpressa* (Motschulsky) lays 500 eggs on leaves. In general, the number of eggs among the chrysomelids can vary from 8 to 3000 according to the genera and the species, their protection system and their embryogenesis. Edaphic, climatic and trophic factors are also variable according to the race-survival strategy adopted.

A few species of chrysomelids are viviparous or ovoviviparous. The great majority of the species are oviparous, but some species can be geographically viviparous. Viviparity is linked with cold climates. It eliminates the stage in the life cycle when the egg might be exposed. It also occurs in the tropics as a protection against egg parasitoids. Donaciinae, such as *Donacia crassipes* Fabricius, lay round or oval eggs into holes dug in the *Nymphaea* leaves and other aquatic plants. In those holes, the eggs are glued on the abaxial face of the leaves or they are set in rows around each hole (Hinton, 1981). In *Paropsis atomaria* Olivier, the eggs are laid on the terminal shoots of *Eucalyptus* or around the extremity of the sharp-pointed leaves. Packages of eggs are symmetrical (Carne, 1966).

Other *Paropsis* have totally different habits, sometimes in relation to the shape of the very variable leaves of *Eucalyptus* trees. *Paropsis dilatata* Erichson lays its eggs in two rows on the surface of the leaves and the eggs are glued together (Clark, 1930; Selman, 1994b). According to Hinton (1981), who examined the eggs of *P. atomaria*, there are no acropyles and the respiratory system was lacking in the egg. Generally, beetles which live on trees, depositing their eggs on the leaves lack egg respiratory plastron-like device.

Other Chrysomelinae, like *Gastrophysa viridula* Degeer, lay 30–40 eggs glued together on the abaxial side of the leaves of *Rumex* spp. (Jolivet, 1951b). In *Timarcha*, a few big eggs are glued together by a mixture of secretions and regurgitated food (Selman, 1994a). It is possible that some species use also excreta to make primitive ootheca. In some free-living, primitive eumolpines, like *Stenomela pallida* Erichson, the eggs are covered by faeces which form a protective scale similar to *Hornibius grandis* (Philippi and Philippi) (Jerez, 1996). In the last genus, the female also constructs egg cases from faeces. The faeces are soaked with a secretion and the female adds wooden fibres to the construction. The same happens with *Bohumiljania* in New Caledonia.

In *Altica lythri* Aubé, during egg-laying, the styles of the females spread distally and ventrally around the ovipositor to feel the leaf surface. The sensilla of this ovipositor feel the topography and the pubescence of the leaf (Phillips, 1977a,b, 1978). This feeling device determines the behaviour of the female and the choice of the substrate when laying eggs.

The eggs of *Iscadida* spp. in South Africa (Jolivet, 1995) are dropped one by one by the female, from her resting place, on to the bushes of the food plant, *Rhoicissus* spp. This is similar to the behaviour of female clytrines. They throw down their eggs, which are covered with excreta and look very much like seeds. Many phasmids use the same technique. With clytrines or phasmids, the eggs are taken over by ants, which carry them into their nest. In the first case, it is in order to rear them, and in the second case, it is in order to eat the capitulum, the equivalent of the seed elaiosome. It seems that the round and smooth eggs of *Iscadida* are not taken away by the ants. What happens to the eggs later on is still unknown. We have not seen the larvae outside rearing cages. They are hairy, active and probably not soil-dwelling. They must climb back over the bushes. Among the *Syneta* (Synetinae) from the Northern Hemisphere, the eggs are also dropped down one by one from trees. The larvae are radicicolous and probably dig into the soil immediately after hatching.

With the Galerucinae, there is often (e.g. in *Coelomera*) a simple ootheca similar to the one made by female *Timarcha*. These ootheca are made from a secretion and contain 60–80 or 100 eggs. Those eggs are well protected from the predators and parasitoids when they are laid inside the *Cecropia* trees' internodes. They remain, however, more vulnerable when they are laid (in case of certain species) outside, under the folioles. In this case, we often find empty eggs above the ootheca as they have been emptied by bugs or ants.

Among other galerucines, as in *Diabrotica* spp. and related genera, eggs are laid in the soil, around 2.5–10 cm deep, sometimes glued to the roots or debris. In dry conditions, beetles lay eggs in soil crevices up to 35 cm deep. There are differences in behaviour between the different species of *Diabrotica*. Many *Galerucella* also lay eggs in the soil or on the trunks and stems above the soil. These species are not root-feeders.

Other galerucines, like *Pyrrhalta humeralis* (Chen), lay eggs on their host, *Viburnum awabuki*, only in rather peculiar places (near the bud, at the extremity of the stem) and only on young stems. *Galeruca tanaceti* (L.) and most other *Galeruca* make a primitive ootheca glued to grasses, a little like that of *Timarcha* or *Coelomera*.

Species of *Coelomera* of the *C. ruficornis* Baly group lay eggs inside the stems of the *Cecropia* trees, the host plant. The female pierces the prostoma at the point of least resistance, which has already (or not) been pierced by the young queen of *Azteca* to lay her own eggs. *Azteca* ants are the normal associated hosts of *Cecropia*, a myrmecophilous plant. The *Coelomera* female introduces her abdomen inside the stem and lays her eggs there. Then she closes the opening with her anal gland secretions. It takes the *Azteca* queen about two hours to pierce the prostoma, but the galerucine takes about half a day to do the same job. *Coelomera* females always choose young stems of *Cecropia* to lay eggs with internodes not yet occupied by ants or their eggs. Larvae will come out after hatching, reopening the former prostoma, which is often replaced by a cicatricial tissue. Other species of *Coelomera*, like *Coelomera lanio* Dalman, lay eggs under the leaflets far away from places normally patrolled by ants. *Coelomera helenae* Jolivet which feed on myrmecophobic *Cecropia* where there are no ants or prostomas, have no way to penetrate into the internodes, and so lay their eggs outside.

The Criocerinae lay their eggs in rows or irregularly over the stem. Several species, like *Lema trilineata daturaphila* (Kogan and Goeden, 1970), can lay up to 2700 eggs. Among Megalopodinae in America (Selman, 1994a) the eggs are laid 2–3 mm before the end of the stem of the host plant (a Solanaceae or an Asteraceae) inside a hole dug by the female itself. The female cuts just before the top of the stem and then, 10 cm below this, she makes a circular or spiral incision around the same stem with successive bitings. This incision is made to reduce the sap access around the egg.

According to Doguet (1994), alticine eggs are laid in small groups in the soil at the base of the host plants for root-feeders, and on the abaxial side of the leaves for miners or phyllophagous species. Sometimes, the female digs a small hole into the petiole (*Phyllotreta rugifrons* Küster). Eggs can be covered with excreta (*Altica*, *Dibolia*). Among *Blepharida* spp. the eggs are laid in groups and glued together with faecal matter (Furth, 1982b).

The hispines sometimes lay their eggs between the folds of the leaves (*Brontispa*) and, in some cases, one by one on the abaxial side of the leaf (*Promecotheca*). These eggs are sometimes contained in a dish of a foamy substance. Among other mining species, the eggs are laid in groups of 2 or 6 into small cavities dug into the epiderm of the leaf, below and near the median vein or near the end, or sometimes on the vein. These egg groups are also covered with a foamy secretion.

It may be noted that, as in the case of certain South American rodents, when there is not enough food, developing eggs can be resorbed into the ovaries (*Leptinotarsa decemlineata*) (Selman, 1994a). It must also be pointed out that some chrysomelids avoid laying eggs in highly infested areas. For instance, the females of *Crioceris duodecimpunctata* (Linné) avoid asparagus berries already occupied by larvae of the same species. Avoiding infected food is a phenomenon well known among Lepidoptera. The *Heliconia* spp. never lay eggs on parasitized *Passiflora* or even on plants harbouring false eggs on their leaves.

OVIRUPTORS OR EGG BURSTERS

In many chrysomelids, there is a pair of deciduous *ruptor ovi* on the meso- and metathorax and on the abdomen of the nearly-hatching larva. Each one is a short and strong spine or symmetric tooth on one of the dorso-lateral thoracic or abdominal tubercles. When the young larva is about to hatch, it often produces two longitudinal slits in the eggshell. The spines generally cut a slit through the chorion and embryonic membranes at the anterior pole or a longitudinal slit in the dorsal region of the egg (Cox, 1994a), which helps in the emergence of the first instar larva. Bertrand (1924) was a pioneer in the study of these spines. According to Hinton (1981), the second slit facilitates the final emergence. In *Paraphaedon tumidulus* Germar hatching lasts about 20–30 minutes.

In Chrysomelidae, oviruptors have been examined by Cox (1994a) among 15 subfamilies. They are missing among Donaciinae, Galerucinae, Hispinae and Cassidinae. They are found on different segments according to the subfamily and they have been found on all the segments of the body, except the abdominal segments 9 and 10. Among Bruchidae, *ruptor ovi* are present in the whole family. The value of *ruptor ovi* as a generic character is negligible (Cox, 1994a). However, they can be used to separate the Paropsina. They are not a good specific character but they can be used to differentiate between certain subfamilies. Steinhausen (1978) used their position as a means by which to separate larvae of Galerucinae from Alticinae. Recently, Cox and Windsor (1999a,b) studied the oviruptors in Megascelinae and Aulacoscelinae.

Other aspects of egg biology such as ovoviparity, parasitism and hibernation are discussed in other chapters.

Larvae and larval biology

The larvae of the Chrysomelidae vary from campodeiform to eruciform. These larvae either live freely on the plants (Chrysomelinae, some Galerucinae and Alticinae), or they are leaf-miners (Hispinae, many Alticinae), or they live in fresh water (Donaciinae) or fresh or brackish water (*Macroplea*), or they are stem-borers (Megalopodinae), or even gallicolous (Sagrinae), at least at the end of the larval life or during pupation, or they are internal or external root-feeders (many Alticinae, Galerucinae, all Synetinae, most of the Eumolpinae).

Little is known of the larval biology of Orsodacninae (probably bud-diggers), Megascelinae (almost definitely root-feeders, but primitive Eumolpinae can be free-living), and Aulacoscelinae (definitely internal feeders, either root-feeders or stem-borers). The new subfamily of Palophaginae has a very peculiar larval biology. Larvae and adults live in the male strobiles of *Araucaria* trees in Queensland, Argentina and Chile, together with the Nemonychidae. They probably exist elsewhere where there are native *Araucaria* and related plants (New Guinea, New Zealand, Norfolk Island, New Caledonia, southern Brazil). They are difficult to find and to study. *Araucaria* have disappeared from Africa and probably the Palophaginae with them. Palophagine larvae feed on *Araucaria* pollen and live inside the central marrow of male strobiles, eating the stems of the sporophylls. In the younger cones, the larvae feed only on pollen and one at a time, together with the weevils. Dorsal and ventral ambulatory ampullae facilitate moving among the sporophylls. The larva pupates in the ground after 5–6 months and hatches after 30–50 days.

Zeugophorine larvae (*Zeugophora* and *Pedrilla*) are miners into young tree leaves (*Salix*, *Populus*, *Juglans*, *Corylus*, *Betula*, *Santalum*, etc.). *Aulacoscelis* (Aulacoscelinae) and *Megascelis* (Megascelinae) neonate larvae have been described recently by Cox (1998a) and Cox and Windsor (1999a,b). *Aulacoscelis* feed on *Zamia* and other cycad fronds and *Megascelis* on various Fabaceae, including *Stizolobium* or *Mucuna*. *Sagra* larvae dig into the stems of young plants, mostly arbustive Fabaceae, then when they reach their final size, they build a cocoon and pupate in the middle of the gall induced by the larva. Eclosion is often a synchronous event covering a large area after the first big rains in Africa. Nothing is really known of the biology of the archaic gondwanian species and genera of Sagrinae. *Megamerus* also seems to pupate in a gall in Madagascar and it lays eggs around stems. However, the life history of *Atalasis* in Argentina is not well known. It is also a gallicolous species but we do not know exactly if the pupation is in the ground or in the gall. *Atalasis* is a specialist on Malvaceae. Other archaic Sagrinae probably have stem-boring larvae which are more than likely also galligenous in the last instar, but we are only hypothesizing. Some genera probably deviate from this pattern. Adults in Australia frequent flowers and eat pollen. They are also attracted by light.

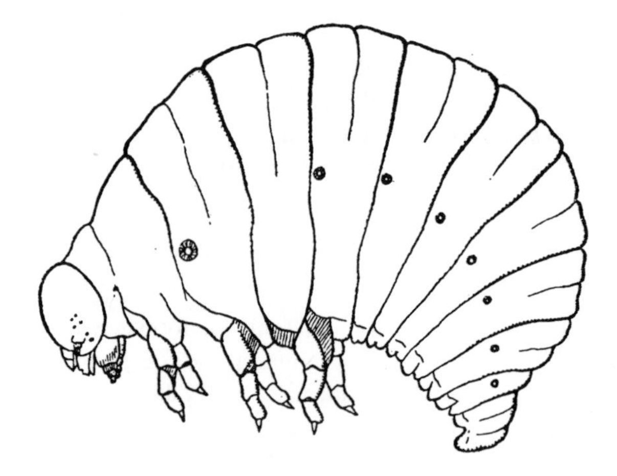

Figure 5.4. *Timarcha tenebricosa* Fabricius. Larva 3rd stage (after Chen, 1934a).

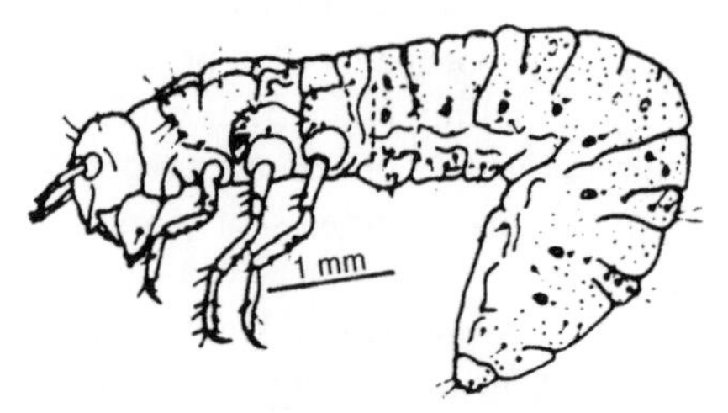

Figure 5.5. *Oomorphus concolor* (Sturm) (Lamprosomatinae). Larva (after Kasap and Crowson, 1976).

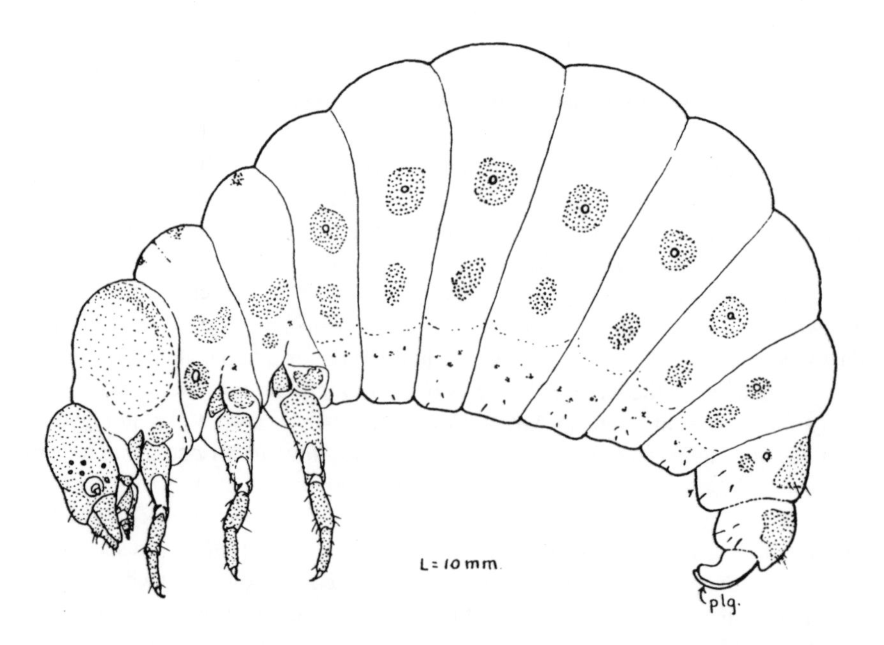

Figure 5.6. *Leptinotarsa decemlineata* (Say) (Chrysomelinae). Larva. 3rd instar (after Peterson, 1960).

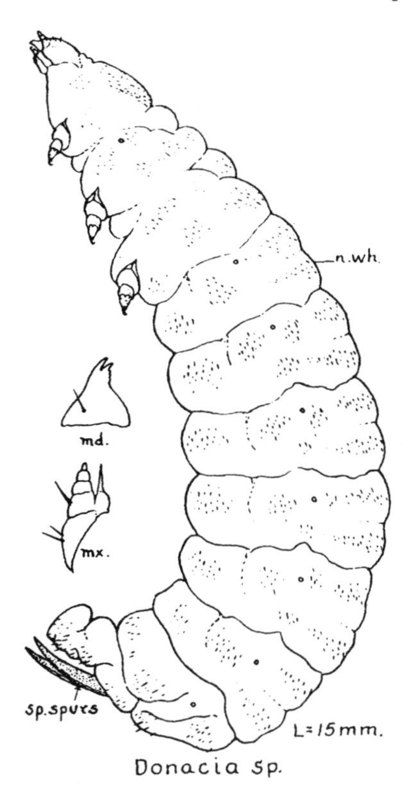

Figure 5.7. *Donacia* sp. (Donaciinae). Larva with spurs (Donaciinae) (after Peterson, 1960).

Donaciine larvae feed on roots and immersed stems, extracting sap and oxygen from them. Several genera (*Macroplea* and *Neohaemonia*) are totally aquatic, and others are mostly terrestrial at the adult stage. *Macroplea mutica* (F.) lives in brackish water, in the Baltic Sea for instance, on various marine plants like *Zostera* and on others which are not especially halophytic, like *Myriophyllum*. This larva is the only chrysomelid larva living in the sea. It has been also mentioned from other European seas. More research is needed.

The cerambycid behaviour of Megalopodinae has been noted before. When the egg is laid inside the stem of the host plant, a herbaceous Solanaceae or Asteraceae, the larva hatches and bores inside. The larva is not galligenous. *Agathomerus* and *Megalopus* larvae in South America feed on internal tissue, like sagrines. They are yellow and covered with setae which allow them to move easily within the galleries. These larvae have a maximum length of 26 mm. Pupation takes place in the soil (Tella, 1952). Their biology is now well known in America (Monros, 1954a), in China (Yu Peiyu and Yang, 1994) and in South Africa (Schultze, 1996). See also Santos (1980, 1981a,b) and Jolivet and Hawkeswood (1995).

In Criocerinae, larvae are either leaf-miners on monocots (Commelinaceae, etc.) such as *Neolema*, or covered with excreta and free, such as *Lilioceris* and *Plectonycha* spp. Pupation takes place inside a cocoon on the soil at the base of the plant (*Oulema*, *Crioceris*, *Lilioceris*), or on the leaves (*Plectonycha*). The cocoon can be simple or rather elaborate. The larvae are often solitary, but can be gregarious as for *Quasilema latipennis* Pic in Argentina. They live together in a foliar mine and move together towards the next leaf when the first one is exhausted. Some South American criocerine larvae even gather together in a cycloalexic ring.

The larva of *Ortholema abnormis* Heinze is gallicolous and the larva of *Neolema* spp. shows a tendency to spermatophagy into the ovules of Commelinaceae. *Crioceris* spp. tend to eat *Asparagus* fruits and can be a similar reddish colour. Schmitt (1988) reviewed the biology of Criocerinae in detail.

Synetinae larvae have been described by Kurcheva (1967) (*Syneta betulae* F.) and various other authors. These larvae feed on the rootlets of various trees. *Syneta* spp. drop their eggs on the ground and so there is no special protection, as for *Iscadida*. The eggs hatch rapidly and the larvae dig into the soil. See also Edwards (1953), Moznette (1916), Wilson and Moznette (1913–1914), Verma and Jolivet (2000).

Clytrinae, Cryptocephalinae, Chlamisinae and Lamprosomatinae larvae are case-bearers and the scatoshell is mostly made of excreta (as in the first two subfamilies), or of excreta mixed up with agglutinated particles. Clytrine larvae retract into the shell in case of danger. They eat ant eggs, cadavers, pieces of wood and any detritus they can find in the ant nest. They can feed on any insect body and are practically omnivorous. Cryptocephaline larvae are generally free-feeding on plant detritus at the base of the host plant, on decaying leaves and even rotten wood (Stiefel, 1993). Some rare species live with or near the ants. Chlamisine larvae feed on leaves, bark, or lichen. Some species also feed on stems, berries, or detritus. A larval scatoshell is a rather complex structure and often looks like a caterpillar or bird excreta. Sometimes the camouflage is sophisticated. Outside their mimetic, toxic or cryptic protection, chlamisines have other means of defence at the adult stage: rapid flight, thanatosis, etc. It may be briefly pointed out here that the adult suture connection between the two elytra is very peculiar and has sometimes been compared to a zipper. That does not prevent them from flying away rapidly in case of danger. Lamprosomatinae have also a case-bearer larva different from the previous ones. They lead a hidden life on or at the base of the host plants. See the excellent review on the biology of the Camptosomata (with the Lamprosomatinae) by Erber (1988).

Eumolpinae have almost all root-feeder larvae. Literature on larvae and adults has been cited by Jolivet and Hawkeswood (1995). There are, however, exceptions in the primitive genera: *Stenomela*, *Bohumiljania* and *Hornibius*. In Chile, several genera of eumolpines have been studied by Jerez and Ibarra-Vidal (1992), Jerez (1996, 1999). Larvae of *Hornibius* (*Hornius*) *grandis* (Philippi and Philippi) dig as the first instar into *Nothofagus* buds, the Antarctic beech. Later on they will feed on the leaves. Pupation takes place in the soil near Poaceae roots, as for most of the eumolpines. *Stenomela* is specialized on Myrtaceae in Central Chile. Adults feed on the stems and the leaves of the host plants whilst the larvae feed only on foliar lamina. These genera are actually the only known external-feeding larval eumolpines with *Bohumiljania* in New Caledonia. It is possible that other primitive genera might behave in the same way. Ferronato (1986, 1988, 1999) has described in detail the biology of the eumolpine species attacking cacao leaves in Brazil. While the adults eat cacao leaves, larvae feed indifferently on cacao or Poaceae roots. As is the case with other endogeous larvae, eumolpine larvae are blind.

Chrysomelinae larvae normally feed openly on the leaves of herbaceous plants, bushes or trees. Their toxicity protects them from predators, but not against all and not against parasitoids. Certain chrysomeline larvae take to the ring disposition when resting (cycloalexy). This is the case with many *Platyphora*, *Plagiodera*, *Gonioctena*, *Paropsis*, *Proseicela*, *Eugonycha*, *Labidomera* (*L. suturella* Chevrolat) and several others. The system seems efficient against predators. This means of defence is discussed in another chapter (Chapter 8) and is shared with some cassidine, criocerine and other larvae. Some larvae, such as *Phola* (*Chalcolampra*), eat *Vitex* sp. stems (Verbenaceae) and build nidiform shelters with excreta on or between the branches. Larvae live hidden with half of the anterior body in the nest and the extremity of the abdomen exposed. Actually, this case seems unique among Chrysomelinae, since *Allocharis* larvae, a related genus, are free on *Veronica* spp. in New Zealand. Those of *Chalcolampra* are also free-living on *Parahebe* sp. (Scrophulariaceae) in Australia.

Cox points out that Entomoscelina and Phyllocharina larvae are generally densely pubescent and have ventral pseudopods on apical segments. We have often observed yellow larvae of *Phyllocharis* very active on *Clerodendrum* plants. Similarly, densely hairy larvae of *Iscadida* move very quickly on the ground.

Timarcha larvae are well known and have been studied by Paterson (1930, 1931), Chen (1934a), and various recent authors (Cox, 1982). Other papers on chrysomeline larvae have been published in China, India, Japan, Russia and Korea. *Microtheca* larvae are the only ones to produce a light cocoon in the ground (Jolivet, 1950b, 1951b).

Galerucine larvae can be root-feeders (*Diabrotica*, *Aulacophora*, *Luperus*, *Phyllobrotica*, *Platyxantha*, *Craniotectus*, *Ceratoma*). *Cerotoma* larvae feed on the nitrogen-fixing root nodules of Leguminosae. Many galerucine larvae are polyphagous and can thus become pests of many crops. *Exosoma* larvae feed inside the Liliaceae and Amaryllidaceae bulbs in a semiliquid medium. Several *Monoxia* larvae are miners. Larvae of *Galerucella*, *Galeruca*, *Agelastica*, *Arima*, *Monocesta* and *Lochmaea* are free-living on the host plant along with the adults. Pupation takes place in the soil, inside the buds, on the leaves or on the trunks.

According to Doguet (1994), alticine larvae pass through three instars and a prenymphal instar. Rarely, chrysomelid larvae have more than four instars. Often, they have only three, but more than four are also known. Alticine larvae are miners inside leaves (*Sphaeroderma*, *Dibolia*, *Febra*, *Mantura*, *Mniophila*, *Argopistes*, several *Phyllotreta*, *Argopus*), fruits, petioles, flowers, buds (*Chaetocnema*, *Psylliodes*, *Clitea*), or root-feeders (*Phyllotreta*, *Longitarsus*, *Podagrica*, *Nisotra*, *Aphthona*, several *Chaetocnema*, *Xenidea*, *Epitrix*, *Neocrepidodera* (=*Asiorestia*), *Crepidodera*, *Batophila*, *Clavicornaltica*, several *Psylliodes*, *Erystus*, several *Altica*). Other larvae are free-living (*Podontia*, *Blepharida*, many *Altica*) on tree leaves or herbaceous plants. Many of them are external feeders on roots, while others such as the larva of *Longitarsus echii* (Koch) eat the central part of the roots. With the alticines it is a common phenomenon for the leaf-mining larvae to move towards the stems and from the stems towards the roots or vice versa (Smith, 1986). Phyllophagous, free-living or mining larvae are often cyphosomatic in shape, whereas root-feeding and endogenous larvae are always orthosomatic. Endogeous larvae are generally completely blind and totally whitish and depigmented, except for the head, the prothorax, and the ninth abdominal segment.

Mining or free-living larvae are always more pigmented over the head. As for galerucines and alticines, the antennae are mono- or bisegmented and ocelli are often reduced or absent. Larval characteristics are rather similar in the two subfamilies. However, as mentioned earlier, Steinhausen (1978) discovered differences between the two groups in the position of the oviruptor. Among miners, the legs are small and well separated. The abdomen is composed of 9–10 segments. Often, the tenth segment is disc-shaped and used as a pseudopod. The length of alticine larvae varies from 4–14 mm, according to the species. *Podontia* larvae are free and covered with a bright black film, chiefly made of excreta. They feed on the side of the leaves and can reach a large size. Pupation takes place in the soil within an earth cell. A *Podontia* adult jumps in case of danger or drops to the ground, rather rare behaviour among alticines and they are practically the only alticines to present reflex immobilization. *Blepharida* larvae (Furth, 1982a) measure up to 14 mm long and are yellow or green with coloured bands on the body. They have a viscous cuticle like a slug and faecal threads attached to the anus. As for *Podontia*, they feed on the sides of the leaves, like some Chrysomelinae, also rather exceptional behaviour among alticines.

It is always difficult to separate alticine and galerucine larvae (Steinhausen, 1985, 1994). Many of them have no ocelli but some alticines do have them: *Dibolia*, *Mantura*, *Podagrica*. Sometimes there are much regressed.

Hispine larval biology is relatively well known (Maulik, 1919) and experimental rearing techniques have been devised, for instance among the *Coelaenomenodera*, miners in oil palms in Africa (Mariau, 1988).

Hispine larvae can live between the appressed leaves of palm trees or *Heliconia* and related (Zingiberales), all monocots (Strong, 1977–1982). Some adults and larvae of *Cephaloleia* even can live inside the water of *Heliconia* bracts and are more than subaquatic (Jolivet, 2002, in print). Some species modify their habits during development from mining to free-living on leaves. The differences between hispine and cassidine larvae were described by Böving and Craighead (1930), Chu (1949) and Peterson (1960). The eighth pair of abdominal spiracles is well developed. It is positioned dorsally among the Hispinae and is vestigial among the Cassidinae. The absence of the furca on the tergum of the eighth abdominal segment is also peculiar to Hispinae. Hispine larvae generally measure 5–10 mm but they can be bigger. The body is depressed, orthosomatic, with lateral projections on the abdominal segments. The antennae are trisegmented. There are 4, 5 or 6 ocelli visible, but sometimes they are in total regression and are replaced with a black spot. Hispine larvae are lightly coloured, white, yellow, or green, except for the head, the prothorax and the eighth abdominal segment which are more strongly pigmented. The legs are generally trisegmented and end with a tarsungulus with a claw and a bifurcated pulvillus. Several larvae (*Octotoma*) do not have segmented legs. The abdomen has 8 visible segments. The number of larval instars is variable and can reach a maximum of 6 in *Brontispa*. In the tropics, the number of generations per year can reach 6, as in *Dicladispa*.

Hispine larvae can be divided into four main groups:

(1) Those which live between folded leaves of monocots (Poaceae), between the folded folioles of new fronds, at the base of palm tree petioles (Arecaceae) or at the base of palm crowns. There are also several species and genera of Cephaloleini living inside the water-filled bracts of *Heliconia* plants. These are Cephaloleini, Botryonopini, Callispini, Leptispini and several others.

(2) Those which dig into the stems of herbaceous or semiligneous plants (Anisoderini).
(3) Those which mine the leaves (Coelaenomenoderini, Hispini).
(4) Those which are free-living. These are the minority.

Many species are miners, a great number live between the young leaves and only a small number of them are stem-borers or free-living. One exception, *Oediopalpa* spp. (Oediopalpini), demonstrates behaviour which is similar to the Cassidinae, since the larva of this American genus builds a shield made of excreta and exuviae at the rear of its body and uses it as a means of defence. Intermediary genera are often recorded in both the subfamilies, Cassidinae and Hispinae, in the same way as for Galerucinae and Alticinae. Having intermediary genera is normal if one believes in evolution, but it is not a good enough reason for amalgamating both subfamilies.

In India, *Leptispa pygmaea* Baly has a larva which lives on the leaves with the adult. *Brontispa froggatti* Sharp larvae from Vuanatu (New Hebrides) lives with the adult on the opened buds of *Cocos nucifera* L. (Arecaceae) (Maulik, 1919).

A larva of *Dicladispa armigera* (Olivier) eats an average of 132.4 mm of a rice leaf during its life (Budhraja *et al.*, 1979). In large quantities, the larvae can be a serious pest of rice. Gonophorini larvae are either miners or free-living on the leaves.

Andrade (1981) has studied the biocoenosis of the *Cecropia* trees in Brazil. One *Metaxycera* and two *Sceloenopla* larvae mine *Cecropia* leaves. The female *Sceloenopla maculata* (Olivier) lays 8 eggs grouped into a small cylindrical and bevelled ootheca measuring 2.5 × 1.5 mm. This ootheca is inserted into the mesophyll on the abaxial surface of the leaf. The larvae (up to 8) hatch after 14 days of incubation. They mine but do not cut the median veins of the foliar lobes and after several moults reach 10 mm in length. Legs are vestigial and the anterior extremity is strongly sclerotized. Larval development requires 45 days and the pupal stage lasts only 2 days. One chalcid parasitizes the ootheca.

In certain cases, several species can be found inside one mine, *Arescus* and *Chelobasis* in *Heliconia* and *Calathea* (Maulik, 1937).

As previously mentioned, many hispine miners leave their mine when the food is exhausted and penetrate into a new leaf to start a new mine. For more details about hispine biology see Jolivet and Hawkeswood (1995).

Cassidinae larvae are generally free on the leaves on the upper or lower leaf surface. They carry their old exuviae and faecal matter over their backs. The shield, efficient against the predators, is carried on the furca which is the prolongation of the ninth abdominal tergum. Some other species, like *Hemisphaerota cyanea* (Say), carry a faecal thatch on their backs, which is efficient against most, but not all, predators (Eisner and Eisner, 2000). Some species are miners. The larvae which carry the furca belong to the genera *Aspidomorpha*, *Laccoptera*, etc. but others carry only a bifid extension (*Oocassida*). The structure of the extension varies a lot in the neotropical region and several species do not carry a furca or an extension. Excreta carried by the larvae vary in shape and are often characteristic of the species and genera. Several larvae, such as *Basilepta*, are naked and without any special protection. Some gregarious larvae of Stolaini are given efficient maternal protection, not always linked with a dorsal shield and in conjunction with cycloalexy, as among certain Chrysomelinae and Galerucinae. As mentioned earlier, some American larvae carry

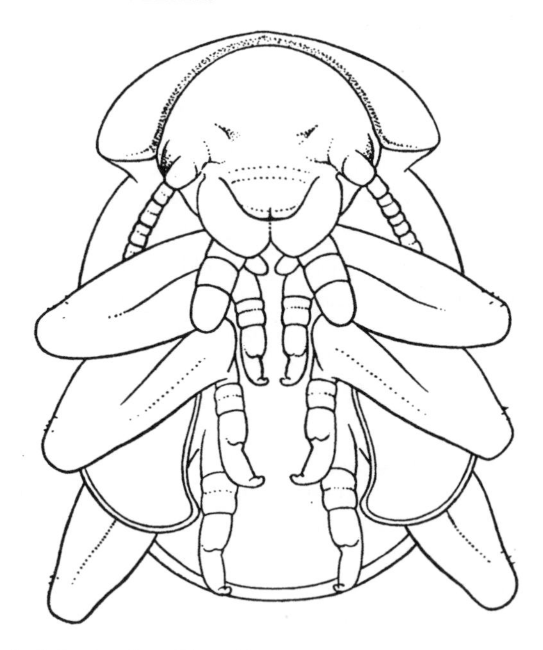

Figure 5.8. *Timarcha balearica* Gory (Timarchinae). Ventral view of the pupa (after Jolivet, 1953).

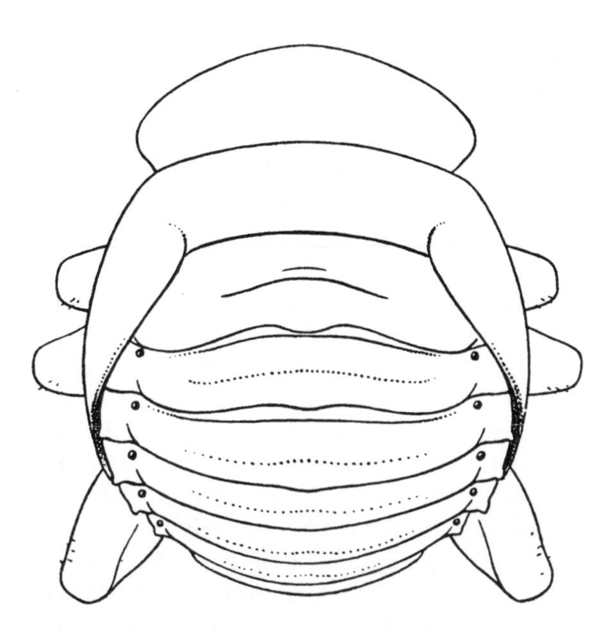

Figure 5.9. *Timarcha balearica* Gory (Timarchinae). Dorsal view of the pupa (after Jolivet, 1953).

curious balls of excremential threads (*Hemisphaerota*, *Dorynota*). For the defensive use of the faecal shield, see Eisner *et al.* (1967).

In the last instar, cassidine larvae measure 6–15 mm. Several species are large, depressed and oval and are covered with spines along the sides. Others are elongated.

Larval general body colour is whitish, yellow, orange or green. The antennae are bi- or trisegmented and there are 5–6 ocelli. The abdomen, as in Hispinae, is made of 8 normal segments with 9–10 strongly reduced. The final tube (tenth segment) deposits the excreta over the furca. The spiracles of the eighth segment are vestigial or absent. There are generally 5 larval instars.

It is the dorsal excremential shield or the thatch ball which protect the larva. Pupation takes place under this excremental parasol in *Hemisphaerota* or *Dorynota* and this shield protects the larva against the ants, dryness and rain. Many larvae can move their dorsal shield against predators (Eisner *et al.*, 1967).

Among Cassidinae living on the trees (*Cordia*, *Tabebuia*), the larvae feed on the abaxial side of the leaf and the adults on the adaxial side. The larvae chew the parenchyma and not the superior epiderm and the veins. Adults, by contrast, eat the shoots or foliar limbs by digging irregular holes.

Roughly speaking, outside Donaciinae, the larvae of which are aquatic and fix themselves with respiratory hooks to submerged stems, chrysomelid larvae are all terrestrial, with sometimes special adaptation to accidental immersion. Larval stages vary from 3–6 instars, often followed by a short pre-nymphal period. Pupation takes place at the inferior (abaxial) surface of the leaves, in the debris at the base of the plant or in the soil. Miners among them often pupate inside the mines and can hibernate there. Case-bearing larvae seal the opening of their cases and pupate inside the scatoshell. A good account of chrysomelid larvae is given in Peterson's (1960) and Stehr's books (1991). See also Lawrence (1990), Lawrence and Britton (1994) and Lawrence *et al.* (1999a,b).

Pupae

The pupae of leaf beetles are exarate, i.e. with appendages free. There is not a great deal to say about pupal biology. Whether they are protected or not by a cocoon, or if they live in an earthen cell or are practically free, they are slightly mobile when they are disturbed.

Hinton (1981) made a special study of insects pupating on the leaves in exposed places and devoid of a cocoon. Those species often retain final stage cuticles attached to their posterior abdominal segments. Hinton found out that, when disturbed, the pupa of *Chrysomela tremula* Fabricius was able to release droplets of a toxic liquid stored in cuticular reservoirs from the remains of the larval skin. Some other chrysomelids, such as *Plagiodera versicolora* (Laicharting), behave in a similar way.

Humid soil promotes pupation in *Altica subplicata* LeConte larvae, as in most of the chrysomelids (Rickelmann and Bach, 1991). Dry soil delays pupation, as we can see with eumolpine larvae. Before pupating, *Chrysomela* larvae walk actively, then they attach themselves upside down on a leaf or on a stem, generally from 30 cm to one metre above the ground. The pupating chrysomelid cuts the larval skin up to the metathorax and pushes it posteriorly until the 5–6 first abdominal segments are exposed. When the larva moults, the reservoirs of thoracic and abdominal glands keep their contents. When the larva is stimulated, it contracts the dorsal longitudinal muscles of the abdomen which move nearly at a right angle to folds in the larval skin. The abdomen pushes against the larval cuticle and the fluid of the reservoirs is ejected, producing a big droplet at the extremity of the larval tubercles. When the pupa reverts to its resting position, the reservoirs reabsorb the fluid to be used against aggressors.

Coccinellid pupae have 'gin-traps', or '*pièges à loups*', which push back or pinch mites or small insects (Hinton, 1946, 1951). Such traps seem to be missing among chrysomelids, but we cannot be sure of this.

By contrast, *Cassida* pupae can have spines which can be projected against a potential aggressor by appropriate movements of the abdomen. There are also two projections on the larval abdomen (furca) which can be used in this way. Free pupae are very often aposematic, while pupae hidden in cocoons are neutrally coloured (white or yellow).

The *Plagiodera versicolora* pupa produces a sound when the ventral part of the head rubs against the leaf. This sound can have a defensive value exactly like the sound produced by some sphingids (*Acherontia atropos* (L.) for instance). Pupation in *Timarcha* takes place in the soil, as with many other chrysomelids. Cox made a detailed study of the morphology of chrysomelid pupae (1996b). However, the pupae of Orsodacninae, Megascelinae and Aulacoscelinae are undescribed, since we know only the neonate larvae. Cox (1998b) discussed the significance of pupal setae, urogomphi, position of pupation, and 'silken' cocoon construction in comparison with Curculionidae, Cerambycidae and Bruchidae. Pupal morphology and chaetotaxy seem to be good characters for chrysomelid classification at subfamilial and even tribal level.

6

Ecology

A typical leaf beetle feeds and oviposits on the leaves of plants. The larva, on hatching, feeds on the leaves and then pupates either on the leaf itself or drops to the soil beneath and pupates there. Chrysomelids are mostly oligophagous, though some primitive groups (and rare evolved genera) tend to be polyphagous.

However, this large family of beetles presents many interesting deviations from the typical life pattern. The larvae of some chrysomelids are leaf-miners (some Alticinae, most Hispinae, some Criocerinae, and all Zeugophorinae), root-feeders (many Eumolpinae, some Galerucinae, Synetinae, Donaciinae and perhaps also Megascelinae), stem-borers (Megalopodinae, Sagrinae and some Alticinae), gall-producers (Sagrinae) and myrmecophilous, feeding on ant eggs, dead ants and detritus in ant nests (Clytrinae). In addition to leaves, adults may feed on flowers, pollen and young shoots (e.g. Orsodacninae, Clytrinae, some Galerucinae). In exceptional cases, even entomophagy and cannibalism are known.

In this chapter we shall deal with the life of chrysomelids in special environmental conditions (in water, desert, etc.), how they coexist in the same habitat (niche separation), and how they overcome adverse environmental phases (diapause).

Aquatic and subaquatic leaf beetles

Under the aquatic category are placed those leaf beetles which, at some stage of their life cycle, live submerged in water. Only Donaciinae qualify for this category. Subaquatic leaf beetles are those which live at the edge of or close to water. They are capable of surviving prolonged submergence in water. They include some members of Alticinae, Galerucinae, Hispinae and Chrysomelinae.

AQUATIC CHRYSOMELIDS

The larval stages of all Donaciinae are wholly aquatic. Adults of *Macroplea* and *Neohaemonia* also live in water. Some species of *Donacia*, *D. malinowskyi* Athrens and *D. hirticollis* Kirby too spend some time in water, but they often emerge from the water and take to flight. Other species are more aerial.

Macroplea mutica F. is the only truly marine chrysomelid and is mostly known from the Baltic Sea.

A donaciine larva respires through its cuticle. Houlihan (1969) believes that, when the larva has grown, cuticular respiration is not enough to meet its needs. The larva is subcylindrical, whitish and with curved legs. Arising from its abdominal end is a pair of spiny processes. They are elevations of the body surface around the eighth abdominal spiracles. The spiracles are thus carried on tips of the processes. For respiration, the processes are inserted into air-holding spaces in plant tissue. In this way, oxygen is obtained from the rhizomes, roots and stems of aquatic plants. For respiration in *Donacia* see Hoffman (1940) and Varley (1939).

When feeding, the larva damages plant tissue with its mandibles. The galea and lacinia of its maxillae are so shaped as to facilitate the sucking of the plant juices by means of the pumping action of the pharynx.

The adult *Donacia* has an elongated body with metallic colours such as blue, green, reddish, violet, black, etc. It respires by biting into plant tissue, but most of its respiration is through a plastron of hydrofuge hairs on the venter.

Mann and Crowson (1983d) described the internal anatomy of Donaciinae. They pointed out that there is lack of association of the terminal parts of the Malpighian tubules with the rectal wall, i.e. there is no cryptonephridic arrangement. Obviously this is due to the aquatic habits of these beetles. It would be relevant to point out here that Poll (1932) correlated an absence of the cryptonephridic arrangement in *Donacia* with its aquatic habit. In *Donacia* and *Plateumaris*, there are six tubules in two sets, a set of four long tubules and another of two shorter tubules. Among the four long tubules, two belong to each side and those belonging to the same side join together to form a long loop. Thus, there are two long loops of Malpighian tubules. In *Plateumaris*, the short tubules join distally the loops of their respective sides (*Figure 6.1*). In a typical chrysomelid, too, six Malpighian tubules are in two sets of 4 and 2, and the three tubules of each side join distally before entering the rectal wall (see Chapter 9: Cryptonephridic arrangement of Malpighian tubules). Thus, the situation in Donaciinae seems to have derived from the typical chrysomelid arrangement by the withdrawal of the tubules from the rectal wall, as enhanced water conservation is not their ecological need (*Figure 6.1*).

Some species of *Donacia* enter water and lay their eggs on the leaves of submerged plants, but some other species of *Donacia* and species of *Plateumaris* oviposit on aerial leaves of floating aquatic plants.

Pupation occurs in cocoons, attached to submerged parts of aquatic plants. The development stages of *Donacia* species found around Moscow were studied by Bienkowski (1996). The full-grown larvae of *D. thalassina* Germar and *D. clavipes* F. make cocoons attached to the living roots of the host plant. The internal cavity of the cocoon is connected to air cavities in the roots. This allows the full-grown larva or pupa or freshly ecloded adult, within the cocoon, to pass one winter (about nine months) in diapause inside the cocoon. The cocoons, made by full-grown larvae of *D. dentata* Hoppe, are attached to tubers of the host plant as all roots die out in autumn. Hence, such an air connection (as in *D. thalassina* and in *D. clavipes*) between the interior of the cocoon and plant tissues is not possible, and eclosion of the adult and emergence from the cocoon occur in late summer itself. Thus, one diapause period is less than in the two earlier mentioned species of *Donacia* (there is diapause in the 2nd/3rd instar in all the three species). Egg–to–egg life history of *D. thalassina* and *D. clavipes* requires three years, but that of *D. dentata* is completed in only two years.

The dispersal of *Donacia* is through periodic migratory flights. During a migratory period they can be readily collected in light. How the aquatic adults of *Macroplea* and *Neohaemonia* disperse is not clear.

Donaciinae have a wide distribution, but they are totally lacking in South America

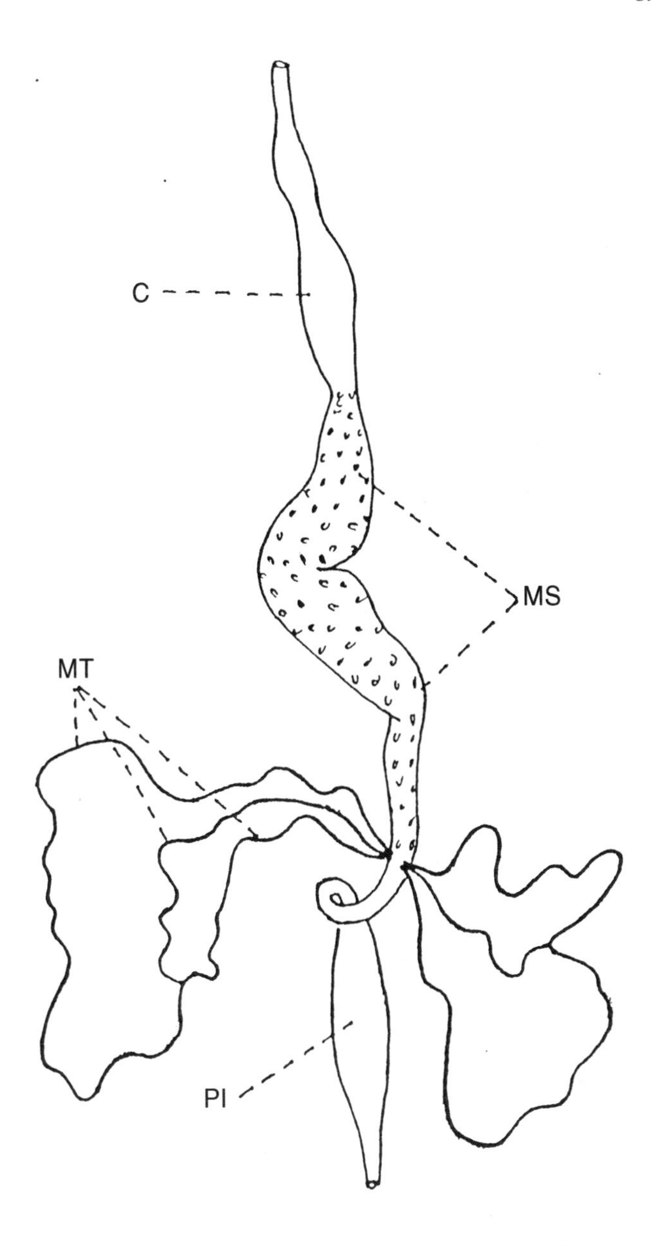

Figure 6.1. Alimentary canal of *Plateumaris discolor* (Panz.) (after Mann and Crowson, 1983d). Note the absence of a cryptonephridic arrangement and that the three Malpighian tubules of each side join together distally (C = crop; MS = mesenteron; MT = Malpighian tubules; PI = hind-gut).

and Central and South Australia, as well as in northern New Guinea, for no obvious reason as suitable habitats are not lacking in these parts. In fact, there are some unsolved riddles about the distribution of Donaciinae which have been pointed out in Chapter 4.

SEMIAQUATIC CHRYSOMELIDS

A number of chrysomelids live either on vegetation close to or at the brink of water, or on the emergent leaves of aquatic plants. To survive in this situation, they must run the risk of falling into the water.

One of us (P.J.) has authored a detailed review on our present knowledge of semiaquatic leaf beetles for presentation in the Fifth International Symposium on the Chrysomelidae (Jolivet, 2002, in print).

Unfortunately, submergence-survival adaptations have been only scantily studied in leaf beetles. In the related family Curculionidae, this aspect of life has been better investigated. Among weevils, subaquatic adaptations include swimming movements, the presence of pellicular wax in the skin and of small scales on the cuticular surface which hold a film of air on the body surface on submergence, acting like a respiratory plastron. Angus (1965) studied the swimming movements in *Bagous limosus* Gyllenhal. He found that these included paddle-like movements of the fore-legs, while the remaining two pairs of legs performed the usual walking movements. Buckingham *et al.* (1986) examined 17 species of *Bagous* in Florida, and found that the adults of some species have a respiratory plastron. Roudier (1957) noted that the weevil *Phytobius leucogaster* Marsham may remain safely under water for quite a long period. During submergence it attaches to a water plant with an air bubble around itself.

Subaquatic chrysomelids mostly belong to Chrysomelinae, Galerucinae and Alticinae. Whatever observations are available go to show that these beetles, living in proximity to water, have adaptations comparable with those of subaquatic weevils. Leaf beetles, however, do not have cuticular scales like weevils, but they may possess a plastron of tiny hairs. The galerucine *Agelastica alni* L. and several species of the galerucine genus *Sastra* in New Guinea have a hairy venter, and are resistant to immersion due to their light weight. The chrysomeline *Phaedon cochleariae* F., which live on emergent aquatic plants, can survive prolonged immersion and are often seen walking or swimming on the surface of the water. *Galerucella nymphaeae* L. has a setose venter. *G. nymphaeae* and *G. sagittariae* Gyllenhal are two very closely related species. The former species lives on aquatic plants with leaves floating on water, while the latter on terrestrial or semiaquatic plants. Larvae of *G. nymphaeae* have tiny papillae on the cuticle which can retain a film of air like a plastron, on submersion in water. Such papillae are lacking in the other species (*G. sagittariae*) (Nokkala and Nokkala, 1998; Nokkala *et al.*, 1998). Larvae and adults of the latter species do not tolerate submersion and, on falling into water, they drown. On the other hand, larvae and adults of *G. nymphaeae* survive in water for a long period. Kauffman (1970) pointed out some semiaquatic adaptations in *G. nymphaeae*. The adult of this species is capable of walking on the surface of water, and in the larva the tenth abdominal segment is beneath the ninth, and acts as a pygopod. When the pygopod is inflated with blood pressure, it can act as a vacuum cup, increasing adherence to the leaf surface. The alticine *Pseudolampsis guttata* LeConte lives on the emergent leaves of the floating fern *Azolla* in Florida. This flea beetle has setose venter and hairy legs (Buckingham and Buckingham, 1981; Buckingham *et al.*, 1986). When immersed in water, it carries an air bubble around itself.

Hispinae are mainly terrestrial but in Central America some Hispinae associated

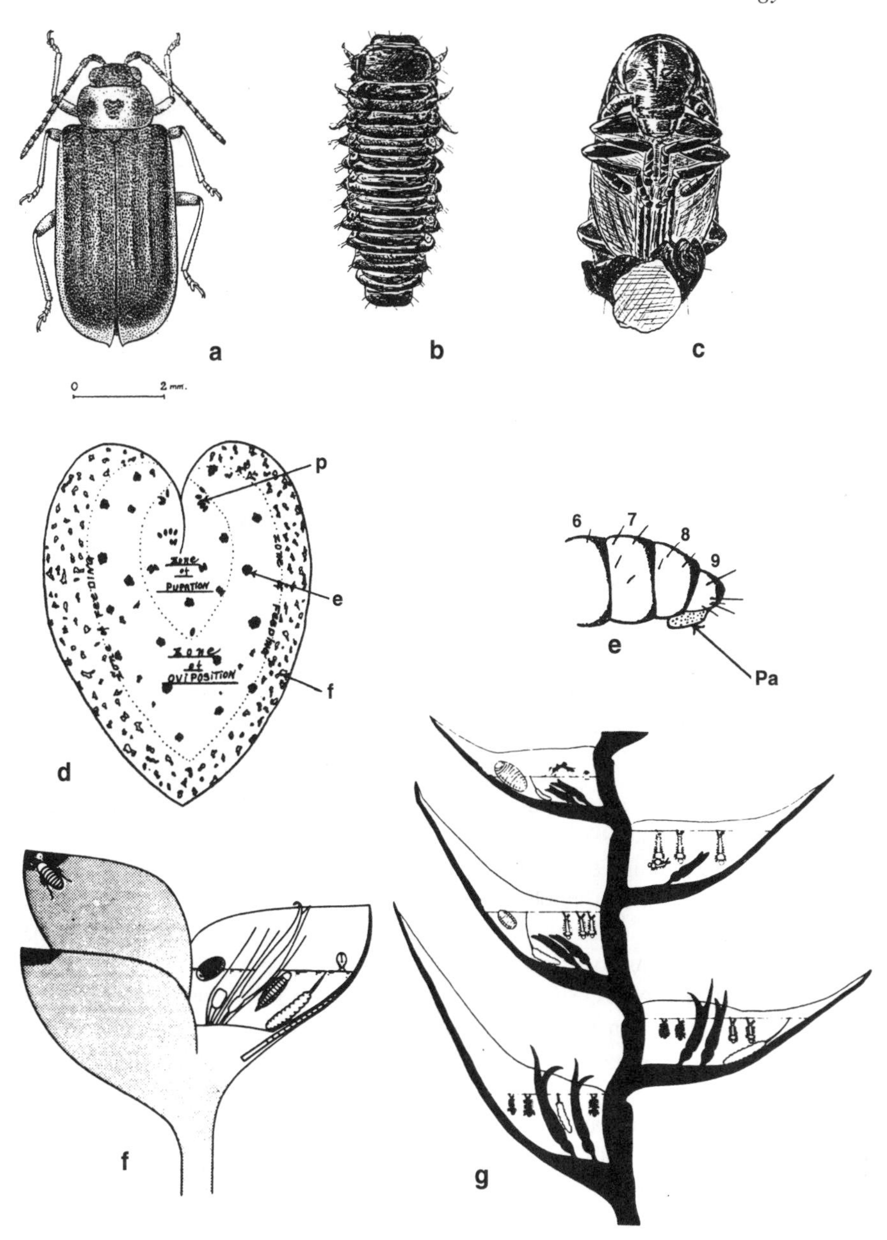

Figure 6.2. Subaquatic Chrysomelidae. (a) *Galerucella nymphaeae* (L.) adult from France (after Balachowski, 1963) on Nymphaeaceae; (b) *Galerucella nymphaeae* (L.) larva (after Laboissière, 1934); (c) *Galerucella nymphaeae* (L.) pupa with remains of larval skin (after Laboissière, 1934); (d) waterlily leaf with three zones of microhabitats for *Galerucella nymphaeae* (L.) (p = pupa; e = eggs; f = feeding area) (after Kaufmann, 1970); (e) *Galerucella nymphaeae* (L.) larval abdomen and the larval hold-fast organ (pygopode) (after Kaufmann, 1970) (6–8 mm); (f) stylized view of *Heliconia imbricata* (Kuntze) Baker from Costa Rica showing dissected bract with inquiline insects. The water-penny larva of *Cephaloleia puncticollis* (Baly) is on the edge of the bract entering the water (after Seifert and Seifert, 1976a,b); (g) diagrammatic representation of the insects living in the water-filled floral bracts of *Heliconia aurea* from Rancho Grande nature reserve, Venezuela. *Cephaloleia neglecta* Weise larvae are visible on the top left two bracts (after Seifert, 1982).

with the plant *Heliconia* are subaquatic and others almost aquatic. For example, hispines of the genus *Cephaloleia*. *Cephaloleia puncticollis* Baly and *C. neglecta* Weise live in floral bracts, which hold a considerable amount of water. Even in a dry season, a bract may hold as much as 6 ml of water; hence, bract-living species are more than subaquatic. Crowson (1994) and Farrell (1998) have found some morphological and molecular relationships between the two subfamilies of leaf beetles: Donaciinae and Hispinae. *C. puncticollis* have a setose abdomen, which is considered as an adaptation for the semiaquatic environment. By contrast, species of *Cephaloleia* living in the rolled leaves of *Heliconia* live in a humid environment but cannot be considered as subaquatic. The larvae of the bract-living species of the genus have not been adequately studied (Jolivet, 2002, in print). *C. puncticollis* larvae, after developing for some time inside the bracts, eventually move to young leaves to avoid competition.

Jolivet examined a number of species of *Cephaloleia* and noted that 13 species had setose venter, and 17 species were without setae on the ventral surface. He points out that a number of fully terrestrial galerucines, having no relationship with an aquatic/semiaquatic environment, have setose venter. Hence, he infers that certain morphological features, such as seta covered venter, may prove a preadaptation in taking to aquatic or semiaquatic life.

Adaptations for life in deserts

Chrysomelidae primarily inhabit humid areas, where vegetation provides them with nourishment. But some of them have invaded deserts and semidesert areas (Medvedev, 1996b). Unfortunately, detailed and precise studies on such leaf beetles are lacking (Cloudsley-Thompson, 1996, 1999, 2000).

Ghilarov (1964) pointed out two ways in which insects could invade deserts, namely ecological and morphophysiological.

First, let us consider the ecological way. Some insects which have not developed any special adaptations for desert life, have penetrated into deserts along rivers, irrigation canals and oases. Some insects live hidden in soil or penetrate into plant tissues. They may get transferred to desert areas along with some plants and with soil around their roots. Two species of *Dicladispa* (Hispinae), the larvae of which are leaf-miners, are found in the deserts of Saudi Arabia, though they are not common. Some Alticinae and Clytrinae have been recorded on plants in irrigated zones (Jolivet, 1997). *Chlamisus*, some Eumolpinae and some Cryptocephalinae have also been collected in green patches in Saudi Arabia. *Donacia* occurs in ponds in Mongolia. All the desert-living leaf beetles mentioned in this paragraph do not show any special adaptations for desert life, but some desert-dwelling chrysomelids do show morphophysiological adaptations of this nature. There have only been a few studies on such adaptations (Lopatin, 1999). *Timarcha* has been looked at from this viewpoint. *Timarcha* does well in the cold and green plains of Middle Europe as well as in arid and semiarid areas of North Africa. In the Libyan desert, it occurs up to 80 km from the border.

European *Timarcha* demonstrate certain preadaptations which could be helpful for extending its range into arid areas. The main preadaptative feature is its apterism. The absence of wings has left a considerable cavity beneath the elytra. Spiracles open into

a

b

Figure 6.3. Subaquatic Hispinae. (a) *Heliconia latispatha* Bentham (Heliconiaceae). They harbour adults and larvae of *Cephaloleia puncticollis* Baly inside their water-filled bractea. Matagalpa, Nicaragua, May 1994; (b) *Cephaloleia puncticollis* Baly (Hispinae). The beetle is red against the red background.

this cavity. According to Slobodchikoff and Wismann (1981) and Roskaft *et al.* (1986), this situation reduces loss of water through transpiration. Slobodchikoff and his team have found that there is a significant correlation between the rate of water loss and the depth of the subelytral chamber.

There is another advantage of the subelytral chamber in *Timarcha*. The cavity allows rapid expansion of the abdomen, and this permits the beetle to drink a large quantity of water when a temporary source is available, such as a rainwater puddle.

In most species of *Timarcha*, the elytra are fused along the suture. As Zachariassen (1977) pointed out, fusion of elytra is a useful adaptation for resisting dehydration. Nocturnal forms among Timarchini (e.g. *Americanotimarcha*, *Metallotimarcha*) do not have complete fusion of elytra. *Timarcha tenebricosa* Fabricius and *T. goettingensis* L. show a zipper-like arrangement or coaptation (*Figure 6.4 (1)*) at the suture. This should also help in the retention of moisture in the subelytral chamber.

Jolivet (1957–1959) referred to modifications in *Timarcha* for life in arid/semiarid areas as 'timarchization'. This phenomenon includes hardening and fusion/coaptation of elytra, enlargement of epipleura, atrophy of wings, wing muscles and of the related apodemes, increased fecundity, heavier eggs, reduction of the metasternum, and reduction of the metathoracic stigmata. Such changes are seen also in some other genera (*Elytrosphaera*, *Iscadida* etc.). Comparable changes are seen also in some galerucines (*Galeruca*, *Arima*), which live in dry, mountainous areas. Obviously, these are cases of convergence under similar environmental conditions. Loss of wings is clearly not a new phenomenon among *Timarcha*, or among *Meloe* spp, as there is no evidence of wings in either larvae or pupae.

Chen and Wang (1962) studied desert modifications in the chrysomelids of Sinkiang, which is traversed by two mountain chains. Between the mountains are desert plains and the slopes of the mountains are bare, covered with stones and of the nature of a semidesert. *Oreomela* (Chrysomelinae) is found in the desert plains and demonstrates considerable timarchization. According to Chen, *Chrysochares* (Eumolpinae), *Ischyronota* (Cassidinae), *Cryptocephalus astracanicus* Suffrian (Cryptocephalinae) and *Diorhabda elongata deserticola* Chen (Galerucinae) are also found in desert plains. On the arid slopes of mountains occur *Crosita*, *Cystocnemis*, *Chrysolina* (all three Chrysomelinae), *Galeruca* and *Theone* (both Galerucinae). Chrysomelids found in desert plains have brushy tarsi and a fusiform body shape, adaptations for hiding in sand. On the other hand, the leaf beetles living on dry mountain slopes have a flattened body and flattened tarsi with reduced brushes because they hide under stones.

Chrysomelids in the alpine environment

Beetles are among the dominant forms of high-altitude insects on nearly all the mountains of the world (Mani, 1974). Among these beetles, debris-feeders, carrion-feeders and predators constitute the majority. As Mani points out, 55% of high altitude Coleoptera in the Himalayas are Carabidae. Curculionidae and Chrysomelidae are also included, but they are much less numerous.

Most subfamilies of Chrysomelidae are thermophile, and so naturally do not include high altitude forms. Generally, most of the alpine leaf beetles belong to Chrysomelinae, Galerucinae and Alticinae. Some members of Eumolpinae,

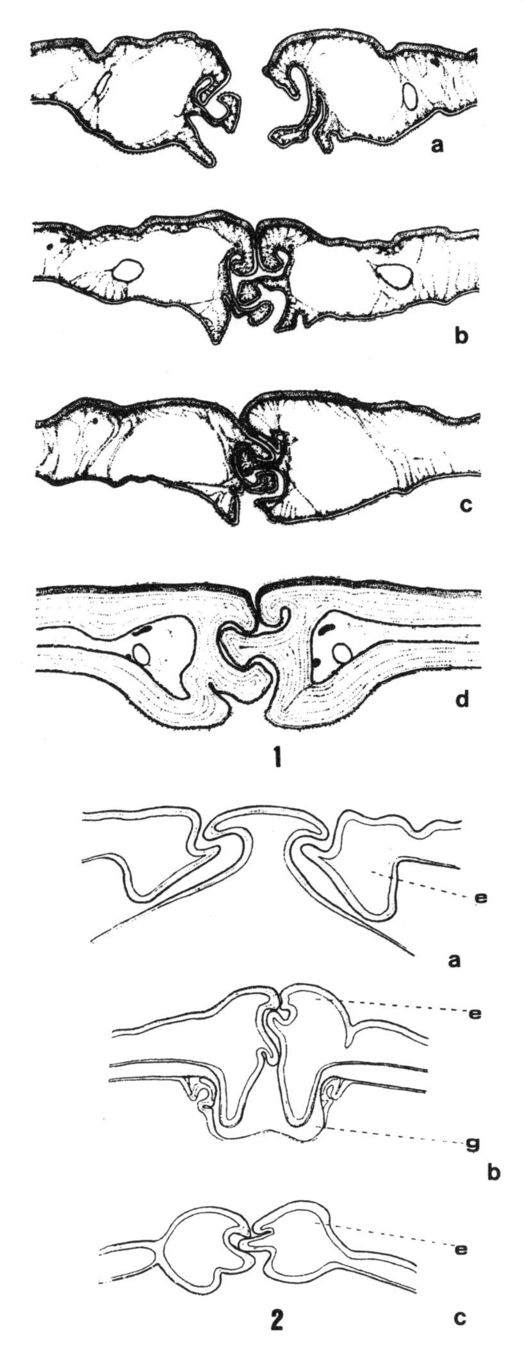

Figure 6.4. Coaptations (after Corset, 1931). (1) *Timarcha tenebricosa* Fabricius (Col. Chrysomelinae) (× 50). Coaptation along the sutural margin of the elytra (line of juncture). (a) before any coaptation; (b) and (c) during coaptation; (d) in adults, when the elytra are fused together; (2) *Cassida viridis* Linné (Col. Cassidinae). Transverse sections of the sutural side of elytra. (a) at the posterior of the scutellum, (b) at the middle of the metathorax and (c) at the first abdominal segment.

Cryptocephalinae, Clytrinae and Cassidinae have also taken to high-altitude life. Hispinae are not found above a certain limit and are generally thermophile insects. Lopatin (1996) points out that high-altitude chrysomelids in the mountains of central Asia include Chrysomelinae (115 species, 65% of the total number of specimens), Galerucinae (19 species, 10.9%), Eumolpinae (15 species, 5.7%), Alticinae (11 species, 6.3%), Cryptocephalinae (10 species, 5.7%), Clytrinae (4 species, 2.3%) and Cassidinae (1 species, 0.6%). The Chrysomelinae have been most successful among the Chrysomelidae in invading high altitudes.

Of the high-altitude Chrysomelinae, *Elytrosphaera* have been extensively studied (Jolivet, 1987b, 1997; Macedo *et al.*, 1998). In the tropics, *Elytrosphaera nivalis* Kirsch has been recorded at 4650 m in Ecuador. The basic metallic colours of this genus are replaced by black, a characteristic seen in the high mountains and volcanoes of the Andes. Generally, in tropical mountains *Elytrosphaera* species occur up to 4200 m maximum, and commonly are found at 1500 m on average. In the Himalayas, the endemic chrysomeline *Apaksha* has been found at 4575 m (Mani, 1974). In Yunnan, China, *Chrysolina zhongdiana* Chen and Wang occurs up to 4300 m. *Chrysolina* spp. exists in Sri Lanka up to 3000 m and in East Africa not above 2000 m. *Oreomela* in Xizang, China has been recorded at 4850 m (Wang and Chen, 1981). *Xenomela* is also mountainous in central and middle Asia.

The highest alpine record is for *Chrysolina zanyana* Chen and Wang in Xizang; it is 5020 m. This record in the Northern Hemisphere suggests the possibility of future captures at higher altitude in the tropics in South America and Africa.

Various species of *Chrysolina* and *Timarcha metallica* Laichharting are found in European mountains, also *Oreina*, *Oreomela* and *Cystocnemis,* and some other Chrysomelinae are found at high altitudes in middle Asia.

In north Yunnan in China, the alticines *Orthespera* and *Microspera* have been frequently collected at altitudes between 3500 and 4300 m (Chen and Wang, 1987; Chen *et al.*, 1987; Wang, 1990). The remarkable endemic chrysomelid of the Himalayas, *Chaetocnema alticola* Maulik, is another alpine alticine (Mani, 1974).

The primitive galerucine *Galeruca* is holarctic in distribution, and it also occurs in central Asia and China. Yang and Yu (1984) have extensively studied *Galeruca* spp. in China. Thirty-six species of this genus are found in China, and most of them are alpine forms, and show the usual high-altitude adaptations (*vide infra*). Some of the species attain the altitude of 4000 m or more. Galerucines of the high mountains of Yunnan in Sichuan include occidental forms, such as *Xingeina*, *Shaira*, *Geinella*, *Geinula*, etc. (Chen *et al.*, 1987). They also exhibit the expected alpine adaptations.

High-altitude adaptations affect a number of morphological features (Yang and Yu, 1994). Behaviour is also affected by the alpine environment. Major adaptations include the following:

1. Reduction or absence of hindwings is intimately connected with adaptation to extreme environment in alpine regions, polar regions and desert/semidesert conditions. This modification not only affords protection against air blast but also, due to the reduction of wing muscles and related apodemes, allows production of larger and more eggs/larvae, the latter in cases of viviparity. In many high-altitude forms, the elytra are fused, highly convex and cover a subelytral cavity. The layer of air between the body and elytra functions as a temperature

and humidity buffer against harsh environmental conditions (Lopatin, 1996). Another possibility is that a brachypterous/apterous condition is accompanied by reduced elytra (brachelytry). *Galeruca*, for example, shows reduction of elytra, so that a last few segments remain uncovered (*Figure 6.5*). In South America, the apterous *Elytrosphaera* shows fusion of elytra, as in *Timarcha*, which may also occur at middle-to-high altitudes (*Metallotimarcha*). However, lowland-frequenting female galerucines of the genus *Metacycla* in Mexico exhibit apterism and brachelytry. The yellow bands on the abdomen coincide with those of the reduced elytra. This is evidently in correlation with reproduction since the male is winged and has complete elytra. As in *Agelastica* in Europe, the female abdomen becomes enormously distended before egg-laying.

2. Generally, high altitude species show blackening or darkening of the body colour. Dark bronze, metallic blue, green and violet and highly melanistic colours are common. *Elytrosphaera*, for example, usually has brilliant metallic colours. But alpine species like *E. nivalis* and *E. melas* exhibit an obvious darkening of the body colour. The former species shows a pattern of sinuous metallic lines on the pronotum and elytra which is almost covered by a general darkening. In the Andes, cassidines and many other insects tend to grow darker and darker with an increase in altitude. Lopatin points out that melanization in high-altitude forms is specially obvious in many species, the near allies of which, on lowland, are pale coloured. Melanization helps in the absorption of solar radiation, resulting in body warming, and at the same time it decreases passage of UV into the body (Lopatin, 1996; Macedo *et al.*, 1998).

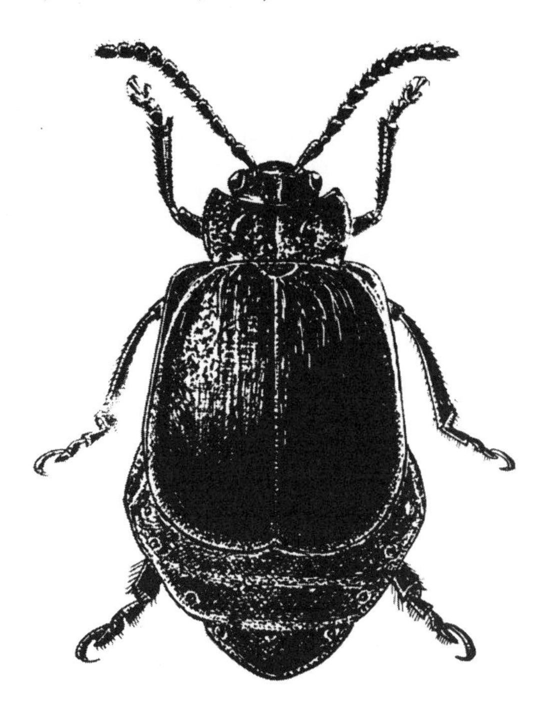

Figure 6.5. *Galeruca monticola* Kiesenwetter from high Pyrenees, showing reduced elytra which leave some terminal abdominal segments exposed (after Laboissière, 1934).

3. For thermoregulation, alpine leaf beetles remain active when it is sunny. When the sky is overcast or it is windy, and during the night, they are immobile and remain concealed under blades of grass or cushions of alpine weeds (Lopatin, 1996).
4. In order to take advantage of short-lived favourable periods for development, alpine leaf beetles have a long duration life history of two years or more, including diapause in various stages.

"One of most important features of high-altitude insect fauna is the high level of endemism. Of 115 species of Chrysomelinae in altitude, the endemics constitute 109. For Galerucinae, the ratio of endemics/non-endemics is 19/15, for Eumolpinae 15/14, for Alticinae 11/6, for Cryptocephalinae10/3, for Clytrinae 4/2, and for Cassidinae 1/1" (Lopatin, 1996). This situation suggests that most alpine species faced strong selection when they tried to invade the high-altitude environment and, as a result, underwent considerable modification. It also seems to be true that some forms, like species of *Timarcha*, with their well-sclerotized bodies and apterism, were preadapted to invade the subalpine environment.

While we are becoming acquainted with leaf beetles in the alpine environment, it would be relevant to know something about aeolian derelicts. Certain substances found on the plains are lifted by warm air currents to the upper layers of the atmosphere, where they may get deposited on the slopes of high mountains. Such aeolian derelicts have been described in the Alps, the North American mountains and the Himalayas (Mani, 1974). Besides dust particles, pollen and small dried fruit, such airborne material includes spiders and insects. Among such insects are included even heavy-bodied species like wasps, cerambycids and locusts. It would therefore not be surprising to come across the frozen and well-preserved bodies of plain-dwelling leaf beetles at high altitude in mountains. Furthermore, leaf beetles often migrate over mountain passes carried by the wind, as in Panama (*Aulacoscelis* spp.) and in Venezuela (Rancho Grande, Panchuelo Pass). This is probably the way in which many species became established in highlands during the different geological periods. Mountains can also become places of sanctuary for Pliocene species which were formerly distributed in the plains, like *Chrysolina staphylea* L. on the top of Mount Halla in Cheju-do Island, Korea (Jolivet, 1975). This beetle can be considered as subarctic, even if it is present in some middle European plains, as it only survives in the mountains in northern Italy and Spain and in the Arctic in Asia.

Leaf beetles in the polar region

A small number of chrysomelids have become adapted to life in the polar regions. There are only a few studies on such chrysomelids. Two important papers, recently published, are by Silfverberg (1994) and Chernov *et al.* (1994). Another significant contribution is by Medvedev and Khruleva (1986) and Khruleva (1994), who have studied in detail the life cycle of the arctic *Chrysolina subsulcata* Mannerheim. More recently, the leaf beetle fauna of the Arctic has been reviewed by Medvedev (1996a). Let us also mention Medvedev (1998a), who recorded the first member of the Chrysomelidae in the Severnaya Zemlya archipelago.

LEAF BEETLES IN THE ARCTIC

The arctic zone extends roughly from the latitude 65° to the northern limit. In Canada, however, the southern limit of the Arctic is close to 60°. The northern limit of the distribution of Chrysomelidae is between 75° and 77°. Severnaya Zemlya is situated even farther north, between 78° and 79°N. This is the northern limit for a *Chrysolina*. Curculionids have penetrated even deeper northward, e.g. in Greenland, from where all chrysomelids have recently disappeared.

Only a small number of chrysomelid species are found in the Arctic. Silfverberg (1994) puts the number at 12 and Chernov *et al.* (1994) cite 20. More have been described by Medvedev (1996a, 1998a). It is not possible to decide on the exact number of these species for two reasons. The taxonomy of the arctic leaf beetles remains to be investigated satisfactorily, and synonymization of some species is likely. The other reason is that the arctic zone has so far been inadequately researched. Studies in Russia, Siberia, parts of Canada and Alaska will have to be carried out more extensively.

Most of the arctic leaf beetles belong to Chrysomelinae. Some Cryptocephalinae and Galerucinae are also included in the arctic fauna.

A list of some of the better-known arctic species of Chrysomelidae and brief details of their distribution are given below. The account is based on Silfverberg (1994), but more data are given by the other authors mentioned above.

Chrysolina subsulcata (Mannerheim) (Chrysomelinae).
Distribution: Along the Siberian coast, Siberian islands and Alaska (*Figure 6.6*).

Chrysolina septentrionalis (Menetries) (Chrysomelinae).
Distribution: Siberian coast of the Arctic Sea and some Siberian islands.

Chrysolina caurina (Brown) (Chrysomelinae).
Distribution: North Alaska.

Chrysolina bungei (Jacobson) (Chrysomelinae).
Distribution: East Siberian coast of the Arctic Sea and some Siberian islands.

Chrysolina cavigera (Sahlberg) (Chrysomelinae).
Distribution: Arctic coast of Siberia, Wrangel Island and along north coast of Alaska.

Chrysolina arctica (Medvedev) (Chrysomelinae).
Distribution: Wrangel Island.

Chrysolina hudsonica (Brown) (Chrysomelinae).
Distribution: Coast of Hudson Bay and north Newfoundland.

Ophraella arctica (LeSage) (Galerucinae).
Distribution: Arctic coast of Canada.

Phratora hudsonica (Brown) (Chrysomelinae).
Distribution: Alaska, Canada about the latitude 60°. In fact, it is a north boreal species extending its range to the arctic zone.

Cryptocephalus krutovskyi (Jacobson) (Cryptocephalinae).
Distribution: NW Siberia and Japan. Another boreal species, with range extending into the arctic.

Pachybrachis amurensis (Medvedev) (Cryptocephalinae).
Distribution: western Mongolia. Its range also extends into the Arctic.

Galerucella nymphaeae (Linné) (Galerucinae).
Distributed widely in Europe and Siberia. Also found in Siberian tundra. It is evident that this species needs water to breed and cannot survive on ice.

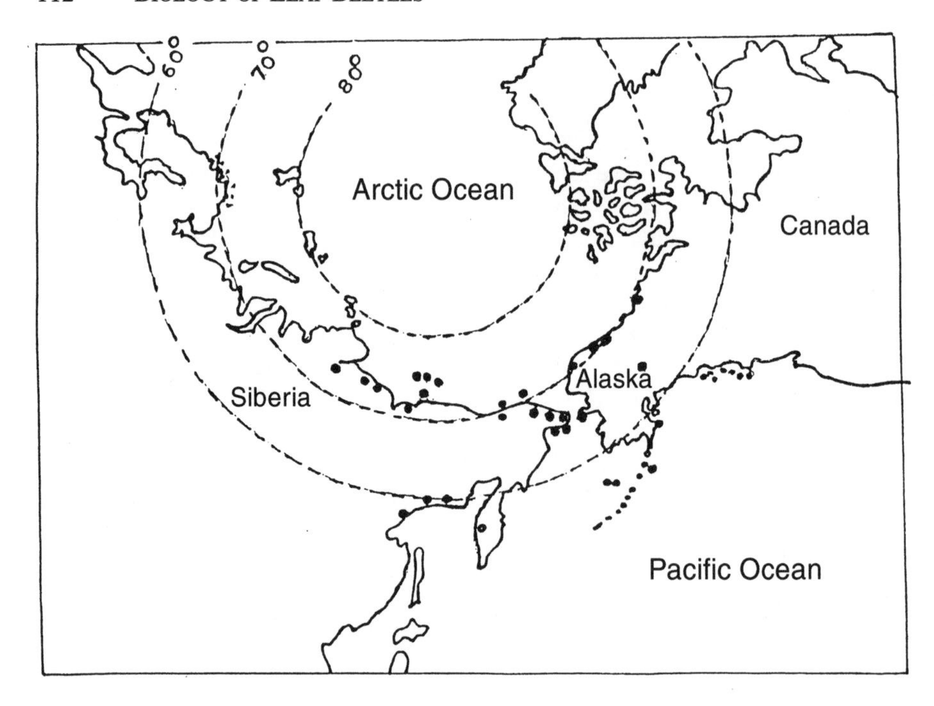

Figure 6.6. Map to show the distribution of *Chrysolina subsulcata*. Based on Silfverberg (1994). The outline map is only approximately correct. The solid black dots denote the sites of collection.

For a fuller account of the arctic species of Chrysomelidae and their ecology refer to Silfverberg (1994) and to various papers by Russian authors including Medevedev (1996a, 1998b).

It is remarkable that there are no chrysomelids in Greenland, when there are a few in Spitzberg. As evidenced by fossils, Chrysomelidae were richly represented in Greenland between quarternary glaciations (Bocher, 1988, 1989). The absence of leaf beetles from this large island of the Arctic is obviously due to recent extinction during the last glaciations, and even more recently around the 14th century when the temperature dropped extensively. Recolonization on crops in the south is always possible with global warming. Also of interest is that, although a number of fossil chrysomelids have been found in Greenland, none is *Chrysolina*. However, the identification of remains is vague, meaning it is difficult to differentiate between *Chrysomela* and *Chrysolina* and much depends on the taxonomic knowledge of the palaeoentomologist. *Chrysolina* fossils have been collected elsewhere in the arctic region. A lack of suitable nourishment was probably responsible for the fact that there are no *Chrysolina* among the past chrysomelid residents of this large land mass (Jolivet, 1997).

Another area which is interesting in terms of chrysomelid distribution in the Arctic is Wrangel Island, a small island located north of the easternmost part of Siberia. It probably separated from the continent about 12 000 years back. It is a relic of the Siberian tundra, with a rich vegetation of herbaceous species still surviving. Its fauna includes 7 chrysomelid species, obviously due to the availability of required nourishment. As usual, the curculionid fauna is richer with 11 species (Khruleva and Korotyaev, 1999). However, *Chrysolina* species in the Arctic became practically polyphagous to survive and are not restricted to a few plant families like the temperate and tropical species.

It is important to note that no chrysomelid species has a circumpolar distribution, but some species do have a very wide distribution. For example, as noted earlier, *Chrysolina cavigera* and *C. subsulcata* range from Siberia to Alaska.

Among *Chrysolina* species of the Arctic some researchers have suggested species 'bunches', which have been referred to as subgenera (Kontkanen, 1959). This suggests that there was a multiple arrival of leaf beetles in the arctic zone. It is evident that the thermophilic and steppic *Timarcha* never entered these cold areas but probably existed in central Asia before the Pliocene. It was eradicated at that time from southern Canada and the Baltic States and Scotland in Europe, except in a few areas. For instance, the western side of Vancouver Island was never covered with ice and has retained one *Timarcha*. In view of the small number of species of arctic leaf beetles, it is obvious that there was poor speciation after entry into the zone. Nonetheless, arctic species had to develop some adaptations for survival in these frozen areas, and we shall consider those adaptations in the following part of this section.

Morphological adaptations

As pointed out by Chernov *et al.* (1994), arctic Chrysomelidae have a tendency of reduction in body size. A maximum size is seen in some species of *Chrysolina. C. cavigera* for example can reach 9 mm in length.

As in the case of alpine forms, in arctic leaf beetles too the body colour tends to be dark. Black is the dominant colour, while some show dark blue, green and copper colours (in the Pribilof Islands with *Chrysolina*). It is believed that a black/dark body colour facilitates the absorption of heat. But, as Jolivet (1997) pointed out, the dark colours may not necessarily be an adaptation for the arctic environment. *Timarcha*, for example, is black in the warm steppes as well as in frozen regions. There is an exceptionally dark red example in Oregon, but this species is nocturnal. Two species in the Balearic Islands and eastern Spain are of a brilliant metallic blue colour, perhaps as a protection against the sun's radiation. However, all the species of *Metallotimarcha* are metallic red and mostly nocturnal in the Central European mountains. As we can see, it is difficult sometimes to associate beetle colours with environment.

Some arctic species of *Chrysolina* (not all) are apterous or brachypterous, and are incapable of flight. This degeneration of wings leads to a loss of flight muscles and a reduction of related apodemal structures. This affords an advantage in that the production of larger eggs/larvae (K-strategy) becomes possible. Another advantage resulting from a loss of wings is the formation of a subelytral cavity, which acts as a layer of thermal insulation, and helps to stabilize the body temperature. Chernov *et al.* (1994) cite the case of *Basilepta ovulum* Weise, living in the cold desert of Tibet. It has a subelytral cavity, which is about 50% of the total volume of the insect. The arctic Chrysomelidae have the subelytral cavity constituting about 5 to 11% of the total volume. Elytra are sometimes fused, or nearly fused, along the suture in these forms. It is interesting to note that similar modifications (i.e. brachypterous/apterous condition and related changes) have helped in the adaptation to different adverse environmental conditions, such as desert, alpine environment and polar conditions.

Chernov *et al.* (1994) pointed out also that in the arctic *Chrysolina*, *Phratora* and *Hydrothassa*, the elytral surface shows striations or ridges, which condition is also seen in high mountain forms.

Another arctic modification, mentioned by Chernov *et al.* (1994), is that the tarsal brush is greatly reduced. This the authors have taken as an indication that the arctic forms have derived from high mountainous forms, but this view is debatable. One obvious possibility is that similar environmental adversity, namely extreme cold, has led to parallel changes in alpine and arctic forms.

Physiological and developmental adaptations

Some arctic species of *Chrysolina* and *Gonioctena* are viviparous or ovoviviparous. In viviparity, well-grown larvae are born. In ovoviviparity, the eggs laid are large and with nearly completely developed larvae, which hatch out soon after oviposition. This condition shortens the developmental history outside the mother's body.

Viviparity/ovoviviparity is not an adaptation seen only in arctic forms. Alpine forms and those living in the tropics also show this adaptation. In tropical forms, the shortening of the postembryonic development period has the advantage of reducing the chances of entomophagy (oophagy) by parasitoids. It also reduces the possibility of predatory attack on eggs. In alpine and arctic environments, brief and relatively favourable conditions may be taken advantage of by short larval development. Viviparity/ovoviviparity in the Arctic is associated with a reduction in fertility. *Chrysolina subsulcata*, for example, lays only 3–9 large eggs. *Gonioctena pallida* L. is viviparous in the Arctic, and produces only 22–24 larvae.

Another arctic adaptation is rapid but flexible larval development. Larval development is not continuous and is interrupted by a variable number of diapauses. The total development from egg to adult may take 7 or 8 years.

Chernov (1978) studied the development of *Chrysolina septentrionalis* and *C. subsulcata* both in the field and the laboratory. The larvae enter into diapause at least twice. There may be some additional larval diapauses. Khruleva (1994) has pointed out that in *C. subsulcata* there are a minimum of two hibernations in the adult stage. In this species Chernov observed (1978) that the life cycle at 4 to 5°C takes in total 6 years from egg to adult.

Another adaptative feature of arctic chrysomelids is cold resistance. As noted by Jacobson (1910), the larvae of *C. subsulcata* are still active at −3 – −4°C. Khruleva (1994) observed that in *C. subsulcata* copulation occurs at 0°C on melting snow.

Arctic leaf beetles are very flexible when it comes to food choice. They generally show oligophagy but, if need be, this can be extended until it amounts to polyphagy. For example, *Chrysolina septentrionalis* and *C. subsulcata* belong to the same subgenus, *Arctolina* (Kontkanen, 1959). *C. septentrionalis* feeds on *Armica iljini* (Asteraceae) and *Delphinium middendorfi* and *Ranunculus borealis* (Ranunculaceae) (Chernov, 1973). These host plants seem to be the basic food choice for *Chrysolina* (*Arctolina*). But in Wrangel Island *C. subsulcata*, according to Chernov (1973), derives nourishment from *Carex lugens* and *C. stans* (Cyperaceae), which are abnormal hosts. Caryophyllaceae are normally not attacked by species of *Chrysolina* which have their favourite and specialized hosts generally selected by subgenera (Jolivet and Petitpierre, 1976a). However, *C. septentrionalis* is known also to feed on *Silene repens* (Caryophyllaceae) for survival in the Arctic. Feeding on Ranunculaceae is interesting and is shared with several temperate species of *Chrysolina*. Ranunculaceae, with Magnoliaceae and others, are very primitive among the flowering plants and this could mean that some of the old feeding habits (atavism?) were re-emerging. In the Pribilof Archipelago, in the Bering Sea, on the Alaska coast, the local *Chrysolina subsulcata* is also polyphagous (Jolivet, 1992). Certain species have been mentioned as feeding even on dwarf *Alnus* and other low trees (Betulaceae).

In conclusion, it may be pointed out that, though it seems likely that some leaf beetles could extend their range to the arctic zone because of their preadaptations for such an environment, the flexibility of their life cycle and their modifiable oligophagy suggest that some adaptations did develop in response to arctic conditions. Medvedev (1996a) is of the opinion that the majority of arctic leaf beetle fauna evolved in the east Siberian tundra.

CHRYSOMELIDS IN THE ANTARCTIC

Leaf beetles have been much less well studied in the Antarctic than in the Arctic. There is only one chrysomelid in the Falkland Islands. At the extreme point of Chile and Argentina, chrysomelids are extremely rare. However, one species of apterous Chrysomelinae with small eyes, *Brachyhelops hahni* Fairmaire, probably hides under stones at the base of plants and lays very few big eggs (Brendell *et al.*, 1993). It has been described from Isla Bertrand, Tierra del Fuego. As usual, curculionids are present in the subantarctic islands when leaf beetles seem absent. Many of the weevils are cryptogam-feeders (bryophytes, lichens) and this strategy probably evolved in response to the Pleistocene glaciations (Chown and Scholtz, 1989).

As remarked on by Chernov *et al.* (1994), the chrysomelid diversity gradually decreases from the tropics towards the Arctic and the same is true towards the south, that is towards the Antarctic. There are only a few leaf beetle species in Patagonia and Argentina, as mentioned above. They are small and dark species. There are no chrysomelids in the subantarctic and volcanic islands, though there curculionids and carabids present considerable diversity. There are chrysomelids in southern Australia, Tasmania and southern Africa. They are rather diversified but we cannot think of them as antarctic fauna, even if some are related in some ways to *Brachyhelops* (Daccordi, 1994, 1996).

Chrysomelids in the canopy

Canopy here means the canopy of tropical rainforests. This habitat still remains inadequately studied from the point of view of arthropod communities.

The main problem arising in the investigation of canopy fauna is that of collection techniques. One method is to set up bamboo scaffolding (or a metallic tower like in Barro Colorado Island, in Panama). The investigator climbs up the scaffolding to make observations and to collect specimens, but this process disturbs the foliage of the tree so much that we can expect at best only sporadic observations and a scanty collection. Paulian used this technique in the Ivory Coast forests (or at least in what was still forest at that time) (Jolivet, 1997). In the dense rainforest of the tropics, various methods have been tried, for example, long wooden passages (Rancho Grande, Venezuela, French Guyana), heavy cranes (Panama, Venezuela, Queensland and Borneo) or a system of hoists. None of these can be readily moved or positioned among the trees without disturbing the forest canopy. Balloon airships have been tried but, again, trees and insects were greatly disturbed. The best system for simple observation seems to be with the crane but, of course, many insects will escape discovery with this method.

The most satisfactory technique developed so far for statistical study of canopy insects and other arthropods is to fog the tree crown with a solution of pyrethroid, so that the arthropods are instantly killed/immobilized and fall on to the trays held underneath. This technique was developed by Erwin (1983), who initiated research on canopy insects. Using this method, studies on canopy arthropod fauna have been carried out by Erwin (1982, 1983) and Farrell and Erwin (1988) in tropical America, by Stork (1987, 1988) in Borneo, by Basset (1996) and Basset and Samuelson (1996) in Papua New Guinea, and by Wagner (1998) in Uganda. Basset and Samuelson (1996) also collected specimens by hand picking, foliage beating, branch clipping, and intercept flight traps, in addition to pyrethroid knockdown. The only objection to the use of the pyrethroid knockdown method is that, in view of so many new species among canopy forms and of lack of trained taxonomists, the concept of morphospecies will be used, and this may lead to unreliable data.

The main inferences about canopy-dwelling Chrysomelidae which have been reached by these studies include the following:

1. It is obvious that the Chrysomelidae which reach the canopy must be efficient flyers. Such Chrysomelidae are Megascelinae, Alticinae, Galerucinae, Cryptocephalinae, Hispinae and small Eumolpinae and they are included in the chrysomelid canopy fauna. Heavy-bodied and poor-flying leaf beetles, like Chrysomelinae, are generally not among the canopy forms. Small chrysomeline species, like those of *Plagiodera*, are exceptions. Everywhere in the New and the Old World, Chrysomelidae and Curculionidae (not taking into account the large ant biomass) are dominant among insects in the canopy, from the point of view of the number of individuals as well as species (Jolivet, 1997). Basset and Samuelson (1996) point out "In terms of species richness, Chrysomelidae were the most prominent leaf-feeding insect family on the foliage of the study trees, followed by Curculionidae (leaf-feeding species only)....". These authors found that, in their canopy chrysomelid collection in New Guinea, Galerucinae dominated in terms of the number of species, but Eumolpinae were more prominent in terms of individuals and biomass. The sequence of chrysomelid subfamilies (in decreasing order) was: Eumolpinae–Galerucinae–Alticinae–Cryptocephalinae. Farrell and Erwin (1988), working in Peru, reached a different conclusion. According to them, the sequence of dominance among canopy chrysomelids in terms of species number was: Alticinae (70 species), Eumolpinae (29 species), Galerucinae (16 species). According to Wagner (1998), the order of numerical/biodiversity dominance was variable among the four tree species chosen for study, but Eumolpinae, Galerucinae and Alticinae were obviously prominent among the chrysomelid subfamilies from all these viewpoints.

2. Wagner (1998) included some swamp areas among the forest patches he studied and noted that, in swamp forest canopy in Africa, Chrysomelidae presented a greater diversity. Among the trees selected by him for investigation, most were 7–16 m tall, that is their crowns belonged to the middle canopy stratum. But one tree, *Cynometra*, was exceptionally tall, and its crown could be taken as a part of the upper canopy. On this tree, a particularly diverse chrysomelid fauna was recorded. This situation suggests that habitat structure influences the composition of an insect community.
3. Efforts have been made to find out the host specificity of canopy leaf beetles, that is whether they are polyphagous, oligophagous or monophagous, but a clear picture of this is still to emerge. Basset (1994) believes that about half the canopy forms were polyphagous. Wagner (1998) reported that the number of leaf beetle species confined to one tree species in East Africa was very small, only 2.98% of the total number of species in his collection. Basset and Samuelson (1996) separated living chrysomelids from their collection in New Guinea, and kept them in jars with leaves of 10 different tree species, including the species from which the insect had been collected. They found that out of 134 species, 36 species actually fed in the jars, out of them 24 were specialists (i.e. they fed only on leaves of the tree from which they had been collected), 9 were generalists (i.e. they fed on leaves of three or more species) and, in the case of three, no inference could be drawn. It must also be considered that a number of the insects found on tree tops are just 'tourists' and not specialist or even generalist feeders.
4. Wagner (1998) found that the number of chrysomelid individuals varied from 14–100 per tree, and that there was a significant positive correlation between canopy volume and the number of leaf beetle individuals collected from it. This correlation could be noted in each of the four tree species included in his study in east Africa. Wagner (1999) also observed that chrysomelid communities in rain forests are characterized by a high number of rare species.

In all the studies on canopy chrysomelids so far, only morphospecies have been considered and details of chrysomelid taxonomy have not played a part. This is because of the obvious difficulty in involving specialists on all the different subgroups in such a study. It is quite likely that many of the species collected in such a project turn out to be new to science. Forests are disappearing fast. There is therefore a real fear that a fairly large number of leaf beetle species may be wiped out before they could be discovered, described, and recorded. In recent research done in Panama, around 25% of the curculionids recorded in dry and wet forest canopies were new to science (H. Barrios, pers. comm.).

Chrysomelidae in caves

Subterranean caves are completely dark and contain no vegetation. Hence, phytophagous/phyllophagous forms would not be expected to be found in such a habitat. But phytophagous homopterans have been reported feeding on roots projecting from the walls and ceiling of caves, especially in Madagascar. The homopterans have reduced eyes, and are thus modified for cave dwelling. There are a few reports of galerucine larvae and adults (*Aulacophora* sp.) from Indian and Sri Lankan caves

(Abdulali, 1948). They too seem to be root-feeders, but do not appear to have any special modification for cave life.

Niche separation among leaf beetles

'Niche' is a basic concept in ecology. The 'niche' of a species means the place occupied by it in its habitat. Elton (1927) defined 'niche' as an animal's place in the biotic environment, its relation to food and enemies. Most ecologists, however, define 'niche' as the place occupied in the environment with reference to space, food, time or other environmental variables.

Root (1967) introduced the phrase 'niche exploitation pattern', which is more meaningful than 'niche'. This phrase refers to the population response of the species in question to the different variables in the environment.

Sympatric populations of related species show various degrees of niche separation. Niche separation minimizes interspecific competition and allows for a fuller utilization of resources.

There are very few studies on niche separation among chrysomelids. Oligophagy and monophagy result in niche separation among many sympatric chrysomelid populations but, in some cases, closely-related populations may share the same food plant. Niche separation in one such case was studied in considerable detail by Verma and Shrivastava (1985).

The case studied by Verma and Shrivastava is of two cassidines, *Aspidomorpha miliaris* F. and *A. sanctae-crucis* F. They live on the leaves of the same host plant, *Ipomoea fistulosa* (Convolvulaceae). Both the cassidines have similar active periods, which are from July to November and mid-January to mid-February. The authors noted a number of environmental variables in relation to these species. *A. miliaris* feeds mostly on submarginal or deeper parts of the leaf lamina, and adults of the other species eat mostly the marginal parts of leaf. *A. miliaris* occurs on host plant bushes mostly in more humid situations, while *A. sanctae-crucis* is specially numerous in drier conditions. The authors (Verma and Shrivastava, 1985, 1986) suggested a modification of a niche separation model, described by Hutchinson (1957). This modified version of the model (*Figure 6.7*) shows population response of the two cassidines to these two environmental variables.

In the modified Hutchinsonian model, the two axes have been assigned to the two variables, the vertical axis showing the distribution of feeding in the marginal and submarginal parts of the leaf, and the horizontal axis showing the number of individuals per 100 sq m of weed area in more or less humid situations. The broken lines, crossing the axes at right angles, are lines of niche demarcation with reference to the two variables. The one crossing the vertical axis shows demarcation between marginal and submarginal parts of leaf lamina. In a leaf 75 mm wide at base, where it is the widest, an arbitrary 10 mm wide area from the margin has been taken as the marginal area, and the rest is submarginal. Fed-away areas in 6 leaves have been examined and measured for both species, and distribution of feeding on either side of the line of niche demarcation has been shown in the model. Similarly, 16 different areas of weed growth, some more humid, being close to a water body, and others less humid, were examined for the presence of the two species. The distribution of their population densities (as number of individuals per 100 sq m of weed growth) is shown

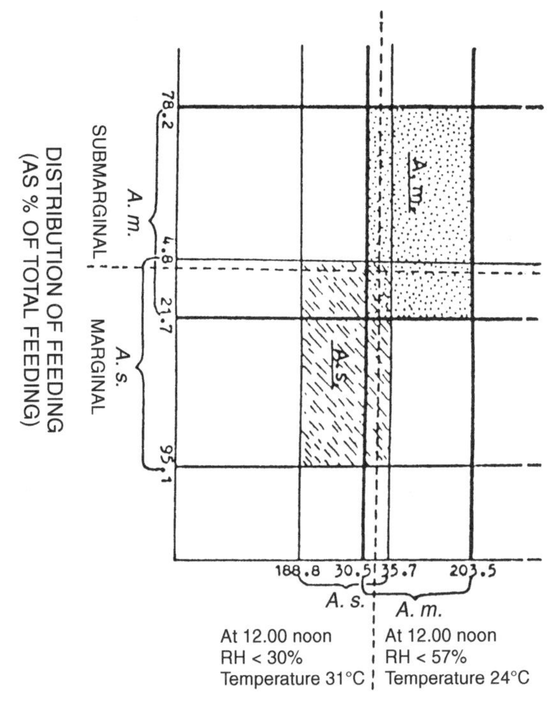

Figure 6.7. Modified Hutchinsonian model showing niches of *Aspidomorpha miliaris* and *A. sanctae-crucis* with reference to two environmental variables. A description of the model is given in the text (from Verma and Shrivastava, 1986).

on either side of the arbitrarily chosen line of demarcation between more and less humid areas along the horizontal axis. On each axis, the broken line has been taken as marking 0 and values have been shown in ascending order on either side of this line.

The niches for the two cassidine species presented in this way show only 3.34% overlap.

The following matter describes another example of niche separation. As was pointed out in the section in this chapter on semiaquatic chrysomelids, in Central America some hispines live in the floral bracts of *Heliconia*, e.g. *Cephaloleia puncticollis* (Baly) and *C. neglecta* Weise (tribe Cephaloleiini). *Xenarescus monoceros* (Olivier) (tribe Arescini) also lives as young larvae in the water-holding bracts of *Heliconia* but these larvae, due to competition with *Cephaloleia neglecta*, leave the bracts after they (the larvae) have grown to some extent, and move on to rolled leaves. Some species of *Cephaloleia* also feed on the rolled leaves of *Heliconia*. Jolivet (2002, in print) points out "Leaf-feeding *Cephaloleia* feed near the edges and Arescini tend to feed away from the edges, probably to avoid competition".

In the First International Symposium on the Chrysomelidae (1984), Selman (1985) presented a paper on the colour separation among paropsine chrysomelid species living on eucalyptus in Australia. He pointed out that several species of these beetles live together on the same host plant. During oral presentation of the paper he mentioned that he could note several instances of niche separation among the chrysomelids. If these cases of niche separation could be studied in some detail, some fresh aspects of the biology of the leaf beetles may come to light.

At the end of this section it must be mentioned that an overlapping of niches to some extent is not unexpected. An abundance of food in the habitat may be responsible for the overlapping of niches of those populations which exploit the same food source (Ananthkrishnan and Viswanathan, 1976).

Diapause

Many insects are known to enter a period of dormancy under the influence of climatic and other environmental changes. This state of dormancy is referred to as 'diapause'. In many instances, diapause is clearly a device to bridge over an adverse environmental phase.

Diapause means a stage in development when changes and growth become retarded or fully arrested, and severe environmental conditions can be withstood. The intensity of changes in diapause is variable. This will be discussed in the following subsection.

For an extensive and very useful review on diapause in Chrysomelidae, see Cox (1994c). This section is a brief account based on this review.

TYPES OF DIAPAUSE

Diapause may occur at any stage in insect development. We can distinguish between egg or embryonic diapause, larval diapause, pupal diapause and adult or imaginal diapause. In adult diapause, there is a repression of development of the reproductive cells in the gonads. Hence, it is also referred to as reproductive diapause. In Chrysomelidae, diapause is known to occur in the egg stage, later larval instars, in the prepupa, and in the adult. As in Coleoptera, pupal diapause is unknown in Chrysomelidae.

Müller (1966) classified different types of diapause according to the factors responsible for inducing or terminating them. Thiele (1973) presented a *résumé* of Müller's scheme.

Müller's scheme is not easy to apply as the factors which determine the onset and end of a diapause are very variable and, in most cases, have not been adequately investigated.

Another scheme for the classification of diapause types was presented by Mansingh (1971). It is based on the intensity of changes in diapause, and is more practical in application. In this scheme, the following different diapause types were described:

1. Quiescence. This is developmental retardation in response to short-term, non-cyclic, sudden and unexpected changes in one or more environmental factors. Such changes may appear in any part of the year. Termination of this type of dormancy is relatively simple and immediate when the environmental factors return to normal. It is probably experienced by all species, and no physiological preparation is needed for entering this dormancy phase.
2. Oligopause. This is developmental retardation or arrest in response to adverse long-term and cyclic environmental conditions, which appear more or less

regularly at a particular time of year. Locomotor and feeding activity become obviously reduced. Some physiological preparation seems necessary for entering this sort of dormancy. If the environmental changes inducing this type of dormancy are not severe, oligopause may be avoided. In other words, oligopause is facultative.

3. Diapause. The main difference between oligopause and typical diapause is that the latter includes a genetic anticipatory mechanism, and it sets in considerably before the environmental adversity appears, or, when it appears, it is much milder than the one necessitating diapause. In other words, this kind of diapause is obligatory. Its termination is a lengthy process.
4. Aestivation. While winter cold is one of the factors relating to the other types of diapause, aestivation is induced by summer heat. Aestivation may also be referred to as 'summer dormancy' and varies in intensity. Instances of aestival diapause may be comparable to quiescence, oligopause or typical diapause.

With reference to environmental temperature, we may distinguish two types of diapause: hibernal diapause (related to winter conditions) and aestival diapause (related to summer conditions).

SOME INSTANCES OF DIAPAUSE AMONG CHRYSOMELIDAE

Details of diapause are known in only a small number of species. The phenomenon has been studied mainly in economically important species. A few examples of chrysomelid diapause are described briefly here to illustrate different types of diapause.

Mediterranean species of *Timarcha* exhibit what has been called 'aestival pseudodiapause' (Cox, 1994c). During a dry summer, adults of such species, e.g. *Timarcha balearica* Gory, become inactive in dry regions but, if a summer is humid due to rain, this does not occur (Jolivet, 1965). In Morocco during the summer, the insects become active immediately after rain and the consequent revival of food plants (Cox, 1994c). Obviously, this is a case of aestival quiescence.

In central India, the cassidine *Aspidomorpha miliaris* (Fabricius), which is a multivoltine species and lives on the weed *Ipomoea fistulosa*, undergoes imaginal diapause both in summer (mid-February to mid-July) and in winter (December to January). In the former diapause, movements and feeding are greatly reduced in the adult. There is no copulation or egg-laying. The development of body fat can be so extensive that, in dissection, the alimentary tract and other internal organs may be difficult to locate. The ovaries can be much regressed (Bhattacharya and Verma, 1982). The ovariole has a smaller germarium and only one or two small egg chambers (*Figure 6.8*). In section, the last egg chamber shows an oocyte breaking up into fragments, and also vitellophagy by follicular cells. The testes also suffer reduction in size, but not as much as the ovaries (compare dimensions of diapause and non-diapause gonads given in Mohan and Verma, 1981 and Bhattachrya and Verma, 1982). Later investigations have shown that testis follicles also undergo considerable regression in the summer imaginal diapause, though a compensatory increase in the thickness of the *tunica externa* of the testis follicle makes the testis look not much smaller than a non-diapause testis (*Figure 6.9*). It may be noted that the *tunica externa*

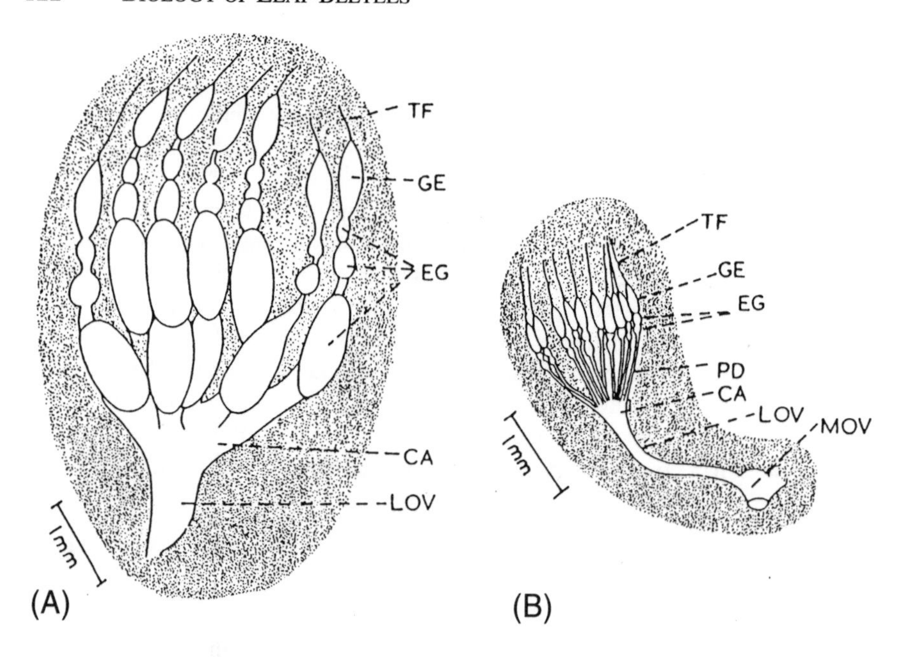

Figure 6.8. *Aspidomorpha miliaris.* (A) an ovary of a non-diapause adult female; (B) an ovary of a summer diapause adult female (from Bhattacharya and Verma, 1982) (CA = calyx; EG = egg chamber; GE = germarium; LOV = lateral oviduct; MOV = median oviduct; PD = pedicel; TF = terminal filament).

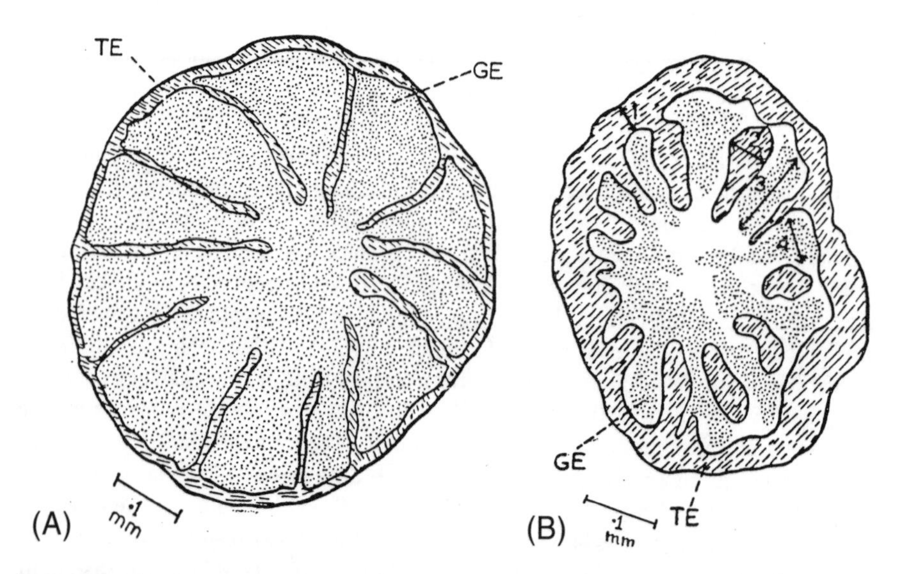

Figure 6.9. *Aspidomorpha miliaris.* (A) section of a testis follicle of a non-diapause adult, approximately parallel to the flat surface of the follicle; (B) similar to (A) but of a summer diapause adult (from Verma, unpublished) (GE = germ cells; TE = *tunica externa*).

is made up of a fat body-like tissue, and, in diapause, body fat generally shows hypertrophy.

In winter imaginal diapause in *A. miliaris,* the repression of gonads is almost as much as in the summer diapause. Hypertrophy of body fat is also almost comparable. Feeding is reduced, but much less than in summer diapause. In fact, it is only a little less than in non-diapause individuals. Winter diapause individuals are almost as active as non-diapause beetles. Thus, while feeding and activity are not much affected in winter diapause, gonads become much repressed. This situation has been referred to as 'gonotropic dissociation' (de Wilde and de Loof, 1973).

In *A. miliaris,* the summer diapause, which is the more marked of the two adult diapauses, is terminated soon after the start of the rainy season, and normal feeding and oviposition are resumed. The rainy season in central India is from the middle of June to September, during which period is concentrated about 80% of the annual rainfall. In the laboratory, summer diapause beetles have been kept under rainy season-like conditions (temperature 30 to 40°C, photoperiod 14 hours and a relative humidity of 90 to 95%) in May. Such adults showed complete recovery from diapause. Copulation started and eggs were deposited in typical oothecae (Verma, unpublished). These observations suggest that the summer diapause in *A. miliaris* is facultative and of the nature of aestival oligopause.

The observations of Nakamura and Abbas (1989) also support the inference that the summer diapause in *A. miliaris* is facultative. These authors studied the biology of this beetle in Sumatra on the same plant as in India, namely *Ipomoea fistulosa* (= *Ipomoea carnea*). The island has a humid equatorial climate. In such a climate it remains warm and humid throughout the year, as it rains all the year around. The authors noted "In contrast (from observations of Bhattacharya and Verma, 1982), in Padang where no distinct dry season occurred, the two species (*Aspidomorpha miliaris* and *A. sanctae-crucis* (Fabricius)) were active throughout the year, although abundance of adult beetles and oviposition activity declined occasionally during the present study". Thus, *A. miliaris* does not undergo any diapause in Sumatra.

An example of obligatory diapause is found in adults of *Chrysomela collaris* L., which eclode in August, do not have mature ovaries even in September, and show reduced oxygen consumption. This condition continues if they remain at 21°C. Their reproductive diapause is over if they are kept at 0°C for 19 weeks and then returned to 21°C (Meidell, 1983). Oviposition starts within 10 days after the adult females are returned to 21°C. The diapause in this case seems to be in anticipation of the approaching winter and is well before the period of climatic adversity starts.

Cox (1976), after Abeloos (1935, 1937a,b) in France, has described obligatory egg or embryonic diapause in *Timarcha tenebricosa* (Fabricius) in England. When eggs are kept at 19°C, a fully formed embryo is developed in 42 days, but hatching will not occur if the eggs continue at this temperature. For successful hatching, a chill treatment is necessary. If eggs with completely formed embryo are chilled to 2°C, if they remain at this low temperature for 30 days and are then taken to 20°C, 14 days of exposure to this higher temperature leads to hatching. *Timarcha goettingensis* (Linné) in France (Abeloos, 1933b, 1937a,b, 1938a,b, 1941) has a much more complex system of diapause in winter and shows an inconstant embryonic diapause. *Chrysolina hyperici* (Forster) and *C. quadrigemina* (Suffrian) used for the biological control of St John's wort (*Hypericum perforatum* L.) in New Zealand enter their summer diapause at the beginning of January (Schöps *et al.*, 1996). *C. hyperici* overwinters in the egg stage and *C. quadrigemina* in the more vulnerable larval stage. So, the last one is more successful in controlling the weed in countries with mild winters since its larvae feed and grow during that period.

Hormonal control of diapause in Chrysomelidae

As with other insects, for Chrysomelidae there is evidence to show that diapauses are under hormonal control. In *Galeruca tanaceti* (Linné) there is a low level of activity in the neurosecretory cells of the brain during imaginal diapause, and there is also reduced activity in the corpus cardiacum–corpus allatum complex. When diapause is terminated, there is a distinct rise in the activity level of these endocrine centres (Siew, 1965).

In *Oulema melanopus* (Linné), the adult diapause can be terminated and oviposition may be induced by the topical application of farnesenic acid, which is a juvenile hormone mimic (Connin *et al.*, 1967).

De Wilde and his team demonstrated the role of the juvenile hormone in the termination of adult diapause in *Leptinotarsa decemlineata* (Say). They found that feeding on mature foliage tends to induce the diapause. If *corpora allata* from beetles feeding on young leaves are transplanted into adults feeding on mature foliage, egg production is promoted, even when the females continue to feed on old leaves (de Wilde *et al.*, 1969).

7

Biogeography. Island Faunas

Chrysomelid beetles are distributed almost throughout the world, except in the Antarctic, where only a few insects survive and, of course, no beetles. The other exceptions are Greenland and the subantarctic islands, where only certain weevils and carabids live. Leaf beetles are missing from most volcanic islands, but were introduced to some by human activity or by storms. In this last case, only lightweight species are present but even they can have produced endemics if they have been around for a very long time. In Greenland, leaf beetles which were abundant in the Tertiary and early Quaternary, disappeared with the glaciations. A few survived in the northern Siberian archipelagos, Alaska and a few islands of the Eurasian and American Arctic. A few species, probably galerucines and alticines, may exist in Spitzberg but so far we do not have precise knowledge of this. Strand (1942) and Kangas (1967) mentioned one *Gonioctena* sp., found on *Salix* (as well as staphylinids, cucujids, cryptophagids and curculionids). There must be other species but the coleopteran fauna in Spitzberg is 'exceedingly scanty'. Weevils survive in Greenland and leaf beetles will probably be reintroduced some day with crops. These would be common species feeding on Brassicaceae or Apiaceae. Glaciers are melting there and warming seems general.

Some interesting biogeographical theories have been put forward in the last century (Joleaud, 1939; Jeannel, 1942; Croizat, 1961; and many more). The theory of land bridges seems totally unrealistic and Jeannel imagined too many basaltic connections between volcanic islands and the mainland. However, Jeannel, despite the criticism of other entomologists and arachnologists, was a forerunner in promoting the Wegener theory of continental drift. Nowadays, this theory of plate tectonics has been accepted everywhere and remains the only plausible explanation for the actual distribution of insects throughout the continents. Millot's criticism, based on the distribution of spiders in Madagascar, is unacceptable because we are dealing here with passively distributed arthropods. The distribution of large, ancient terrestrial apterous beetles, like *Timarcha*, carabids or *Meloe*, is important since they could not have been affected by storms. A man-made invasion of millipedes or tenebrionids in areas such as the Cabo Verde archipelago is another, more recent phenomenon.

We should also mention here that Pathak and his team (Pathak *et al.*, 2001) collected and studied airborne insects above the Bay of Bengal, the Arabian Sea and the southern open part of the Indian Ocean, 30 to 700 km away from the nearest land mass in wind direction. Insects can rise to great heights with thermal currents and may drift long distances. Pathak and his group found 8 species of Chrysomelidae in their collections. The body length of the collected leaf beetles measured 1.2–6.2 mm.

The chrysomelids probably split from a common 'trunk' along with the cerambycids and the bruchids in tropical conditions during the Jurassic. It is very difficult to pinpoint their exact origin. Botanists believe that the Angiosperms developed slowly in tropical mountainous areas and then spread over the planet before and during the Trias. Jeannel's theory (1942), which was largely based on assumptions, placed the origin of many beetle genera in what he called Angaria, i.e. Central Asia, and he believed that glaciation destroyed their many groups. In fact, Europe and many other areas endured hot climates at various periods. It is impossible to guess the origin of a given group. Transatlantic migrations could have taken place over ancient land mass connections. Madagascar has been linked in this way with Africa during the Mesozoic, and Australia with New Guinea comparatively recently. Many changes occurring during the Mesozoic and the Tertiary were responsible for fauna exchanges and island speciation. During its hot periods, the Antarctic was also an important migration corridor, but no beetle fossils have yet been discovered.

Very few papers have been published on the biogeography of chrysomelids.

Timarcha is an archaic genus, a relic from the Jurassic. It is possible that it had wings at that time but then rapidly lost them. A proof that this is an ancient apterism is that the pupa itself is apterous. This is rare among beetles and shared by only a few species, including *Meloe* and certain tenebrionids. Rüschkamp (1927) claimed to have found vestigial wings (scales) among rare specimens of *Timarcha tenebricosa*. Its nervous system is one of the most primitive among the chrysomelids.

Timarcha probably originated in the central Asian steppes and *walked* towards Europe, North Africa and North America. This occurred during the end of the Mesozoic and in the Cenozoic and was certainly a very slow process. Quaternary glaciations seriously reduced the primitive distribution as the genus is thermophilic and cannot survive beyond 50° north. By contrast, *Chrysolina*, becoming apterous and ovoviviparous or viviparous, lives and reproduces between 78° and 79° north in the polar islands of Siberia. The transatlantic migrations probably took place during the Eocene. *Timarcha* did not survive in the Baltic States, Denmark, Scotland, Canada, except at a few places north of Idaho, in British Columbia, and Vancouver Island, which was not totally covered with ice during the glaciations (the western part was spared). *Timarcha* survives today in western areas of the USA such as Oregon, Washington, Idaho, California, Montana and probably other places like Colorado. Since it is nocturnal, it will have largely escaped collection. It has disappeared, however, from the rest of the USA, including the Appalachian mountains, where its host plant grows and good climatological conditions exist. We think that it survived on the Pacific US coast because this area had a much warmer climate than the east during the early Quaternary. The absence of the genus in Japan, Siberia, Korea and China is probably due to the Pleistocene glaciations which impoverished central Asia. Although *Timarcha* penetrated 80 km down into the oases of the Libyan desert, its actual distribution (absent in Egypt, in Sinai, in southern Libya and in the Syro-Lebanon area) is certainly due to the desertification of the area in the recent past. One isolated area where it does exist is Cyrenaica. Information on its recent presence in western Syria (*T. amesthystipes* Chevrolat or *T. laevigata* (L.)) probably results from a confusion between Tripoli (Libya) and Tripoli (Syria) (Jolivet, 1996b).

Timarcha was very probably circum-Mediterranean, African species being in the south and south-east and European species in the north and north-east (Turkey). The *T. olivieri* group, *T. metallica* subgenus survives in Anatolia, and *Timarcha* probably reached the Iranian border. *T. olivieri* and related groups are European, as are all forms from Turkey, Caucasia and Azerbaidjan. *Timarcha* is in decline everywhere because of urbanization, road building, forest destruction and the use of insecticides but it has been extinct for a long time in the Maltese islands which were linked with Sicilia during a great part of the Cenozoic (*T. melitensis* Weise). It has survived in the larger western Mediterranean islands (Majorca, Minorca, Corsica, Sardinia, Sicilia) and in few small islands surrounding Sicilia (Egades), Corsica and Sardinia. It has disappeared (or never existed) in the islands of Iviça, Formentera and Cabrera, and it is declining quickly due to urbanization in the big islands of the Balearic group. It has survived in small islands off the eastern coast of Spain and in some Atlantic and Channel Islands (Chausey, Guernsey, Jersey, Aurigny, etc.). The islands of the eastern Mediterranean (Crete, Cyprus, Aegean Islands, Rhodes) have lost their *Timarcha*, population, if, indeed, they ever had one. *Timarcha* never reached the Elburz mountains in Iran, and the prospects for its presence in north-western Iran are rather poor. The genus has also disappeared from the western part of Turkey (Izmyr) and is restricted to medium-altitude mountains in central and northern parts of the country. With climatic warming, the potential geographical range of summer and winter temperatures suitable for *Leptinotarsa decemlineata* may extend into Britain as far as the north of England (Joffree and Joffree, 1996). A similar scenario is possible for *Timarcha*, but a wingless species is not so mobile.

Old World species of *Timarcha* feed exclusively on *Galium* and various Rubiaceae, on *Plantago* and on other plants such as members of the Brassicaceae, Scrophulariaceae, Dipsacaceae and Asteraceae for some rare species. In the New World, they feed on Rosaceae (*Rubus* and *Fragaria*) and also (actually in study, Poinar *et al.*, 2001) on various Ericaceae (Jolivet, 1989a). This seems to reinforce the old description of *Vaccinium* and *Erica* (Ericaceae) as a food plant of *Metallotimarcha* in Europe and so, after all, there is some truth in the old reports (Kuntzen, 1919; Jolivet, 1999a). A common chemical denominator may exist among the food plants selected by *Timarcha*. This remains to be discovered. Phylogeography and speciation in leaf beetles, including *Timarcha* species have been studied by Gomez-Zurita *et al.* (1999, 2000a,b) by applying a cladistic analysis using mitochondrial DNA. It seems that there is a significant relationship, for example among the *Timarcha goettingensis* complex, between genetic structuring and geography. According to the authors, oligophagy, mountainous habitat and apterism are factors which have encouraged speciation in these beetles. This is true for European, north-west African and American species but remains to be researched in Africa and North America, where the species are probably more numerous than is actually known.

Another apterous tropical American genus, *Elytrosphaera*, is related to *Leptinotarsa*. Its distribution was studied by Jolivet and Vasconcellos-Neto (1993). The subgenus *Elytrosphaera* s. str., which contains large species, is distributed on the south-east of the Brazilian plateau at altitudes varying from 700 to 1700 m. One exception, the black species of the oriental side of the Andes in Bolivia (*E. melas* Jolivet), is found at 2000 m. The subgenus *Elytromena* Motschulsky which is often elegantly coloured, sometimes in zig-zags, seems restricted to the mountains of southern Mexico, Ecuador, Columbia and Bolivia. It reaches the snow level in Ecuador and one species, *E. nivalis* Kirsch which appears to be black, is actually metallic. *E. cartographica* Bechyne reaches 3550 m and *E. nivalis* 3700–4150 m on the volcanoes. The last

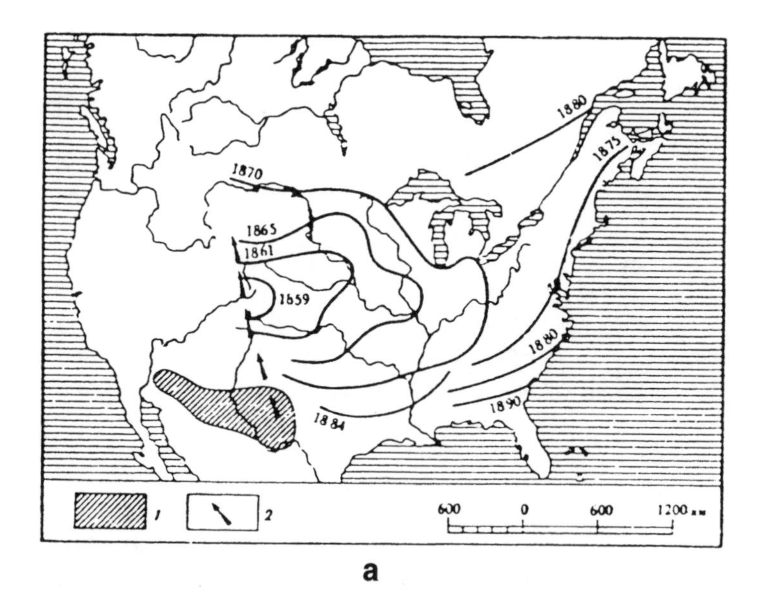

a

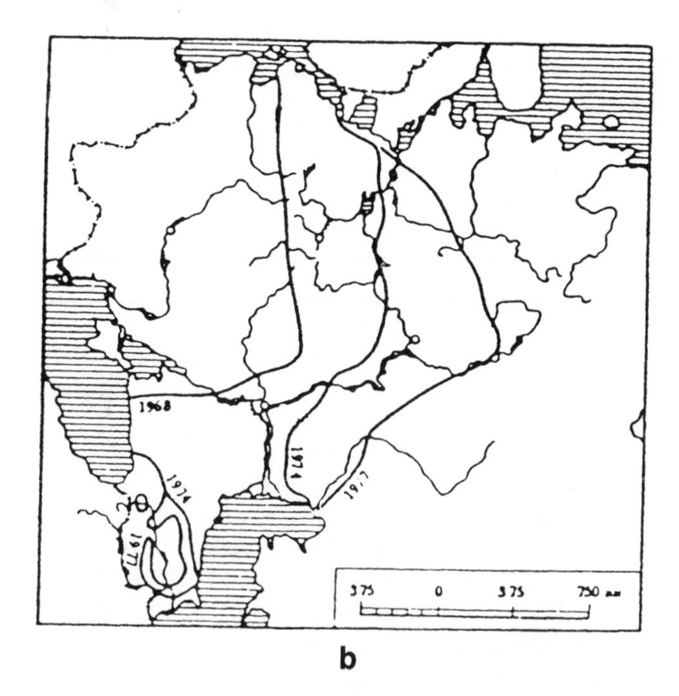

b

Figure 7.1. Biogeography. (a) Progressive invasion by the Colorado Potato Beetle (CPB) in the USA from a theoretical focus in Mexico (hatched area) (after Sokolov, 1981); (b) advance of CPB towards Asia from 1968 to 1977 and the passing of Ural. The CPB has now reached the gates of China (after Sokolov, 1981); (c) successive invasions of Europe by CPB since 1871 and the definitive invasion from Bordeaux (hatched area) in 1922 (after Sokolov, 1981); (d) distribution of the genus *Timarcha* (Chrysomelinae) in Europe, Northern Africa and the Middle East.

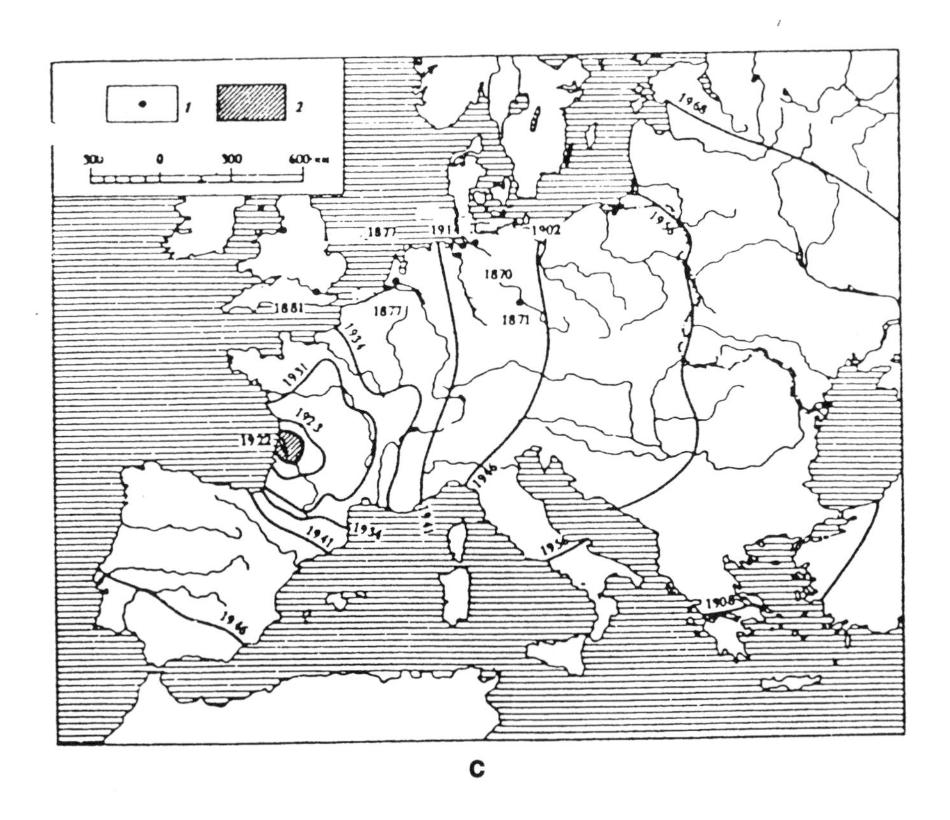
1
2
0
1968
1877
1914
1902
1870
1871
1881
1877
1934
1931
1923
1922
1941
1934
1946
1956
1908
1966

c

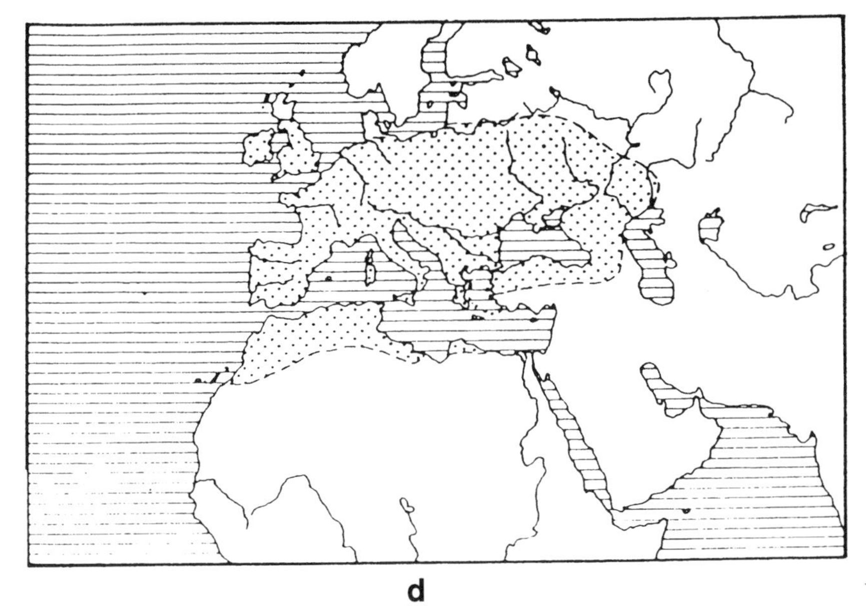

d

species is black in colour. This characteristic is, like apterism, an adaptation to high altitude. The genus *Elytrosphaera* is now endangered, due to the destruction of its natural forest habitat on the Brazilian plateau. It feeds on various Asteraceae (*Adenostemma*), which are also in decline through forest destruction, and on Solanaceae (*Solanum*). Plant and beetles have disappeared, forest fires destroying what was left. We do not know for sure on what *Elytromena* species feed on the volcanoes, but the host plant is probably a member of the Solanaceae.

The biogeography of Alticinae was studied by Scherer (1988). Although the hypotheses offered by him are always questionable, his ideas remain reasonable. Scherer starts with an estimate of 8000–10 000 species and 500 genera of Alticinae, most of them actually described. As Alticinae have probably evolved from the Galerucinae and as the two groups are sister groups, it would probably have been better to consider the phylogeny of both subfamilies together. Scherer deals in his paper only with Alticinae.

Scherer starts from the Mesozoic, 100 million years ago with the breaking of the Gondwanaland. It is in tropical America that the greatest diversity exists. More than 200 genera are known there, against 65 in Africa. Almost all American genera are endemic, except *Chaetocnema*, *Epitrix*, *Terpnochlorus* and *Longitarsus*, which have a large distribution around the world. This rich South American fauna presumably had a very long isolation, probably 100 million years. India has also been isolated for a long time and has 60 endemic genera and 20 more, distributed in the southern islands. The reason for an impoverished Australian fauna seems to be the cold climate during the Palaeocene when Australia was farther south. The impoverishment of the European fauna is also attributable to the cold climate during the Pleistocene which destroyed many genera belonging to various families. Scherer thinks that host plants are responsible for the gradual speciation of certain species. He also believes that, before the end of the Jurassic, alticines already had the enlarged femora which allowed them to jump, as this phenomenon was widespread throughout the continents at this time. Alticines and galerucines were already separated at that early period, and in spite of this, cladists tend to amalgamate them in one subfamily.

The case of the archaic Sagrinae, with its biggest concentration in Australia, is a good example of Gondwanian distribution. The remaining species are found in Madagascar, Argentina, and north-east Brazil. The genera *Coolgardica*, *Ametalla*, *Mecynodera*, *Atalasis*, *Neodiaphanops*, *Pseudotoxotus*, *Polyoptilus*, *Diaphanops*, *Duboulaya*, *Carpophagus* and *Megamerus* are very primitive. They have cerambycid characters and even bruchid ones, in the case of *Carpophagus, Megamerus* is distributed in Australia, Madagascar and in north-east Brazil. In the last part of its range, it is represented by a very small species: *Megamerus alvarengai* Monros, which probably hatches, after the annual heavy rains, from a pupa enclosed in a gall. These genera are probably all gallicolous and stem-borers, but there may be some exceptions. Madagascan *Megamerus* seems to construct an ootheca around the stem of a plant but we cannot be certain as no reliable observations have been made on these Sagrinae. Sagrinae adults eclode in the humid season, all together in an area, after the first rains. The adults of many genera, including *Megamerus*, are attracted by light traps but nothing more is known of their peculiar biology.

In Argentina, *Atalasis sagroides* Lacordaire bores inside Malvaceae stems. It produces galls and probably pupates inside the stem. Megalopodines, also stem-borers, go into the ground to pupate. They are not gallicolous. It seems strange that no primitive sagrine survived in South Africa. This must be due to extinction, although climatic conditions very similar to the ones prevailing in Australia and South America seem to have existed in Southern Africa for a very long time. *Araucaria* trees vanished also from South Africa and with them the specific fauna (Palophaginae), while the subfamily survived in Australia and South America. The distribution of the Stenomelini (Eumolpinae) in western Chile and New Caledonia, which feed on the Myrtaceae, seems really surprising (Jolivet, 1957). As far as we can tell, the biology seems similar in New Caledonia, and Fauvel, when he discovered the species in 1907, named it *Stenomela caledonica*, a name that Jolivet kept when he described the wings. Unfortunately, part of the Fauvel manuscript (1907) was lost and never published. Jerez (1995, 1996) described the biology of this Gondwanian relic in Chile. Monros (1958) described the New Caledonian genus as a *Bohumiljania* and the species becomes *Bohumiljania caledonica* (Jolivet).

The chrysomelid fauna of islands follow MacArthur–Wilson laws (1967). Roughly, the fauna of oceanic or continental islands is proportional to their surface area and inversely proportional to their distance from the adjoining continent. Continental islands have a tendency to lose their fauna progressively because of anthropic factors. For the oceanic (volcanic) islands, colonization results from elements disseminated at random and proportional to the time elapsed (Simpson, 1965b). The oldest island contains the largest quantity of insects if its surface area is large enough to keep and feed the species. The biodiversity of the flora also has a part to play in the survival of certain species which are very specific about their host plants Diversity in volcanic islands is evidently very small, colonization being carried out only by winds and in the tropics by floating rafts, a rather slow and exceptional means of dispersal. Human plant introductions also play a part in the diversity of archipelagos such as the Cabo Verde islands where phytosanitary control is nil. It is evident that only lightweight forms are transported by winds and that these introductions are by chance and are not related to an abundance in the country of origin. Polyphagous or parthenogenetic species have more chance to adapt and maintain themselves in new areas than monophagous or bisexual ones. To reproduce, a newcomer must arrive with both sexes, or as a fertilized or parthenogenetic female and find an adequate food plant.

In the Mascarenes group, La Réunion, Mauritius and Rodriguez, the chrysomelid fauna is impoverished compared with Madagascar. It also consists only of lightweight species (Jolivet, 1979). It is a disharmonic fauna: Criocerinae, Cryptocephalinae, Eumolpinae, Galerucinae, Alticinae, Hispinae and Cassidinae. There are no heavy-weight Sagrinae or Megalopodinae and no Chrysomelinae. This fauna certainly originated from Madagascar. The small number of species, introduced through time, adapted themselves to local plants, which sometimes differ greatly from the original ones. Often being polyphagous, Eumolpinae adapted to the local flora, but Sagrinae and Chrysomelinae, being heavyweight and poor flyers, are missing here, as in most of the oceanic islands. Clytrinae are also missing. Their complex cycle with the ants and the problem of their weight are other obstacles in the way to reaching and colonizing volcanic islands. There are, however, exceptions such as Pantellaria, near Sicily. Clytrinae are rare even on continental islands (Madagascar, Australia, New Guinea), although they are good flyers. Perhaps adaptation to ants is more recent than previously thought.

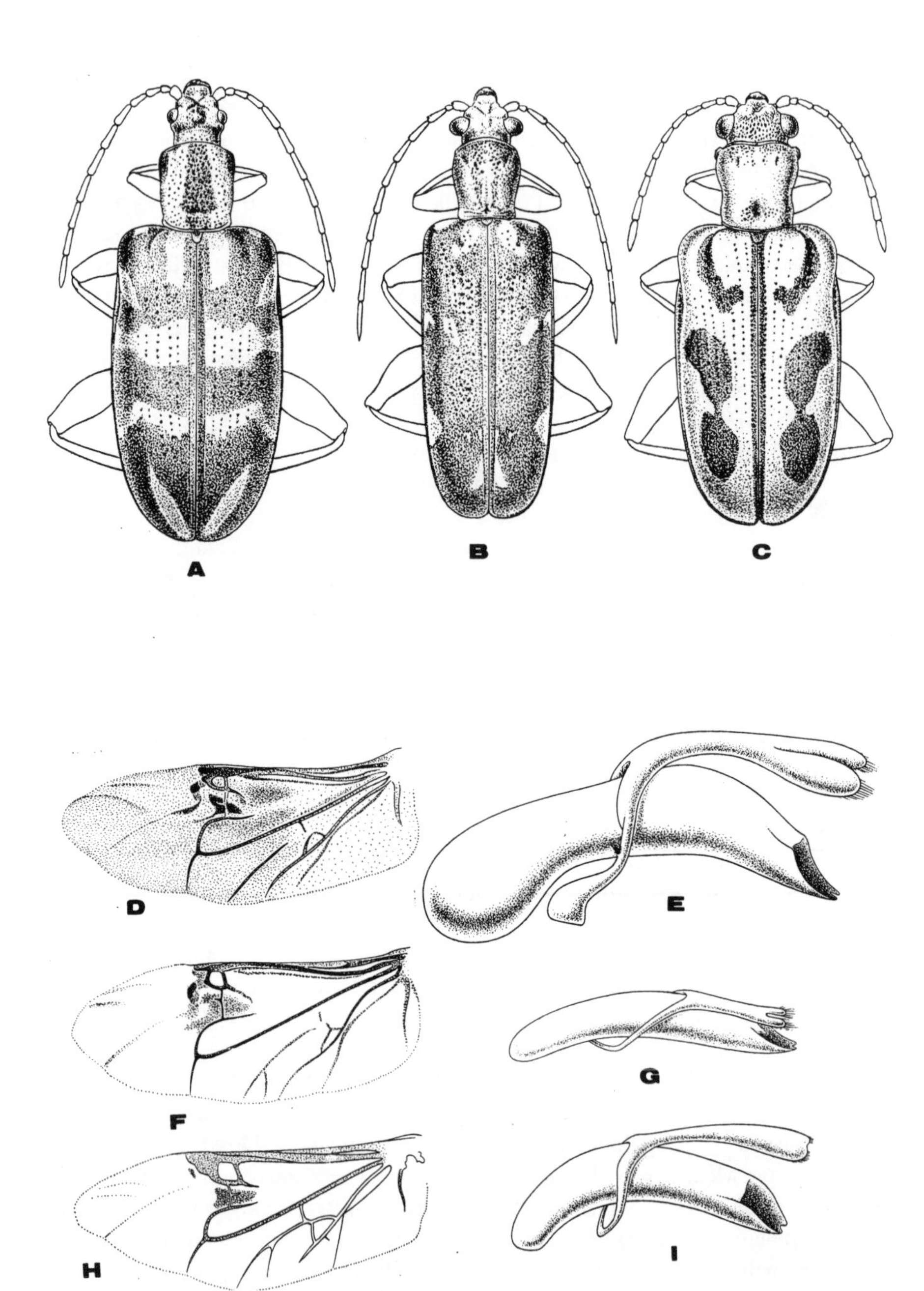

Figure 7.2. Three Gondwanian Megamerini Sagrinae. (A) *Mecynodera balyi* Clark, 1864, from Southern Australia; (B) *Megamerus madagascariensis* Chapuis, 1878, from Madagascar; (C) *Atalasis sagroides* Lacordaire, 1845, from Argentina; (D) *Macynodera balyi* Clark, hindwing (× 6.5); (E) Id. aedeagus (× 45); (F) *Megamerus madagascariensis* Chapuis, hindwing (× 4); (G) Id. aedeagus (× 14); (H) *Atalasis sagroides* Lacordaire, hindwing (× 10); (I) Id. aedeagus (× 36) (after Jolivet, 1995).

The differentiation of the Eumolpinae in the Mascarenes is very important. It can be explained by an evolution and an 'explosion' starting with two or three ancestors and can be compared with the 'explosion' of weevils of the genus *Cratopus* Schönherr in the whole of the Mascarenes. Eumolpines as alticines are small species, quickly adaptable to a new host. It is likely that some alticines (*Chaetocnema confinis* Crotch), hispines and rare cassidines have been imported in recent years on plants brought by man. *C. confinis* came from North America, where it is bisexual. It feeds on very common cultivated plants (*Ipomoea aquatica* and *I. batatas*) and has invaded Madagascar, Mauritius, the neighbouring African continent and is now invading south-east Asia and Pacific regions such as Taiwan, Japan, India, Vietnam, Thailand, Palau, Moorea, Hawaii, etc. It has also been introduced in the Galapagos Islands. *C. confinis* will soon be a pantropical species. Its food plant is everywhere, the beetle is lightweight and easily transported by wind, and, an enormous advantage, the female can become parthenogenetic.

It is, however, surprising that in only 10 million years, eumolpines, alticines and weevils of the genus *Cratopus* evolved into so many species in the Mascarenes. The fauna composition is similar to the Seychelles and richer than that of the Cabo Verde archipelago, another disharmonic population. It is possible that the Seychelles were linked temporarily with the African continent. This did not happen with the Mascarenes. In volcanic islands like the Mascarenes, it seems that a genus is imported every 450 000 years. Diversification was more apparent for the two subfamilies imported first. Many of those species (*Cratopus*) are actually extinct with the use, in Mauritius mainly, of insecticides. In a way, Mauritius (1865 sq km), the less diversified topographically, is the oldest and was probably the cradle of the endemism. In contradiction of the MacArthur–Wilson island theory (1967), it is the midddle-sized island Mauritius, more distant from Madagascar than La Réunion, which shows the bigger diversity. The contradiction is obvious: La Réunion is bigger (2512 sq km) and more diverse in topography and vegetation, though the island is younger.

Mauritius is 10 million years old, La Réunion 3 million, and Rodriguez, the smallest (109 sq km) is the youngest (only 1.5 million years old). The biodiversity on these islands is in proportion to the age.

Another example of volcanic islands is found in the Cabo Verde archipelago. Colonization occurred there with the help of winds and, for some groups (Tenebrionidae), by man's intervention. This does not explain, however, the enormous endemicity there among certain beetle families. However, there is practically no endemic species among the chrysomelids. The fauna there is in greater disharmony than in the Mascarenes (Jolivet, 1985; Scherer, 1986; Geisthard and Von Arten, 1992). Roughly, there are 4 Criocerinae, 1 Cryptocephalinae, 1 Cassidinae, 1 Galerucinae and 11 Alticinae. We can see that Hispinae, Chrysomelinae and Clytrinae are missing. These last two subfamilies are characteristic of continental islands. Some of these species, namely some crop Alticinae, may have been introduced with plants since they seem to be recent arrivals.

Oceanic islands, smaller in surface area and more isolated, are poorer in leaf beetles and some have none at all. Saint Helena contains only three species of *Longitarsus*, all endemic, and Pascua Island in the eastern Pacific has only one species of *Diabrotica*, a crop pest imported from neighbouring Chile. The Hawaiian Islands have no native leaf beetles, except rare ones imported with the crops. Tristan da Cunha, in the southern Atlantic, has only one species, one alticine, *Stegnaspa trimeni* Baly, probably imported by man or winds from South Africa (Voisin, 1980; Jolivet, 1998c). As we can see, most of the isolated islands in the eastern Pacific have no original fauna, and the fauna from New Guinea towards Fiji and Samoa diminishes progressively from west to east to contain only alticines and eumolpines at the end, but endemics do exist there. Clipperton Island, for instance, in the extreme east, has only 4 beetles: one tenebrionid, one coccinellid, one staphylinid and one cicindelid (Sachet, 1962). However, the survey was so superficial that we could expect there to be more. No leaf beetle seems to inhabit this islet. As it is uninhabited, there are no crops to attract leaf beetles and no plants have been imported.

The Galapagos Islands, volcanic in origin but with lava emersions (stepping stones), seem to have allowed for a more easy penetration of continental species and a small local and interesting fauna has survived. American entomologists have adopted Jeannel's theory about basaltic bridges with the continent. All the insects of this archipelago were imported from Ecuador during the past but endemics had time to differentiate, as in Cape Verde for example (Van Dyke, 1953).

In Galapagos we can find one eumolpine, *Metachroma labiale* Blair, one galerucine, *Acalymna limbatum* (Waterhouse), and four alticines, *Docema galapagoensis* (Waterhouse), *D. darwini* Mutchler, *Longitarsus galapagoensis* (Waterhouse), *L. lunatus* (Waterhouse). All are lightweight insects, easily dispersed by storms. The North American alticine, *Chaetocnema confinis* Crotch, which feeds on *Convolvulus* and *Ipomoea* spp., may also be added. It is spreading all over the tropics of the New and the Old World. The insect, in its bisexual form, is common in 40 states in the USA and in 7 provinces in Canada. It extends its distribution through the tropics, thanks to its ability to become parthenogenetic. Evidently, it has been recently imported into the Galapagos.

Sometimes we get some surprises. One alticine, *Argopistes gourvesi* Samuelson, has been discovered recently in Tahiti, an island supposed to be devoid of chrysomelids (Samuelson, 1979). *Argopistes* spp. are associated with Oleaceae (privet, *Olea*, lilac, etc.) and an importation is always possible. However, their distribution through the Pacific islands is a well known fact. All other Tahiti and Moorea chrysomelids (Chabrol, 1995) have been imported: 2 Alticinae, 1 Cryptocephalinae and 1 Hispinae. *A. gourvesi* could be an authentic endemic.

The Azores, another oceanic archipelago in the Atlantic, has been populated in the same way as the Cabo Verde archipelago, also a volcanic group. Its fauna resembles a European fauna, with several South American imports like *Calligrapha polyspila* (Germar). There, we find one *Cryptocephalus*, three *Chrysolina* (rather surprising for oceanic islands), including *C. banksi* (F.), heavyweight and brachypterous species, with winged forms unable to fly, and 12 Alticinae (Borges, 1990–1992). As in all oceanic islands, the majority of species are lightweight forms, which can fly. Rare, heavier forms could have been imported in the past by human intervention or by means of floating rafts (rather an unlikely theory). The Azores cannot be compared with the Canary Islands or Madeira with a high endemism and a pure Mediterranean fauna. Madeira has 36 species of leaf beetles, including eleven endemics (Erber, 1984). *Chrysolina* spp. are endemic in both the Canaries and Madeira, but both faunas are slightly disharmonic, even though they were probably once connected with the continent. The Balearic Islands are also disharmonic, even though they are continental islands connected during the Pliocene with Spain and probably other Mediterranean islands.

The Balearic fauna is very rich on the two main islands, Majorca and Minorca, but is less so on Ibiza, namely with the loss of a *Timarcha*, but still with *Chrysolina banksi* (Fabricius). Formentera has lost *Timarcha* and even *C. banksi* (Sacares and Petitpierre, 1999). Cabrera has a very poor fauna, consisting only of alticines (Jolivet, 1953). Biodiversity there is well in proportion with the size of the islands, but probably also with the history of land connections. It is notable that Isabel II, a small island east of Melilla, on the Moroccan coast, has retained *Chrysolina banksi* but has no *Timarcha*. It is difficult to say if *C. banksi* was imported or was native there (Palmer *et al.*, 1999) as in the Azores or in the Canaries. It seems that some individuals can fly exceptionally but the beetle is generally apterous.

A continental island like Greenland was populated in the past with an abundant chrysomelid fauna. It now has several curculionids but no chrysomelid, even in the south. The country was depopulated by glaciation during the Pliocene. Even the north had leaf beetles during the Pleistocene. One chrysomelid has been seen in Spitzberg but never captured. It is probably a species of *Phratora*, *Gonioctena* or *Chrysomela* living on the dwarf *Salix* which grows there.

Iceland is purely volcanic and was never connected with a continent. All the chrysomelids found there came with storms or as a result of human activity: *Chrysolina staphylea* (L.), *Phratora polaris* Schneider, *Phaedon veronicae* Bedel (Larsson, 1959; Larsson and Gigja, 1959). *Timarcha tenebricosa* F. was imported recently, probably with salt from Italy, but evidently did not survive. Beetles can also be imported in animal fodder, like one species of the *T. goettingensis* complex in Gran Canaria which probably originated from Catalonia.

Almost all the islands of northern Siberia have chrysomelids: Novaya Zemlya (3 species), Novosibirskie Islands (2 species), Severnaya Zemlya (1 species), Wrangel Island (6 species). The northern limit of their distribution is therefore between 72° and 77°N (Medvedev, 1996a, 1998a). The biology of these insects is very peculiar and their activity very short during the 'summer' in this area. The complete cycle of species like *Chrysolina subsulcata* (Mannerheim) or *C. cavigera* (J. Sahlberg) can last 5 to 6 years, hibernation being at all stages (Medvedev and Khruleva, 1986; Khruleva, 1994, 1996). Several *Chrysolina* can be found in North America. These include *C. subsulcata* in the Pribilof Archipelago, *C. caurina* Brown in the extreme north of Alaska and *C. cavigera* in Alaska (Jolivet, 1992). The Pribilof Islands lie between 56° and 58°N and *Chrysolina* are not present in Alaska or Canada above 70°N. Siberia and the Wrangel Islands have chrysomelids above the 70° limit. There are no chrysomelids in north-eastern Canada. All these arctic areas contain what is left from a much warmer period. Severnaya Zemlya is between 78° and 79°N and represents the most northern point of the distribution of the group. All the arctic *Chrysolina* from Pribilof to the Siberian islands have a much wider trophic spectrum than that of the holarctic species living in a cooler climate. They accept the leaves of dwarf trees (*Betula*, *Salix*) and low plants, including *Carex* and dicot families. The diet of *Chrysolina* further south is normally greatly restricted.

During the hot spells of the Pleistocene, Perry Land, north of Greenland, was populated with a rich fauna of chrysomelids (Böcher, 1988, 1989) such as *Hydrothassa*, *Graphops*, *Chrysolina*, *Chrysomela* and *Galeruca*. Spitzberg also lost a part of its fauna during the Pleistocene glaciations. The Faroe Islands have an interesting fauna with a northen holarctic *Chrysolina, C. staphylea* (Linne).

Madagascar (Paulian, 1961) was separated from south-east Africa during the Cretaceous. Its chrysomelid fauna is relatively poor considering the surface area of the island. It has, however, certain Gondwanian endemics such as the genus *Megamerus*, shared with Australia and north-east Brazil. Many endemics appeared from South African genera, mostly among Chrysomelinae, Alticinae, Galerucinae, Eumolpinae, Cryptocephalinae and Criocerinae. Clytrinae are rare, as in all islands, even continental ones. The presence of Chrysomelinae, which separated early on from African stock, with three original genera, is evidence of an old link with the nearby continent. Unfortunately, we do not know anything of the biology or host plants of these leaf beetles. Sagrinae and Megalopodinae have African affinities, but the two species of *Megamerus* present there also suggest close Australian affinities. The most interesting endemics are found among Eumolpinae, Galerucinae and Criocerinae. Generally speaking, little is left of the old Gondwanian status of the island, at least as far as the chrysomelids are concerned.

Donaciinae probably reached Madagascar from the east coast of Africa where they are still abundant. Cassidinae and Hispinae often have African or Asian affinities, but their differentiation is important. It is likely that Indonesia influenced this fauna, and Madagascar certainly played a part in the original colonization of the Mascarenes.

The Falkland archipelago (Malvinas or Malouines) (Robinson, 1984) at the south of the American continent in the southern Atlantic includes more than 200 islands with a total surface area of about 16 180 sq km. Their position is at the same latitude as the extreme south of Tierra del Fuego and at about 650 km from the Patagonian coast. The islands have no trees, but only sheep pastures (*Poa* spp.). Among the beetles there is a strong degree of endemicity and, because of the strong winds, many species demonstrate apterism and brachypterism, all the endemic species being apterous. There are endemic genera among the curculionids and several other families. Only one chrysomelid is mentioned, a galerucine, *Luperus marginalis* Allard, evidently imported. This is a far cry from the rich endemicity of southern Argentina and Chile. In all the subantarctic island carabids, curculionids are present, but chrysomelids are missing. The reason for this absence is difficult to explain. Many of these islands are volcanic and received passive colonization through the ages. This is not due to climate, since in the continental islands south of Patagonia, *Brachyhelops* (Brendell *et al.*, 1993) and various chrysomelids have survived and have adopted a hidden life at the bases of the plants or under stones and are perfectly adapted to a rough climate. They became K-strategists and lay only a few big eggs. It is likely that several alticines occur on subantarctic islands, but they have not yet been captured.

In brief, small alticines and eumolpines have been the last to colonize the Pacific oceanic islands towards the east. The disappearance of other families has happened progressively from west to east, starting with New Guinea and ending with Samoa. There is a notable exception in the existence of a *Plagiodera* (Chrysomelinae) in Samoa. Heavyweight genera are missing in the east and it is not just a coincidence if alticines and eumolpines can survive there on host plants which are different from their original ones. Their adaptability is greater than that of Chrysomelinae and other groups. Cassidines survive normally on the Pacific islands on Convolvulaceae (*Ipomoea* spp.), even on the beaches.

Socotra island (3580 sq km) at the south of the Arabian peninsula and on the north of the African horn (225 km east of Cape Gardafui) was part of continental Africa which was separated during the Tertiary. Its characteristics are crystalline rocks and cretaceous and tertiary deposits. Its flora is original and includes the only tree cucurbit, *Dendrosicyos socotrana*. Are there *Aulacophora* spp. feeding on them? It is evident that there must be an original fauna of chrysomelids on this island which is related to those of Somalia and Arabia but we do not have any precise data. A paper by Gahan (1903) describes only an eumolpine (*Eryxia socotrana* Gahan) from Socotra and the neighbouring islands. On the Arabian peninsula itself, several rather interesting leaf beetles, mostly from the central mountains, have been listed in surveys over the years. The fauna of the Arabian peninsula, together with the Emirates, includes Palaearctic and Ethiopian elements (Medvedev, 1996b), and roughly 155 species. It is likely that the Emirates will be found to have some more species. Middle Eastern elements will be present because of the proximity of Iraq, Iran, Syria and Lebanon. It is likely that only adaptable species would have survived the desertification of the area and tolerated the rough environment. Some areas have not been properly surveyed and much remains to be discovered, particularly in hilly regions. Many subfamilies are missing and Socotra is not included in the surveys. We know that 2 Criocerinae, 1 Chlamisinae, 25 Cryptocephalinae, 16 Clytrinae, 25 Eumolpinae, 5 Chrysomelinae (including 2 *Crosita* and 2 *Colaphus*), 21 Galerucinae, 55 Alticinae, 3 Hispinae and 2 Cassidinae occur in Arabia. It is evident that there are many galerucines and alticines on Socotra. Clytrines and chrysomelines are probably absent in this dry insular area, even though there are several permanent streams in the mountains.

Winds, storms and typhoons can easily carry lightweight species. Human introduction is evident in the Cabo Verde islands, Mascarenes, Azores, Iceland, Galapagos, etc., but it could be an old introduction (Iceland) or a more recent one (Cabo Verde). Finally, this human 'contribution' to the biodiversity concerns only a few species. These are mainly omnivorous tenebrionids and carabids, cassidines and alticines linked with sweet potatoes, Brassicaceae, etc. Several islands, such as Juan Fernandez in the Pacific, have developed their own alticines, the genus *Minotula*, derived from a continental genus. The diversification of *Minotula* species took place on several host plants: *Boehmeria* and *Solanum*. An introduced species can 'explode' into several endemic species on a variety of plants or by topographic isolations. In Tahiti, the recently discovered (1979) *Argopistes* certainly arrived passively, some time in the past, originating from a common trunk in the western Pacific archipelagos. The genus *Argopistes* has a very large distribution: Japan, China, India, New Guinea, tropical Africa, Madagascar, Mexico, Cuba, etc. It also comprises 16 endemic species spread from Bouin Islands, Fiji, Vuanatu (New Hebrides) and probably they arrived late in the Tahiti area after having colonized all the intermediate lands. All these species are leaf-miners on Oleaceae leaves and the host plant is present everywhere.

Many islands, islets or even rocks of continental origin, like Chausey Islands near Granville, France, have retained in part their original fauna from when they were linked with the continent. A *Timarcha* exists in Chausey and over all the islets, whatever their size. Some other archipelagos, by contrast, have lost a great part of their original fauna. One example is Malta, which is a simple extension of Sicily and where the genus *Timarcha* has been eliminated (Jolivet, 1996b). It does not seem that

T. melitensis Weise is really Maltese. The genus probably disappeared with the intensive agriculture and urbanization of the area which occurred during the Phenician period and the Roman invasion. The fauna of both islands (Malta and Gozo) is disharmonic and regresses slowly by extinction.

Apart from *Timarcha*, *Oreina*, *Chrysolina* and some others, few biogeographical studies have been conducted on Chrysomelidae. The absence of *Donacia*, south of Cuba and in Central and South America, remains a mystery as the species have reached South Africa and Madagascar. *Donacia* is a good flyer and readily takes to migratory flights to colonize new areas. It is possible that one day it will recolonize the neotropical zone. South-east Asia, Indonesia, New Guinea, and even northern Australia, as well as Sri Lanka, have become colonized. Why is there this surprising gap in the distribution? Why in New Guinea, does only the Fly River area possess *Donacia*? This chrysomelid does not exist in the highlands, in the Madang and Weewak area or in the north-east. There are slow rivers, ponds, lakes and swamps all over the country: why are there no *Donacia* in these? It is possible that predatory fish or insects control the situation. Certainly piranhas or other South American fish would eat *Donacia*. There is no clear answer.

Sri Lanka is rich in endemics and in beetles in general. *Chrysolina*, characteristic of a continental island, are localized in the mountains. In the tropics (Ethiopia, Sudan, Kivu, Indonesia, etc.) *Chrysolina* is always to be found in middle altitudes from 1500–2000 m since the genus belongs to a cold holarctic group. It is missing in tropical America but represented there by vicarious genera. The genus, according to the taxonomists, feeds on low plants (with the exception of a certain polyphagy on dwarf trees in the Arctics), while the related genus *Calligrapha* feeds both on low plants and trees. In Cheju-do (Quelpart), a great continental island, south of the Korean peninsula, *Chrysolina staphylea* (L.) survives on the top (alpine zone) of Mount Halla at 3000 m. *C. staphylea* is a Pleistocene relic of the former link between Korea and the island. It survives on the top of a volcano but has completely disappeared from the surrounding plains and from the near mainland. On the other hand, the island is rich in chrysomelids and several *Chrysolina* have also survived in low altitude (Jolivet, 1975).

Daccordi (1994, 1996) recently tried to establish a biogeography of the sub-Gondwanian chrysomelids as to whether they moved via Antarctica or not. According to him, the origin of all chrysomelids is Gondwanian. It is certainly true for *Phaedon* and related genera from Australia and New Zealand. Daccordi gives also a Gondwanian and transantarctic origin for *Brachyhelops* (Patagonia), *Gasterantrodes* (South Africa), and *Ethomela* (Australia), all dating from the end of Jurassic. The same author sees also a transantarctic (amphynotic) origin between *Araucanomela* (South America) and *Novocastria* (Australia) associated with *Nothofagus* and between *Gavirga* (tropical America) and *Paropsimorpha* (Australia, New Guinea). It is certain that *Brachyhelops* (Patagonia) looks like *Allocharis* (New Zealand), the host plant being *Veronica*. The host plant of *Brachyhelops* is unfortunately unknown (see also Bechyne and Bechyne, 1973).

In conclusion, this review of the distribution of chrysomelids does not really clarify the origin of the family at the end of the Jurassic. It probably happened like an explosion of species and genera under a tropical climate during the Cretaceous. The Chrysomelinae migrated according to the connecting and breaking of tectonic plates and several genera remained concentrated in the southern zone of Gondwana, while some others migrated north (such as *Timarcha*, probably via central Asia). All archaic Sagrinae remained tropical, but *Sagra* had in the past a much bigger distribution to the north, in what is now Europe. The rest is only hypotheses which will be, we hope, supported later by fossil findings.

Many of the endemics have unfortunately disappeared with deforestation (Haiti, Cuba, Brazil, Madagascar). Many of the biogeographical theories are based only on specimens in the collections, since the beetles have disappeared in nature. In our temperate countries, *Timarcha* are depleting alarmingly (Baleares, Morocco, Tunisia, Djerba, Normandy and the Paris area in France, Turkey). This is due to a number of things, including careless use of insecticides and fungicides, urbanization, asphalting and cementing, the building of hotels along the coasts, the construction of ski trails in mountains, and of roads separating populations and forcing some to destruction. The fragmentation of the habitat is dangerous for apterous species.

According to Bechyne and Bechyne (1971a,b), in Peru there are about 4000 species on a surface area of 1 000 000 sq km, while in Europe there are only 700 species over 10 000 000 sq km. This is based on excellent sampling in Europe and practically none in Peru (several dozen stations only). Our great biodiversity, which still remains to be adequately described, is fast losing its members and we are doing little to save it. If whatever remains of it can be saved, it will prove valuable in the understanding of the history of our planet.

On the subject of new invaders (Jolivet, 1998c): besides *Leptinotarsa decemlineata* (Say) invading North America, Europe and, slowly, Asia (it has now reached China) from its origin in central Mexico, there are quite a few species going beyond their original territory. They adapt to new and related plants or they find their original food in the new areas. In this context may be mentioned: *Chaetocnema confinis* Crotch (Alticinae) from North America, where the species is bisexual, invading the whole of Africa, south-east Asia and the Indian Ocean and the Pacific land masses in a parthenogenetic form. It has reached New Caledonia and the Philippines, and perhaps even Southern Japan. It feeds mostly on *Ipomoea* (Convolvulaceae). *Epitrix hirtipennis* (Meslheimer) (Alticinae) from North America, is now found in Italy, Turkey and eastern Europe. It feeds on tobacco and other Solanaceae. *Diabrotica virgifera virgifera* LeConte (Galerucinae), also from North America, is found in Serbia, Croatia, Hungary, Romania and further east. Though normally a cucurbit-feeder, *Diabrotica* has adapted to maize and has become polyphagous. *Ophraella communa* LeSage, from North America, has recently invaded Honshu Island in Japan (Suzuki and Nakamura, 1999), where it feeds on *Ambrosia artemisiaefolia* (Asteraceae).

Exchanges have been made deliberately in Europe, Australia, South and East Africa, Hawaii, New Zealand and North America. For instance, *Chrysolina hyperici* (Foerster), *C. quadrigemina* (Suffrian), *C. varians* (Schaller) (Chrysomelinae), all native European species, have been introduced more or less successfully in several places to fight the introduced *Hypericum*. *Longitarsus jacobeae* (Waterhouse) (Alticinae) has been introduced from Europe to the US for use against *Senecio jacobaea* L. and many species of *Agasicles*, *Aphthona*, *Psylliodes*, *Altica*, *Longitarsus* (Alticinae), *Galerucella*, *Galeruca*, *Diorhabda* (Galerucinae), *Cassida*, *Stolas* (Cassidinae), *Lema* (Criocerinae), etc. have been introduced from one continent to another for biological control. Some have survived, some were eliminated by other factors, such as climate, birds, and other predators (Fornasari, 1996; White, 1996). However, some European chrysomelids came from Europe accidentally, and became pests in the US, like *Plagiodera versicolora* (Laicharting) (Chrysomelinae) (Wade, 1994). Uninvited guests crossed the ocean both ways. Some have been more active than others, and *Chaetocnema confinis* seems to be the most active of all.

8

Defence Strategies

Being, in general, poor flyers and because they are well exposed when feeding on the host plant, chrysomelids are very vulnerable to predation and parasitic attack. Naturally, they have evolved a variety of devices and mechanisms for their defence. A sharply-defined classification of defence strategies among Chrysomelidae is not possible. A convenient classification, with diffuse and overlapping separation between categories, is as follows:

1. Use of waste material for defence.
2. Chemical defence.
3. Behavioural defence:
   (a) Cycloalexy.
   (b) Gregariousness.
   (c) Thanatosis/feigning death.
   (d) Reflex bleeding.
4. Structural defence:
   (a) Body shape and adhesive hairs of cassidines.
   (b) Spines of hispines.
   (c) Jumping mechanism in alticines.
   (d) Mimicry, aposematism and camouflage.
   (e) Change of colour.
   (f) Stridulation.
5. Parental care and subsociality.
6. Concealed feeding.

Use of waste material for defence

The term 'waste material' includes excreta, faecal discharge and cast skin or exuvium. Olmstead (1994) wrote a very useful review on the use of waste for protection by leaf beetles. The following account of this protective stratagem is mainly based on this review. According to Vencl and Morton (1999), faecal shields (Criocerinae, several Alticinae) and parasols (Cassidinae) have direct chemical links with larval diets and the plants on which they feed. They are used to repel or deter invertebrate predators and as a means of protection against rain and sun. Larvae transform the host precursors found in the plants before their incorporation into the shields.

The females of some Chrysomelinae, Galerucinae, Alticinae, Cassidinae and Hispinae defecate on their eggs after laying them. The eggs thus lie concealed under the faecal deposit, which may also act as a repellent to enemies. *Timarcha* spp. cover their eggs mostly with stomach regurgitations, mixed up with plant tissues. This structure can be referred to as a pre-ootheca (Jolivet, 1997). Some *Galeruca*, *Phratora*, etc. add secretions to their faecal deposits on the eggs.

The larvae of many cassidine beetles retain their faecal discharge at their hind end, forming a protective shield. The larval abdomen may be curled upward and forward to cover more or less of the body. The cassidine larval faecal shield varies considerably in form and structure.

In some cassidines, the cast skin or exuvium is retained at the hind end of the larva, and is mixed with bits of faecal matter. In this way, the exuvium contributes to the formation of a protective shield. In *Aspidomorpha miliaris* (Fabricius), for example, the larva carries a chain of exuviae from previous instars, containing a darkish material. Among some South American cassidines, exuviae and a ball of excrement form a cover for the larvae. For instance, in *Hemisphaeronota* and *Dorynota*, the larvae take cover under a complex network of excremental filaments covering all the abdomen (Böving and Craighead, 1930; Jolivet, 1997) (*Figures 8.1* to *8.3*).

Besides the repulsive substances in the faecal discharge, secretions from externally opening glands may also be added to the protective shield, to improve its defensive quality.

In Camptosomata, an assemblage of subfamilies (i.e. Clytrinae, Cryptocephalinae, Chlamisinae and Lamprosomatinae), faecal waste is used for constructing a larval case, which offers protection throughout the larval life. In these subfamilies, the larva is bent upon itself on its venter, and becomes U-shaped, so that mouth and anus are close together. Faecal discharge is mixed with glandular secretions, it is manipulated with mandibles, and is arranged as a tubular case around the body. Often, particles of various nature from the surroundings are incorporated in the case; this provides the advantage of camouflage to the larva and probably gives strength to the structure. Similarly, larvae of fresh- or saltwater Trichoptera assemble small pieces of sand, diatoms or coral algae on their silky larval cases. Before pupation, the case of the Camptosomata is made to adhere to the substratum using some more excreta.

Field observations suggest that the protective shield, made of waste, affords adequate protection against mandibulate predators, but is not quite as effective against predators with piercing and sucking mouth parts, such as heteropteran bugs. Moreover, it offers only partial protection against active and vigorous predators, e.g. paper wasps.

There have been some experimental observations attesting to the protective value of shields and covers made of waste associated with chrysomelid developing stages. Olmstead (1994) mentioned two such experiments. Egg masses of the hispine *Macrorhopala* spp. are covered by the mother's faecal discharge. If the faecal cover is removed, predators readily destroy the eggs. Covered eggs are not, however, attacked by predators. The larval case of *Chlamisus* (Chlamisinae) has been found to be an effective protection against ants and other predators. When the cases were removed, the larvae were readily attacked by coccinellids, which did not attack the encased larvae. Recently, Eisner and Eisner (2000) have studied the defensive use of faecal thatch by larvae of *Hemisphaerota cyanea* (Say). The system seems quite efficient, though some beetles do succeed in feeding on the larva.

A unique case among Chrysomelinae, *Phola octodecimguttata* (Fabricius) from China, has been described from Chen (1964–1985). The larvae are stem-feeders on *Vitex cannabifolia* (Verbenaceae) and build tubiform nests with excrement on or

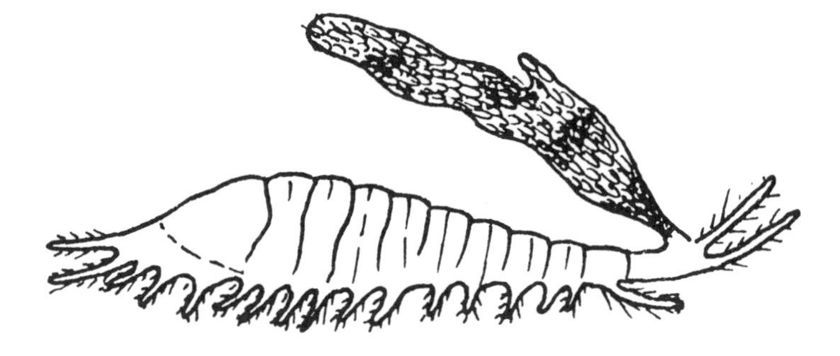

Figure 8.1. *Oocassida cruenta* (Fabricius) (Cassidinae) larva with its protective shield made of excrementitious material (after Maulik, 1919).

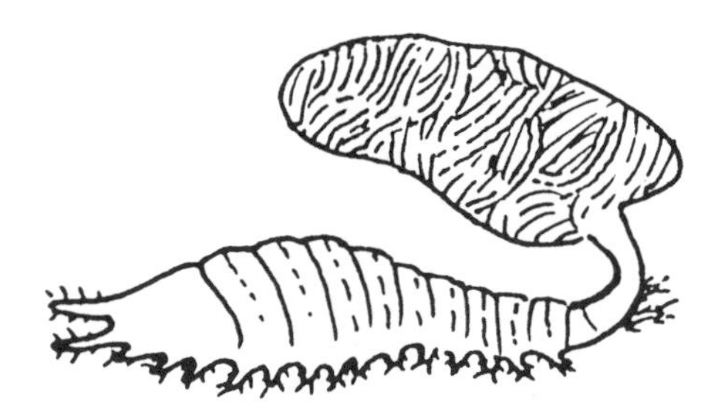

Figure 8.2. *Laccoptera quadrimaculata* (Thunberg) (Cassidinae) with excrementitious shield at the caudal end (after Maulik, 1919).

between the branches. The larvae live partially hidden, with the anterior half of the body in the nest and the tip of the abdomen exposed (Jolivet, 1997). In South Africa, the adult of *Iscadida* spp. (Chrysomelinae) feeds on various Araliaceae (*Rhoicissus*) and several other plants (Jolivet, 1995; Grobbelaar and Jolivet, 1996). The insect drops the eggs onto the ground from the leaves of its food plant, in the same way as some *Cryptocephalus* do (Owen, 1999). What happens to the larvae later on is not known. Whether they climb back onto the food plant or they feed on the ground, like *Cryptocephalus* larvae, on fallen leaves or debris, remains unobserved. The *Iscadida* larvae remain free and do not produce a shield or a faecal case. The larva seems protected by its abundant hairy cover. As for the encased larvae of *Cryptocephalus* spp., the shell remains free in the case of *C. coryli* (L.) but, in the related species *C. sexpunctatus* (L.) in England, its shell is attached by silken material to an object such as a dead leaf (Owen, 1997), possibly an extra protective measure (*Figures 8.4* and *8.5*).

A protective cover or shield (scatoshell) made of waste may, in a way, impose a disadvantage too. The substances included in such a cover or shield may behave as a kairomone and invite parasitoids. Possible adaptations of parasitoids to chrysomelid pheromones and allomones are discussed by Hilker and Meiners (1999). Little is known about the use of waste in defensive strategy outside the family Chrysomelidae.

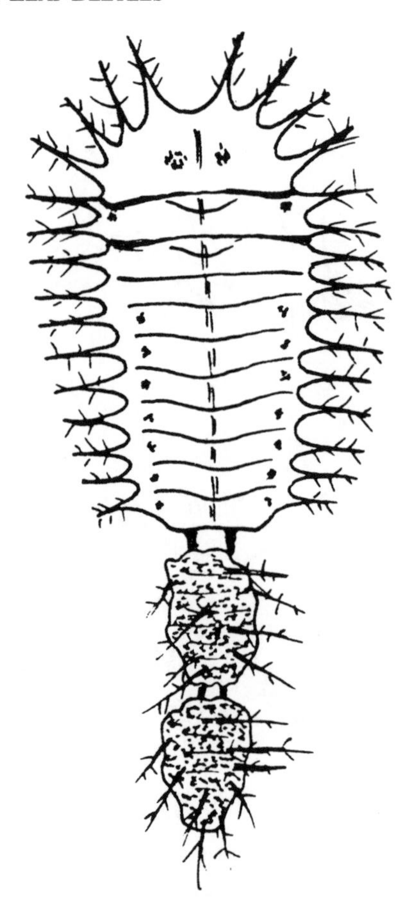

Figure 8.3. Larva of *Aspidomorpha miliaris* (Fabricius) (Cassidinae) with a chain of cast skins at the caudal end (after Maulik, 1919).

Chemical defence

Many leaf beetles protect themselves from predation by releasing toxic chemicals at the body surface. The present knowledge about these chemicals has been summarized by Pasteels *et al.* (1988a,b, 1994).

The toxic compounds present a great diversity, which seems to be due to the variety of chemicals available to the beetles from host plants and the number of ways in which the beetle's body can synthesize toxic chemicals.

The toxic substances released by adult leaf beetles include derivatives of isoxazolinone, nitropropanoic acid, steroids, amino acids, pyrrolizidine alkaloids, cucurbitacins, etc. The last named compounds are bitter-tasting, oxygenated, tetracyclic triterpenes. *Gastrophysa cyanea* Melsheimer produces a defensive secretion composed of chrysomelidial and an enol lactone (Blum *et al.*, 1978).

The toxins produced and released by leaf beetle larvae include salicylaldehyde, methylcyclopentanoid monoterpenes, cardenolides, benzaldehyde, HCN, cucurbitacins, anthraquinones and anthrones (Blum, 1994).

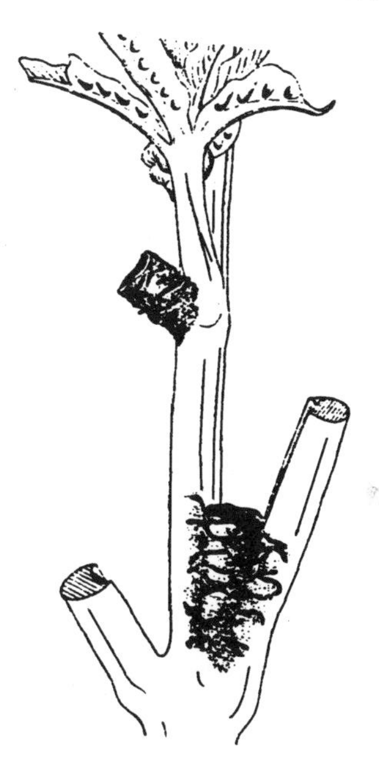

Figure 8.4. *Phola octodecimguttata* (Fabricius) (Chrysomelinae) larva on its case on a twig of *Vitex quinata* A.N. Will (Verbenaceae) (after Chen, 1964).

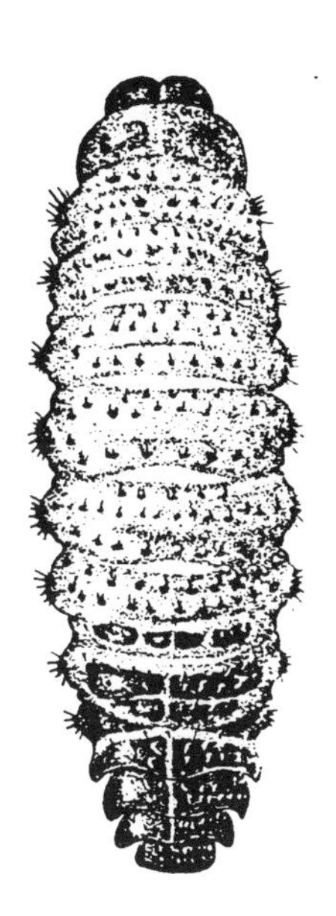

Figure 8.5. Larva of *Phola octodecimguttata* (Fabricius) (after Chen, 1964).

Toxins in many cases are released through certain exocrine glands. Such defensive glands are known to be present in adults, as well as larvae. Pronotal and elytral glands have been noted in all the Criocerinae and Chrysomelinae which have been examined, and they are also present in some Alticinae and Galerucinae. On mechanostimulation,

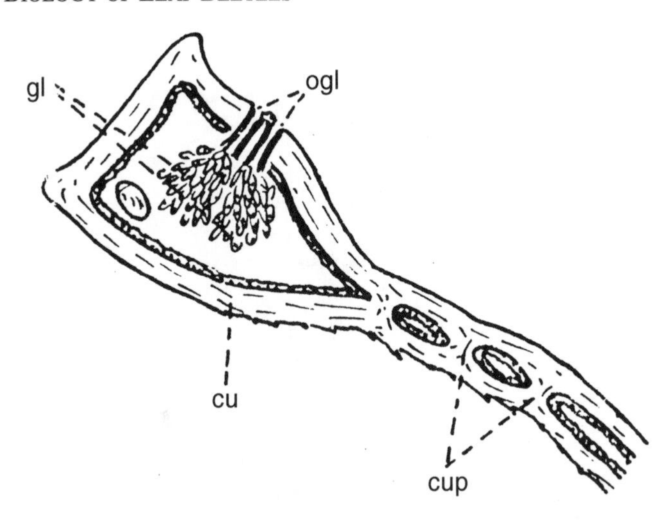

Figure 8.6. Transverse section of an elytron of *Chrysomela tremulae* Fabricius (Chrysomelinae) (after Jolivet, 1997) (cu = cuticle; cup = cuticular pillars connecting the two faces of the elytron; gl = glands secreting defensive fluid; ogl = openings of the defensive glands).

Figure 8.7. Larvae of *Aspidomorpha miliaris* (Fabricius) (Cassidinae) showing cycloalexy (photograph by H.V. Ghate, Pune, India).

the glands give out their secretions. Pasteels *et al.* (1988a) describe the discharge of defensive secretion from these glands in this way: "After disturbance, the secretion oozes out from the gland pores and accumulates in marginal grooves of elytra and pronotum, as well as in more or less defined pronotal and elytral depressions, constrictions and concavities. These contours certainly help to retain the secretions on the insect" (*Figure 8.6*).

Another way of releasing toxins is through reflex bleeding, seen mostly in galerucines. Some alticines and chrysomelines also show this phenomenon. On disturbance, the thin membranous skin, at the tibio-femoral joints or near the mouth (or at both places, as for example in *Timarcha*), becomes distended and ruptures or exposes fine and preformed orifices where droplets of blood appear. The exuded blood rapidly clots and turns into a sticky mass. Sometimes, the blood is partly reabsorbed through the orifices. Besides providing a mechanical protection, the bleeding brings to the surface the toxins present in the blood.

Still another scenario is that toxins simply accumulate in the blood and are released when the insect is attacked and wounded by a predator. The wounded insect may get killed, but the experience with the blood toxins will make the aggressor learn to avoid the kin of the prey. This selection advantage has been referred to as 'kin selection' by Pasteels *et al.* (1983).

Evidence as to the defensive value of these compounds is mostly indirect and is based on observations of their effect on general predators. There are practically few studies on predators actually attacking leaf beetles in nature (Cox, 1996a). Ferguson and Metcalf (1985), for example, have found that, if certain galerucines are reared on a diet rich in cucurbitacins, they are rejected by the Chinese praying mantids, which, however, readily consume the galerucines reared on a cucurbitacin-free diet. The mantids are general predators.

The nature of toxins, produced for chemical defence, may help with the classification of chrysomelids as far as grouping the genera into subtribes, but not beyond this level in higher classification of leaf beetles (Pasteels, 1993; Pasteels *et al.*, 1994). For detailed information on chemical defences in leaf beetles, see papers by Pasteels and his associates (Pasteels *et al.*, 1988a,b, 1990, 1992, 1994, 1996; Pasteels and Rowell-Rahier, 1991; Duffey and Pasteels, 1993; Rowell-Rahier *et al.*, 1995; Dobler *et al.*, 1996; Hartmann *et al.*, 1997, 1999; Soetens *et al.*, 1998).

Behavioural defence

CYCLOALEXY

Cycloalexy is a defence stratagem used by certain insect larvae. According to Jolivet *et al.* (1990), cycloalexy is "the attitude adopted at rest by some insect larvae, both diurnal or nocturnal, in a tight circle where either the heads or ends of the abdomen are juxtaposed at the periphery, with the remaining larvae at the centre of the circle. Coordinated movements such as the adoption of threatening attitudes, regurgitation, and biting are used to repel predators or parasitoids". Similar behaviour is shown by some semisocial vertebrates.

Among Chrsyomelidae, this phenomenon is known in some Chrysomelinae, Galerucinae and Cassidinae (Vasconcellos-Neto and Jolivet, 1988). Among Coleoptera, besides certain Chrysomelidae, cycloalexy has been recorded in the curculionid *Phelypera distigma* (Boheman) (Jolivet and Maes, 1996). Stray cases of this defence stratagem have been observed in Hymenoptera, Diptera and Neuroptera (Vasconcellos-Neto and Jolivet, 1994). From Africa, only two cases are on record (Heron, 1992, 1999). The phenomenon is rather common in America and sometimes associated with parental care. From India, the only known case is of the cassidine *Aspidomorpha miliaris* (Fabricius) on *Ipomoea carnea* Jacq. (Maulik, 1919; Verma,

1992, 1996b). When at rest, larvae of this cassidine form, on the undersurface of leaves, compact oval groups with heads towards the centre and hind ends forming the periphery of the group. When such a group is disturbed, the larvae curve their abdomens upward in a seemingly threatening manner. Larvae of a closely related species, *A. sanctae-crucis* (Fabricius), on the same *Ipomoea carnea*, however, do not show such gregarious behaviour, perhaps because they are protected from excremential filaments on the abdomen (Verma, 1992) (*Figure 8.7*).

Nakamura and Abbas (1987), while describing the ecology of *Aspidomorpha miliaris* on *Ipomoea carnea* in Sumatra, point out that the larvae of this leaf beetle live in compact groups, and show low mortality. But there is a steep rise in death rate when the larvae pupate. This situation lends further support to the well-founded hypothesis that cycloalexy is for larval defence. It may be noted here that the larvae of *A. miliaris* undergo some dispersion before pupation. In the New World, the cassidines showing cycloalexy often also show parental care, which affords protection not only to eggs and the larvae, but also to pupae.

As has been noted above, the larvae in cycloalexy, in the face of approaching danger, demonstrate coordinated movements/activity. This seems to be due to an alarm pheromone (Jolivet, 1997). The concept of cycloalexy in insects has been established and developed by Joâo Vasconcellos-Neto and P. Jolivet (1988, 1994). Certainly, cycloalexy is a form of gregarism and pheromones may be involved. Eisner and Kafatos (1962) studied a pheromone-promoting aggregation in an aposematic distateful insect and it is well known that male crucifer flea beetles (*Phyllotreta cruciferae* (Goeze)) also produce an aggregation pheromone (Peng and Weiss, 1992; Peng *et al.*, 1999).

Recently, Weinstein and Maelzer (1997) have studied cycloalexy in Australia among *Perga dorsalis* larvae (Hymenoptera Pergidae). The behaviour of these larvae is very similar to that of chrysomeline larvae. A subgroup of about 20% of the larvae preferentially occupied the outer positions in the resting colony and appeared to lead the foraging expeditions. Leaders were quick to regain outer positions if removed and placed in the centre of the colony. Polyethism, or division of labour among members of a colony, exists here and should be verified among *Coelomera* larvae and others showing cycloalexy. Small colonies are non-viable and subject to attack by predators.

GREGARIOUSNESS

Cycloalexy, shown by some chrysomelid larvae, is a kind of gregariousness. Many members of the leaf beetle family demonstrate group formation. It may be seen in the egg stage, in the larval stage, or in the adult stage. Gregariousness in the larval stage may or may not show features of cycloalexy.

Many chrysomelids deposit their eggs in an aggregate which is often not provided with any protective cover. Gregoire (1988) has reported that in the hispine *Cephaloleia consanguinea* Baly uncovered eggs are laid in a group. Such a group is open to attack by parasitoids, and a certain percentage of eggs are destroyed, but for an egg in a group the probability of being parasitized is smaller than if it were isolated. In the case of larvae too, gregariousness has a positive influence on survival.

When eggs are laid in a group, the larvae hatching out from them will be gregarious to begin with and will, in many cases, have a tendency to remain so. Perhaps a pheromone is responsible for keeping them together.

That the larvae of some chrysomelids remain in a group is not just because the eggs from which they have emerged were in a group, but because of an inherent gregariousness. This situation is well illustrated by the case of *Platyphora*, species which are mostly viviparous. Jolivet (1997) points out that, in this case, larvae are given birth to one by one at quite long intervals and, since the species has a gregarious tendency, the larvae move rapidly into an aggregate.

Some leaf beetles show polyspecific aggregates, for example *Phratora laticollis* + *Phratora vitellinae* (Jolivet, 1997). It seems that in such cases the aggregate pheromone is not very species specific and the species are closely related.

Adult leaf beetles may also show gregariousness. In the mountainous region of Panama, two alticines, *Macrohaltica jamaicensis* (Fabricius) and *Macrohaltica* sp., form large aggregates, each consisting of thousands of individuals, on the giant leaves of the plant *Gunnera chilensis* Lamarck (Gunneraceae) (Eberhard *et al.*, 1993). Flowers and Tiffer (1992) pointed out that the alticine *Hypolampsis* sp. forms aggregates on the leaves of *Desmodium* and *Cordia* spp. Evidence of gregariousness in a population of flea beetles (probably *Macrohaltica*) from the Greater Antilles has been obtained from a single piece of Dominican amber (Poinar and Poinar, 1999). Individuals of insects like *Timarcha*, from the Mediterranean region, rest in groups during the night or during cold spells under plants like *Thymelaea hirsuta* L. (Thymelaeaceae). They readily aggregate when resting and they disperse during daytime activity (Jolivet, 1999a). Flowers (1991) points out aggregations in various Cassidinae, not necessarily on their food plants, in Costa Rica during the dry season. The insects are not exactly in diapause but merely quiescent (dormancy). Some Endomychidae are known to rest by thousands each year on the same palm tree species in Panama, quiescent but still slowly moving (Roubik and Skelley, 2001), and a pheromone is involved exactly as for *Phyllotreta cruciferae* (Goeze) on oilseed rape, *Brassica napus* L. The male *Phyllotreta* is known to produce an aggregation pheromone when in numbers (Peng and Weiss, 1992; Peng *et al.*, 1999).

It should be noted that in adult gregariousness there is nothing to suggest cycloalexy. Adult aggregates may be formed also as a prelude to diapause or migration.

THANATOSIS/FEIGNING DEATH

Collectors of leaf beetles often notice that, when they approach a beetle with their fingers, the insect becomes immobile, folds up its appendages against its body, and may drop off from the host plant to the ground. It then becomes very difficult to locate it among the litter on the ground. After a period of immobility, which varies with species, the insect moves about again. This feigning death has been referred to as thanatosis. The defensive value of thanatosis is obvious.

The physiology of thanatosis has not been satisfactorily worked out, despite some intensive research from Rabaud (1919). It seems to be of the nature of reflex action, i.e. it does not appear to be by 'choice'. But how is it that mechanical disturbance by man or a would-be predator induces thanatosis, but that a breeze does not?

Occurrence of thanatosis is variable, even within a group. For example, most alticines, on being disturbed, jump and fly away, but species of *Podontia* and *Blepharida* turn to thanatosis (Jolivet, 1997). These are big species for Alticinae and have a peculiar biology.

REFLEX BLEEDING

Reflex bleeding is a defensive behaviour shown by some insects. It is seen in a number of chrysomelids, and has been briefly referred to under the section on chemical defence. On stimulation by a prospective predator, drops of blood come out through the intersegmental membrane in some parts of the body, where the membrane is specially weak or is provided with minute preformed pores. The blood of chrysomelids often contains toxic substances, and reflex bleeding is a way to release defensive compounds (Seguy, 1967; Pasteels *et al.*, 1988a,b). The blood of *Timarcha* contains anthraquinones and is very toxic to vertebrates (Jolivet, 1997). Various papers by Hollande (1911a,b, 1926) reviewed the phenomenon. According to Andersen *et al.* (1988), *Diabrotica virgifera virgifera* LeConte and *D. undecimpunctata howardi* Barber produce cucurbitacins through metabolic transformation in their bodies. Cucurbitacins are present in the blood of these leaf beetles, and that is why they are rejected by the Chinese praying mantis after being bitten and wounded (Ferguson and Metcalf, 1985). These species of *Diabrotica* are known to show autohaemorrhage (Andersen *et al.*, 1988).

In reflex bleeding, blood discharge usually occurs at tibiofemoral joints and/or from the prebuccal region. *Timarcha* will bleed at both these sites. Hollande (1911a,b), who carried out an initial survey of this phenomenon among Chrysomelidae, noted tibiofemoral haemorrhage in *Timarcha*, *Chrysolina* (example *C. herbacea* (Duftschmid)), *Oreina*, *Chrysomela*, *Galeruca*, *Galerucella*, *Altica* and *Exosoma*, and prebuccal haemorrhage in *Agelastica alni* (L.), *Sermylassa halensis* (L.), *Exosoma lusitanica* (L.) and *Timarcha* spp. In the nocturnal species among the *Timarcha* (*Americanotimarcha* in the USA and, to some extent, *Metallotimarcha* in the middle European mountains), reflex bleeding is almost useless and is reduced or absent. Blood is frequently partly reabsorbed after display. Part of it coagulates. The production of blood is limited and absorption of water is needed to allow abundant autohaemorrhage. Among cassidines, reflex bleeding has been reported in *Stolas* (*Botanochara*) *impressa* (Panzer). *Leptinotarsa* spp. and *Megalopus* spp. are also known to show reflex bleeding (Jolivet, 1997). In fact, this defensive behaviour is quite widespread among leaf beetles (Jolivet, 1997).

In *Timarcha*, prebuccal haemorrhage occurs through a pair of orifices situated on the arthrodial membrane between the bases of maxillae and mandibles. During discharge, the membrane becomes distended due to an influx of blood. As a result, the orifices become exposed, and blood spurts out. In this chrysomelid, as noted above, there is also tibiofemoral haemorrhage. This bleeding occurs through a preformed orifice in the ligament and the arthrodial membrane at this location. Flexion between femur and tibia is necessary to expose the orifice for the release of blood. A drop of blood comes out and seals the orifice.

With the chrysomelids that show reflex bleeding, the phenomenon is confined to the adult. It is not normally seen in the larva, except with *Coelomera*, where the adult does not bleed in this way. *Coelomera* species live on the ant plant, the *Cecropia*, and coexist with the aggressive *Azteca* ants. They need extra protection, like cycloalexy, autohaemorrhage, toxicity, etc. Wallace and Blum (1971) studied reflex bleeding in two species of the galerucine *Diabrotica*. According to them, the bleeding is shown only by the larva and not by the adult in *Diabrotica undecimpunctata* and *D. balteata*

LeConte. In the larvae of these galerucines, the sites of autobleeding are the arthrodial membrane between the head and the prothorax, and the intersegmental membrane between the last two abdominal segments. Reflex bleeding occurs only at the site stimulated, and results in considerable fluid loss, up to 13% of body weight. The exuded blood very rapidly clots, and this is specially effective against ants, which get caught up in the congealing sticky blood. Andersen *et al.* (1988) find that adults of *Diabrotica undecimpunctata howardi* and *D. virgifera virgifera* will show autobleeding, both buccal and tibiofemoral, if the thoracic venter is lightly pressed.

The fluid exuded in autohaemorrhage is actually blood. Hollande (1911a,b) examined the fluid exuded in *Timarcha* microscopically, and noted the presence of blood cells in it. Matsuda (1982) studied microscopically and chemically the autohaemorrhage fluid of *Gallerucida nigromaculata* Baly, and confirmed that it was blood.

Blood colour is variable among leaf beetles. It may be different in different species, and also between the two sexes of the same species (Jolivet, 1997). There is a possibility of confusing the discharge of secretion from certain exocrine glands with autohaemorrhage. Careful assessment of the situation is therefore necessary.

Reflex bleeding in Chrysomelidae may be accompanied by some other defensive behaviour or devices. In *Timarcha*, for example, prebuccal haemorrhage is sometimes accompanied by stomach regurgitation. A similar situation may be observed in young adults of *Leptinotarsa decemlineata* (Say) (Deroe and Pasteels, 1977).

Meloe, which is wingless and metallic blue like *Timarcha*, shows a similar liquid discharge. Thus, in these two cases, autohaemorrhage is accompanied by aposematism: blue black beetle on a green background. In the highlands of the Tchiaberimu, in the Kivu mountains, the adults of *Meloe* are abundant on the *Galium* plants, exactly like *Timarcha* were in the middle European lowlands. There is really a convergence in the behaviour of the two genera and both are becoming rarer and rarer due to pollution and destruction of habitat. In some instances, reflex bleeding, stomach regurgitation, aposematism and thanatosis are independent phenomena. *Pimelia* spp. (Tenebrionidae) and *Timarcha* spp. in North Africa are certainly mimetic, probably a case of Müllerian mimicry, but *Pimelia* does not show reflex bleeding, only stomach regurgitation which is probably associated with toxic glands.

Structural defence

BODY SHAPE AND ADHESIVE HAIRS OF CASSIDINES

A flattened body, a hard dorsum provided with well-sclerotized pronotum and elytra, the elytral explanate margins, a head which in most forms is covered by the pronotal explanate margin and flattened and adhesive tarsi provide an effective protection to cassidines. Olmstead (1996), discussing cassidine defences, described these defensive structural adaptations in this way: "As the name suggests, tortoise beetles can retract all appendages beneath the elytra and pronotum when disturbed". The pronotum and elytra are often heavily sclerotized and may be difficult to penetrate with an insect pin (*Desmonota variolosa* (Weber)) (Jolivet, 1994c). In many species, the elytral and pronotal margins are explanate, further covering the body. Additionally, cassidines have tarsal pads that strongly grip the substratum, making them difficult to remove (Windsor *et al.*, 1992). In fact, if they are literally ripped from the host plant, the tarsal pads often remain behind on the plant surface (Olmstead, personal observation). These attributes seem to well defend most species of cassidines as relatively few enemies have been recorded for adults. In *Table 8.1* in Olmstead (1996), far fewer parasitoids and predators have been listed for cassidine adults than for the eggs, larvae and pupae of these leaf beetles. The caudal process of larvae of some *Aspidomorpha*, like *A. deusta* (Fabricius), holds a movable faecal shield thought to be important in larval defence (McBride *et al.*, 2000).

The flattened tarsal joints of tortoise beetles are provided with a dense brush of bristles on their under-surface. The tarsal bristles in cassidines are bifid and spatulate, and the tarsal claws are sharply pointed and hook-like (Mann, 1985). Jolivet (1997) has extensively cited Eisner's study (1971) on *Hemisphaerota cyanea* (Say) to describe the defensive role of the tarsi. This small, blue cassidine species live on the bush tree palmetto (*Sabal* sp.) in the southern part of North America, notably Florida. In the event of an attack by an ant, the beetle holds tightly on to the leaf surface, like a limpet, and it is practically impossible for a predator to dislodge it.

In *Hemisphaerota cyanea*, the tarsal joints are flattened, and are like large pads, the under-surface of which are provided with a dense pad of bristles with their terminal portions enlarged into an adhesive cushion. When adhesion is needed, the bristles become covered with an oily secretion, which is produced by special glands opening between the bases of the bristles. In normal locomotion, that is when there is no fear of attack, the legs are placed lightly onto the substratum, only a few bristles of each tarsus come into contact with the substratum, and tiny oily droplets do not appear on the bristles. When an attack is likely, the beetle presses the three pairs of tarsi on to the substratum and some 60 000 oily droplets appear on the tarsal bristles. Thus, the cassidine becomes tightly stuck to the leaf surface and is secure. The insect must sacrifice some of its valuable oily secretion to achieve this security. But, according to Eisner, the sacrifice is moderate. This author estimated that the oily droplets produced by the three pairs of tarsi, and collected on a glass surface, represented 0.005% of the total body weight. The secretion consists of hydrocarbons, which can be readily synthesized by the insect from waxes present in its nourishment. Eisner has shown experimentally that the cassidine, when stuck to the leaf surface in preparation of an approaching attack, can resist a weight of 3 gm, which is quite considerable in view of the size of the insect.

Using a light microscope and SEM, Stork (1983) studied the adherence of tarsal setae to glass in eight species of beetle. His material included a cassidine (*Cassida flaveola* Thunberg). He inferred that secretions do help adhesion, but also suggested that molecular adhesion is also important here.

HISPINE SPINES

A subfamily closely related to Cassidinae is Hispinae. Most hispine beetles rely on an armour of spines for their protection, instead of flattening the body. This protective adaptation is equally effective. The big palm- or *Heliconia*-frequenting American species are, however, flat and smooth and have warning colours which must be meant to convey natural toxicity. Palm-frequenting species also exist in Asia.

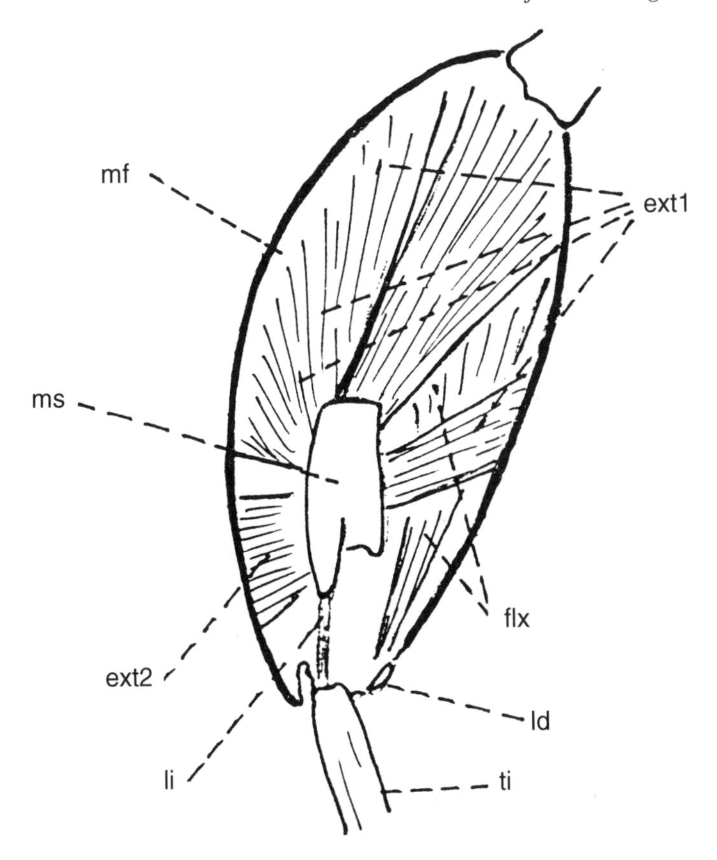

Figure 8.8. Metafemur of an alticine in section (after Furth, 1982a) (ext1 = primary tibial extensor fibres; ext2 = secondary tibial extensor fibres; flx = tibial flexor fibres; ld = Lever's triangular sclerite; li = ligament; mf = metafemur; ms = metafemoral spring; ti = tibia).

JUMPING MECHANISM IN ALTICINES

Alticine chrysomelids can jump like fleas; hence their common name, 'flea beetles'. The jumping ability obviously helps them escape from predators. Those alticines which have not lost their flying capacity may use their wings following a jump, instead of landing on a surface. Thus, jumping and flying away can be combined. Apterous alticines can only jump and not fly, but they are also very efficient in escaping predators.

The distance covered by the jump of a flea beetle is considerable. Furth *et al.* (1983) and Furth (1988) have reported that *Blepharida sacra* (Weise), whose body is 7 mm long, can cover nearly 70 cm in a jump, that about 100 times its body length. Some other flea beetles, like *Longitarsus* spp., are even more efficient jumpers.

The jumping capacity of alticines is due to metathoracic legs, which present an interesting modification for this purpose. The femur is especially swollen and enlarged. The tibial extensor muscles, lodged in the capsule of the hind femora, are greatly developed. They find insertion on a sclerotic structure lying free, floating in a distal part of the femur capsule. This structure has been referred to as the 'Maulik's

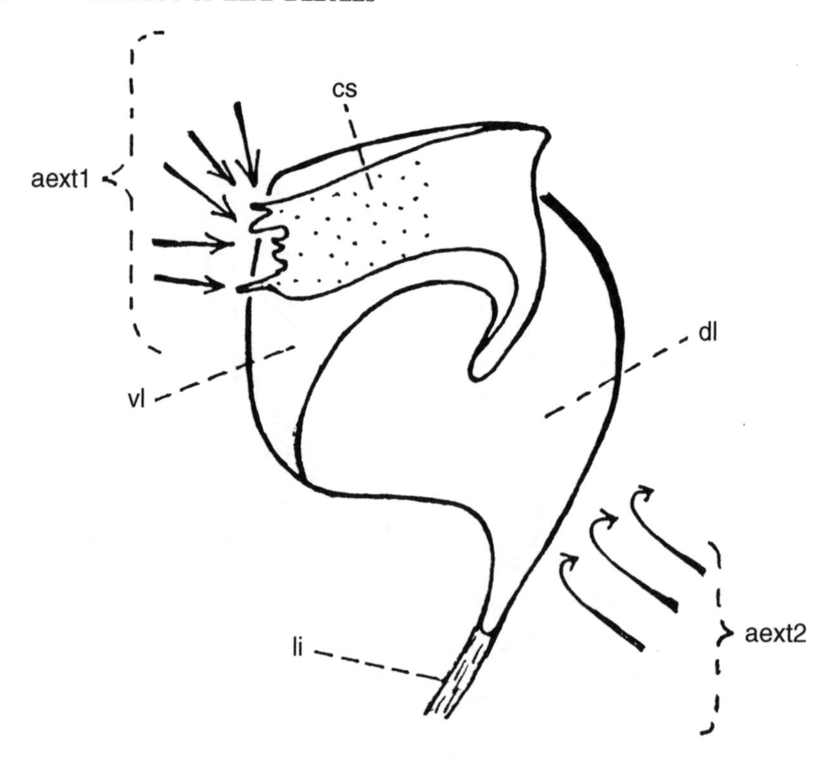

Figure 8.9. Metafemoral spring of *Psylliodes* sp. (Alticinae) (after Furth, 1982a) (aext1 = attachment site for primary tibial extensor fibres; aext2 = attachment site for secondary tibial extensor fibres; cs = cuticular sheet into which the ventral lobe of the metafemoral spring continues and on which the primary tibial extensor fibres end; dl = dorsal lobe of the metafemoral spring; li = ligament; vl = ventral lobe of the metafemoral spring).

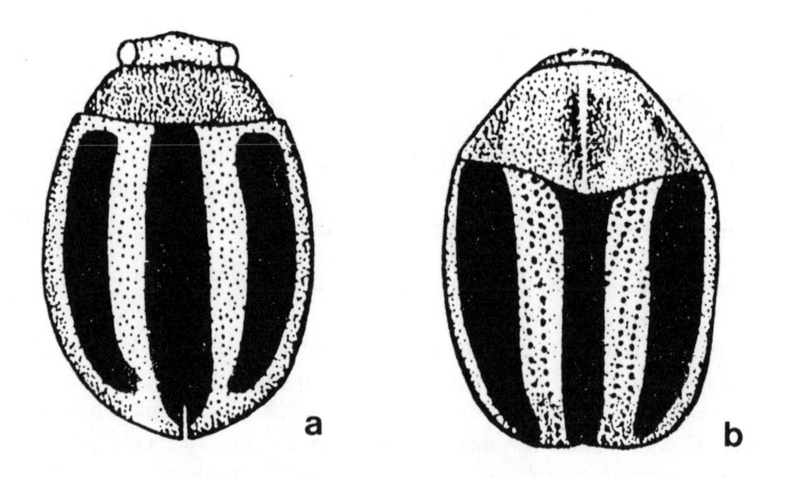

Figure 8.10. A case of aposematism (after Verma, 1996a). (a) *Brumus suturalis* Fabricius (Coccinellidae) (× 20); (b) Id. *Cryptocephalus ovulum* Suffrian (Cryptocephalinae) (× 20).

Figure 8.11. Adult of *Exema indica* Guérin (Chlamisinae) looking like a caterpillar dropping (after Verma and Vyas, 1987).

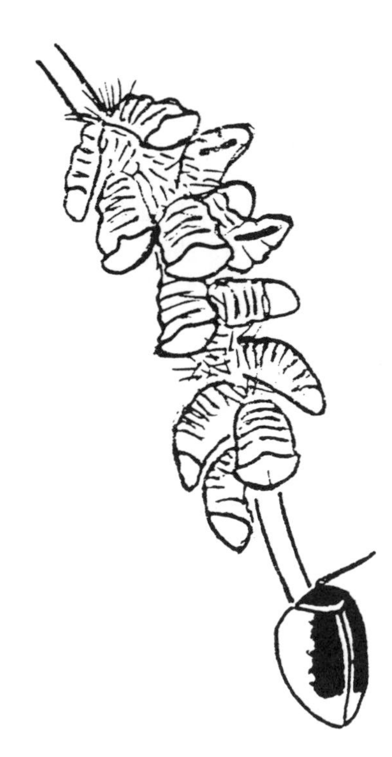

Figure 8.12. A female of *Eugenysa cascaroni* Viana (Cassidinae) guarding her progeny, a group of pupae (based on Jolivet, 1997).

organ' after the name of the discoverer. Furth (1982a) called it 'metafemoral spring' in view of its location and function. The metafemoral spring is like a curved plate, somewhat S-like in a cross section, and made up of alpha-chitin and protein molecules. The tibial extensor fibres are in two groups and have been referred to by Furth (1982a, 1988) as 'primary and secondary tibial extensors'. The primary extensor is the larger and its fibres are attached along the lower edge of the metafemoral spring. The secondary fibres attach to a dorsal part of the spring. A tough ligament-like strand connects the ventral lower end of the spring with the postaxial part of the proximal end of the tibia. The tibial flexor fibres have their distal tendon attached close to a triangular sclerite in the femoro-tibial arthrodial membrane on the preaxial side. This sclerite, called the Lever's triangular plate, is drawn into the femoral capsule on contraction of the tibial flexor (*Figures 8.8* and *8.9*).

The alticine's jumping mechanism was first explained by Barth (1954). The explanation, as modified by Furth (1982a) and Furth and Suzuki (1992, 1998), is given here. Contraction of the extensors of the tibia brings about deformation and dilatation of the elastic and chitinous metafemoral spring. This produces an energy reserve in the spring. When this energy reserve reaches a certain threshold value, a catch is undone, and the spring returns to the original shape, releasing the reserve energy. This brings about an abrupt extension of the tibia, and the consequent jump. The nature of the catch is not clear and so the mechanism of the alticine jump is not adequately understood.

Some alticines, especially the larger ones, use other defensive behaviour/devices in addition to jumping, for example, immobilization reflex and camouflage. It is not clear how a defensive strategy is selected for a particular moment of danger, that is whether the insect will jump away, or take to thanatosis, or feign death and remain immobile for a while and then suddenly take to a leap.

Furth (1980, 1988) pointed out intergeneric differences in the form of the metafemoral spring among a number of alticines and, on the basis of these differences, divided the alticine genera into 6 morphogroups. As far as a natural classification of alticines into tribes is concerned, however, these groups do not seem to be of much importance (Jolivet, 1997).

MIMICRY, APOSEMATISM AND CAMOUFLAGE

These adaptations are quite common among leaf beetles and their classification is rather complex. The classification followed here is as according to Jolivet (1993, 1997). Balsbaugh (1988) proposed a rather complex interpretation of mimicry among the Chrysomelidae. Recent papers by Hespenheide (1976, 1996) and Staines (1999) complete the picture.

Batesian and Müllerian mimicry

In Batesian mimicry, the model is protected through some defensive device and the mimic, by copying the body colour pattern and shape of the model, takes advantage of the defence preparedness of the latter. In Müllerian mimicry, on the other hand, both the model and the mimic have their own defence adaptations and, by resembling each other, each can take advantage of the defence device of the other and thus becomes better protected. As Dodson (1968) puts it, Müllerian mimicry is like a 'double insurance' for either of the two participants. It also minimizes the number of patterns predators must remember. Some species of *Cephaloleia* (Hispinae) are homochrome with their substrate, the water-filled floral bracts of *Heliconia* spp. in Central America (Jolivet, 2002, in print). Mimetic complexes, probably Müllerian, have been recorded in 17 *Cephaloleia* species involving Chrysomelidae, Cantharidae, Lampyridae, Cleridae and Lycidae (Staines, 1999).

Dodson (1968) points out that Müllerian mimicry is much more common than Batesian. It is so among chrysomelids, too. In a number of cases, a chrysomelid will mimic a coccinellid which is more toxic than itself (Jolivet, 1997). There are different opinions as to whether this situation can be taken as Batesian or Müllerian mimicry, but it seems more appropriate to think of it as a case of Müllerian mimicry. Wickler (1968) cited many such cases. Verma (1996a) pointed out a notable Müllerian mimicry between the chrysomelid *Cryptocephalus ovulum* Suffrian and the coccinellid *Brumus suturalis* Fabricius, the two being aposematic (*Figure 8.10*).

Gahan (1891), as cited by Jolivet (1997), noted a remarkable similarity in the colour pattern of *Lema* and sympatric *Diabrotica*, both chrysomelids. He showed how ten species of *Lema* closely resembled *Diabrotica* spp., all from tropical America. It is believed that *Lema*, through this aposematism, takes advantage of the toxicity of *Diabrotica*. Besides, it is known that *Lema* species are generally toxic. This aposematism, therefore, seems to be of the nature of Müllerian mimicry. Gahan (1913) reported another mimetic complex involving Hispinae and several families of beetles.

Peckhamian mimicry

This is the reverse of Batesian mimicry. An aggressor mimics a phytophagous species, and thus finds it easier to commit aggression (*Figure 8.13*).

Hemipteran nymphs, which are carnivorous and suck their nourishment from some chrysomelid adults, may resemble their prey. Chrysomelids of the genus *Mesoplatys* are fed upon by the nymphs of the bug *Afrius purpureus* Westwood in Africa. The predator is very similar to the prey in colour pattern and body shape. Also in Africa, the larvae of *Mesoplatys cincta* Olivier, a chrysomeline, are attacked by the carabid *Cyaneodinodes ammon* Fabricius, which also resembles the adult of the prey (Jolivet, 1993). There are many cases in the literature of Peckhamian mimicry between chrysomelids and Lebiidae, their predators (*Figure 8.13*).

Wasmannian mimicry

Social commensals and parasites may resemble their host so as to facilitate the association. This resemblance has been called 'Wasmannian mimicry'. The adults of the clytrine *Hockingia curiosa* Selman and the cryptocephaline *Isnus petasus* Selman, which are myrmecophiles in East Africa living in the stipules of thorns of *Acacia* trees, mimic the host ants, and also produce, through their trichomes, a secretion which is attractive to the ants (Selman, 1962; Jolivet, 1997). Eumolpines of the genus *Syagrus* also live inside the thorns with these beetles and ants and apparently are not mimetic.

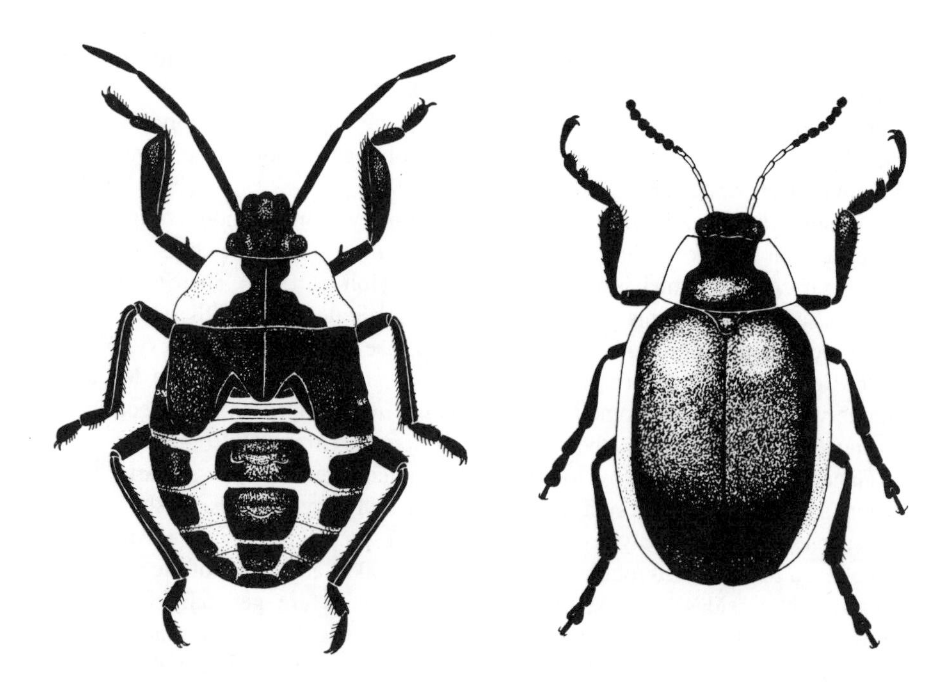

Figure 8.13. Nymph of a bug *Afrius purpureus* Westwood (left). It tends to resemble its prey, the chrysomelid *Mesoplatys cincta* Olivier (right) (after Jolivet, 1993).

Camouflage and homochromy

Into this category are put those insects which resemble, in colour and shape, some part of the host plant, soil or some other inert object in the surroundings. This makes them inconspicuous, and they may escape detection. Larvae of some *Platyphora*, *Eugonycha* and of some Chlamisinae pull out the hairs of the host plant and attach them to their own bodies for camouflage. Adults of Chlamisinae and the case-bearing larvae of Cryptocephalinae may resemble insect faecal masses (Verma and Vyas, 1987). The case-bearing larvae of Lamprosomatinae may resemble the thorns on the stems of tropical plants in America (*Figure 8.11*).

Hispines use their spines for defence. Hispines are mostly drab coloured, but in central America, some of the *Cephaloleia* spp. which live inside the bracts of various *Heliconia* are red in colour, matching the crimson coloured bracts of the host plant (Jolivet, 2002, in print). Those cases are different from those reported by Staines (1999) among *Cephaloleia* spp. which are cases of Müllerian mimicry. Large species of Hispines which feed on palm trees can also be crimson or black and yellow, but these are pure cases of aposematism (*Mecistomela*, *Coraliomela, Alurnus*). All these species are flat and do not have spines. They develop between the folded leaves of palms and other monocots.

Small cassidines are often green, or somewhat transparent. Obviously, such cases are instances of camouflage.

Change of colour

Some leaf beetles change their body colour. Such changes are especially marked in cassidines. For a useful review on the colour changes in tortoise beetles, see Jolivet (1994c).

The colour changes may be rapid, taking place in the course of a few minutes, or they may occur with a change of season or age or, in some cases, after death. Mature individuals of the North American tortoise beetle, *Metriona bicolor* (Fabricius), will change body colour from gold to brick red in less than one minute when disturbed. *Charidotella ventricosa* (Boheman), a cassidine from Panama, changes from gold to orange in one minute with a change of humidity, or on disturbance. *Aspidomorpha tecta* Boheman passes from golden to cupreous red in 2–3 minutes on being disturbed. *Cassida murraea* Linné, a European species, is green soon after emergence in July. It turns yellow-brown in autumn, and it is brick red at the end of the winter diapause. The Indian species, *Oocassida pudibunda* (Boheman) and *Cassida pusillula* Boheman, shiny green in life, turn dull brown soon after death.

Body coloration may be due to some pigments in the epidermis (chemical colour), or it may be due to interference and reflexion by certain structural features of the cuticle (structural colour). In some instances, both chemical and structural features are involved in producing the body colour.

When changes in colour are seen in a short period, as in *Metriona bicolor* in life and in some cassidines on death, the body colour is mainly structural. However, some green chemical pigments can resist death in some species, and in many brightly coloured, tropical species structural colours can persist after death (*Desmonota*). There are no rules. In some structurally coloured species, there are horizontal thin lamellae in the endocuticle, the interference effect of which changes with their degree of hydration (Hinton, 1973). When, however, body colour changes with age or season, it is due to changes in the pigment contents of the epidermis.

Colour changes may have some protective value but, in the absence of experimental evidence, this cannot be proved. These changes may serve other purposes, too. The green colour of young imagines of *Cassida murraea* may be helpful in camouflage, but the post-hibernation brick red colour may be also a declaration of sexual maturity, a case of aposematism. The brilliant golden colour of many cassidines may make them look like water drops in sunlight, and thus may be of defensive value. More observations and experiments are needed in this area.

It may be mentioned here briefly that nuptial colour changes are widespread among some chrysomelids (*Zygogramma*, *Calligrapha*, *Disonycha*, *Omophoita*, etc.) (Knab, 1909), but they are not seen in all the species of these genera. It seems that they act as a sexual signal, not as a means of camouflage. Here also, more research is needed.

STRIDULATION

Some chrysomelids are known to stridulate. The stridulatory organs of these beetles have been studied through SEM by Schmitt (1992a, 1994).

The stridulatory organ of an insect consists essentially of a ‘file’, which, in most cases, is a set of fine ridges on a certain sclerite, and a ‘scraper’, which is generally a set of stiff bristles located on an adjoining sclerotic area. The scraper is made to rub against the file to produce a chirping sound.

Schmitt (1992a) made out the following types of stridulatory organs of Chrysomelidae on the basis of the location of the file and scraper.

Mesoscutopronotal

The file is located on the middle of the mesonotum. The stiff posterior edge of the pronotum acts as a scraper. Such stridulatory modifications are known in some Cerambycidae, Zeugophorinae, Megalopodinae, Palophaginae and Orsodacninae. Schmitt (1992a) notes a close similarity in these organs of Zeugophorinae and Megalopodinae on one hand and of some Cerambycidae on the other.

Elytroabdominal

In Criocerinae, the eighth abdominal tergite carries the file, which consists of close transverse ridges made up of fused cuticular denticles. The scraper consists of conical denticles located on the apical sutural angle of each elytron. Sound is produced by contracting and extending the abdomen and the consequent movements of the file against the scraper.

Verticopronotal

This is known in Hispinae and Cassidinae. The file is located on the vertex of the head, and consists of fine transverse and parallel ridges. A single ridge on the ventroanterior edge of the pronotum is the scraper. Movements of the head against this produces the sound. Schmitt (1992a) also examined certain cassidines (*Aspidomorpha* spp., *Stolas* spp.) which do not stridulate. He could not find vestiges of the stridulatory ridges. However,*Cassida viridis* Linné, which is also not known to chirp, has such vestiges.

Schmitt (1994) recorded parallel wrinkles on the mesonotum of certain Clytrinae using SEM, but they are very low, and do not seem to be capable of sound production. Tan *et al.* (as cited by Schmitt, 1994) described a file-like modification, which could act as a stridulatory device, on the first abdominal sternite of *Oomorphoides* sp. (Lamprosomatinae).

It is generally accepted that the purpose of stridulation in Chrysomelidae is to disturb an approaching predator (Schmitt, 1992a), but the possibility that it may also be for sexual call cannot be ruled out. One of us (Jolivet, 1988a) observed adults of *Physonota alutacea* Boheman in Amazonia (Ilha de Maracca) chirping by moving their heads up and down. There were numerous individuals in a bush and it was difficult to ascertain if the beetles were communicating amongst themselves or if it was a warning signal.

Parental care and subsociality

Some chrysomelids demonstrate parental care. This obviously helps in the continuity of the race. Cassidinae are foremost in this respect.

Cassidines are known to deposit eggs as egg aggregates or egg masses. Many species (such as those belonging to the tribes Aspidomorphini, Basiptini, and

Physonotini) produce oothecae around their egg masses with the secretions of their colleterial glands (Olmstead, 1996). The oothecae provide a protective cover around the eggs.

Some cassidines, *Chelymorpha* spp., *Ogdoecosta* spp., *Omaspides* spp. and *Stolas* spp., lay eggs at the end of filaments, like the chrysopids (Olmstead, 1996), and this probably makes the eggs difficult to reach by predators.

A number of tortoise beetles defecate on an egg mass after oviposition (Olmstead, 1994, 1996). The excremental cover for an egg aggregate not only conceals the eggs but is believed to provide physical and chemical protection against predators and parasitoids. Hilker (1994) and Olmstead (1994) noted that the excremental deposit on eggs also included secretions from the colleterial and other glands. In the case of *Timarcha*, the deposit includes chemical stomach regurgitation and a variety of secretions.

In some cases, the mother remains close to the progeny beyond the egg stage to protect them. This maternal behaviour has been called 'subsociality' (Jolivet, 1988b, 1997). Subsocial maternal behaviour is known in various species of *Gonioctena* (Chrysomelinae), but not all, in Europe and Japan (Kudo and Ishibashi, 1995; Kudo *et al.*, 1995) in several species of *Labidomera* (Chrysomelinae), including *L. suturella* Guérin-Méneville in Costa Rica, and in five genera of neotropical cassidines, namely *Acromis*, *Omaspides*, *Paraselenis*, *Stolas*, and *Eugenysa* (Windsor, 1982, 1987; Windsor and Choe, 1994; Jolivet, 1997). In all these subsocial New World cassidines, the female, after laying an egg mass, stays close to the progeny throughout the egg, larva and pupal period, actively defending them against predators like ants. In several cases, the female covers the eggs with her body and leaves the progeny only when the adults eclode from the pupae. In the case of *Gonioctena* spp. which show parental care, the species are ovoviviparous and are aggressive against intruders, but the maternal care, effective against non-flying predators, seems ineffective against parasitoids (Kudo and Ishibashi, 1995; Kudo *et al.*, 1995).

It is interesting to note that subsociality has only been observed in neotropical cassidines and still remains to be discovered in the Old World, if indeed there is any. Why? Perhaps the reason for this is that highly aggressive ant species are confined to the New World. Most of the subsocial tortoise beetles of the New World live on lianas and climbers, lay their eggs as egg masses, have gregarious larvae and demonstrate maternal parental care. In *Eugenysa cascaroni* Viana, however, the male parent may stay close and seemingly give company to the female when she is guarding the offspring. In many chrysomelids, like the Colorado Potato Beetle, the male keeps guard over the female to drive away rivals after copulation (Boiteau, 1988b; Jolivet, 1999b). However, in some cases, male care of the offspring cannot be totally excluded. According to Hsiao and Windsor (1999), parental care has a dual origin in America, one (*Acromis* and *Omaspides)* in a clade of Convolvulaceae-feeding Stolaini and the other independently, and possibly more recently (*Eugenysa*), within the Asteraceae-feeding Stolaini. Among Stolaini, one species of *Echoma* has lost larval gregariousness, presumably in connection with flower feeding habits *(Figure 8.12)*.

The laying of eggs in groups or egg masses has a defensive value in several ways, namely:

(i) If the eggs are in a group, some may be left unparasitized if the parasite has a limited egg-laying capacity at one time, or if some eggs in the mass are shielded by others. That some eggs in a cassidine ootheca may escape parasitization has been observed by Becker and Frieiro-Costa (1988).

(ii) The mother may provide a protective cover for the eggs. Such a cover may be an ootheca or a faecal deposit. If eggs are in a group, the female can provide this protection for a number of eggs at the same time.

(iii) In some species, the newly hatched larvae may take to a group defence strategy, such as cycloalexy, and the laying of eggs in a group helps larval aggregate formation.

(iv) In some chrysomelids, as we have already noted, there is subsociality that is parental protection of a group of eggs/larvae/pupae. If eggs are in a group, the hatching larvae readily form an aggregate, which is needed for subsociality. In some subsocial chrysomelids, like *Platyphora*, the female is oviparous and the larvae are deposited separately but they rapidly congregate on their own.

Concealed feeding

Many chrysomelid larvae eat through and burrow into parts of the host plant. Thus, they feed and grow in a concealed situation, and are protected against predators and parasites.

The larvae of Megalopodinae burrow into twigs. Their pupae are formed in soil. Sagrinae larvae also bore into young and less woody stems and form galls. The larvae of some Alticinae, Galerucinae, Hispinae, and Zeugophorinae are leaf-miners. Some of the Alticinae, Galerucinae, Eumolpinae and Synetinae are root-borers. The biology of Aulacoscelinae is little known. In Nicaragua, adults of Aulacoscelinae have been seen to emerge from soil, but larvae can be stem- or root-feeders. The larvae of Megascelinae, due to their close relationships with Eumolpinae, are certainly root-feeders on various Fabaceae and Caesalpiniaceae. The adults appear in numbers on species of *Stizolobium* (= *Mucuna*) (Fabaceae) and they do not stridulate or jump, and do not practise reflex immobilization. The adults seem to be protected by their toxicity and the larvae by their concealed life.

Mines and burrows in plant tissue offer good protection against predators and parasites. An indication of this is that larvae with concealed feeding are not truly gregarious. If such larvae are removed from their mines, they lead a solitary life and do not show tendency to form a group. In fact, feeding in a concealed situation offers such good general protection that there is no need for any additional stratagem. However, with some larvae, for example *Diabrotica*, reflex bleeding from anterior and posterior parts of the body occurs as a protection against ants (Wallace and Blum, 1971).

While concealed feeding offers good protection against predators and parasitoids, it is not enough to prevent fungal and bacterial attacks.

Chrysomelidae are believed to have evolved from certain Cerambycidae. Cerambycid larvae are wood-borers. The larval habit of burrowing into plant tissues is therefore regarded as primitive in the group of leaf beetles.

9

Anatomy

In this chapter we will discuss only those anatomical features which are of special interest, either because they are notable synapomorphies and/or because they help to sort out the phylogenetic relations among the much diversified leaf beetles.

It was pointed out in the Introduction (Chapter 1) that large numbers of fresh specimens are needed for morphological studies and that the techniques necessary are often lengthy and difficult. Hence, our knowledge of anatomical details, being limited to only a small number of species, is patchy, and the need for further studies is clear.

Anatomical features of special interest include:

(1) Wings and wing venation.
(2) Tarsal vestiture and digitiform sensilla on the maxillary palp.
(3) Rhabdoms and rhabdomeres.
(4) Abdomen.
(5) Digestive canal.
(6) Cryptonephridic arrangement of Malpighian tubules.
(7) Ventral nerve cord.
(8) Female reproductive system.
(9) Male reproductive system.

Wings and wing venation

Wings have been important in insect evolution as they have made it possible for these arthropods to invade different land habitats. Wing venation is a complex and well-integrated system and so provides useful taxonomic characters (Mayr and Ashlock, 1991). Hindwing venation has been used in the discussion of relationships among higher taxa in many groups of Coleoptera (Jolivet, 1957–1959; Suzuki, 1994).

FOREWINGS OR ELYTRA

As in all Coleoptera, the forewings, or elytra, of leaf beetles are well-sclerotized, without clear venation and form a protective cover for the hindwings. In some protocoleopteran fossils of Lower Permian (e.g. *Tshekardocoleus* Rohdendorf) the forewings/elytra show megalopteran-like wing venation (Crowson, 1981); this supports the view that coleopterous elytra have been derived from typical insect forewings.

Leaf beetle elytra are variable in colour, texture and hardness. They may be of brilliant, metallic colours as in *Sagra*, many Chrysomelinae and Criocerinae, with varous non-metallic colours, or sombre brown, dark brown or black. In many leaf beetles (especially cassidines and paropsines), the metallic/iridescent colours, seen on live specimens, disappear on death and become replaced with a yellowish brown colour. This change is generally due to structural alteration, namely a loss of hydration and the consequent reduction in thickness of the numerous interference lamellae in the cuticle (Jolivet, 1994c). Stenomelini (Eumolpinae) in Chile and New Caledonia are soft green when alive and become dark metallic when in collections. Physiological colour changes in cassidines, which are controlled by the insect itself, can occur under stress. They last for a few minutes only. The surface of the elytra may be impunctate, finely pitted or obviously punctured, the punctures being irregularly distributed on the surface or arranged in longitudinal rows. The elytra may be fairly flexible and soft, as in Galerucinae, or quite hard.

In apterous forms (*vide infra*) the two elytra may be fused together along the suture.

HINDWINGS AND HINDWING VENATION

A hindwing has the usual Remigium or Remigial region (Rr), Vannus or Vannal region (Vr) and Jugum or Jugal region (Jr). The three regions are separated by the anterior fold (AF) and the posterior fold (PF) (*Figure 9.1*).

Arrangement of veins in the hindwing of Chrysomeloidea was extensively studied by Jolivet (1957–1959) and more recently by Suzuki (1994).

The account of hindwing venation included here is mainly based on Suzuki (1994), who has adopted the terminology of Snodgrass (1935).

Hindwing venation in Chrysomeloidea is of the Cantharid type, the main characteristic of which is a peculiar recurrent vein. Some special features of the hindwing venation in Chrysomeloidea are:

(i) The radius (R) is a thick and well sclerotized vein. In its apical portion it presents a pterostigma (Pa). The Pa includes a small radial cell (Rc), which is formed by R1 and a small cross-vein (r). The other branch of R, the radial sector (Rs) is a lightly sclerotized longitudinal vein.

(ii) Another prominent vein in the hindwing is the median (M). It divides into two branches, M1+2 and M3+4. In Chrysomeloidea the basal part of M1+2 has disappeared, while distally it is connected with M3+4 by a cross-vein (m–m). From this cross-vein, the remnant of M1+2 is seen arching basally. Thus, the characteristic recurrent vein (Rv) is formed.

(iii) The cubitus vein (Cu) divides into two branches, Cu1 and Cu2. Cu2 is always very small, inconspicuous and confined to the basal part of the wing. Cu1 divides into two branches, Cu1a and Cu1b. The basal part of Cu1a has disappeared in all Chrysomeloidea. Cu1a may be connected with Cu1b by a cross-vein (indicated as 'cv' in *Figure 9.1*). Generally, Cu1a bifurcates, and the anterior among the two branches may divide again.

Though there is considerable variation in venation within the large group of leaf beetles, the principal types of venation are derivable from a basic plan (*Figure 9.2*), strengthening the view that these beetles had a parallelophyletic origin, i. e. origin from a common ancestral stock.

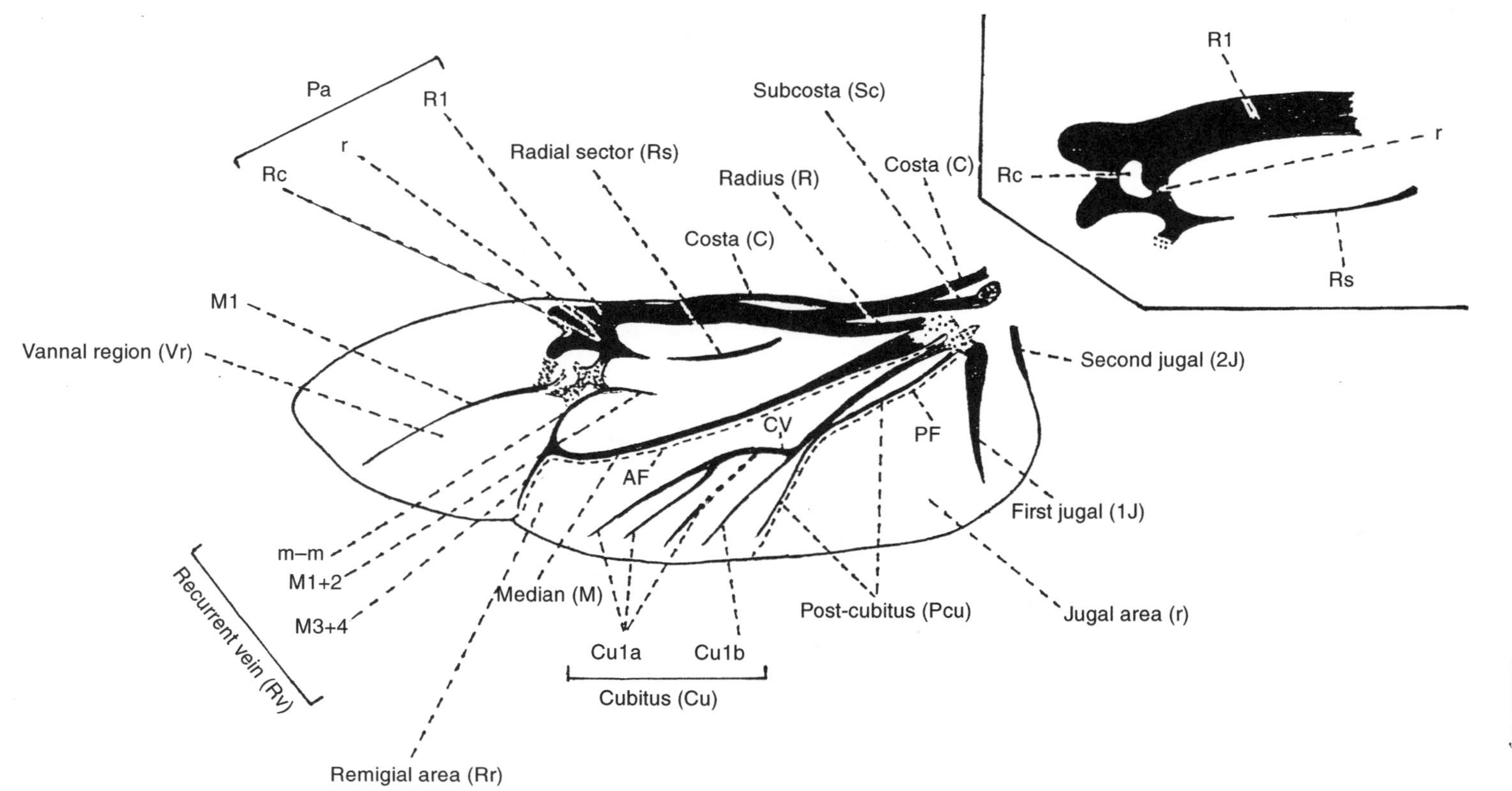

Figure 9.1. Typical hindwing venation for Chrysomeloidea (after Jolivet, 1954). Vein nomenclature following Suzuki, 1994. The inset shows the pterostigma region of the hindwing enlarged.

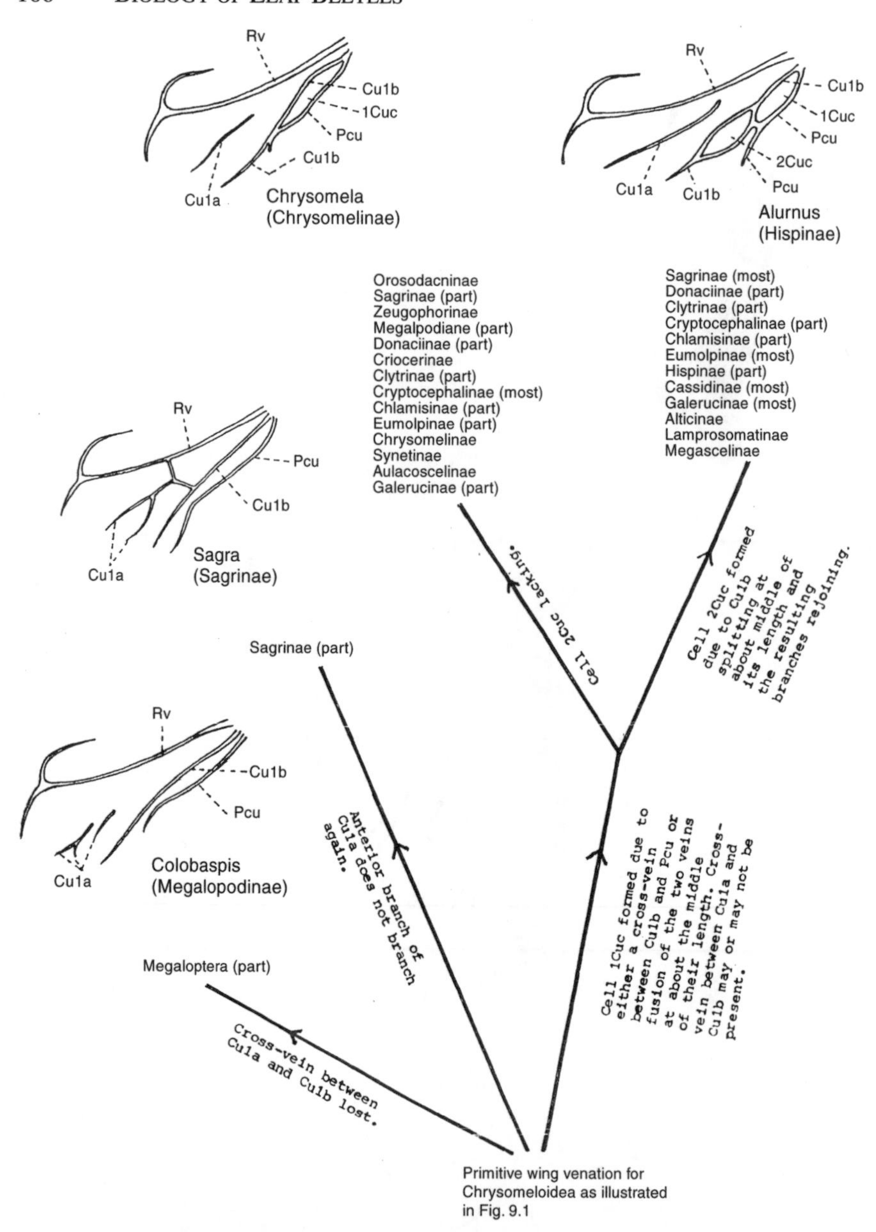

Figure 9.2. Diagram suggesting the derivation of the principal types of hindwing venation among Chrysomelidae from a primitive plan. The arrows do not indicate phyletic lines; they only suggest the direction of the changes in venation (illustration and data after Suzuki, 1994).

FLIGHTLESS LEAF BEETLES

While most leaf beetles have retained their capacity to fly, some have lost it. *Figures 9.3* and *9.4* show examples of apterous, brachypterous, brachelytrous and flightless forms.

According to Vasconcellos-Neto and Jolivet (1998), flight loss is generally due to living for a long time in a stable environment such as highlands, steppic or cold areas

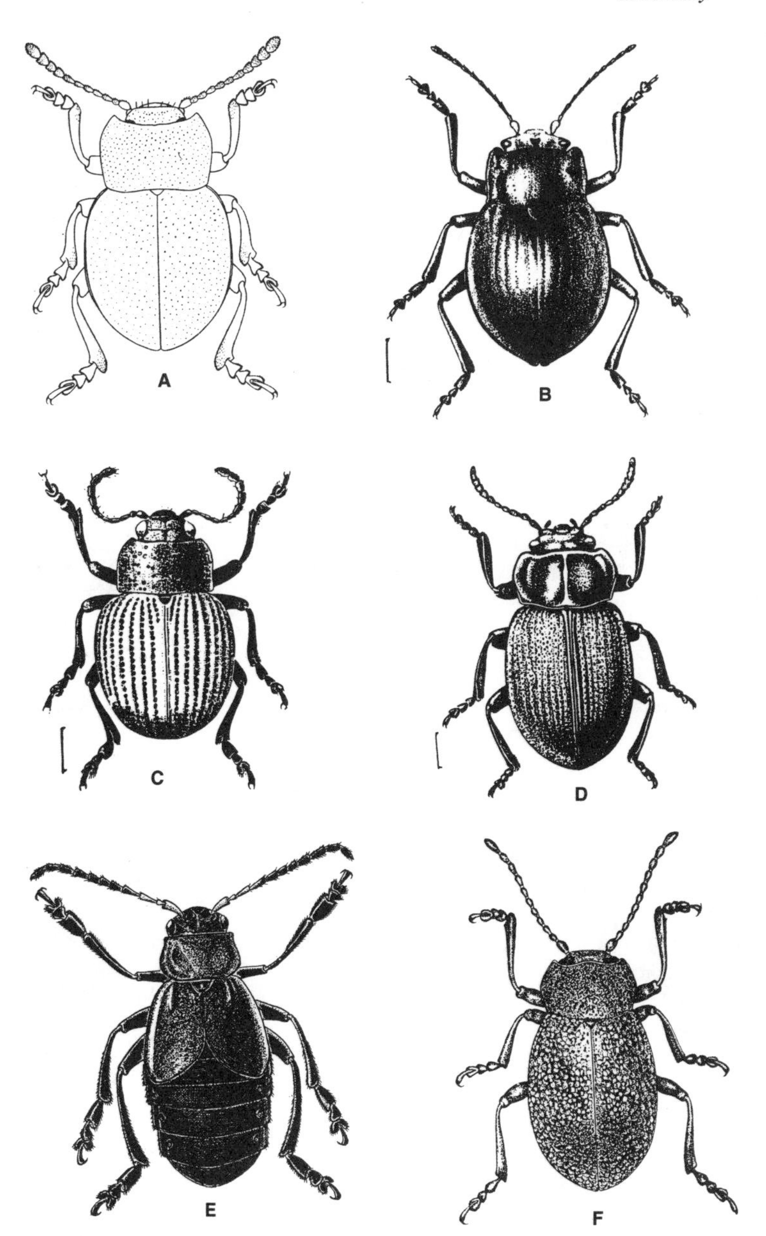

Figure 9.3. Some apterous genera and species. (A) *Timarcha (Metallotimarcha) metallica* Laicharting (Chrysomelinae), feeds on *Galium* spp. and related (Rubiaceae) (after Jolivet, 1997); (B) *Elytrosphaera* (s. str.) *melas* Jolivet (Chrysomelinae), female, Yuangas de Palmar, Bolivia, 2000 m, probably feeds on Asteraceae or Solanaceae (after Jolivet, 1950); (C) *Elytrosphaera* (s. str.) *xanthopyga* Stal (Chrysomelinae) from Viçosa, SP, Brazil, feeds on *Adenostemma brasilianum* (Asteraceae), Atlantic primary forest clearings (after Jolivet, 1997); (D) *Elytrosphaera (Elytromena) nivalis* Kirsh (Chrysomelinae) from Tungurahua, Equator, 3700 m, very probably feeds on montain Asteraceae (after Jolivet, 1997); (E) *Arima marginata* (Fabricius) (Galerucinae), France, apterous and brachelytrous, rather polyphagous but prefer Labiatae and Asteraceae, run on the ground (after Laboissière, 1934); (F) *Timarcha (Americanotimarcha) intricata* Haldeman (Chrysomelinae), female, western USA, feeds on *Rubus parviflorus* Nuttall (Rosaceae) (after Jolivet, 1976).

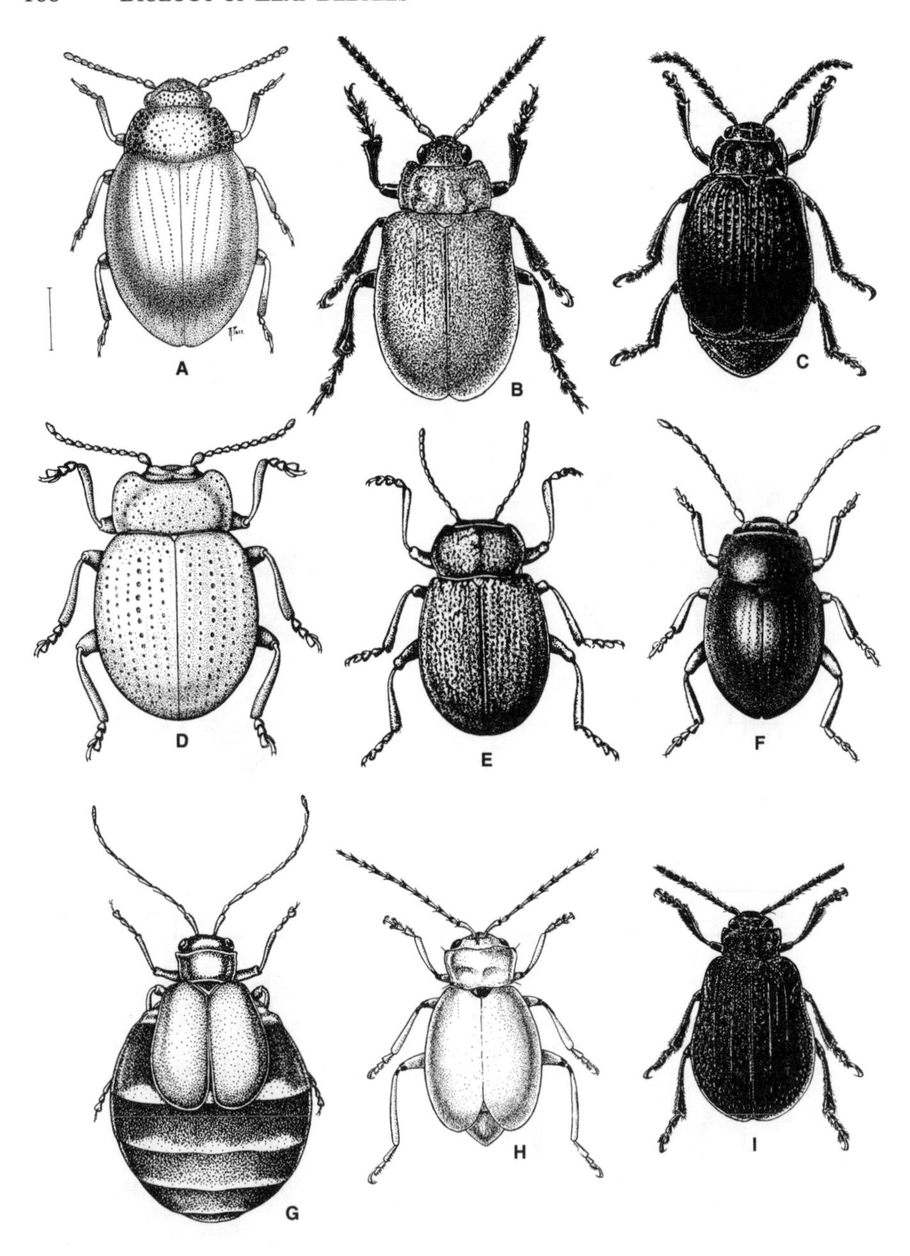

Figure 9.4. (A) *Jolivetia obscura* (Philippi) (Chrysomelinae) from Chile, apterous (after Daccordi, unpublished); (B) *Galeruca (Emarhopa) rufa* Germar (Galerucinae) from Bordeaux, France, do not fly (after Laboissière, 1934); (C) *Glaeruca (Galerima) miegi* Perez (Galerucinae) from Pyrenean mountains, south of France, apterous, slightly brachelytrous (after Laboissière, 1934); (D) *Iscadida stali* Vogel (Chrysomelinae) from Transvaal, South Africa, apterous (after Jolivet, 1995); (E) *Timarcha (Americanotimarcha) cerdo* Stal (Chrysomelinae) from Cannon Beach, Oregon, USA, apterous (after Jolivet, 1997); (F) *Cyrtonastes weisei* Reitter (Chrysomelinae) from Corfu, Greece, apterous (after Jolivet, 1997); (G) *Metacycla coeruleipennis* Jacoby (Galerucinae) from Chiapas, Mexico, female, brachelytrous and apterous (after Jolivet, unpublished); (H) *Mahutia alluaudi* Laboissière (Galerucinae) from Kenya, slightly brachelytrous (after Laboissière, 1917); (I) *Galeruca* (s. str.) *littoralis* Fabricius (Galerucinae) from Corsica, it has wings but does not fly (after Laboissière, 1934).

and islands. If the environment undergoes a change, the flightless condition can be a disadvantage as the insects are unable to find new habitats. Generally, flightlessness helps the female to concentrate on egg-production. Often, flightless leaf beetles lay a few big eggs and are K-strategists (*Brachyhelops*, *Timarcha*, etc.).

Vasconcellos-Neto and Jolivet (1998) have found that a number of neotropical species of the apterous chrysomeline *Elytrosphaera* living at high altitudes (800–4400 m) are fast declining in numbers, and are threatened with extinction. This is mainly due to the destruction of forests on mountain slopes. Similarly, the primitive and apterous genus *Timarcha* is dying out due to the destruction of sand dunes through human activity in Europe, Africa and North America (Jolivet, 1989a,b). *Timarcha* prefers to feed on weeds (*Galium*, *Plantago*, etc.) on sand dunes (Africa, Europe) or on strawberries and brambles on sand dunes and forests (America).

In flightless leaf beetles, both elytra are often welded along the suture, or fastened together by a zipper-like device (tenon and mortise joint) (Corset, 1931). This can be seen, for instance, with *Timarcha* and Chlamisinae. This condition is called 'coaptation' and develops during pupation (*Figure 9.3*).

ELYTRAL BINDING SITES

In Coleoptera there is a device to hold the elytra in place against the body. This device consists of certain patches of a dense arrangement of minute microspicules on the under-face of elytra, and corresponding similar patches on sides of the pterothorax and in some cases on an anterior part of the abdomen. The elytra are lightly bound to the body surface by interdigitation of the elytral and body patches.

The elytral patches were first described by Hammond (1979), and more recently by

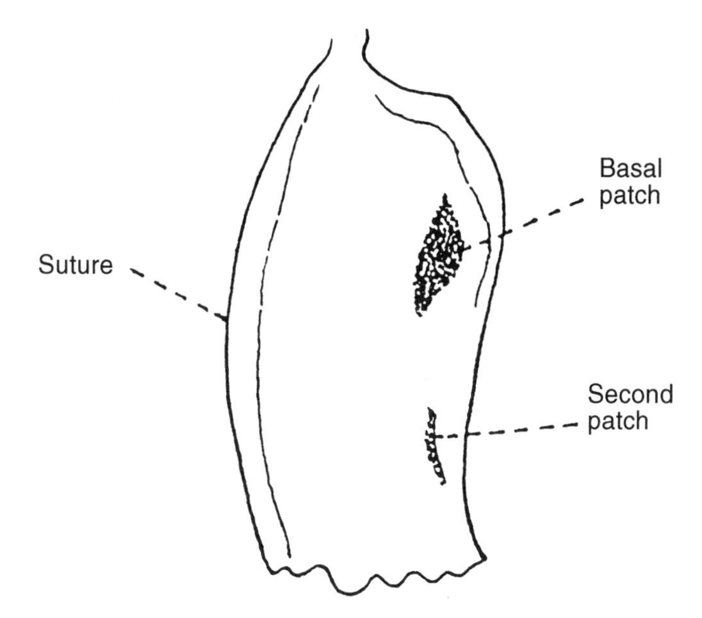

Figure 9.5. Elytral binding sites in *Rhyparida* (Eumolpinae). The figure is based on a SEM photograph of the under surface of the left elytron by Samuelson (1996). Note that the second patch of microspicules is located on a longitudinal strut.

Samuelson (1996) who has studied the patches in Chrysomelidae with SEM. The following account is based mainly on Samuelson's work.

Elytral binding patches are known in all Coleoptera except Archostemata. In Adephaga and Polyphaga there is a basal patch on the under surface of each elytron on or near the epipleural inner wall. The corresponding patch is located on the metathorax (*Figure 9.5*).

In addition, a second elytral patch can be seen in some subfamilies of Chrysomelidae. It interlocks with a patch on the first abdominal segment. This second elytral patch is present in most Sagrinae, certain Chrysomelinae, and most Alticinae. Curiously, it is lacking in Galerucinae. The second patch is located on a longitudinal strut-like elevation on the under surface of the elytron in Eumolpinae, certain Chrysomelinae, many Cassidinae, and also in some Hispinae.

As pointed out by Samuelson (1996), the basal elytral patch seems to be an important apomorphic character, while the second patch is a synapomorphy, which appears to have evolved independently in some chrysomelid subfamilies.

The number and distribution of the elytral binding sites may be useful in distinguishing between clades or tribes in some subfamilies of Chrysomelidae.

Tarsal vestiture and digitiform sensilla on the maxillary palp

TARSAL VESTITURE

The first three tarsal joints of the thoracic legs are covered on their under surface with a dense brush of hairs. This pad of tarsal hairs improves the adherence of the leaf beetle to the plant surface, so that it is not easily dislodged by breeze or predator.

In some chrysomelid subfamilies (Sagrinae, Criocerinae, Donaciinae, Cassidinae and Hispinae) and also in Bruchidae, some or all of the first three tarsomeres are provided with bifid hairs (*Figure 9.6*) (Stork, 1980a,b; Mann and Crowson, 1981; Mann, 1985; Schmitt, 1989, 1996). In Hispinae and Cassidinae, such tarsal setae are on all the first three tarsal segments, but in Bruchidae, they occur on the tarsal segments 2 and 3 only, while in Sagrinae, Donaciinae and Criocerinae only the third tarsal joint is provided with bifid setae. Such bifid tarsal setae are not known in any other beetle taxon (Schmitt, 1989).

The occurrence of bifid setae in all the first three tarsal joints in Cassidinae and Hispinae obviously has the purpose of enhancing the defence value of the tarsi in these

Figure 9.6. Bifid setae on the third tarsomere of *Dactylispa* (Hispinae) (based on a SEM photograph by Schmitt, 1989).

insects. The setae may be spatulate at their tips (Mann, 1985). In the cassidine *Hemisphaerota* found in South USA, the tarsal hairs may be 'moistened' with an oily secretion produced by glands opening between the bases of the hairs (Eisner, 1971). During ordinary walking, only a few tarsal bristles of *Hemisphaerota* come into contact with the substratum, and the hairs are 'dry', i.e. not covered with the oily secretion. When the beetle is attacked by a predator, the hairs become covered with the secretion, and all three pairs of tarsi are pressed down, so that nearly all the bristles are touching the substratum. Under these conditions, considerable force is needed to dislodge the beetle. This defence strategy has already been discussed in detail in Chapter 8 (*Figure 9.6*).

MAXILLARY DIGITIFORM SENSILLA

In a number of beetle taxa, the last joint of the maxillary palp has an arrangement of digitiform and peg-like sensilla (Honomichl, 1980; Mann and Crowson, 1981, 1984). As Honomichl (1980) observed, the sensilla are approximately the same size, irrespective of the size of the beetle.

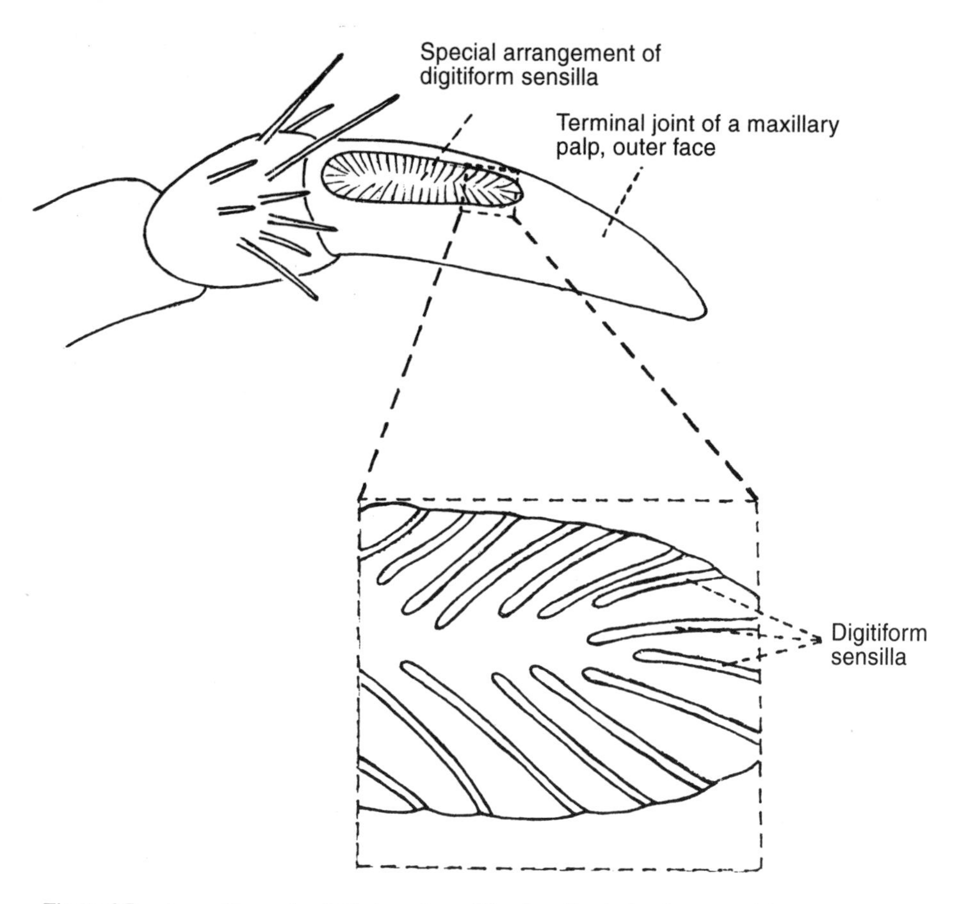

Figure 9.7. A maxillary palp of *Clythraxeloma* (Megalopodinae) showing a special arrangement of digitiform sensilla. A part of the arrangement has been further magnified (based on SEM photographs by Schmitt, 1994).

As regards the function of these sensilla, Zacharuk *et al.* (1977) found that the digitiform sensilla on the labial palps of a larval elaterid show electrical responses to contact and vibrations, but none to chemical stimuli. Mann and Crowson (1984) believe that they are proprioreceptors, and Schmitt (1994) suggests that they function as stretch receptors for the cuticular cover of the maxillary palp.

In some Cerambycidae (e.g. *Callidium violaceum* Linne), Orsodacninae, Megalopodinae and Palophaginae, the digitiform sensilla have a characteristic arrangement on the terminal joint of the maxillary palp (Schmitt, 1994). The sensilla are situated in an elongated area on the outer face in the basal part of the terminal segment of the maxillary palp (*Figure 9.7*). They arise along a U-shaped line and point towards the central axis of the whole arrangement.

In Sagrinae too, a U-shaped arrangement of the maxillary digitiform sensilla was reported by Mann and Crowson (1984). But Schmitt (1994) found that in *Sagra femorata* Drury (which was one of the species studied by Mann and Crowson, 1984), the sensilla in the arrangement were less numerous, the arrangement was rather irregular and was located near the apex of the palpus segment.

PHYLOGENETIC SIGNIFICANCE OF BIFID TARSAL SETAE AND THE DIGITIFORM MAXILLARY SENSILLA

As noted before, bifid tarsal setae are known only in Bruchidae, Sagrinae, Criocerinae, Donaciinae, Cassidinae and Hispinae, and not in any other beetle group. In view of this important synapomorphy, Schmitt (1996) does not doubt that these groups are monophyletic. He believes that, among these groups, the most likely course of evolution is that shown in *Figure 9.8*.

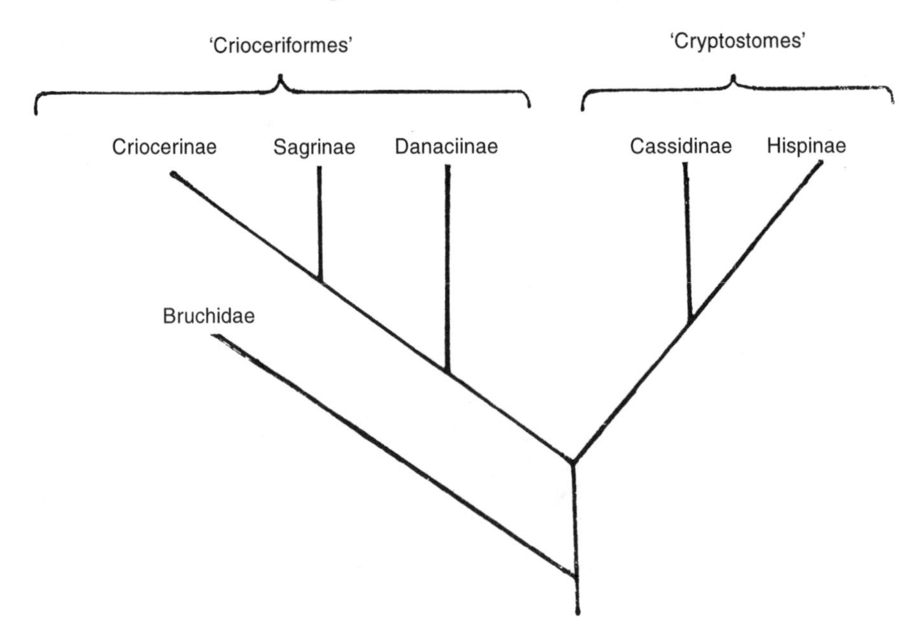

Figure 9.8. Probable phylogenetic relations among chrysomeloid groups with bifid tarsal setae (based on Schmitt, 1989).

The fact that the characteristic U-like arrangement of digitiform sensilla on the terminal maxillary palp joint is shared by some Cerambycidae on the one hand and some primitive chrysomelids (Orsodacninae, Megalopodinae and Palophaginae) on the other, helps to confirm the idea of a basal connection between Cerambycidae aand Chrysomelidae. The presence of mesoscutopronotal stridulatory apparatus in Cerambycidae (part) and in the chrysomelid subfamilies named above (Schmitt, 1994), the common aedeagal structure seen in them (Verma, 1996c) and the presence of dorsal and ventral ambulatory ampullae in the larvae of Megalopodinae and Palophaginae (Schmitt, 1994) give further strength to the notion.

Rhabdoms and rhabdomeres

EM studies on rhabdoms and rhabdomeres by Schmitt *et al.* (1982) reveal that, while perhaps due to functional need, these photosensitive structures present considerable variation. They also exhibit a fundamental similarity.

In Chrysomeloidea, the dioptric portion of an ommatidium includes an acone type of arrangement of vitrellae, surrounded by a mantle formed by two pigmented cells. Proximally, this arrangement is followed by a fascicle of eight retinulae. In this fascicle, two retinulae (7 and 8) occupy a central position, and the remaining (1–6) are arranged peripherally around the retinulae 7 and 8. All retinulae carry on their inner surface a pallisade of microvilli called the rhabdomere. The rhabdomeres together seem to form a refractile rod, the rhabdom.

In all the 108 species of Chrysomeloidea studied by Schmitt *et al.* (1982), the rhabdomeres of the outer retinulae (1–6) are much shorter than those of the central ones (7 and 8). The former are limited in their extent to the distal part of the retinulae.

Two types of rhabdoms are found among Chrysomeloidea, viz. the insula pattern and the ponticulus pattern. In the former type, the rhabdomeres of the central retinulae (7 and 8) are fully separated from the rhabdomeres of the peripheral cells (*Figure 9.9*). In the ponticulus pattern, the central rhabdomeres appear elongated in cross sections, and are in contact with the rhabdomeres of the peripheral retinulae 1 and 4.

Rhabdoms can also be classified as open and closed types. In the open type, the central rhabdomeres are separated from the peripheral ones by larger or smaller distances, or they are only in contact with some of them (*Figure 9.9*).

In both the insula, as well as the ponticulus pattern, the arrangement and direction of microvilli in the rhabdom varies greatly. It can be different among subfamilies, tribes or genera. For details of this variation see Schmitt *et al.* (1982).

All Chrysomeloidea have open rhabdoms, a fact which lends support to the notion of the homogeneity of the group.

Bruchids have only the insula pattern. Cerambycids and chrysomelids show both the insula and the ponticulus pattern. All chrysomelid subfamilies have the insula pattern, except Chrysomelinae and Timarchinae, which have rhabdoms of the ponticulus type. This feature indicates a proximity between Chrysomelinae and Timarchinae, though the two differ in significant ways, e.g. in the tegmen in the male copulatory apparatus and in the nervous system.

On the basis of the arrangement of microvilli in the rabdom structure, Schmitt and associates (1982) have inferred phyletic proximity between Orsodacninae and

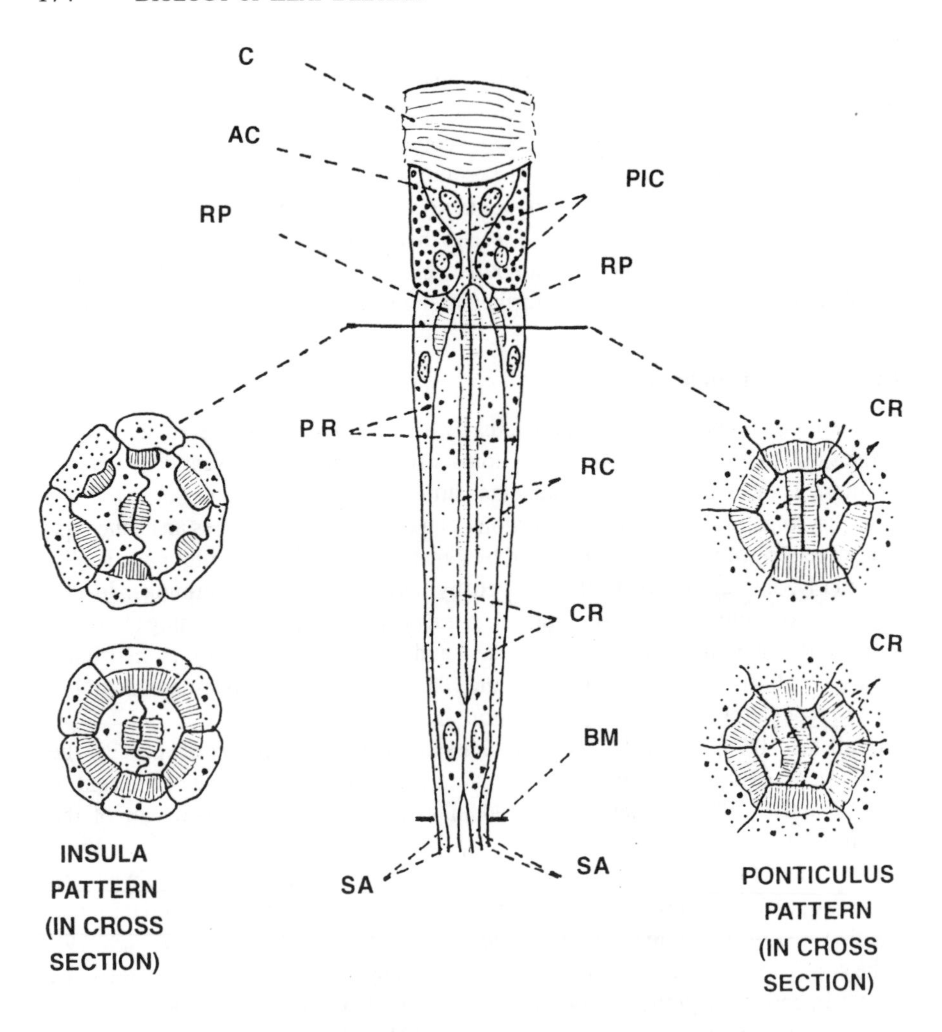

Figure 9.9. The organization of an ommatidium in a chrysomeloid (after Schmitt *et al.*, 1982) (AC = vitrellae in an acone type of arrangement; BM = basement membrane; C = cuticular lens; CR = central retinulae, i.e. retinulae 7 and 8; PIC = pigment cells; PR = peripheral retinulae, i.e. retinulae 1 to 6; RC = rhabdomeres of central retinulae; RP = rhabdomeres of peripheral retinulae; SA = sensory axons).

Megalopodinae. A similar closeness between Clytrinae and Cryptocephalinae and between Galerucinae and Alticinae could also be inferred. Cassidinae and Hispinae, though closely related, have certain differences in the structure of the rhabdom.

The retinula cells are the site of photoreception. Sight among larvae is reduced but adults can distinguish (*Chrysolina* spp.) blue from yellow and orange from violet and green (Schlegtendal, 1934). Several species even seem to be able to distinguish different tones of greens (*Agelastica alni* (L.) and *Chrysolina fastuosa* (Scopoli)). The influence of trap colour on the capture of specimens seems to vary from species to species (Vincent and Stewart, 1986). However, it does seem that yellow or green traps are more attractive than orange, white, blue or red ones (Nielsen, 1988). For further discussion of this topic, see Jolivet (1997).

Abdomen

ABDOMEN IN COLEOPTERA

In all Pterygotes, the first abdominal sternum gets reduced and fused with the metasternum. The second abdominal sternum is also generally lost in late post-embryonic development (Matsuda, 1976). Murray and Tiegs (1935) have shown that in the prepupa of *Calandra* two transverse depressions in the venter of the anterior abdomen indicate the first two abdominal sterna. In the pupal stage, a peg-like process grows forward from the third abdominal sternum, and is received by a corresponding depression on the metasternum. At this stage, the first two sterna of the abdomen come to lie in a groove above the peg, and they become membranous, while muscles related to them undergo degeneration.

At the other end of the abdomen, the 9th segment is usually the last clearly indicated segment. In male Coleoptera, the 10th abdominal segment is not often indicated and in the female it may be included in the ovipositor.

The 9th abdominal segment usually becomes folded in so as to contribute to the formation of the anogenital vestibule, i.e. the mostly membranous chamber at the end of the abdomen, receiving both the anus and the genital opening. The 8th segment may also get invaginated to help in the formation of this chamber.

Jeannel and Paulian (1944) have distinguished 4 types of abdomen among beetles, viz.:

(a) The adephagid type, which is characteristic of Adephaga.
(b) The haplogastrous type, in which the second abdominal sternum survives as a pair of laterally situated triangular sclerites.
(c) The cryptogastrous type, in which the second abdominal sternum is entirely lost through membranization.
(d) The hologastrous type, in which the second abdominal sternum is a distinct and typically developed sternum. This condition is seen in some of Cantharoidea.

ABDOMEN IN CHRYSOMELIDAE

In Chrysomelidae, the abdomen has 7 terga and 5 sterna externally. The first visible sternum belongs to the third abdominal segment. The 8th abdominal segment and some vestiges of the 9th segment are folded in to be included in the anogenital vestibule.

According to Jeannel (1955), the abdomen in Phytophagoidea (= Chrysomeloidea + Curculionidea) is of the cryptogastrous type. While generally in Chrysomeloidea only 5 abdominal sterna may be seen externally, and the second sternum is not represented even by small, triangular and laterally located sclerites, e.g. in *Aspidomorpha miliaris* (F.) (Verma and Kumar, 1972) (*Figure 9.10*) and in *Chrysomela populi* L. (= *Lina populi* (L.)) (Harnisch, 1915), the haplogastrous type of abdomen has been noted in *Galerucella birmanica* Jacoby (Verma, 1969) (*Figure 9.11*) and in *Aulacophora foveicollis* Lucas (Kumar and Verma, 1980).

To summarize, the abdomen in Chrysomelidae has 7 externally visible terga and 5 externally readily visible sterna, and it may be of the haplogastrous or cryptogastrous type. It would be interesting to find out the distribution of the two types of abdomen among the leaf beetles.

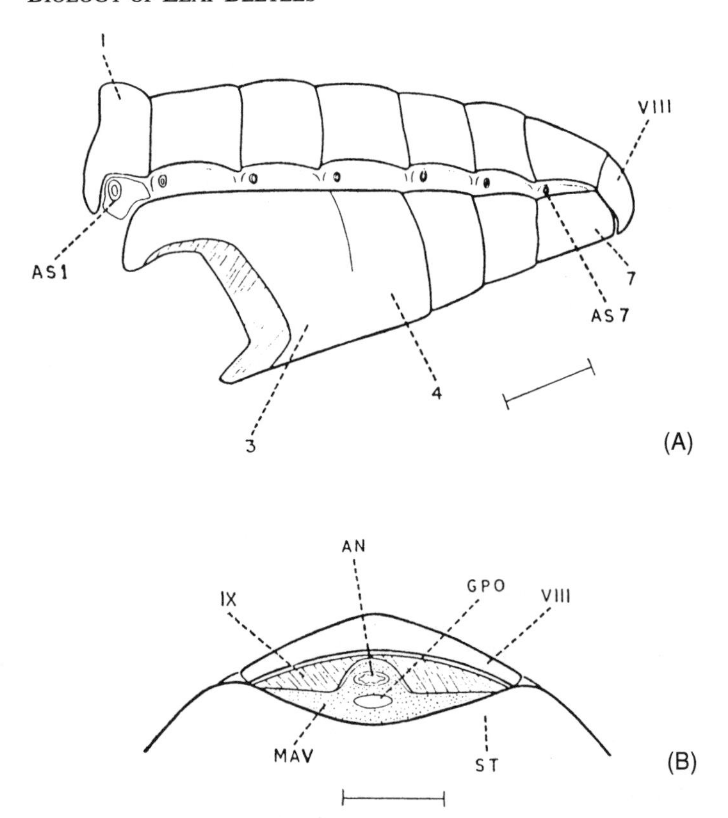

Figure 9.10. Abdomen of adult male of *Aspidomorpha miliaris*. (A) Lateral view; (B) Last externally visible tergum and sternum gaped to show the anogenital vestibule (from Verma and Kumar, 1972) (I to IX = abdominal terga; 3 to 7 = abdominal sterna; AS1 and AS7 = abdominal spiracles 1 and 7 respectively; AN = anus; GPO = opening of the genital pocket; MAV = membranous wall of the anogenital vestibule; ST = sternum), scale denotes 1 mm.

While the terminal abdominal sternites and tergites (i.e. 8th and 9th) are infolded and form the anogenital vestibule, and they may show a varying extent of reduction and modification in relation to the functional need of the aedeagus in the male and formation of the ovipositor in the female, the externally seen abdominal tergites and sternites do not exhibit any special modification, apart from the seventh sternite, which is often quite modified in the male. It may show slight emargination at the hind border to accommodate the base of the aedeagus, when the latter is protruded for copulation. In many galerucines, e.g. *Aulacophora foveicollis*, the last visible sternum is trilobed (*Figure 9.12*). During copulation, a hood-like structure, formed by 8th and 9th terga, moves back, its lateral edges fitting into the deep clefts which separate the three lobes of the 7th sternum. Thus, a tubular passage is formed for protrusion of the aedeagus (Kumar and Verma, 1980). In males of two Indian species of the galerucine *Hoplasoma*, *H. unicolor* Illiger and *H. nilgiriensis* Jacoby, two finger-like processes arise from the second apparent abdominal sternite and extend backward (*Figure 9.13*). The processes are longer in the former species. In addition, in both species the last visible sternite shows a specialized quadrangular and well-demarcated area, another male secondary sexual feature.

Iuga and Konnerth (1963) studied the apical part of the abdomen in some Alticinae,

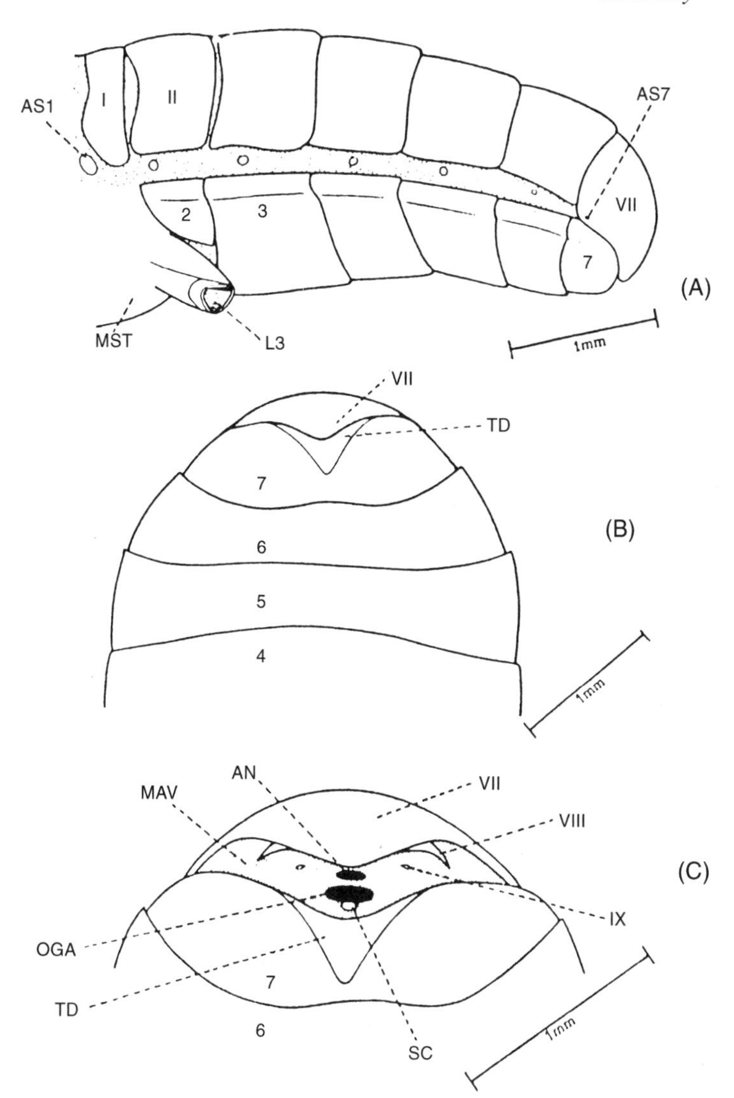

Figure 9.11. Abdomen of *Galerucella birmanica*, adult male. (A) Lateral view; (B) Ventral view of posterior segments; (C) Similar to (B) but last externally visible tergum and sternum gaped to show the anogenital vestibule (from Verma, 1969) (I to IX = abdominal terga; 3 to 7 = abdominal sterna; AN = anus; AS1 and AS7 = abdominal spiracles 1 and 7 respectively; L3 = third leg; MAV = membranous wall of the anogenital vestibule; MST = metasternum; OGA = opening of the genital pocket; SC = sclerite guarding the lower edge of the opening of the genital atrium; TD = triangular depression on the 7th abdominal sternum).

and regarded the spiculum/second spiculum in the male as the 8th abdominal sternum, and the tegmen/first spiculum as the 9th abdominal sternum. These suggested homologies, however, are not in agreement with ontogeny. The tegmen/first spiculum develops as an apodemal ingrowth from the base of the aedeagus or tegmen; hence, in a strict sense, it is a tegminal apodeme (Verma, 1969; Kumar and Verma, 1980; Verma, 1996c). The second spiculum is an apodemal growth from sides of the 9th abdominal segment (Kumar and Verma, 1980).

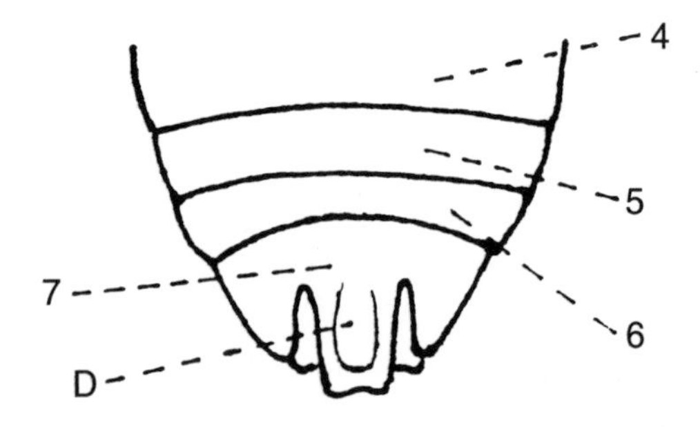

Figure 9.12. Abdomen of *Aulacophora foveicollis*, ventral view (after Maulik, 1936) (D = oval depression on the middle lobe of the three lobed 7th abdominal sternum; 4 to 7 = abdominal sterna).

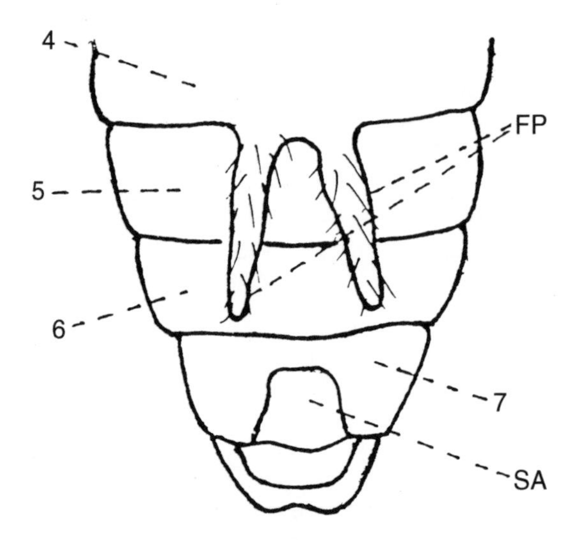

Figure 9.13. Abdomen of *Hoplasoma unicolor*, ventral view (after Maulik, 1936) (FP = finger-like processes, arising from the 2nd visible abdominal sternum and extending backward; SA = quadrangular specialized area on the 7th sternum; 4 to 7 = abdominal sterna).

Digestive canal

REGIONS IN THE DIGESTIVE TUBE

The alimentary system has been studied and described for only a small number of chrysomelids. The following description is based on the accounts available.

The fore-gut is short, without proventriculus or gizzard, and differentiated into an oesophagus and a small crop. In *Syneta*, the crop is comparatively large (Mann and Crowson, 1981). This part of the digestive tube has the usual histological structure of the stomodeum (Shrivastava, 1986). At the beginning of the mid-gut, the stomodeal and the mesenteric walls are invaginated to form the 'oesophageal invagination', which is valvular in nature (Wigglesworth, 1972).

The mid-gut is quite long. Its first part is dilated into a crop-like enlargement, while the rest of it is narrow and intestine-like. The two regions of the mid-gut may be referred to as MS1 and MS2 respectively (*Figure 9.14*).

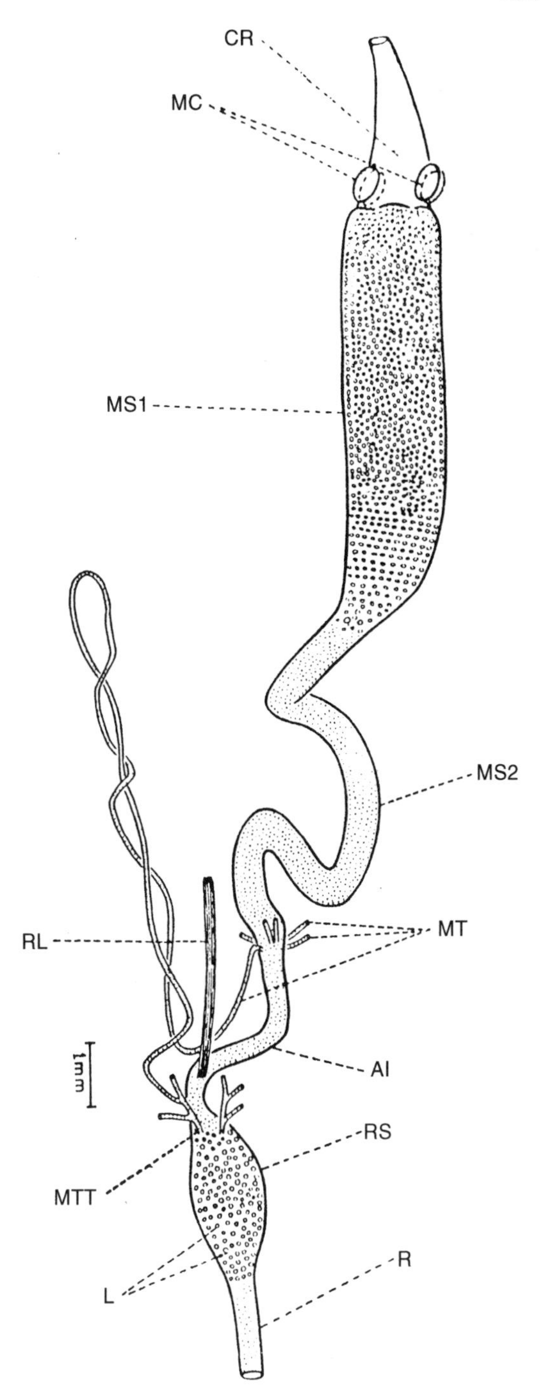

Figure 9.14. Alimentary canal of *Aspidomorpha miliaris*, removed from the insect and uncoiled (from Shrivastava, 1986) (AI = anterior intestine; CR = crop; L = leptophragmata; MC = mid-gut caeca; MT = Malpighian tubules; MTT = common stem formed by union of the three M. tubules of the same side; R = rectal tube; RL = rectal strand; RS = rectal sac; MS1 and MS2 have been explained in the text).

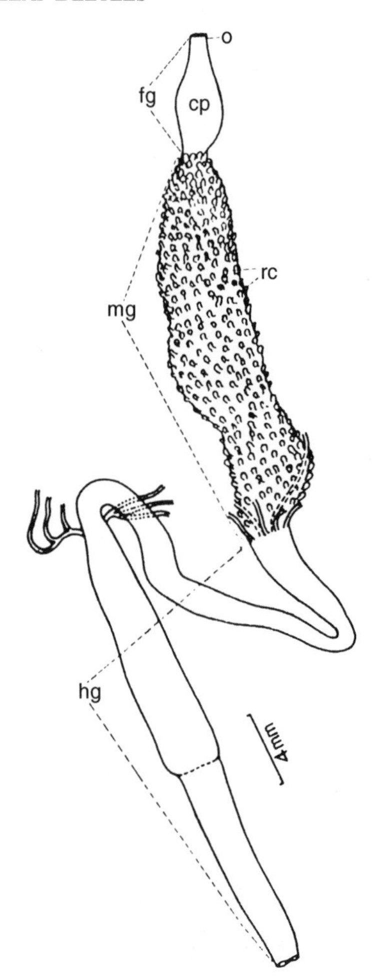

Figure 9.15. Alimentary tract of *Orsodacne* (from Mann and Crowson, 1981) (cp = crop; fg = fore-gut; hg = hind-gut; mg = mid-gut; o = oesophagus; rc = regenerative crypt).

This differentiation of mid-gut into MS1 and MS2 seems to be quite common, not only among Chrysomelidae but also among Curculionoidea. It has been described by Woods (1916) in *Altica*, Wilson (1934) in *Chrysochus*, Khatib (1946a) in *Galerucella*, Shrivastava (1986) in *Aspidomorpha*, Murray and Tiegs (1935) in *Calandra*, Kaston (1936) in *Hylurgopinus*, and Sundman and King (1964) in *Anthonomus*. Such a division of the mid-gut is also characteristic of other Coleoptera such as *Dytiscus*, Rungius (1911). In *Orsodacne* and *Syneta*, however, the mid-gut is nearly uniform in diameter throughout (Mann and Crowson, 1981) (*Figure 9.15*).

The mid-gut wall has the typical histological structure, including cuboid to columnar glandular epithelial lining, often showing a distinct striated border, and groups of dividing smaller regenerative cells or nidi (*Figure 9.16*). At the position of the nidi, an outpocketing or a regenerative crypt is formed by rapid and repeated cell divisions. These crypts may only be small bulges on the outer face of the mid-gut, or

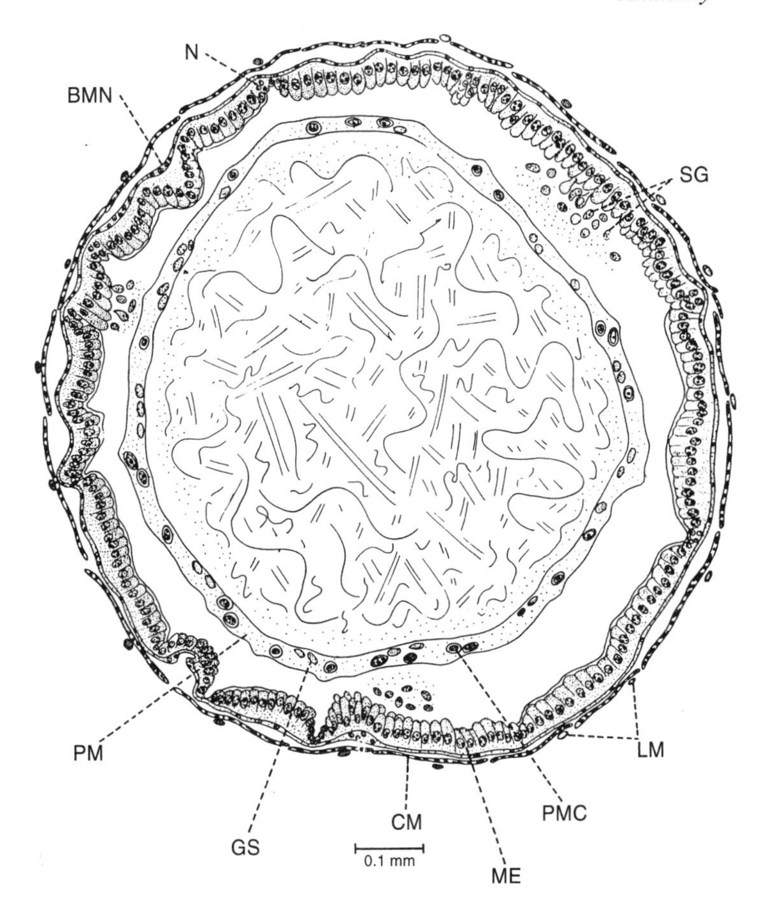

Figure 9.16. T.s. of MS1 of *Aspidomorpha miliaris* (from Shrivastava, 1986) (BMN = nucleated basement membrane; CM = circular muscle fibres; GS = lacunae in pm filled with eosinophilous granular material; LM = longitudinal fibres; ME = mesenteric epithelium; N = nidus; PM = peritrophic membrane; PMC = peritrophic membrane cells; SG = secretory globules formed by breaking off of dilated ends of mesenteric epithelial cells; MS1 has been explained in the text).

they may be quite deep, penetrating the muscular coat and appearing as longish elevations on the surface of the mesenteron.

The epithelial lining of the mesenteron is protected by a peritrophic membrane. In at least one type of leaf beetle, the peritrophic membrane is known to show a special feature, peritrophic membrane cells. These cells will be discussed in a separate section below.

The hind-gut has the usual histological structure and regionation. It is quite long and divided into the anterior intestine and rectum, and the latter is further divisible into the rectal sac and rectal tube. The rectal sac is large and heavy due to reassociation of the Malpighian tubules or the cryptonephridic condition (see section below).

The six Malpighian tubules arise at the junction of the mid-gut and hind-gut, generally in two groups, two tubules a little anterior to the remaining four (see section below). In some chrysomelids, however, e.g. *Orsodacne* and *Syneta* (Mann and Crowson, 1981) (*Figure 9.15*) and *Aspidomorpha miliaris* (F.) (Shrivastava, 1986), all six arise in the same transverse plane, in which case the fact that they belong to two groups is indicated by their relative length and the way they join to form a common stem entering the structure of the rectum. In *Orsodacne*, among the three tubules on each side are included "two of them longer than the third, the longer two becoming attached to each other apically before the shorter one joins them, and thus all three become attached to the side of the hind-gut" (Mann and Crowson, 1981).

GASTRIC CAECA

In many chrysomelids there are a variable number of mid-gut or gastric caeca, also variable in shape. In *Aspidomorpha miliaris,* there are four vesicle-like caeca arising from the anterior end of the mid-gut (Shrivastava, 1986) (*Figure 9.14*).

Mann and Crowson (1983c) recorded gastric caeca in adults of all the 31 species of Eumolpinae examined by them. In addition, they observed the caeca in adults of 8 species of Cassidinae and 5 species of Hispinae. Gastric caeca in the larval stage were noted in 9 species of Sagrinae and in *Megalopus*. Further, the authors point out that in the adults of all the species examined by them, and belonging to the following chrysomelid groups, the gastric caeca were found lacking: Orsodacninae, Synetinae, Galerucinae, Alticinae, Chrysomelinae, Zeugophorinae, and Camptosomata.

In Eumolpinae and in *Megalopus*, Mann and Crowson (1983c) described two sets of gastric caeca, an anterior group of a small number of large caeca and a second set of numerous small caeca spread over a posterior part of the mid-gut. The figure showing the *Megalopus* larval digestive canal included in their paper, however, suggests that the structures, regarded by the authors as posterior gastric caeca, may be regenerative crypts extending beyond the muscular coat. A histological examination of the mid-gut of larva of *Megalopus* or of an adult eumolpine will settle this point.

PERITROPHIC MEMBRANE CELLS

In *Aspidomorpha miliaris*, a peritrophic membrane (pm) surrounds the contents of both MS1, as well as MS2. The pm is of the 'type i' (Wigglesworth, 1972) as it separates by 'delamination' from the inner face of the mesenteric epithelium. The pm of MS1 is considerably thicker than that in MS2, and shows some cellular bodies included in its thickness (*Figure 9.16*) (Shrivastava and Verma, 1982). These cells have been referred to as peritrophic membrane cells (pm cells). They lie in lacuna-like spaces in the structure of the pm. There are no such cells in the pm of MS2. In some sections of MS1 the appearance of some of the mesenteric epithelial cells suggest that the cells become specially basophilic, move out of the epithelial layer, and become included in the forming pm. Thus, the pm cells are mesenteric epithelial cells which have become dislodged and have come to lie in the pm.

In the lacunae occupied by the pm cells, the cells show some changes: the nucleus becomes smaller, eosinophilous granules appear in cytoplasm, and the cell boundary fades away. As a result, the lacuna becomes full of eosinophilous granular material (*Figure 9.17*). Material with a similar appearance is also seen in the lumen within the pm. Shrivastava and Verma (1982) inferred from these observations that there is holocrine secretion by the pm cells, releasing digestive enzymes. This idea is supported by their observations that a patch of pm from MS1

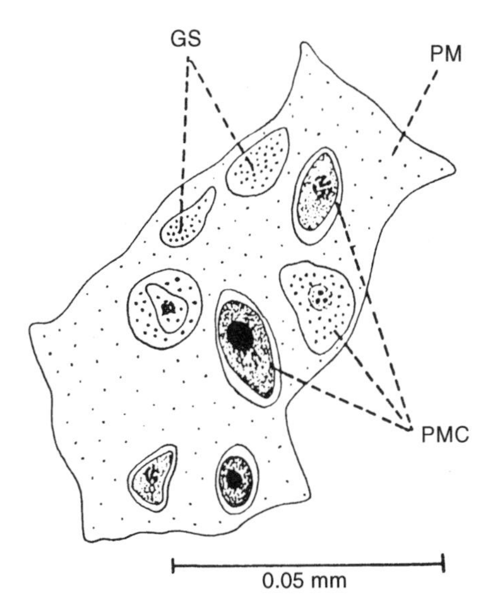

Figure 9.17. A part of pm in a t.s. of MS1 of *Aspidomorpha miliaris* (from Shrivastava, 1986) (GS = lacunae in pm filled with eosinophilous granular material; PM = peritrophic membrane; PMC = peritrophic membrane cells).

is able to digest a starch film, which remains almost unaffected if kept in contact with a piece of pm from MS2.

Through their experiments, Verma and Shrivastava (1989) have inferred that protease and lipase activities are more pronounced in the lumen within the pm of MS1 in *Aspidomorpha miliaris*, while amylase activity is more marked outside the pm in this part of the mid-gut. It seems, therefore, that the mid-gut epithelial cells, dislodged and included in the pm of MS1, are specially efficient in producing proteases and lipases, and those in the general epithelium of MS1 mostly produce amylases.

Are pm cells present in all cassidines? Do they occur in other chrysomelids, too? These questions may be answered only through further investigations.

RECTAL STRAND

Shrivastava and Verma (1983) found a thin strand connecting that part of the anterior intestine, which joins the rectum with the prescutum of the mesonotum (*Figure 9.18*). It is slightly thicker in the female than in the male. Before reaching a median ridge-like thickening in the prescutum, the fibres of the strand separate into two bundles, encircling the aorta and then getting attached to the ridge. In the female, the ovarian suspensory ligaments also join the ridge close to the attachment of the rectal strand (*Figure 9.19*).

The rectal strand is not a tendinous thread. It is contractile, and shows striated muscle fibres in its structure. Shrivastava and Verma (1983) observed some electrical activity in the strand on stretching it between two microelectrodes. They inferred that the contraction of the strand is stretch-stimulated, and that the strand, through its tonus, helps in keeping the rectal sac longitudinal, without folding, and well above the genital tract so that the sac's outer surface, provided with numerous and large

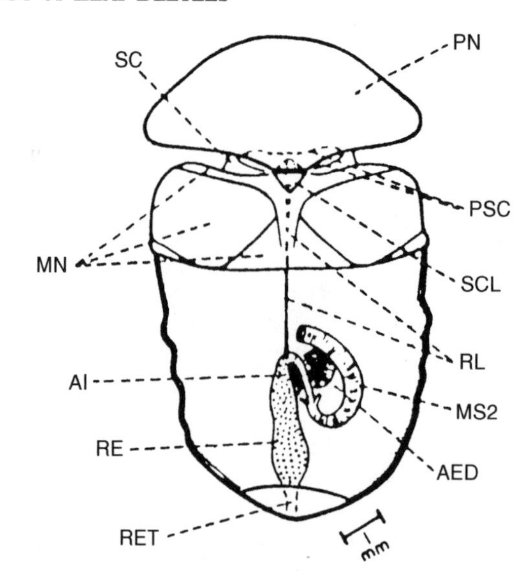

Figure 9.18. *Aspidomorpha miliaris* male, dissected from the dorsal side. Extent of the rectal strand may be seen (from Shrivastava and Verma, 1983) (AED = aedeagus in repose; AI = anterior intestine; MN = mesonotum; PN = pronotum; PSC = prescutum of mesonotum; RE = rectal sac; RET = rectal tube; RL = rectal strand; SC = scutum of mesonotum; SCL = scutellum of mesonotum).

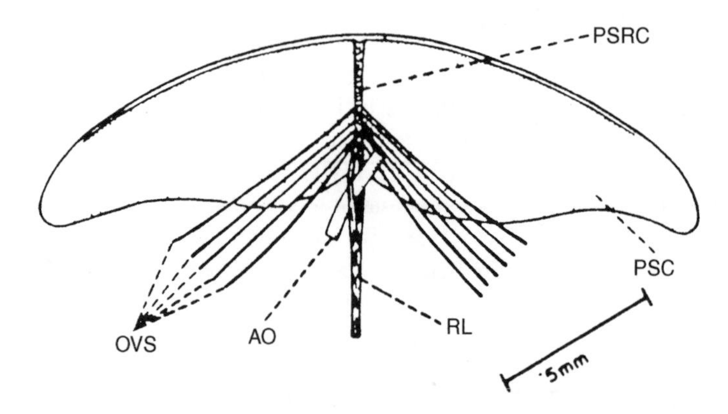

Figure 9.19. *Aspidomorpha miliaris*, prescutum of mesonotum from the ventral side. Ovarian ligaments and rectal strand are seen finding attachment on a median ridge-like thickening on the under surface of the prescutum (from Shrivastava and Verma, 1983) (AO = aorta; OVS = terminal filaments of ovarioles; PSC = prescutum; PSRC = a median ridge-like thickening on the under surface of the prescutum; RL = rectal strand).

leptophragmata, is in full and direct contact with blood. The free bathing with blood of the outer surface of the rectal sac is important for the functioning of the cryptonephridic system.

Shrivastava and Verma (1983) observed the rectal strand in other chrysomelids, viz. *Aspidomorpha sanctae-crucis* (Fabricius), *Oocassida pudibunda* (Boheman), *Cassida* sp., *Platypria erinaceus* (Fabricius), and *Altica* sp. It remains to be discovered which other chrysomelids have such a suspensorium for the rectum.

Cryptonephridic arrangement of Malpighian tubules

The reassociation of the blindly ending distal parts of the Malpighian tubules with the alimentary tube in the rectal region has been referred to as cryptonephridic arrangement. This anatomical association has been interpreted as a system for increasing the reabsorption capacity of the rectum.

The cryptonephridic arrangement of the Malpighian tubules is known in all Chrysomeloidea studied, with the only exception of *Donacia*. Among leaf beetles, the pattern of the cryptonephridic system is remarkably uniform (Saini, 1964). There are six tubules arising in two groups and two tubules arising a little anterior to the remaining four, which form the other group. On each side, three tubules, one belonging to the anterior group, and two to the posterior set, run forward for some distance and then, looping backward, reach the anterior limit of the rectal sac (*Figure 9.20*). Here, the three fuse together to form a common stem, which penetrates into the rectal wall. Within the wall, the stem enters the space between the surface membrane, which has been called the perinephric membrane or the perinephric jacket, and the muscular layer (*Figure 9.22*). Here, the stem immediately divides into three tubules,

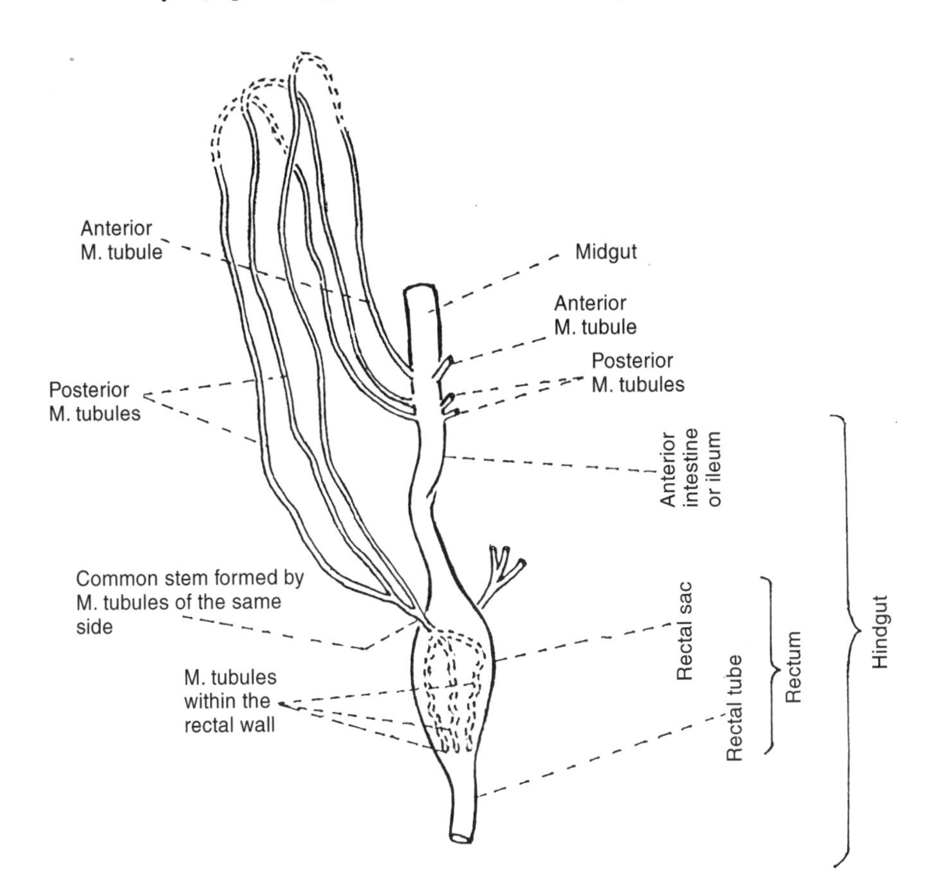

Figure 9.20. Cryptonephridic arrangement of Malpighian tubules in a typical chrysomelid N.B. The ileum or the anterior part of the hind-gut has been twisted clockwise to show arrangement of enclosed M. tubules on one side of the rectum.

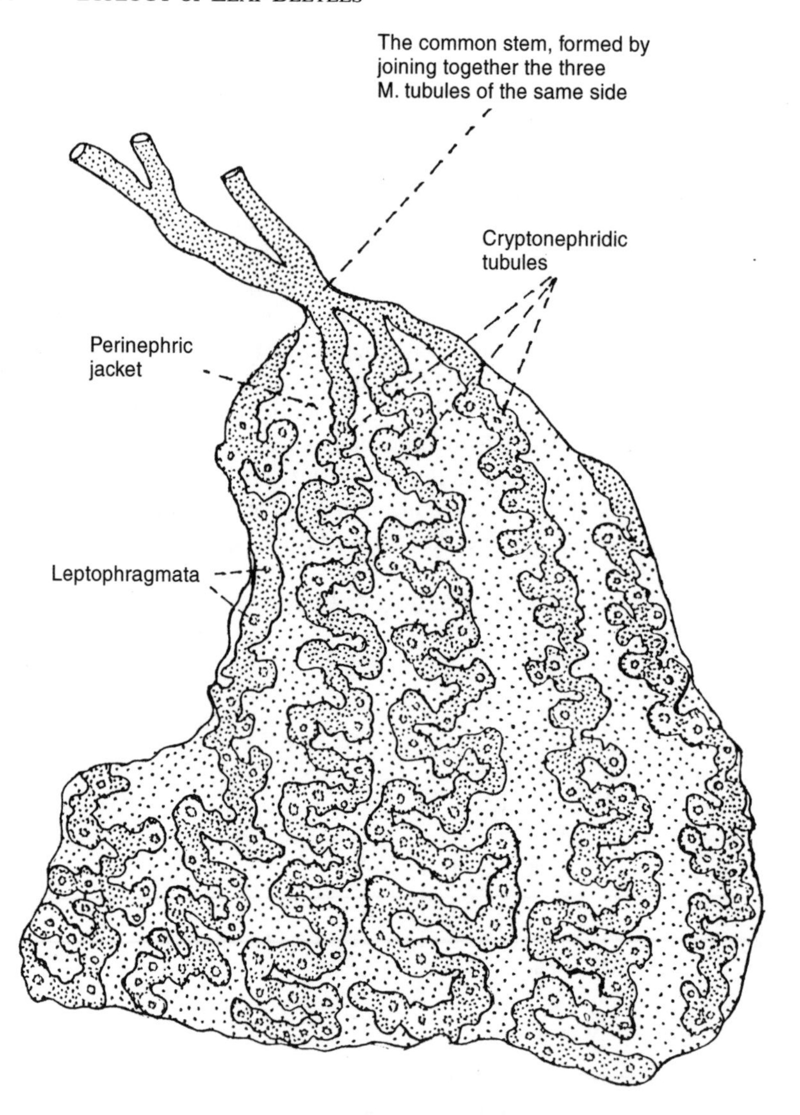

Figure 9.21. Perinephric jacket of the cassidine *Aspidomorpha miliaris*, slit open on one side, removed from around the rectal sac and mounted flat. Note that the common stem, formed by joining together the three M. tubules of one side, divides into three cryptonephridic tubules running a deeply sinuous course posteriorly, closely associated with the perinephric jacket. The three similar cryptonephridic tubules of the other side have become torn anteriorly (after Shrivastava, 1986).

which run a convoluted course posteriorly, remaining adherent to the inner face of the perinephric membrane (*Figure 9.21*). In many chrysomelids, these enclosed tubules are provided with leptophragmata, which are specially thinned out, circular portions of the tubule wall in intimate contact with the perinephric jacket. At the position of a leptophragma, the tubule wall is constituted by a thin, plate-like cell, the leptophragma cell (*Figure 9.22*).

The almost uniform pattern of the cryptonephridic system in chrysomelids is further evidence of the group having descended from a common ancestral stock.

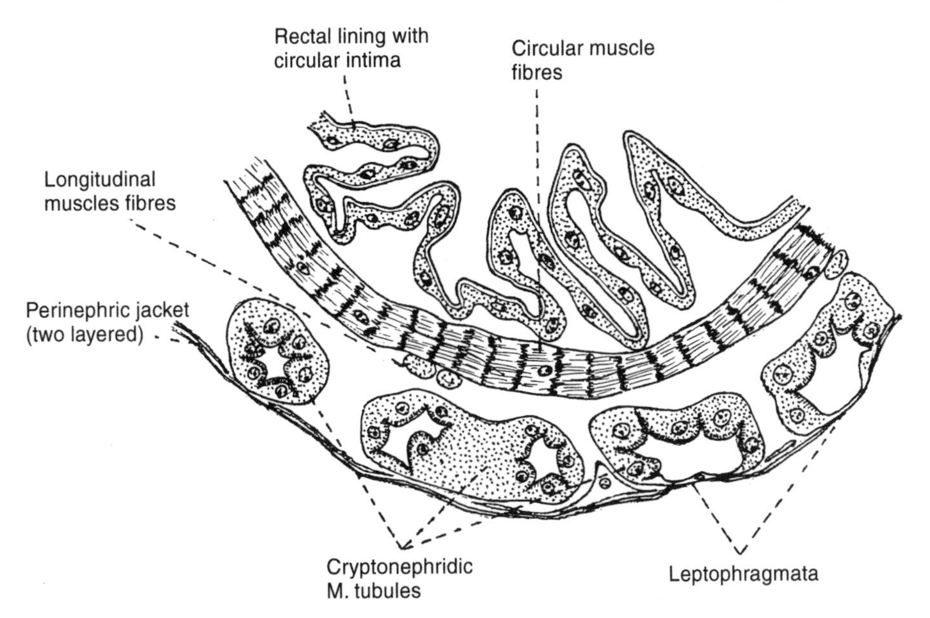

Figure 9.22. Part of a transverse section of the rectal sac of *Aspidomorpha miliaris* (after Shrivastava, 1986).

Heymons and Lühmann (1933), however, gave an account, which deviates to some extent from this pattern, for *Galerucella viburni* Payküll. They found that in this chrysomelid, the anterior two Malpighian tubules do not join the other tubules, and end blindly in the haemocoel. The four tubules of the other group form two lateral stems, which penetrate the perinephric membrane to form three tubules each within the perinephric space.

The facts that the cryptonephridic arrangement of the Malpighian tubules is lacking in the aquatic *Donacia* and that it is poorly developed in the semiaquatic *Galerucella birmanica* Jacoby (Saini, 1964) provide support for the view that this anatomical feature helps with water conservation. Singh (1976) studied excretion in *Aulacophora foveicollis* Lucas, and found that the beetle excretes liquid/semiliquid material, and at times solid. He further reported that on starvation it excretes more in solid form. He attributes the increased discharge of solid excreta to the cryptonephridic system.

Ventral nerve cord

The ventral nerve cord (vnc) part of the central nervous system was comparatively studied in Chrysomelidae by Mann and Crowson (1983a), and the following account is mainly based on their paper.

The most primitive condition of the vnc in beetles includes 8 ganglia in the abdomen, in addition to pro-, meso- and metathoracic ganglia. This condition may be seen in some larviform females among cantharoids. In most beetles, there is varying amount of anteroposterior condensation of the vnc, and a consequent fusion among the ganglia.

In all Chrysomelidae (and also Curculionoidea), the first two abdominal ganglia are fused with the metathoracic ganglion, and the last two have joined to form a

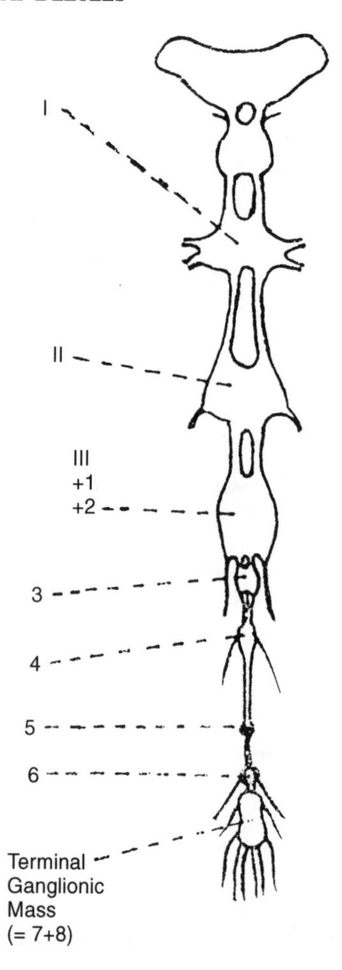

Figure 9.23. Central nervous sytem in *Orsodacne* (after Mann and Crowson, 1983a) (I to III = thoracic ganglia in the vnc; 1 to 8 = abdominal ganglia in the vnc).

terminal ganglionic mass. Thus, there are never more than 5 separate ganglionic masses in the abdominal part of the vnc. The least condensed condition of the vnc may be seen in Orsodacninae (*Figure 9.23*), and in *Timarcha*. But in these chrysomelids the sixth abdominal ganglion is pressed against the terminal ganglionic mass (i.e. ganglia 7 + 8) due to an absence of connectives.

As inferred by Mann and Crowson (1983a) from their comparative study, condensation of the vnc and the consequent fusion of abdominal ganglia took place independently along several different lines among Chrysomelidae. Among Eumolpinae, various degrees of condensation may be seen, and the number of free abdominal ganglia (not taking into account the terminal ganglionic mass) varies from 3, 2, 1 to none. Among Galerucinae, the number varies from 3 to 1. In Cerambycidae too, the condensation of the vnc seems to have occurred independently along certain lines.

It seems that, to some extent, condensation of vnc is related to anteroposterior shortening of the body. Maximum condensation of the vnc is shown by Clytrinae and

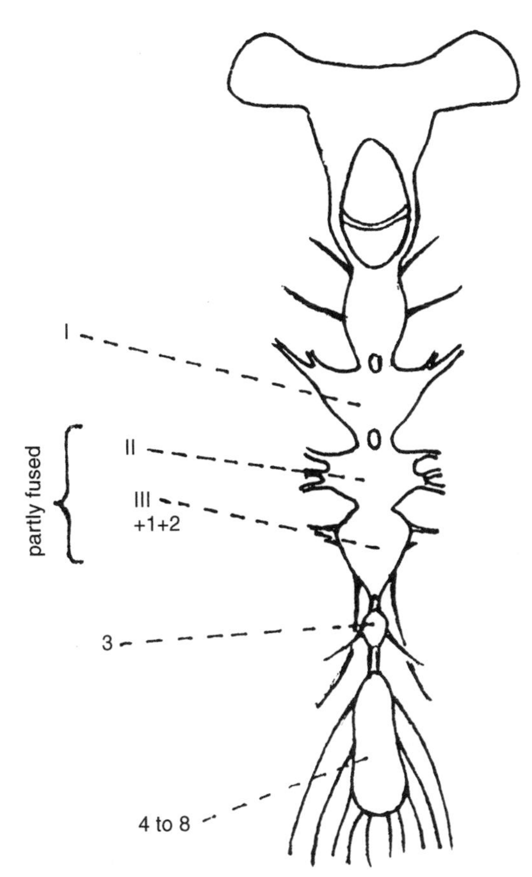

Figure 9.24. Central nervous system in *Cryptocephalus* (after Mann and Crowson, 1983a) (I to III = thoracic ganglia in the vnc; 1 to 8 = abdominal ganglia in the vnc).

Cryptocephalinae, which are compact-bodied with intermediate abdominal sterna showing constriction of the middle. In these chrysomelids, not only does the 4th abdominal ganglion show a tendency of fusion, or actual fusion with the terminal ganglionic mass in the abdomen, but there is also at least partial fusion of the meso- and metathoracic ganglia (*Figure 9.24*).

The structure of the vnc may help in the inference of phylogenetic relations only below the subfamily level among Chrysomelidae.

Female reproductive system

Here, as in the male reproductive section of this chapter, it is useful to divide the system into internal and external genital organs.

INTERNAL REPRODUCTIVE ORGANS IN FEMALE

A pair of ovaries, each a fascicle or bunch of a variable number of ovarioles, are located laterally in the abdominal cavity and occupying a variable part of the cavity,

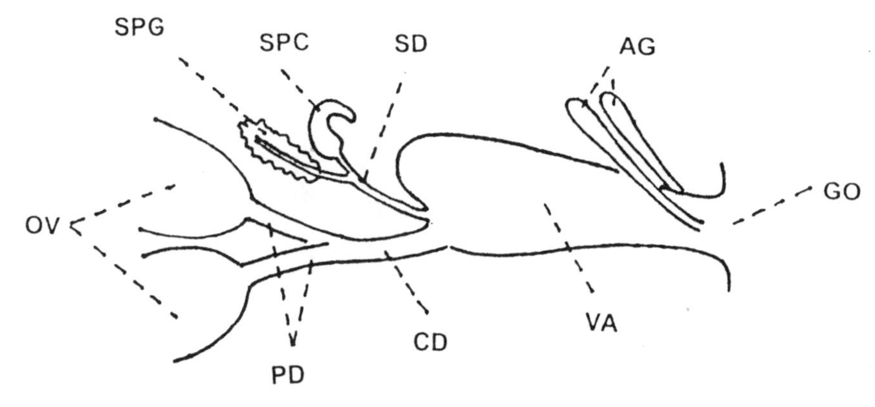

Figure 9.25. Diagram to illustrate internal genital system in a typical female leaf beetle, lateral view (AG = accessory or colleterial glands; CD = common or median oviduct; GO = female genital orifice; OV = ovaries; PD = paired or lateral oviducts; SD = spermathecal duct; SPC = spermathecal capsule; SPG = spermathecal gland; VA = vagina/bursa).

depending upon the stage of growth of eggs in the ovarioles. The calyces of the ovaries continue behind as paired or lateral oviducts, which are short and wide and which join to form the common oviduct (*Figure 9.25*). The common oviduct extends behind into the vagina, without any clear demarcation. Opening into the median duct is the duct of the spermathecal complex (= spermatheca + spermathecal gland + spermathecal duct) (*Figure 9.27*). The point where the spermathecal duct opens is taken as the level of junction between the common oviduct and vagina (Suzuki, 1988). The vagina may be partly swollen to form the bursa, and it opens behind into the anogenital vestibule through the female genital opening.

As in all Polyphaga, in Chrysomelidae the ovarioles are telotrophic or acrotrophic (Matsuda, 1976; Bhattacharya and Verma, 1982). Generally in leaf beetles, the pedicels of ovarioles are united to form a calyx (*Figure 9.26*). But in the genus *Lupesthes* (Eumolpinae), the lateral oviduct is elongated, and ovarioles are arranged in longitudinal rows along it (Suzuki, 1988) (*Figure 9.26 B*).

The number of ovarioles per ovary varies greatly, not only from species to species but also often intraspecifically. Suzuki (1974), Suzuki and Hara (1975) and Mann and Singh (1979) studied this in a fairly large number of species belonging to several different chrysomelid subfamilies. Suzuki (1974) found the number to vary from 6 to 43. But in some cases it may be much higher, for example, in the galerucine *Mimastra oblonga* Gyllenhall it is 73 (Mann and Singh, 1979), and in the alticine *Pseudodera xanthopsila* Baly the number may be up to 114 (Suzuki, 1975).

Suzuki (1974) pointed out that the ovariole number is stable in species having essentially few ovarioles. On the other hand, it tends to vary considerably in species having many ovarioles.

Though often there is considerable intraspecific variation in the number of ovarioles per ovary, in most species this number is constant (Suzuki, 1974). This suggests that the number of ovarioles is primarily determined by the genetic mechanism. Intraspecific variation seems to be largely due to environmental factors, mainly nourishment.

Suzuki (1975) and Suzuki and Yamada (1976) studied the relation between intraspecific variation in ovariole number and variation in body size in five different species of Chrysomelidae. Hindwing length was taken as an index of body size. Only

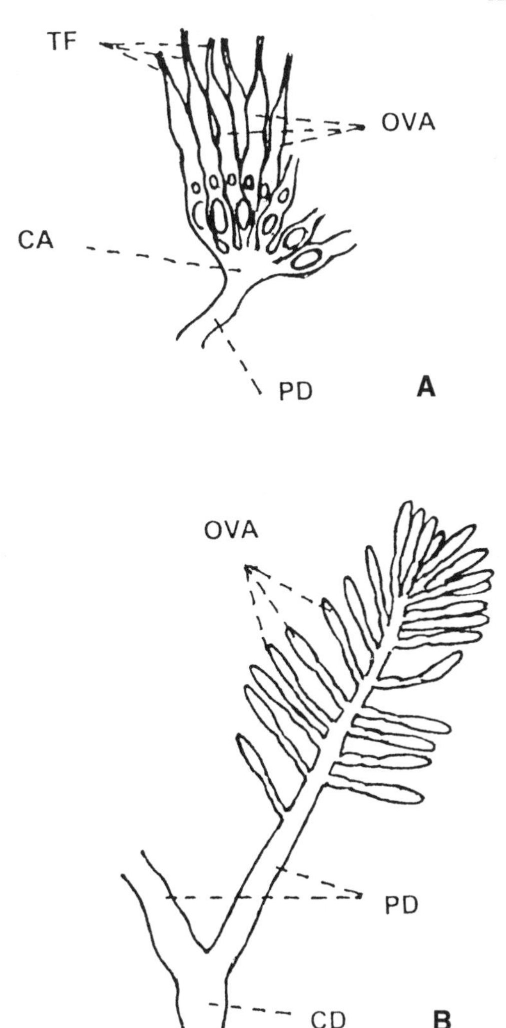

Figure 9.26. (A) Ovary in a typical chrysomelid; (B) Ovary in *Lypesthes* (Eumolpinae) (after Suzuki, 1988) (CA = calyx; CD = common oviduct; OVA = ovarioles; PD = paired or lateral oviducts; TF = terminal filaments of ovarioles).

in one species, *Aulacophora nigripennis nigripennis* Motschulsky, was the correlation between hindwing length and total ovariole number recognized.

The spermathecal complex varies greatly among chrysomelids. This part of the anatomy was comparatively studied by Varma (1955a,b), Pajni *et al.* (1985a,b) and Suzuki (1988). The authors arrived at some important phylogenetic inferences, outlined below.

(a) In Orsodacninae, the spermatheca is simple and horseshoe-shaped with an only weakly-sclerotized lining. The spermathecal duct is thick, short and not sclerotized. There is a united opening for the spermathecal gland and the capsule into the duct. The spermathecal complex opens into a subapical part of the bursa (*Figure 9.28 A*). These features of the spermathecal complex and some special features of

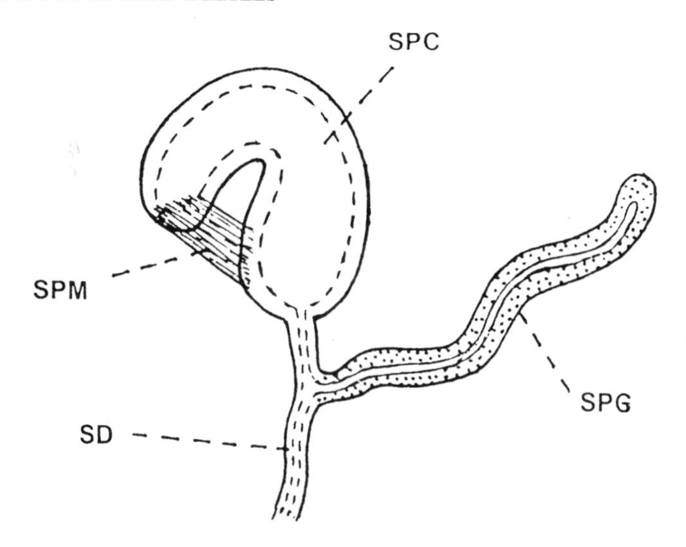

Figure 9.27. Spermathecal complex in a typical chrysomelid (after Suzuki, 1988) (SD = spermathecal duct; SPC = spermathecal capsule; SPG = spermathecal gland; SPM = spermathecal muscle).

the median ejaculatory duct in the male are not known in other chrysomelids. On the other hand, certain cerambycids, particularly Lepturinae, have these features. Suzuki (1988) is, therefore, convinced about the true phyletic relations between Lepturinae and Orsodacninae. In fact, he believes that they were derived from the same ancestors.

(b) Zeugophorinae and Megalopodinae have been taken as having descended from a common ancestor on the basis of the similarity in their spermathecal complex (Suzuki, 1988). The points of similarity are outlined here. The distal part of the capsule is similarly shaped and crane-like; the proximal part of the capsule is narrow, tube-like, forming a spheroidal lump through complex and irregular coiling, and the spermathecal gland in both the groups is very long, rather stout, nearly uniform in diameter and with lateral constrictions at intervals (*Figure 9.28 B*). There are also some notable differences between the spermathecal complexes of the two groups. In Zeugophorinae, the proximal part of the spermathecal capsule branches into three tubes, and there is no such branching in Megalopodinae. In the former subfamily, the spermathecal gland is bifurcated; in the latter the gland is a simple tube.

(c) There are some notable similarities between the spermathecal organ of Orsodacninae and Synetinae. In both the groups, the spermathecal capsule is horseshoe-like with a sharp bend. Its lining is very thin and weakly sclerotized. Inner and outer surfaces of the capsule are entirely smooth, that is without any creases, ridges or folds. In both the groups, the capsule and the gland open into the spermathecal duct very close together, almost by a common aperture. In spite of these points of resemblance, Suzuki (1988) does not infer any direct phyletic relation between the two in view of differences in other parts of their organization, e.g. in their testicular structure.

(d) On the basis of details of their spermathecal complex, the following subfamilies of Chrysomelidae were regarded by Suzuki (1988) as monophyletic in origin:

Clytrinae, Cryptocephalinae, Chlamisinae, Lamprosomatinae, Hispinae, Cassidinae, Megascelinae and Eumolpinae. From Suzuki's descriptions and figures (1988), the following special features of the spermathecal organ in this group can be set out. The spermathecal duct is very long and coiled (*Figure 9.28 C*). The spermathecal gland is short and simple. The proximal part of the spermathecal capsule forms a special outgrowth to receive the openings of the spermathecal duct and the gland. The spermathecal gland opens into the proximal part of the spermathecal capsule forming a sort of nipple or conical prominence projecting

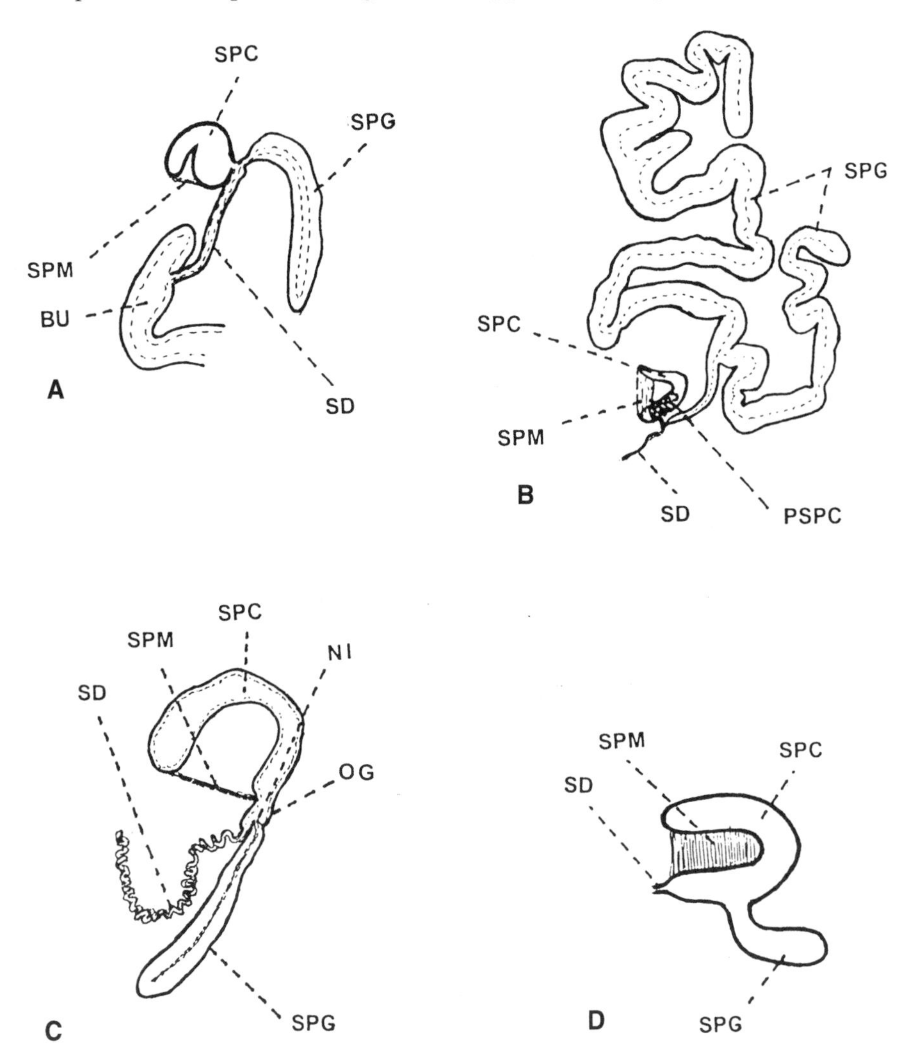

Figure 9.28. (A) Bursa and spermathecal complex in *Orsodacne* (after Suzuki, 1988); (B) Spermathecal complex in *Zeugophora* (after Suzuki, 1988); (C) Spermathecal complex in *Cassida* (after Suzuki, 1988); (D) Spermathecal complex in *Corynodes* (after Pajni *et al.*, 1987a) (BU = bursa copulatrix; NI = nipple-like elevation at the opening of SPG into SPC; OG = an outgrowth of SPC to receive openings of SD and SPG; PSPC = proximal part of SPC, narrow and tubular and irregularly coiled, forming a lump-like mass on the base of the distal part of SPC; SD = spermathecal duct; SPC = spermathecal capsule; SPG = spermathecal gland; SPM = spermathecal muscle).

into the capsule. Some, or all, of these features are presented by the members of these subfamilies. Suzuki (1988) says "The conspicuous coiling in SptD (= spermathecal duct) is a most important characteristic that is generally found in the six subfamilies, Clytrinae, Cryptocephalinae, Chlamisinae, Lamprosomatinae, Hispinae and Cassidinae".

In Eumolpinae, the spermathecal complex is variable. In some species of *Colasposoma*, e.g. *Colasposoma downesi* Baly and *C. lividipes* Jacoby, the spermathecal duct is long and much coiled. In some members of the subfamily, e.g. in *Corynodes* (*Corynodes peregrinus* Fussly, *C. pyrophorus* Parry and *C. modestus* Jacoby), however, the spermathecal capsule is shaped like an incomplete ring and the spermathecal gland arises quite away from the origin of the duct, which is long but with little coiling (Pajni *et al.*, 1987a) (*Figure 9.28 D*). Such variation is also seen in the organization of the male reproductive system in Eumolpinae (Verma, 1996c). Hence, the view of Pajni *et al.* (1987a) that Eumolpinae are a heterogeneous assemblage deserves serious consideration. However, the nervation of the wings presents a constant character.

In Criocerinae too, the spermathecal duct may be fairly long and somewhat coiled, but then there are significant differences from Clytrinae, Cryptocephalinae, etc., namely that the spermathecal capsule is almost C-shaped (Pajni *et al.*, 1985b) and openings of the duct and the gland are considerably separated from each other.

In Donaciinae, as in Criocerinae, the spermathecal capsule is almost C-like, and the openings of the duct and the gland are considerably separated (*Figure 4 a,b,c* and *d* in Mann and Crowson, 1983d). This supports the idea of phyletic relations between the two subfamilies.

(e) Suzuki (1994) examined the spermathecal organ in six different species of Aulacoscelinae, a small neotropical subfamily, and found that this part of the anatomy in Aulacoscelinae is similar to that in Sagrinae and Megascelinae.

There may be a pair of tubular and vesicular accessory or colleterial glands opening into the last part of the vagina. These secrete either the viscid material covering the laid eggs or the substance of an ootheca. Such glands are present in Hispinae, Cassidinae, Eumolpinae and Lamprosomatinae (Suzuki, 1988; Pajni *et al.*, 1987a). Silfverberg (1976) described a single tubular, and more or less convoluted, accessory gland opening into an anterior part of the vagina in *Galerucella nymphaeae* L. This gland is in addition to the spermathecal gland.

In order to draw up a comparison between the female genital organs in leaf beetles and those in other insects, a brief reference to their developmental history would be useful.

The postembryonic development of the female internal genital organs has been studied in a number of Coleoptera, including some phytophagoids, e.g. *Anthonomus* and *Rhagium* (Metcalfe, 1932), *Callosobruchus* (Pajni, 1968), *Calandra* (Murray and Tiegs, 1935) and *Galerucella* (Varma, 1963). From the accounts given by these authors, the following development story of the female internal genital organs in leaf beetles and other phytophagoids can be presented.

Mesodermal ovarian rudiments in the larva send a pair of slender cords extending backward. In late larva, two median ectodermal invaginations appear, one immediately

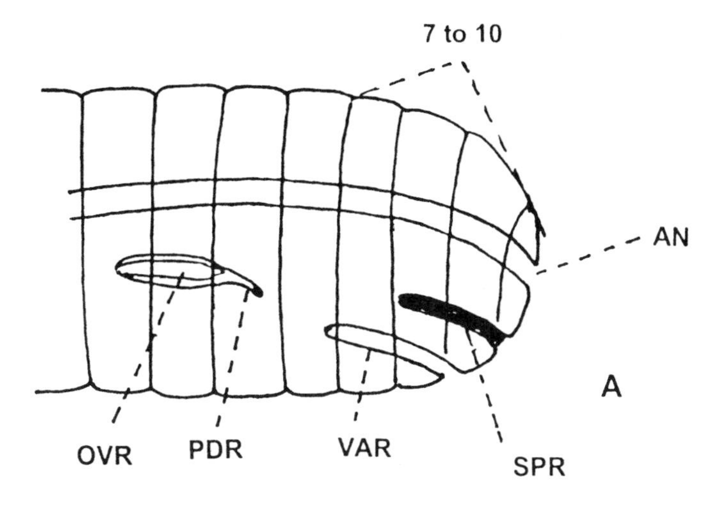

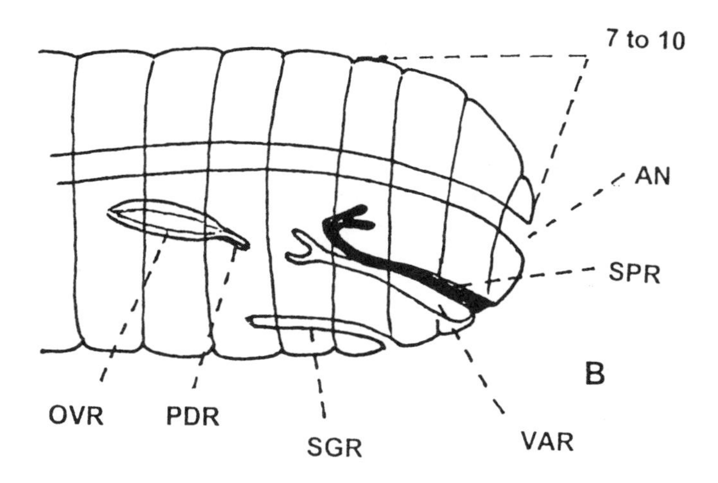

Figure 9.29. (A) Developing female genital organs in a last instar larva, lateral view; (B) Developing female genital organs in a pupa, lateral view (both the diagrammatic figures based on Matsuda, 1976) (AN = anus; OVR = ovarian rudiments; PDR = lateral oviduct rudiment; SGR = spiculum gastrale rudiment; SPR = spermathecal rudiment; VAR = vaginal rudiment; 7 to 10 = abdominal segments).

behind the eighth abdominal sternum, and the other behind the ninth. The latter is called the spermathecal rudiment and the former the vaginal rudiment (*Figure 9.29 A*). From the anterior end of the spermathecal rudiment, the spermathecal complex develops. The vaginal rudiment becomes bifurcated anteriorly to join the mesodermal cords from the ovarian rudiments; thus, paired oviducts are formed, the posterior parts of which are obviously ectodermal in origin. The spermathecal and vaginal rudiments fuse together lengthwise to form a median genital duct (*Figure 9.29 B*). According to Metcalfe (1932), in *Sitodrepa* a spiculum gastrale is formed by an ectodermal invagination in the intersegmental area between the 7th and 8th abdominal sterna in the pupal stage.

FEMALE EXTERNAL GENITALIA

The external genitalia in the female consist mainly of the ovipositor. In Coleoptera, an ovipositor of a gonapophysal nature is entirely lacking. Some of the terminal abdominal segments become modified to help oviposition and are protruded as a vesicular or tubular ovipositor during egg laying.

In Chrysomelidae, 8th and 9th abdominal segments, along with the intersegmental membrane between the two segments constitute the ovipositor. The 7th segment is also modified, but to a lesser extent.

The female external genitalia in *Galerucella birmanica* may be taken as typically chrysomelid. Khatib (1946a) described the terminal abdominal segments in this beetle. The 7th abdominal tergum in *G. birmanica* narrows posteriorly, and its posterior half is bent downward forming a conical downward projection. The 7th sternum is nearly typical, slightly emarginate along its posterior border, and presents in its median posterior part a triangular depression covered with very fine setae. This depressed area is so located as to guide the aedeagus into the vagina. The 8th tergum is a small and nearly semicircular plate, bent inward, so that it lies hidden beneath the 7th tergum, and forms a part of the roof of the anogenital vestibule, which has in its lining both the anus and the opening of the vagina. The 8th sternum is also folded in, and lies immediately above the 7th sternum. It is incompletely divided into two sclerites, as its middle part is membranous. The 9th tergum is a transversely elongated sclerite, separated from the 8th tergum by an ample intersegmental membrane. Situated beneath the 9th tergum is the anus. The lateral parts of the 9th tergum bend downward, and close below its lateral ends are palps on the sides of the vaginal opening (*Figure 9.30*).

During oviposition, the ample membranous area around the sclerites of the 8th and 9th abdominal segments, along with the reduced and modified sclerites, become everted as a short and stumpy or a moderately long and tri-segmented ovipositor (*Figure 9.31*). The basal 'segment' is formed by the tergum and sternite of the 8th abdominal segment, and the middle 'segment' is the ample intersegmental membrane between the 8th and the 9th segment. The 3rd 'segment' is a mostly membranous 9th segment. In many chrysomelids, this 3rd 'segment' includes, in addition to the reduced 9th tergum, the reduced 9th sternum, the latter generally divided into 2 hemisternites or valvifers. The valvifers carry palps, or styli. The palps (also referred to as the vaginal palps, e.g. by Mann and Crowson (1983d)) are soft appendages without any clear joint separating them from the 9th segment sternite; styli, on the other hand, are clearly jointed at base. The vaginal opening, or the female genital aperture, lies between the palps/styli.

Often, the 9th sternum is divided into two hemisternites carrying the styli/palps. In some cases, e.g. in the donaciine *Macroplea*, the hemisternite carries an additional sclerite, the coxite, which in turn carries the stylus (Mann and Crowson, 1983d).

In many chrysomelids, the 8th sternum is drawn into the body like a median or paired apodeme. The paired apodemes join anteriorly to form a Y-shaped structure. The apodemal ingrowth is referred to as spiculum, or spiculum gastrale. The protractors and retractors of the ovipositor are attached onto the apodeme.

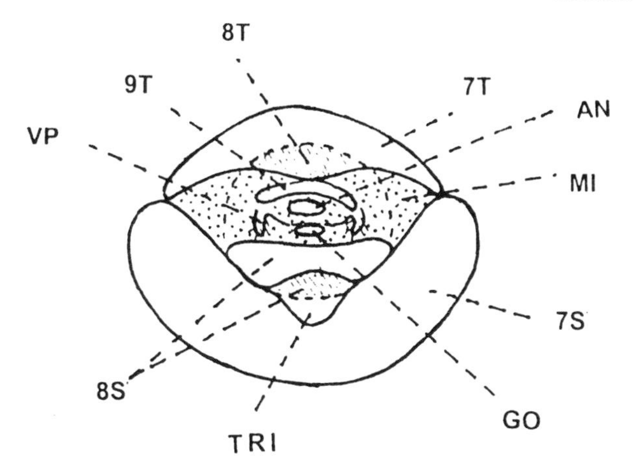

Figure 9.30. Caudal end view of abdomen of female *Galerucella birmanica*. 7th abdominal tergum and the corresponding sternum have been gaped to reveal the structure in the anogenital vestibule (after Khatib, 1946a) (AN = anus; GO = female genital orifice; MI = extensive membranous integument between 8th and 9th abdominal segments forming lining of most of anogenital vestibule and permitting protrusion of the ovipositor during egg laying; S = sternum of the abdominal segment, number of which is appended to it; T = tergum of the abdominal segment, number of which is appended to it; TRI = triangular depression on the last externally visible abdominal sternum; VP = vaginal palp). N.B. 8T is enfolded beneath the 7T. 8S is partly withdrawn and partly covered by 7S.

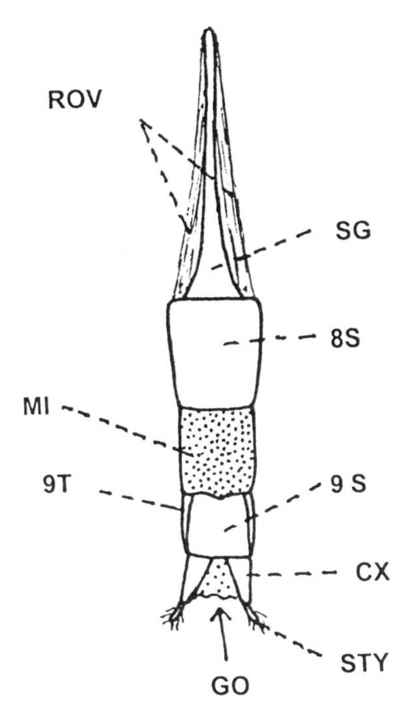

Figure 9.31. Ovipositor in a generalized leaf beetle (CX = coxite; GO = female genital orifice; MI = extensive membranous integument between 8th and 9th abdominal segments forming lining of most of anogenital vestibule and permitting protrusion of the ovipositor during egg laying; ROV = retractors of the ovipositor; S = sternum of the abdominal segment, number of which is appended to it; SG = spiculum gastrale; STY = styli; T = tergum of the abdominal segment, number of which is appended to it). N.B. Sclerites of 8th and 9th segments may present different degrees of reduction and loss.

Tanner (1927) and Iuga and Konnerth (1963) regarded the middle 'segment' of the three 'segmented' ovipositor of chrsyomelids as the 9th abdominal segment, and the 3rd 'segment' as the 10th segment. But Verhoeff (1918), Khatib (1946b) and Matsuda (1976) took the 2nd 'segment' as the extensive intersegmental membrane between the 8th and the 9th abdominal segments. The latter interpretation appears more reasonable, and has been adopted here.

Pajni *et al.* (1987a) compared the ovipositor in certain Indian chrysomelids, and could make out some common features of phylogenetic importance which, in brief, are mentioned here:

(a) In Galerucinae and Alticinae, the spiculum gastrale is long and often dilated at both ends.
(b) In Cryptocephalinae and Clytrinae, the female external genitalia are characterized by complete absence of the spiculum gastrale and broadening of the genital segments (i.e. 8th and 9th abdominal segments).
(c) In Cassidinae and Hispinae, the genital segments are short and broad, and the spiculum is reduced in length.
(d) Criocerinae show considerable variation in their female genitalia. The spiculum gastrale may be lacking, as in *Crioceris malabarica* Jacoby, or there may be a moderately long and apically dilated spiculum, as in some species of *Lema*.
(e) In Eumolpinae there is elongation of the genital segments and of the intersegmental membrane between them, and the spiculum is short.

Thus, from these observations some of the commonly held views on the phyletic relations among chrysomelid subfamilies are supported.

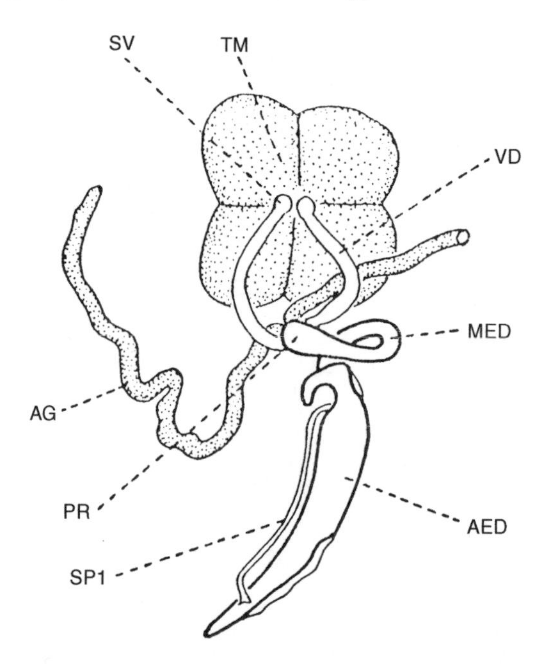

Figure 9.32. Male genital system of *Galerucella birmanica*, ventral view (after Verma, 1969) (AED = aedeagus; AG = accessory genital gland; MED = median ejaculatory duct; PR = 'prostata'; SP1 = first spiculum or tegminal apodeme; SV = seminal vesicle; TM = median testicular mass; VD = *vas deferens*).

Male reproductive system

For descriptive convenience, the male reproductive system can be divided into two parts, namely the internal genital organs and the external genitalia. The former includes the testes and the efferent genital system, along with the accessory genital glands, while the latter consist of the aedeagus and its musculature.

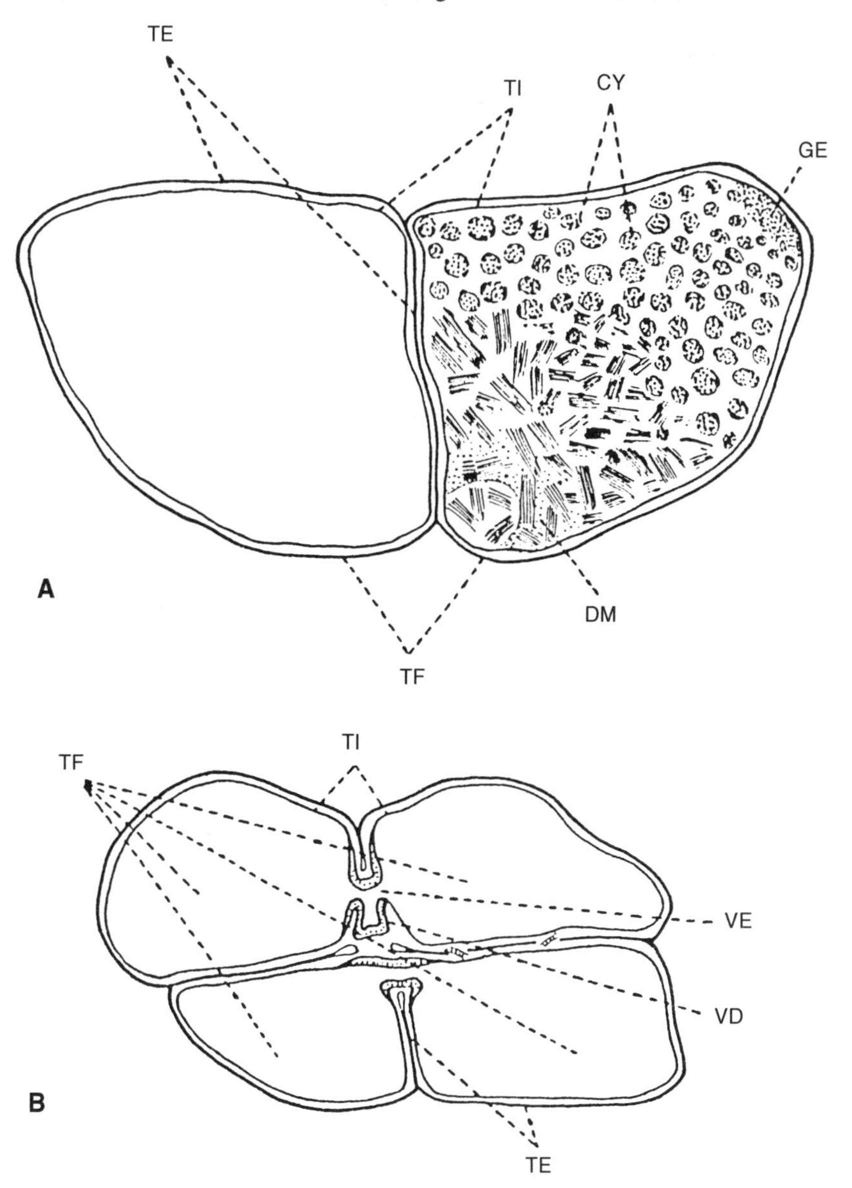

Figure 9.33. *Galerucella birmanica.* (A) Vertical section through two testis follicles of the same side. Contents shown only in one follicle; (B) Horizontal section through the median testicular mass. Germ cells not shown (after Verma, 1969) (CY = developing germ cell cysts; DM = mass of degenerating cells; GE = germarium; TE = *tunica externa*; TF = testis follicle/follicles; TI = *tunica interna*; VD = *vas deferens*; VE = *vas efferens*).

INTERNAL GENITAL ORGANS

The internal reproductive organs in male Chrysomelidae were surveyed by Kasap and Crowson (1979), Mann and Crowson (1983b) and Suzuki (1988). Pajni *et al.* (1985a,b, 1987a) studied these organs in some subfamilies of Chrysomelidae. Verma (1996c) reviewed our knowledge of male genital organs of leaf beetles and tried to infer inter-subfamily relations on the basis of the organization of this part of leaf beetle anatomy. In addition, a number of authors described these organs, in particular chrysomelid subfamilies/species; for their citation, see Verma (1996c), Sharp and Muir (1912), Powell (1941), Zia (1936), etc.

In a male leaf beetle there are two testes, one on each side. Each testis consists of two testis follicles. In this way the chrysomelids are remarkably uniform in their male genital system but in the structure of the testis follicles and in their arrangement in the body, the beetles vary considerably (Verma, 1996c).

From the point of view of their structure, testis follicles of chrysomelids may be placed under two types, viz. non-septate and septate types (Mohan and Verma, 1981; Verma, 1985). These two types were first described by Virkki (1957) among scarabaeoid testes, and Mohan and Verma (1981) extended the classification to chrysomelid testes.

A non-septate follicle is a simple, sac-like structure (*Figure 9.32*), covered on the surface with two sheaths, a *tunica interna*, which is a thin and nucleated membrane, and a *tunica externa*, which is a layer of fat body-like cells (Verma, 1969) (*Figure 9.33*). In the sac on one side is located the germarium, while the rest of the space is filled with groups of germ cells in different stages of spermatogenesis, which develop into sperm cysts. According to Verma (1969), in *Galerucella birmanica* the tracheae,

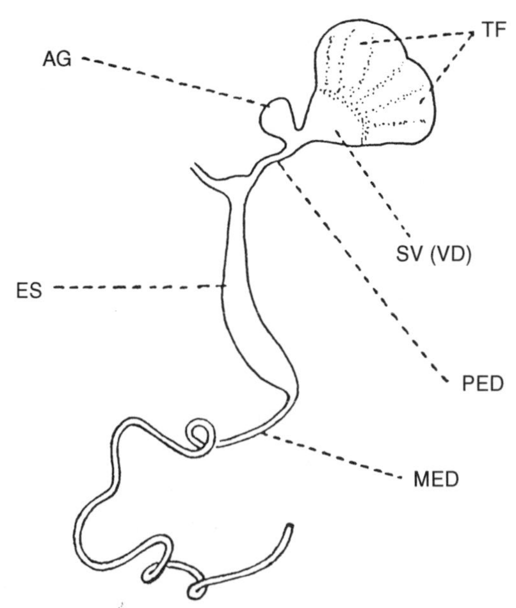

Figure 9.34. Internal male reproductive organs of *Aspidomorpha transparipennis* Motsch (after Suzuki, 1988) (AG = accessory genital gland; ES = ejaculatory sac; MED = median ejaculatory duct; PED = paired ejaculatory duct; SV(VD) = seminal vesicle formed by anterior end of *vas deferens*; TF = testis follicle/ follicles).

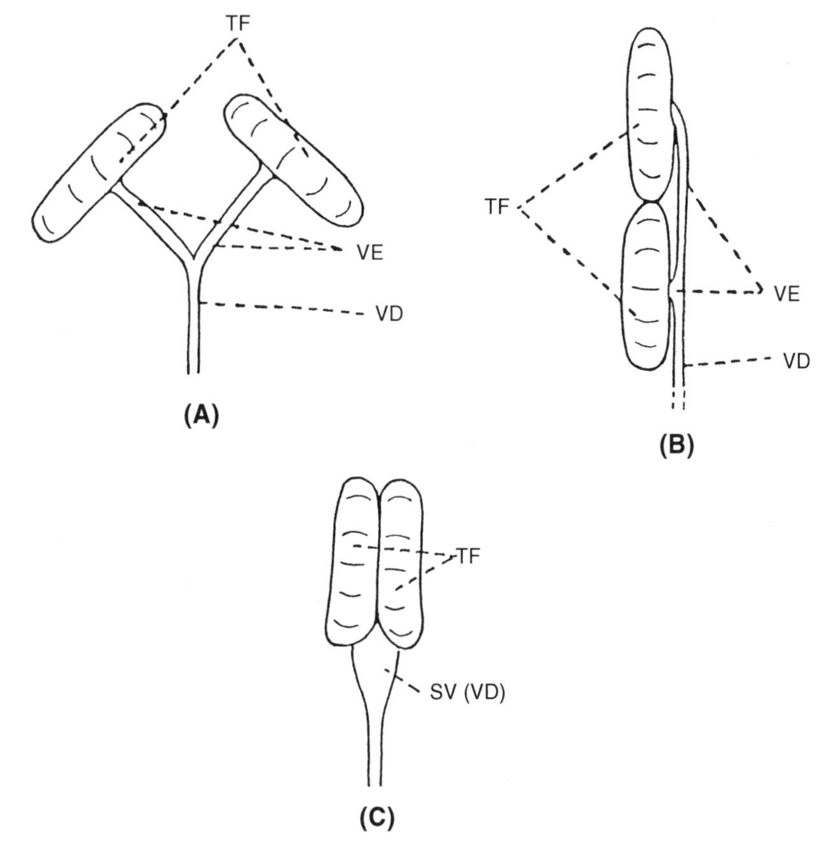

Figure 9.35. Different arrangements of septate testis follicles among chrysomelids. (A) In *Chrysolina aurichalcea* Mannerheim (Chrysomelinae) (after Suzuki, 1988); (B) In *Chrysomela populi* (Chrysomelinae) (after Harnisch, 1915); (C) In *Aspidomorpha miliaris* (Cassidinae) (after Mohan and Verma, 1981) (SV(VD) = seminal vesicle formed by anterior end of *vas deferens*; TF = testis follicle/follicles; VD = *vas deferens*; VE = *vas efferens*).

present in the *tunica externa*, on leaving the tunic, enter among the germ cells, but Bruck (1978), studying the alticine *Alagoasa*, could not locate such tracheae lying free among the developing germ cells.

In a septate testis follicle, the surface covers are similar to those in a non-septate one, but the *tunica interna* is repeatedly infolded to divide the interior of the follicle into a number of radially arranged *loculi* (*Figure 9.37*). The *tunica externa* tissue extends into the septa, formed by the folds of the *tunica interna*. A germarium is located immediately beneath a raised circular area, called the lens of Demokidoff, situated centrally on the flat distal surface of the follicle. This germarium is common for all the *loculi* in the follicle, and gives out cysts of developing germ cells to all of them (*Figures 9.36* and *9.38*). Tracheae are confined to the external tunic tissue. Hence, formation of septa seems to have the function of bringing oxygen (and also nourishment) close to the developing germ cells.

Non-septate testis follicles are known in Galerucinae and Alticinae. In these subfamilies, they show a special arrangement. The two follicles of each testis are fused by their external tunic, and there is a similar fusion between the testes of the two

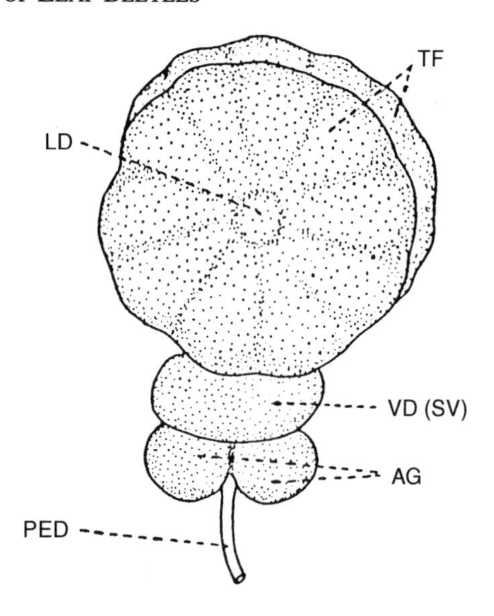

Figure 9.36. Testis of *Aspidomorpha miliaris* with flat surface of one testis follicle facing (after Mohan and Verma, 1981) (AG = accessory genital gland; LD = lens of Demokidoff; PED = paired ejaculatory duct; TF = testis follicle/follicles; VD(SV) = seminal vesicle formed by anterior end of *vas deferens*). N.B. The accessory gland is like an incomplete ring encircling the first part of the paired ejaculatory duct.

sides dorsal to the gut. Thus, there is a four-lobed, sac-like median testicular mass, situated dorsally in the abdominal cavity (*Figure 9.35*).

Typical septate follicles, as seen in Cassidinae and Chrysomelinae, are disc-like or like a flattened summer squash fruit in shape, and are divided into *loculi* by radial grooves (*Figure 9.36*). In some cases, a thick external tunic tissue may cover the radial creases or grooves, corresponding to the septa inside; hence the septate nature of the follicle may not be indicated on the surface. The two follicles of the same side are variously arranged (*Figure 9.35*).

The chrysomelid testis has been investigated histologically by only a small number of researchers (Wieman, 1910a,b; Harnisch, 1915; Verma, 1969; Mohan and Verma, 1981). From the external appearance of the testis follicles and from some details provided by those who have studied male gonads in leaf beetles, it may be inferred that septate testis follicles are found in the following chrysomelid subfamilies: Donaciinae, Sagrinae, Criocerinae, Cassidinae, Hispinae, Clytrinae, Cryptocephalinae, Chlamisinae, Eumolpinae, Chrysomelinae and Synetinae. In the last named subfamily, the *loculi* in the testis follicle are not so numerous, and are loosely held, i.e. not compacted together. Non-septate follicles, as has been said earlier, are characteristic of Galerucinae and Alticinae. In the following subfamilies, the septate nature of the testis follicles is doubtful, and may be confirmed or denied only through careful histological examination: Orsodacninae, Zeugophorinae, Megalopodinae, Timarchinae and Lamprosomatinae. For a more detailed discussion on testis structure in Chrysomelidae, see Verma (1969), Mohan and Verma (1981).

In some beetles, it has been observed that cyst cells, which surround a group of developing germ cells, become separated from fully formed spermatozoa, but

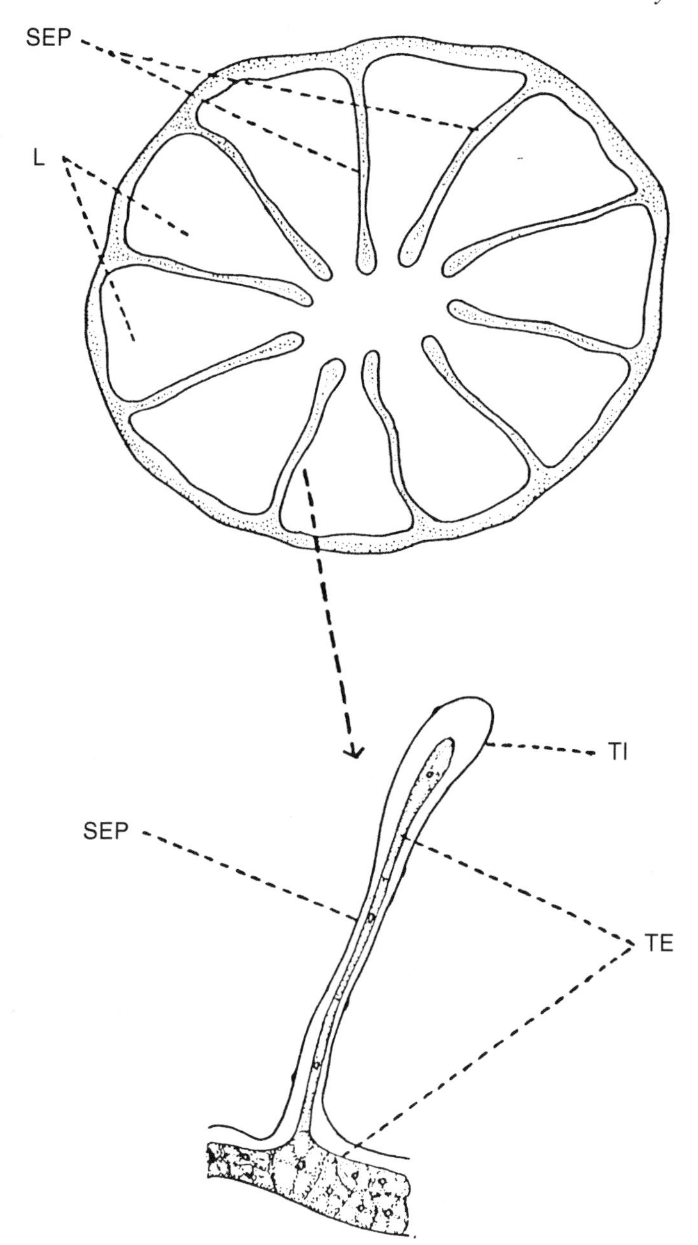

Figure 9.37. *Aspidomorpha miliaris* horizontal section of a testis follicle. Note that septa are formed by repeated infolding of *tunica interna*. Germ cell details have not been shown (based on Mohan and Verma, 1981) (L = loculi; SEP = septa; TE = *tunica externa*; TI = *tunica interna*).

accompany the latter right up to the female genital tract, where they degenerate (Anderson, 1950 in the Japanese beetle, *Popillia japonica* Newman; Landa, 1960 in *Melolontha melolontha* L.; both the beetles are scarabaeoids). In *Galerucella birmanica* (Verma, 1969) and in *Aulacophora foveicollis* (Verma, 1975), however, it has been noted that cyst cells, separating from fully formed sperms, are retained in the testis

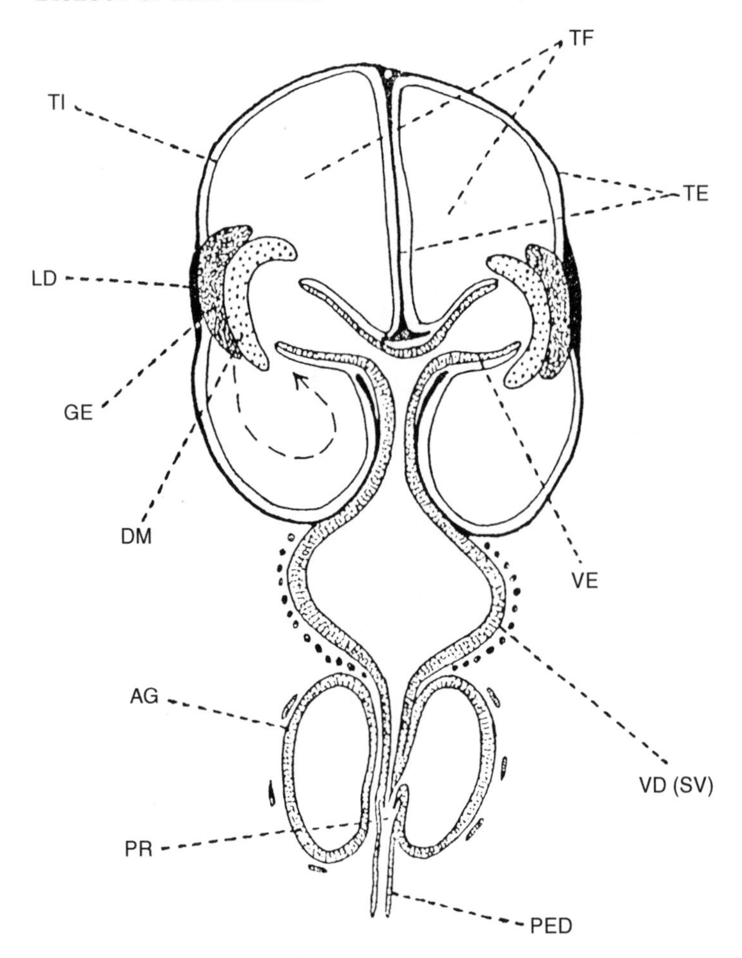

Figure 9.38. *Aspidomorpha miliaris* diagrammatic longitudinal section through the testis follicles of the same side. The curved arrow in one follicle indicates the path taken by cysts of developing germ cells. Germ cell details have been left out (after Mohan and Verma, 1981) (AG = accessory genital gland; DM = mass of degenerating cells; GE = germarium; LD = lens of Demokidoff; PED = paired ejaculatory duct; PR = 'prostata'; TE = *tunica externa*; TF = testis follicle/follicles; TI = *tunica interna*; VD (SV) = seminal vesicle formed by anterior end of *vas deferens*; VE = *vas efferens*).

follicle, and only sperms make their way into the seminal vesicle. In either chrysomelid, the separated cyst cells come to form a degenerating syncytial mass close to the opening of the follicle into the seminal vesicle. A similar situation has been noted in *Aspidomorpha miliaris* (Fabricius) (Mohan and Verma, 1981). In *Aulacophora foveicollis*, there is a basilar membrane consisting of a dome-like arrangement of a whorl of elongated curved cells (*Figure 9.39*) at the opening of the follicle into the seminal vesicle. This arrangement seems to prevent the movement of separated cyst cells into the seminal vesicle, while fully-formed sperms are able to swim across the barrier (Verma, 1975). Mohan and Verma (1981) find that in *Aspidomorpha miliaris* separated cyst cells form this obstruction themselves, and they get pushed up to degenerating syncytial mass, located immediately beneath the germarium (*Figure 9.38*), as new separating cyst cells arrive.

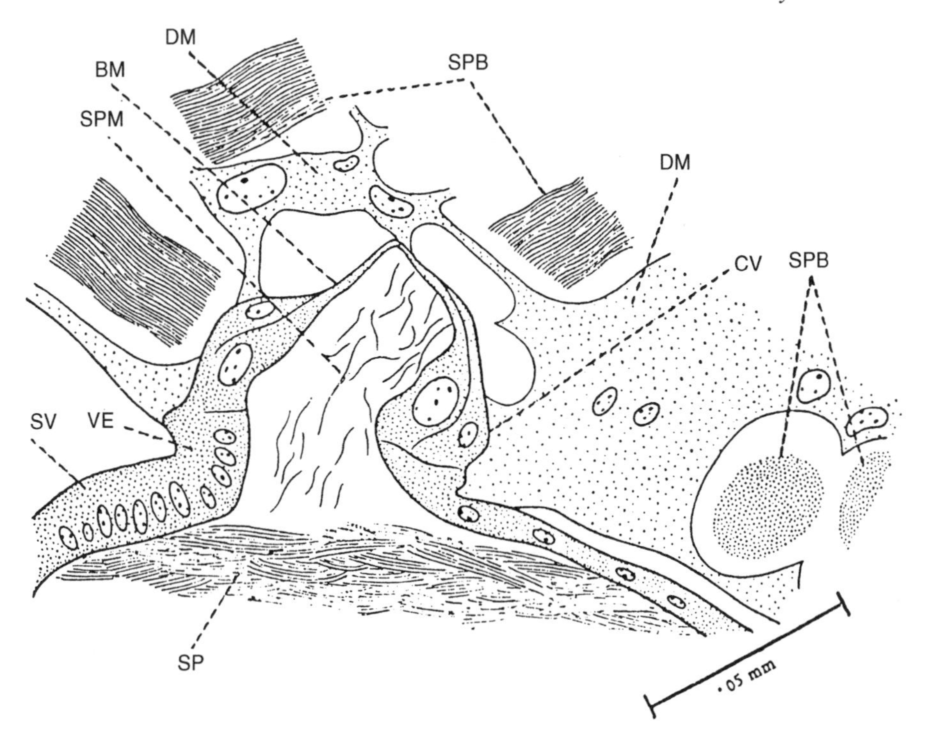

Figure 9.39. *Aulacophora foveicollis* testis section showing a follicle opening into seminal vesicle. Note the tent-like arrangement of elongated cells constituting the basilar membrane (from Verma, 1975) (BM = basilar membrane; CV = elongated cells constituting the basilar membrane; DM = degenerating syncytial cell mass; SP = sperms closely packed in the seminal vesicle, unaccompanied by cyst cells; SPB = sperm bundles; SPM = sperms moving into the SV; SV = seminal vesicle; VE = *vas efferens*).

Wieman (1910a,b) described a degenerating cell mass below the germarium in *Leptinotarsa decemlineata*. He too found that the cyst cells do not accompany sperms to the sperm ducts. Edwards (1961) found a similar situation with the cerambycid *Prionoplus*.

It would be interesting to find out if these histological features occur throughout Chrysomelidae/Chrysomeloidea. This would further attest homogeneity of the group.

The obvious difference between the organization of septate and non-septate follicles is a serious challenge to the concept of homogeneity in the Chrysomelidae. Verma (1996c), however, put forth a hypothesis suggesting the derivation of the non-septate organization from a primitive septate structure like that seen in testis of *Syneta* (*Figure 9.40*). As no intermediate stage is known between the two types of testicular structure, this hypothesis remains to be substantiated.

Male efferent genital system

Verma (1996c) distinguished two different plans of organization of the male efferent genital system among chrysomelids and referred to them as Plan A and Plan B. Plan A is associated with a non-septate testis, and plan B with the septate type. In Plan A, *vasa deferentia* receive very short *vas efferens* from each testis follicle of its side.

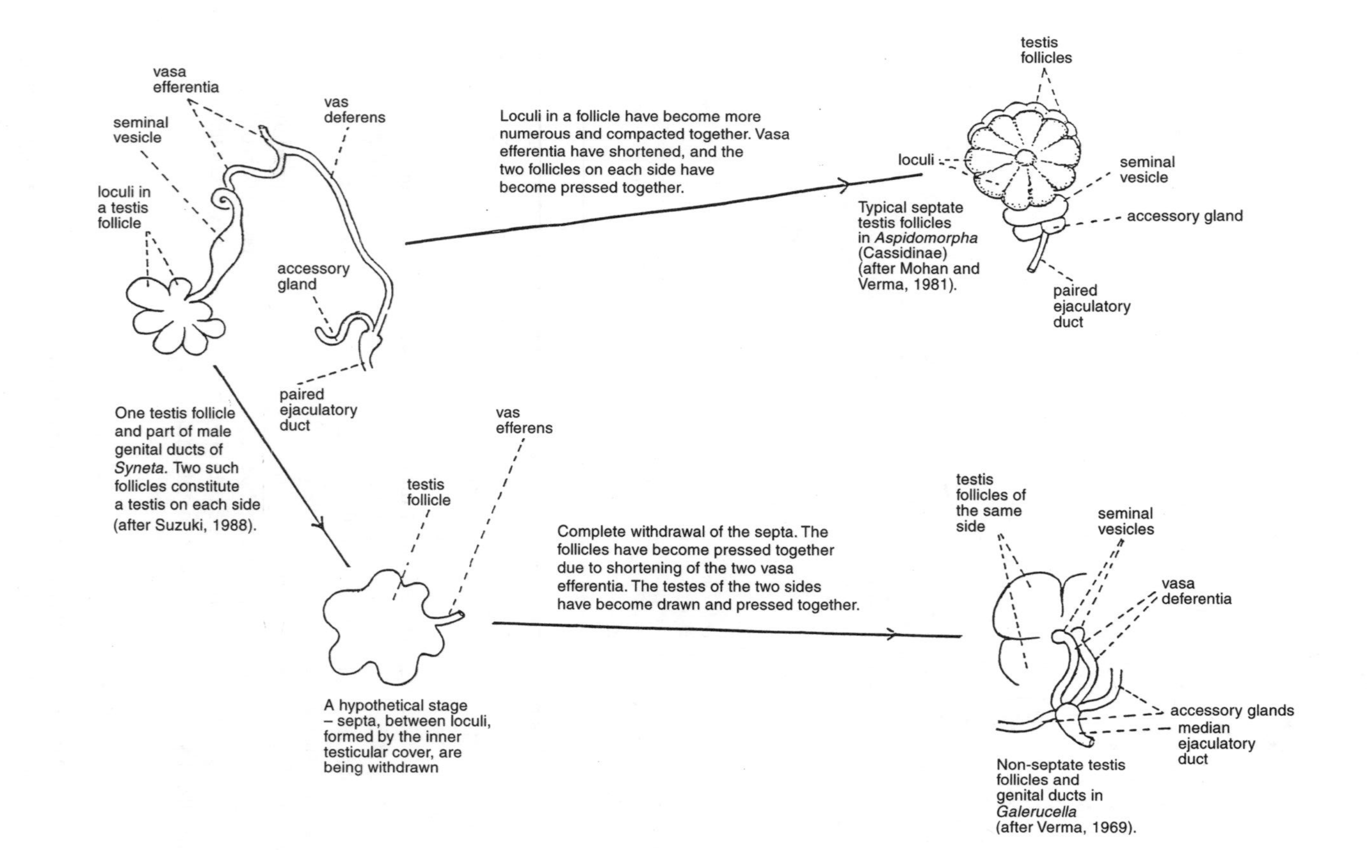

Figure 9.40. Diagram to illustrate Verma's hypothesis (1996c) on the evolution of septate and non-septate condition of testis follicles among Chrysomelidae. It is assumed that phytophaga ancestors had a *Syneta*-like testis (Verma and Jolivet, 2000).

They are of considerable length and open into the anterior end of the median ejaculatory duct. Also opening into the median duct are a pair of long and tubular accessory genital glands. The anteriormost part of the ejaculatory duct may be dilated a little, but the rest of the duct is a short, wide tube with a nearly uniform diameter which runs a winding course to the base of the aedeagus (*Figure 9.32*).

In Plan B, the *vasa deferentia* are very short, and the accessory glands, generally short and either vesicle-like or ring-like, are located immediately beneath the testis (*Figures 9.34* and *9.36*). Both the *vas deferens* and the accessory gland open behind into the anterior extremity of the paired ejaculatory duct. The paired ducts are slender and short, and open posteriorly into the median ejaculatory duct, which is thin, very long and much coiled. An anterior part of the median duct is thick and muscular to form a sperm pump or ejaculatory sac (*Figure 9.34*).

In both Plan A and Plan B the anterior extremity of the median/paired ejaculatory duct receiving *vasa deferentia* and the accessory glands may be dilated a little. This part of the median/paired duct has been often referred to as the 'prostata', though there is no histological indication of a glandular function in it.

A Plan A organization in the male genital system is seen in Galerucinae and Alticinae. Plan B in its typical form is seen in Hispinae, Cassidinae, Clytrinae, Cryptocephalinae and Chlamisinae. Among Eumolpinae, the male genital system seems to be quite variable. In *Corynodes* spp., the two testis follicles of each side are quite separate, and the *vasa deferentia* are considerable in length. The accessory glands are long and tubular, and arise considerably behind the testes. The median ejaculatory duct is long and has an ejaculatory sac. In *Colasposoma* spp., however, the two testis follicles of the same side are closely compacted together, the *vasa deferentia* are small in length, the tubular accessory glands arise immediately beneath the testicular mass, and the median ejaculatory duct is very long, but without a well marked ejaculatory sac.

In Chrysomelinae, Criocerinae and Donaciinae, though the testis follicles are of the septate type and the median ejaculatory duct may have an ejaculatory sac, the *vasa deferentia* are long, and long and tubular accessory glands arise considerably behind the testes.

In Zeugophorinae, Megalopodinae and Timarchinae, as pointed out before, the septate nature of testis follicles remains to be demonstrated, but the median ejaculatory duct is fairly long and has an ejaculatory sac. It is interesting to note that in Megalopodinae, the testes of the two sides are bound together into a median organ, almost in a galerucine way.

In Orsodacninae and Synetinae, the male efferent system is nearly galerucine in its organization.

EXTERNAL GENITALIA

The aedeagus

The aedeagal apparatus is variable among the leaf beetles, but all the variations can be derived from a common primitive plan, which may be seen in the anatomy of Cerambycidae (Verma, 1996c).

We may distinguish the following types among the aedeagi of the leaf beetles:

(i) The cerambycine type

This type has been referred to as *'mode en cavalier'* by Jeannel (1955). The aedeagus is a long sclerotic tube with paired and anteriorly extending, long basal or median struts. The tegmen is an obliquely-placed ring around the aedeagus, with paired parameral appendages dorsally, and forming an apodemal process or manubrium ventrally (*Figure 9.41 (A)*). This type is seen in Cerambycidae, Orsodacninae, Megalopodinae, Zeugophorinae and Timarchinae.

(ii) The bruchine type

This is similar to the cerambycine type, but the basal struts seem to have fused to form a dorsal hood-like structure, and the ventral apodemal part of the tegmen takes the form of a vertical plate of a considerable expanse. The parameral part of the tegmen is in the form of a single or bilobed parameral cap, due to the fusion of the parameres (*Figure 9.41 (B)*). This aedeagal apparatus is found in Bruchidae, Ṣagrinae and Donaciinae.

(iii) The cassidine type

This is similar to the bruchine type, but the dorsal part of the tegmen, including the parameral cap, has been lost. The surviving part of the tegmen is mostly apodemal; hence, one of us (KKV) consistently refers to it as the first spiculum or the tegminal apodeme (Verma, 1969; Verma and Kumar, 1972; Kumar and Verma, 1980; Verma, 1996c). The aedeagus takes the form of a long, curved tube. Its basal orifice, covered dorsally by the basal hood-like forward extension of the dorsal aedeagal wall, is large and anteroposteriorly extensive, covering one fourth to half the length of the aedeagus (*Figure 9.41 (C)*). The median limb, shaped like a vertical keel (but in some cases like a horizontal plate) of the first spiculum or the tegminal apodeme, is confined in its length to the anteroposterior extent of the basal orifice, and the forked posterior portion of the first spiculum is located close to the ventroposterior lip of the orifice. Aedeagi of this type are present in Criocerinae, Hispinae, Cassidinae, Clytrinae, Cryptocephalinae, Chlamisinae, Lamprosomatinae and Eumolpinae. The copulatory apparatus of Chrysomelinae should perhaps also be placed in this category.

(iv) The galerucine type

The aedeagus is a long and ventrally curved sclerotic tube, with an anteroventrally directed and rather small basal orifice. The first spiculum, or the tegminal apodeme, is quite long, extending considerably posteriorly beyond the ventral lip of the basal orifice. Its anterior end curves upward and projects into the basal orifice (*Figure 9.41 (D)*). Such male copulatory organs are seen in Galerucinae, Alticinae and Synetinae.

One significant difference between the cassidine and the galerucine types of aedeagi is that in the latter type there is an ample arthrodial membrane between the forked posterior portion of the first spiculum and the ventral lip of the basal orifice and, therefore, considerable relative movement can take place between the aedeagus and the spiculum during the protraction and retraction of the former. In the cassidine type, however, there is only a limited arthrodial membrane between the two (*Figure 9.41 (C)*) and, therefore, the aedeagus and the spiculum move almost as a unit.

That these four types of chrysomelid aedeagal organs are modifications of the same

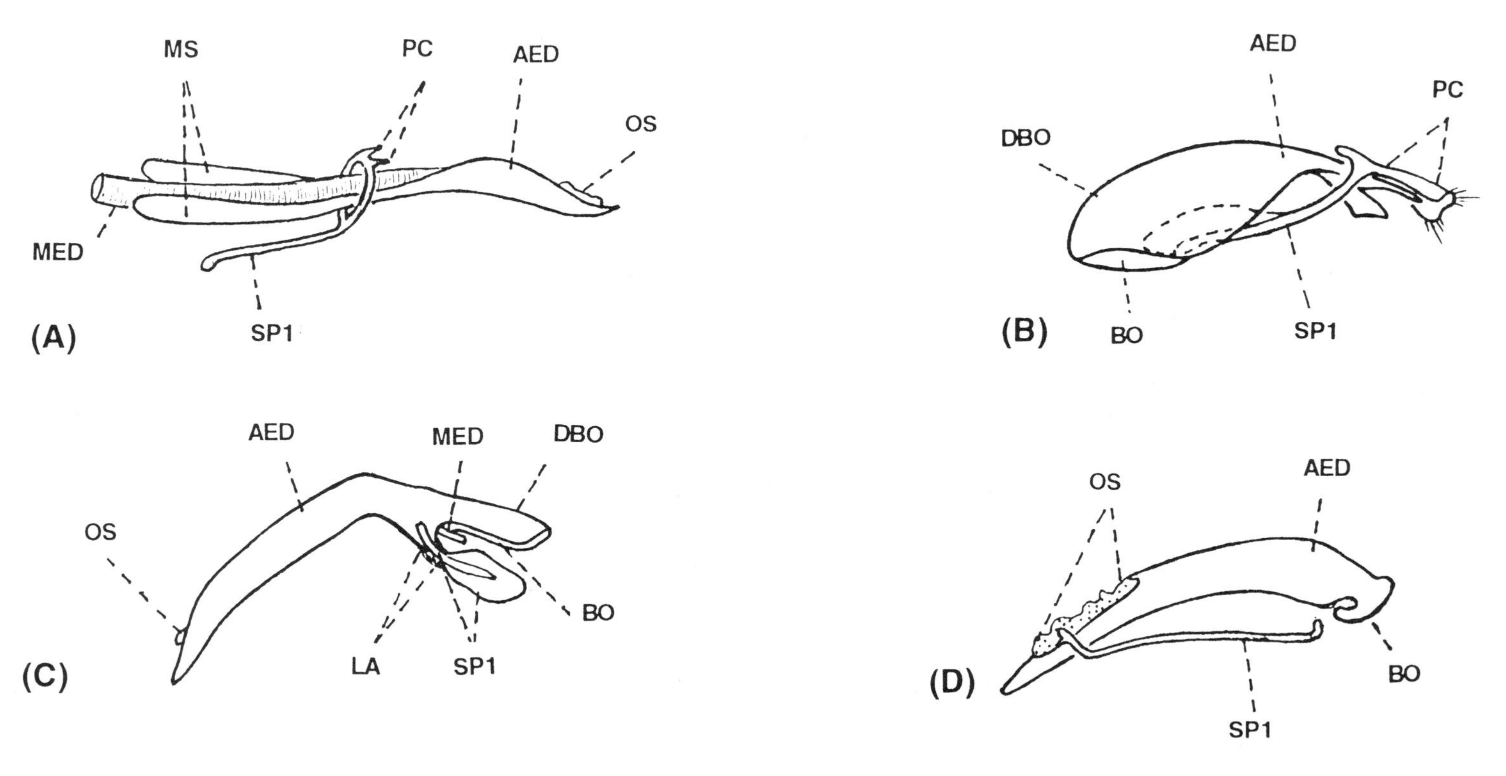

Figure 9.41. Types of aedeagi among Chrysomeloidea. (A) Cerambycine type, ex. *Phyllobius* (after Matsuda, 1976); (B) Bruchine type, ex. *Caryedon gonagra* F. (after Kumar and Verma, 1980); (C) Cassidine type, ex. *Aspidomorpha miliaris* F. (after Verma and Kumar, 1972); (D) Galerucine type, *Galerucella birmanica* Jac. (after Verma, 1969) (AED = aedeagus; BO = basal orifice; DBO = dorsal hood formed by the dorsal wall of the aedeagus extending forward and overhanging the basal orifice; LA = limited arthrodial membrane between the arms of SP1 and ventrolateral lip of BO; MED = median ejaculatory duct; MS = median or basal struts of the aedeagus; OS = ostium; PC = parameres or parameral cap; SP1 = tegminal apodeme or the first spiculum).

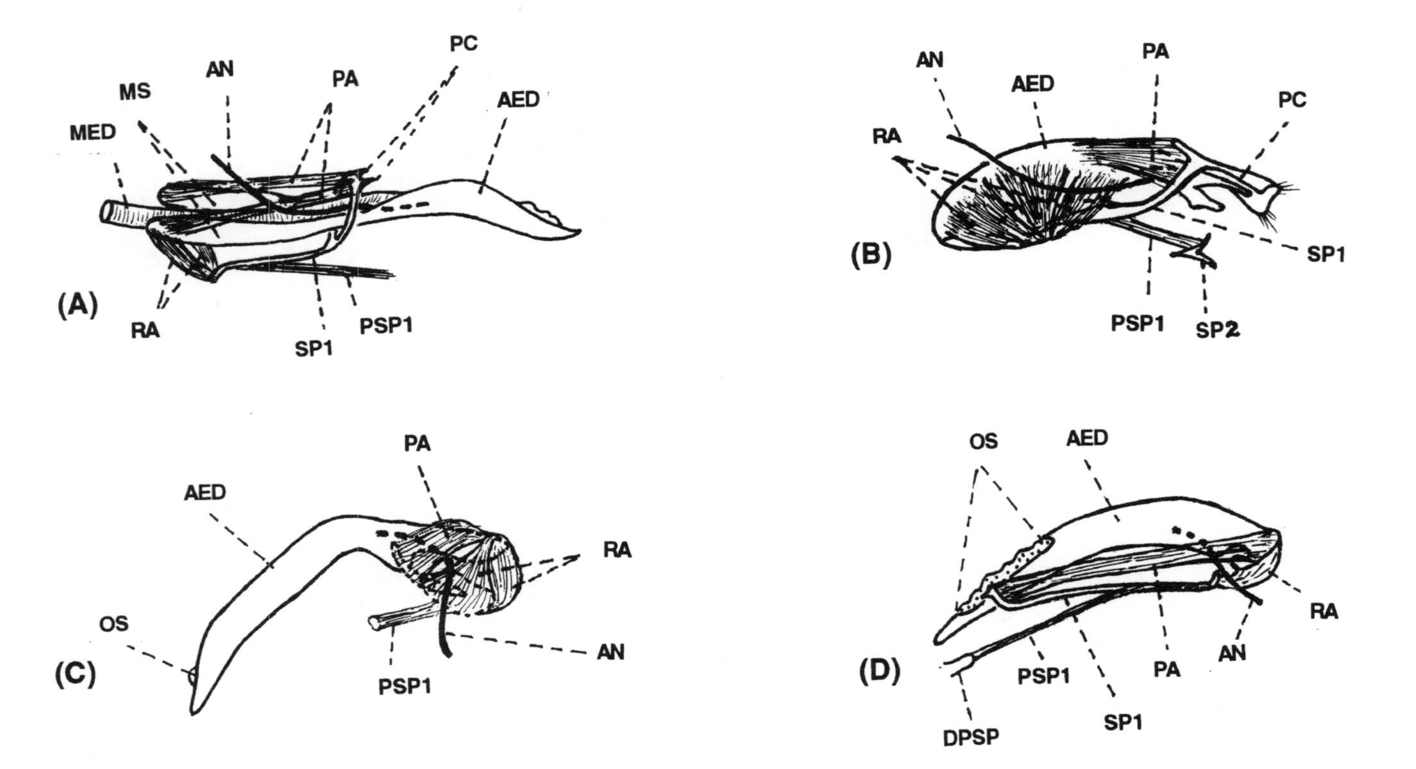

Figure 9.42. Aedeagal apparatus in Chrysomeloidea, including the aedeagal musculature. (A) Cerambycine type, ex. *Phyllobius* (sclerotic parts after Matsuda, 1976. Muscles have been added based on results of Kumar and Verma, 1980, who studied three species of Cerambycidae from the standpoint of aedeagal musculature); (B) Bruchine type, ex. *Caryedon gonagra* (after Kumar and Verma, 1980); (C) Cassidine type, ex. *Aspidomorpha miliaris* (after Verma and Kumar, 1972); (D) Galerucine type, ex. *Galerucella birmanica* (after Verma, 1969) (AED = aedeagus; AN = aedeagal nerve; DPSP = diverticulum of the genital pocket providing attachment to the protractor of the first spiculum; MED = median ejaculatory duct; MS = median or basal struts of the aedeagus; OS = ostium; PA = protractor/protractors of the aedeagus; PC = parameres or parameral cap; PSP1 = protractor of the first spiculum; RA = retractor/retractors of the aedeagus; SP1 = tegminal apodeme or the first spiculum; SP2 = second spiculum).

basic plan is clearly suggested by a comparative study of aedeagal musculature. Muscles connected with aedeagal function are illustrated in *Figure 9.42 (A–D)*. It may be noted that in the cerambycine and galerucine types, the protractors and retractors of the aedeagus perform function as suggested by their names but in the bruchine and the cassidine types, these muscles have become bunched together to form a basal muscular bulb for the aedeagus. This bunching is a result of the shortening of the first spiculum. The muscular bulb seals the fluid in the aedeagus on eversion of the internal sac during copulation, and thus ensures the turgidity of the everted sac. The protraction of the aedeagus in these types is the function of the protractor of the first spiculum. The retraction of the aedeagus seems partly due to the pressing of the abdominal tip against the substratum, and partly due to the protruding organ being repeatedly struck by the legs of the male. Verma and Kumar (1972) separated the male from a copulatory pair of *Aspidomorpha miliaris*, and its legs were amputated soon after the separation, i.e. before the legs could stroke the protruding penis. In order to prevent the insect from pressing the abdominal tip against the substratum, it was held in the air by holding the explanate margin with a small clamp. This experiment was repeated three times, and every time, even after 24 hours, the male could not retract the penis, though the insect was alive and active.

For a detailed description of examples of the four types of aedeagal apparatus, see Verma (1969), Verma and Kumar (1972) and Kumar and Verma (1980).

Though in Chrysomelinae and Criocerinae there is a basal muscular bulb for the aedeagus, the aedeagal apparatus in these subfamilies appear somewhat intermediate between the cassidine and the galerucine types. In Chrysomelinae, the aedeagus presents a regular ventral curvature throughout its length and not an almost ventral angulation near the base, as is seen among those with a typical cassidine type of aedeagus. Moreover, the basal orifice is rather restricted anteroposteriorly, and the middle limb of the first spiculum does not form a deep vertical keel. In Criocerinae, the first spiculum extends considerably behind the posteroventral lip of the basal orifice (*Figure 9.43*).

The sclerotic armature of the internal sac of the aedeagus in Chrysomelidae was comparatively studied by Mann and Crowson (1996). The armature includes sclerites, spicules and flagellum, variously shaped and arranged. In fact, the armature presents a lot of variation. From their study, the authors inferred that the evolutionary development of the internal sac structure had been polyphyletic, and that the following are possible lines of development:

(a) Sagrinae–Donaciinae–Criocerinae–Cassidinae–Hispinae.
(b) Orsodacninae.
    (i) Megalopodinae–Zeugophorinae–Megascelinae–Lamprosomatinae–Chlamisinae–Clytrinae–Cryptocephalinae.
    (ii) Eumolpinae–Synetinae–Galerucinae–Alticinae.
(c) Aulacoscelinae–Timarchinae–Chrysomelinae.

The spicula

There may be one or more spicula associated with the aedeagus in the male external genitalia of Chrysomelidae. The association is both anatomical and functional.

The spicula are paired or unpaired apodemal rods, situated on the outer face of the

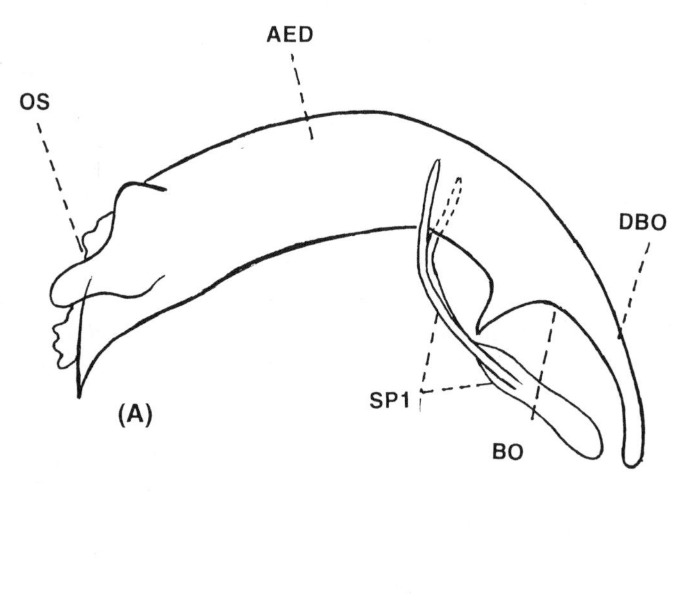

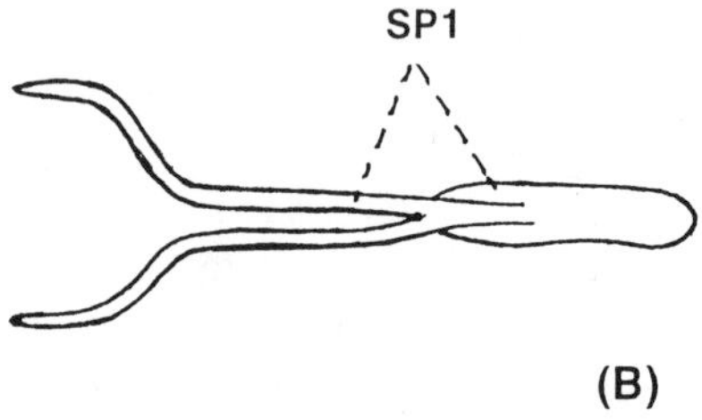

Figure 9.43. Aedeagus of *Lema semifulva* Jac. (Criocerinae). (A) Lateral view; (B) Dorsal view of only the first spiculum. Its median vertical keel-like portion has turned under cover-glass pressure and has become almost horizontal (AED = aedeagus; BO = basal orifice; DBO = dorsal hood formed by the dorsal wall of the aedeagus extending forward and overhanging the basal orifice; OS = ostium; SP1 = tegminal apodeme or the first spiculum).

genital atrium, and are of variable homology and origin (Verma, 1969; Kumar and Verma, 1971; Matsuda, 1976). The first spiculum is an apodemal ingrowth from the tegmen, and is present in all the leaf beetles. A second spiculum may also be present, in which case the protractor of the first spiculum arises on it. This muscle, running anteriorly, becomes inserted on the median limb of the first spiculum. Often, the origin of the muscle on the second spiculum is asymmetric, indicating the direction of '*retournement*' of the aedeagus (*vide infra*), e.g. in *Aulacophora* spp. and *Colasposoma* spp. (Kumar and Verma, 1980). In some Chrysomelidae, there is no second spiculum, in which case the posterior attachment of the protractor of the first spiculum is on a small ventral outpocketing of the genital atrium; this may be seen in *Galerucella* sp. (Verma, 1969) and *Aspidomorpha* sp. (Verma and Kumar, 1972). This part of the genital atrium may represent one of the last abdominal segments, membranized and incorporated in the genital pocket. In *Aeolesthes* sp.(Cerambycidae) there are: (i) a second spiculum, which is clearly constituted by apodemal ingrowths from the sides of the reduced ninth abdominal tergum, and (ii) a third spiculum, connected with the eighth abdominal sternum (Kumar and Verma, 1980).

'Déversement' and 'retournement' of the aedeagus

The terms '*déversement*' and '*retournement*' were introduced by Jeannel (1955) and they refer to certain orientational changes in the aedeagus. One of us (KKV) has adopted these French terms and has been using them for about thirty years as their meaning is clearly defined and no suitable English equivalents are available.

'*Déversement*' means the inclined or tilted orientation of the ventrally curved aedeagus in repose in the abdomen. This inclined orientation is undone every time the aedeagus is protruded for copulation, and the aedeagus returns to it when it is withdrawn into the abdominal cavity. Thus, it is a reversible phenomenon.

In *Galerucella birmanica*, the aedeagus in repose is 'deversed' to the left (Verma, 1969). The undoing of the '*déversement*' during protrusion of the aedeagus seems to be partly due to the median anterior attachment of the protractor of the first spiculum, and partly to the horizontal slit-like opening of the anogenital vestibule between the last externally visible abdominal tergum and sternum. The terminal part of the aedeagus is dorsoventrally flattened and plate-like; hence, when it is being pushed through the horizontal opening during protrusion, it becomes symmetrical in orientation. The reappearance of '*déversement*' on retraction of the aedeagus is partly due to the position of the aedeagus among the abdominal viscera and partly to the asymmetric and twisted tracheal supply reaching the aedeagus base. The twisted nature of the tracheal supply for the aedeagus is a consequence of the phenomenon of '*retournement*' (*vide infra*). For a detailed discussion of '*déversement*' refer to Verma (1969).

Balazuc (1948) described the tilted orientation of the aedeagus in adult *Donacia* spp. The tilting is usually to the left, but in a small number of cases it is to the right. The author referred to this asymmetric orientation as 'inversion'; it is clearly '*déversement*'.

Iuga and Konnerth (1963) recorded the rightward tilting of the aedeagus in repose in the alticine, *Phyllotreta nigripes* (Fabricius).

Harnisch (1915) illustrated and discussed the leftward tilting of the aedeagus at rest in *Chrysomela* (*sens stricta*) *populi* Linné.

'*Retournement*' means the rotation of the aedeagus about its longitudinal axis through about 180° during development. '*Retournement*' is an irreversible developmental change, while '*déversement*' is reversible and functional.

'*Retournement*' is known in Silphidae and Staphylinidae, and probably occurs also in Oedemeridae (Jeannel, 1955). Heberdey (1928) observed it taking place in a dytiscid beetle. Verma (1958, 1969) was first to describe it in a chrysomelid. For '*retournement*' see also Kumar and Verma (1971), Verma and Kumar (1972), and Kumar and Verma (1980).

The direction of '*retournement*' is, in most cases, clockwise when looked at from behind. Rarely is it anticlockwise. In this phenomenon, the rotation involves the aedeagus and the genital pocket. This change of orientation produces a number of morphological effects; the nervous and tracheal supply reaching the aedeagus base becomes twisted (*Figure 9.44*) and the median ejaculatory duct takes a coiled course to the aedeagus base (*Figure 9.45*). In those chrysomelids which have a very long and coiled median ejaculatory duct, for example in hispines and cassidines, this coiled course (resulting from '*retournement*') of the median duct is not very obvious.

'*Retournement*' of the aedeagus is due to degeneration of a member of a pair of

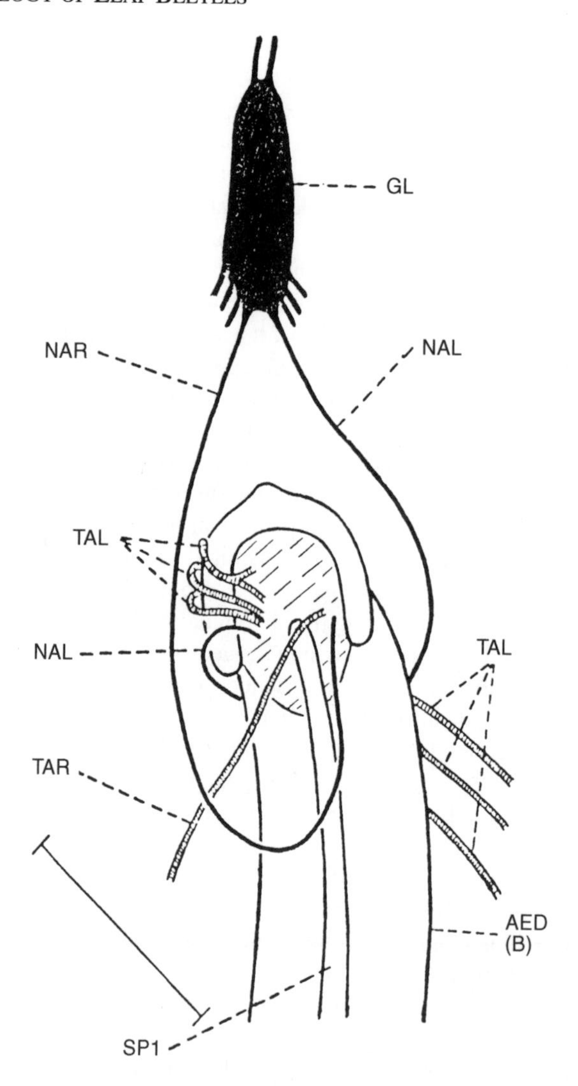

Figure 9.44. Nervous and tracheal supply of the aedeagus in *Galerucella birmanica*, ventral view. The scale indicates 0.5 mm (after Verma, 1994) (AED(B) = basal part of the aedeagus; GL = last abdominal ganglion (compound); NAL = aedeagal nerve of the left side; NAR = aedeagal nerve of the right side; SP1 = spiculum/first spiculum; TAL = tracheae arising on the left side and supplying the aedeagus; TAR = tracheae arising on the right side and supplying the aedeagus).

muscles getting attached to the dorsal surface of the genital tube (= aedeagus + genital atrium). The surviving member of the pair exerts a unilateral pull on the dorsal surface of the tube; as a result, the tube rotates, bringing the dorsal surface to the ventral side. The paired muscles are the protractors of the first spiculum, or the tegminal apodeme. They arise from a short ventral out-pocketing of the genital atrium, close to the opening of the latter into the anogenital vestibule or from the developing second spiculum, and then run forward and upward, finding attachment to the genital tube (*Figure 9.46*). Generally, the right member of the pair survives, and the left degenerates; the consequent *'retournement'* is clockwise. But rarely it is the left member

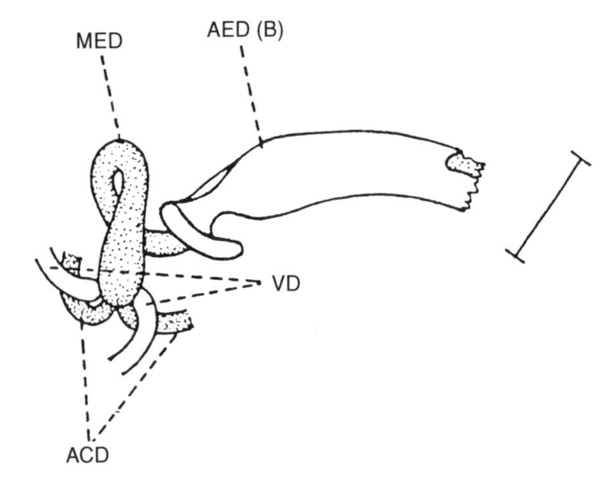

Figure 9.45. Aedeagus and median ejaculatory duct of *Galerucella birmanica*, lateral view (after Verma, 1969). The scale indicates 0.5 mm (ACD = male accessory genital glands; AED(B) = basal part of the aedeagus; MED = median ejaculatory duet; VD = *vasa deferentia*).

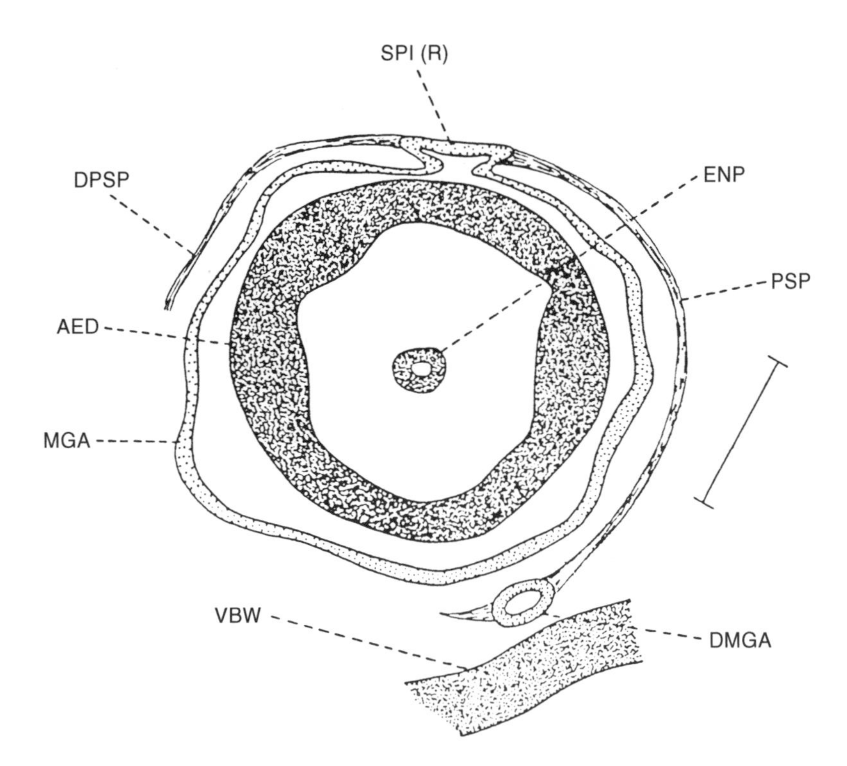

Figure 9.46. Diagrammatic t.s. of the developing aedeagus in a young male pupa of *Galerucella birmanica* to explain mechanism of *'retournement'* (after Verma, 1969) (AED = aedeagus/aedeagus rudiment; DMGA = diverticulum from the floor of a posterior part of the genital atrium; DPSP = degenerating protractor of the spiculum; ENP = endophallus; MGA = membranous walls of the genital atrium; PSP = protractor of the spiculum; SP1(R) = spiculum rudiment; VBW = ventral body wall).

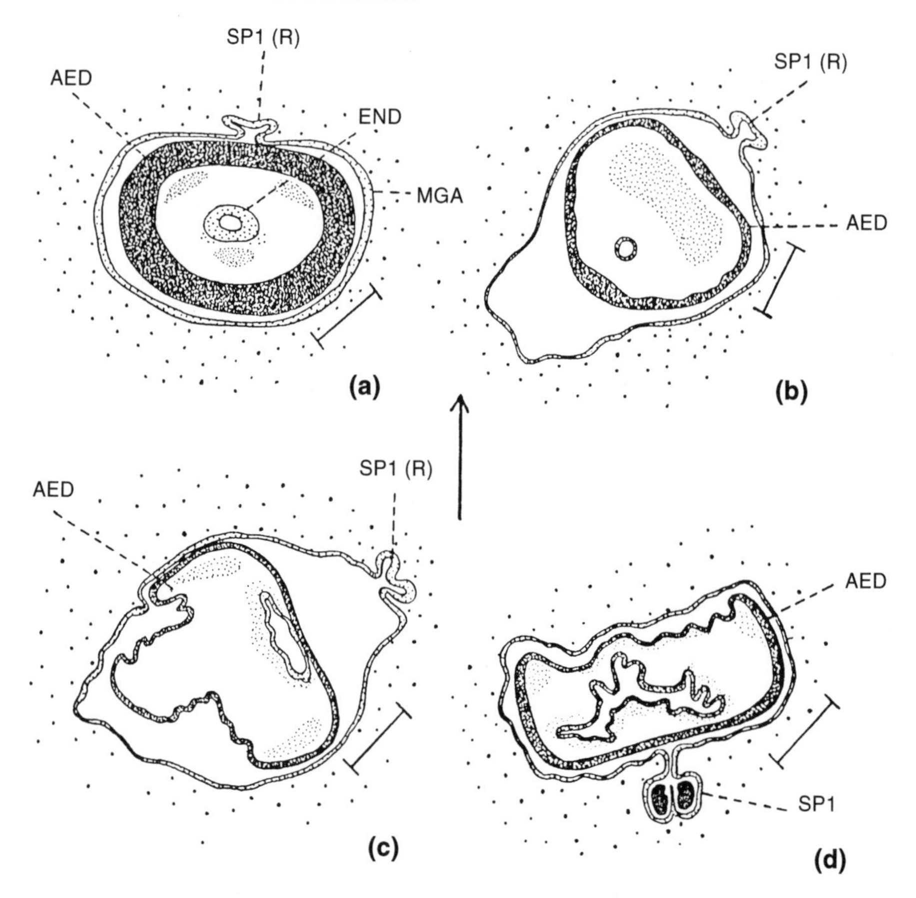

Figure 9.47. Progress of *'retournement'* as seen in transverse sections of pupa/adult of *Galerucella birmanica*. (a) Pupa nearly 2 days old; (b) Pupa nearly 4 days old; (c) Pupa nearly 5 days old; (d) Well sclerotized adult (from Verma, 1994). The scale indicates 0.1 mm. The arrow indicates dorsoventral direction for all sections. Posterior face of the section is upper in all cases (AED = aedeagus/aedeagus rudiment; END = endophallus; MGA = membranous walls of the genital atrium; SP1 = spiculum/first spiculum; SP1(R) = spiculum rudiment).

which survives and the right degenerates; then the resultant rotation is anticlockwise. This mechanism of *'retournement'* has been observed by Verma (1969) in *Galerucella birmanica* (*Figure 9.47*) and by Verma and Kumar (1972) in *Aspidomorpha miliaris*.

That *'retournement'* has occurred (and also its direction) is suggested by the obviously twisted nature of the tracheal and nervous supply to the aedeagus (Kumar and Verma, 1971). In some cases, the asymmetric origin of the protractor of the first spiculum on the second spiculum is also evidence of the aedeagal turning, e.g. in *Aulacophora foveicollis* (Kumar and Verma, 1980) (*Figure 9.48*). Such an asymmetric origin of the muscle may also be seen in *Chrysomela populi* (Harnish, 1915). The asymmetric attachment of the muscle on the tegminal apodeme in the alticine *Longitarsus pratensis* (Panzer) (Iuga and Konnerth, 1963) also suggests clockwise *'retournement'*.

The occurrence of *'retournement'* has been inferred from its morphological effects in all of the 20 species of Chrysomelidae, 6 species of Bruchidae and three species of

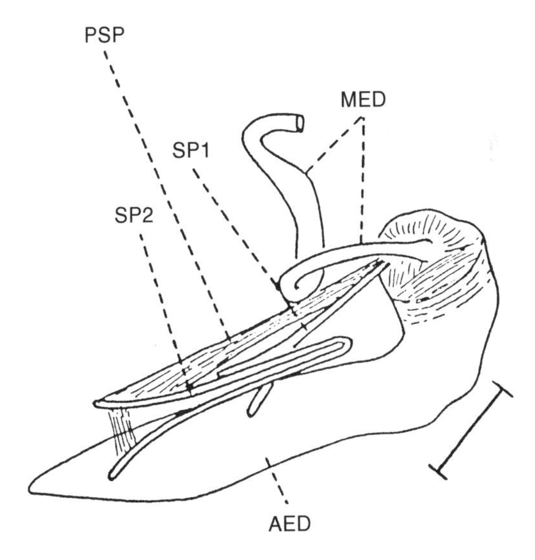

Figure 9.48. Copulatory apparatus of *Aulacophora foveicollis*, ventrolateral view. Note asymmetry in the origin of the protractor of the spiculum (after Kumar and Verma, 1980). The scale indicates 0.5 mm (AED = aedeagus/aedeagus rudiment; MED = median ejaculatory duet; PSP = protractor of the spiculum; SP1 = spiculum/first spiculum; SP2 = second spiculum).

Cerambycidae, examined from this standpoint (Verma, 1994). It seems, therefore, that the phenomenon is universal in Chrysomeloidea. This seems to be another indicator of homogeneity in this assemblage of the three large families.

We do not know the exact functional significance of *'retournement'*. According to a hypothesis suggested by Jeannel (1955), it is an adaptation for the riding mode of copulation. During intromission in this mode of copulation, the ventrally curved aedeagus on protrusion gets inverted in its distal part. One way to ensure that the original dorsal surface of the aedeagus comes in contact with the dorsal surface of the female bursa, and the ventral surface of the penis with the ventral wall of the bursa is by *'retournement'* of the aedeagus. Other ways for achieving this among beetles, according to Jeannel (1955), are: (i) *'l'inversion de l'édéage'*, which means the transposition in the development of the features of the dorsal surface of the aedeagus to the ventral surface, and vice versa, (ii) *'la migration de l'orifice apical'*, i.e. the ostium-bearing region of the aedeagus is twisted, so that the ostium gets shifted to the dorsal surface of the organ.

It has been inferred that *'retournement'* is under endocrine control in *Aspidomorpha* (Kumar and Verma, 1978) and in *Callosobruchus* (Tiwary and Verma, 1989).

For more detailed discussion on *'retournement'*, see Verma (1994).

Crowson and Crowson (1996) tried to explain the loss of the dorsal part of the tegmen in most Chrysomelidae. They say "I (perhaps R.A. Crowson) had once suspected a connection (of loss of the dorsal part of the tegmen) with the phenomenon of *'retournement'* of the aedeagus". But when he noted that *'retournement'* took place even in Chrysomeloidea with the parameral part of the tegmen intact, he changed his opinion and offered another hypothesis, that ".... loss of the dorsal part of the tegmen is related to the problems of copulation on an unstable leaf surface rather than a firm stem or ground as in *Timarcha*". This second hypothesis is not satisfactory either, as

Timarcha copulates on a climbing stem of *Galium* or on the leaves of *Plantago* in the Old World, and on the leaves of *Rubus* in the US. Perhaps the loss of the dorsal part of the tegmen is one of the evolutionary tendencies among Chrysomelidae. The situation that few genera of Chrysomelidae have ring-like tegmen without the dorsal parameral appendages, which are present in *Orsodacne* or *Timarcha*, lends support to this evolutionary tendency concept. A ring tegmen exists exceptionally among several Chrysomelinae (Daccordi, pers. comm.). However, this ring has nothing in common with the plesiomorphic tegmen of the *Timarcha* and has no tegmental cap.

A dimorphism in the male copulatory organ was described in *Arthrochlamys bebbianae* Brown (Brown, 1944). Dimorphism, or even polymorphism, of the male copulatory organs may exist in a species and this should not lead us to describe two or more species, if individuals are otherwise identical. Are several species of *Chaetocnema*, all feeding on Convolvulaceae and seemingly differing only in the shape of the extremity of the aedeagus, products of such a confusion? This case remains to be investigated closely. Some species of *Timarcha* also show polymorphism of the aedeagus.

10

Reproduction

The primary genital organs, or their parts which are often used in taxonomy (aedeagus, internal sac, spermatheca), usually provide us with good information but sometimes, as for example in *Timarcha*, they are so variable as to be of no help in the separation of species (Stockmann, 1966). However, there are some groups of species among the *Timarcha* in which it seems that characteristics of the aedeagi are sufficiently stable to work as taxonomic references. The same is true for spermatheca.

In certain cases, as with some of the Galerucinae, aedeagus characteristics are extremely different from one species to another. Take the example of the galerucine genus *Monoxia*, for instance, and chiefly the species *Monoxia puncticollis* Say, where the aedeagus is four times longer than in *Monoxia debilis* LeConte and very different in shape (Jolivet, 1959). It fills the whole abdomen in the case of *M. puncticollis*. Both species feed on Chenopodiaceae, while others prefer Asteraceae.

The spermatheca has a rather complex structure and various experiments have been carried out with a cassid, *Chelymorpha alternans* Boheman. Rodriguez (1994b) cut the spermathecal muscle of a virgin female of this cassid and obtained the following results: the number of transferred spermatozoa is not affected, but the operation reduces the fertility and produces many infertile (trophic) eggs.

It seems likely that, during multiple matings, certain males could empty with their flagellum the spermatheca of a female and replace its contents with their own spermatozoa (Windsor, pers. comm.). This could be the reason for the long flagella of certain neotropical species, but the function of the flagellum seems to be more as a titillatory organ than anything else. Removal of sperm has been found among several tenebrionids but remains to be proven among cassids or other chrysomelids (Gage and Baker, 1991; Gage, 1992; Jolivet, 1999b).

The secondary sexual characters among chrysomelids are generally the body size, females being bigger and longer than the males (Chrysomelinae, Galerucinae, etc.). However, it is not always the case, since among the Clytrinae, Megalopodinae, for instance, the prominent head, the lengthening of the forelegs and the prominent mandibles in the males make them very different from and bigger than the females. Patay (1937b) described small differences between the sexes in *Leptinotarsa decemlineata* (Say) in the last tergites and sternites of the abdomen. In brief, size provides the easiest method of sex identification in chrysomelids.

In *Chrysolina aurichalcea* (Mannerheim), sexual dimorphism (except a little for the size) is practically non-existent, but, among the Camptosomata, females have an abdominal depression used during the making of the scatoshell. Such a depression is missing, or remains small, in the male. While the female is losing her wing muscles (compensated in ovary development), the male often retains them (*Chrysolina*, *Oreina*).

There are obvious differences in the tarsi of *Timarcha*, which are enlarged in the males and narrower and with a furrow in the females. This difference facilitates copulation for the male as its forelegs can firmly grasp the female elytra. A similar, though less marked, difference is found in other chrysomelids, with variations in the shape and the furrows of the last abdominal sternites. In several rare cases, for instance in *Chrysolina* (*Caudatochrysa*) *angusticollis* (Motschuslky), the pygidium is strongly elongated in the female and takes the shape of an ovipositor, a characteristic already observed in the subgenus *Anopachys*. *Chrysolina angusticollis*, like others of the species from Asia, largely resembles a species of *Timarcha*. Some of these have been described as *Timarcha*. In such cases, however, the tarsi are not so enlarged among males.

Sacchi and Busardo (1935) studied in Italy variations in the size of the male and female of two *Chrysolina*: *C. fastuosa* (Scopoli) and *C. americana* (L.). The differences are really small, but they exist. The male of *Oulema melanopus* (Linné) has a narrower cephalic capsule and shorter elytra than the female.

Many chrysomelid or galerucid females demonstrate physogastry before being ready to lay eggs (*Gastrophysa*, *Xenomela*, *Oreomela*, *Agelastica*, *Galeruca*, etc.). In certain cases, as for the Mexican *Metacycla*, the differences between the sexes are enormous: physogastry, apterism and brachelytry in the females, whereas the males have complete elytra. Apterism, or functional apterism, in the female is generally useful as it helps in concentrating energy for reproduction.

Other differences between the sexes are sometimes encountered but are not easily distinguishable. The difference in size is the most constant character, together with the enlargement of the male tarsi, which allows the male to grasp the female without slipping. There are also certainly physiological differences between the sexes in terms of the haemolymph composition, their concentration of toxins, etc. Some chrysomelines have a differently coloured haemolymph in male and female (*Gonioctena*) (green and yellow). Among the Clytrinae, the highly enlarged forelegs of the male seize the females like forceps and allow a quasi vertical position during mating.

It may also be noted that several viviparous species, but not all, have a regressed spermatheca, or none at all (Bontems, 1988). The detailed structure of the internal sac of the aedeagus can also provide important details for distinguishing between related species, and among *Timarcha* it makes it easier to differentiate between the subgenera (Stockman, 1966).

Reproduction and mating

There are only a few published works on the reproduction and mating of chrysomelids. The first was by Gadeau de Kerville (1900) and several papers appeared recently, namely Medvedev (1962), Medvedev and Pavlov (1988), Jolivet (1999c), Bienkowski (1999a), etc. The whole picture is summarized in Jolivet (1999d). Many papers have been published on mating in other species (Thibout, 1982; Windsor, 1987; Boiteau, 1988b; Jolivet, 1999b, etc.). Mating between unrelated or related groups (Cerambycidae, Bruchidae, Curculionidae, Meloidae, etc.) has also been abundantly reported and can be used for comparison. However, surprises can be expected when studying the Galerucinae with their complex antennae and abundant pheromone production and with the Alticinae which demonstrate in some cases a kind of lek behaviour.

Interspecific matings are frequent among related species feeding on the same plant. *Chrysolina polita* (Linné), for instance, mates often with *Chrysolina herbacea* (Duftschmidt) or *C. graminis* (Linné) on *Mentha aquatica* L. or *M. rotundifolia* L. Such matings remain sterile, at least most of the time, except perhaps in the case of *C. coerulans* (Scriba), this species being closely related to *C. herbacea* (Jolivet, 1952b). Interspecific matings are almost unknown among *Timarcha* spp., even among sympatric species like *T. goettingensis* (L.) and *T. tenebricosa* (F.), but aberrant matings have been noted in rearing-cages, even between larvae and adults. Observations in captivity are generally unreliable and need to be confirmed by observations in the field.

It is likely that the enormous alticine spermatozoa, sometimes longer than the insects themselves, among Oedionychina (Virkki and Bruck, 1994) carry proteins to be used by the future embryo, but they are probably also used as vaginal or mating plugs (Ladle and Foster, 1992). In *Drosophila hibisci* Bock, sexually mature males cannot copulate with previously mated, young females (Polak *et al.*, 1998) and here the plug may be a deterrent. With Oedionychina, there is no special plug apart from the large spermatozoa. The formation of large spermatozoa could also be a kind of sperm economy. To Bacetti and Daccordi (1988) there are relationships, indicated by the spermatozoa, between Bruchidae, Chrysomelidae and Curculionidae. According to them, leaf beetles are not polyphyletic, a view which is different from that of most cladists and taxonomists. What is definite is that chrysomelids arose from the cerambycids with the parallel branch of Bruchidae. Dallai *et al.* (1998) separate Cerambycidae, Chrysomelidae and Bruchidae on the basis of spermatozoa ultrastructure.

Since Gadeau de Kerville wrote the first paper on beetle mating habits (1900), there have been many others. Observations have been made on the male fight for possession of the female among Lucanidae, Dynastidae, Cerambycidae, Tenebrionidae, etc. A complex system of rivalry exists between strong and weak males according to size, the development of the cephalic or thoracic horns, the mandibles, etc. In some tenebrionid species, like *Bolithoterus cornutus* Panzer (Brown and Siegfried, 1983), the males vary much in size and horn length. The relative size of the horn is an important factor in the acquisition of a female among males of equal size.

In *Doryphora punctatissima* (Olivier), males use their sternal horn as an aggressive weapon during rival encounters between males (Eberhard, 1981) on the host plant, *Prestonia isthmica* Woodson, an Apocynaceae (*Figure 10.1 F*). In this species, both males and females have horns. The use of the horn by the female is not clear. Males, and (rarely) females, compete for a mate among chrysomelids and cerambycids and among representatives of other families.

Thibout (1982) described the sexual behaviour of the Colorado Potato Beetle (*Leptinotarsa decemlineata* (Say)) and its control by juvenile hormone. He noted that a young virgin female, not yet fed, is not receptive and rejects the male (*Figure 10.1 A*). An old, well-fed female has an inflated abdomen and copulates immediately after the male antennae vibrate. An undisturbed female does not stop feeding during mating.

In the Colorado beetle, males are attracted by the pheromones of the females after 10 days from the emergence of the latter (Edwards and Seabrook, 1997). Male choice is often due to the female showing her preference. Both sexes remain polygamous. Males are aggressive when trying to gain access to a female (Thibout, 1982; Szentesi, 1985). Their respective size is not of much importance, while size is important among other beetles like tenebrionids or scarabeids. Boiteau (1988b) observed that males, being fundamentally polygamous, guard the female once they have acquired one. He also studied the use of the sperm by the female. A female mated three times will lay three times as many eggs than a female mated once and her spermatheca will be fuller. It may be noted that three copulations are necessary to fill the spermatheca of the Colorado beetle female. Extra copulations result in sperm transfer, and cryptic female choice occurs. Certain matings remain sterile. The sperm of the last male to copulate always takes precedence. Females tend to be less and less receptive in a series of matings. The guarding of the female by the male is important, since it allows him to have permanent access to her, offering repeated matings and protecting the male investment (Thornhill and Alcock, 1983). These observations of the male guarding the female in the Colorado beetle are interesting since this has not yet been observed among other chrysomelids. We think that, in some rare cases in South America, males may guard offspring along with the females, but this has still to be confirmed. According to Tallamy (1994), females seek to acquire sufficient, high quality nutrients to fuel oogenesis, while males seek to father as many offspring as they can. Remaining with the female guarantees the male permanent access to his mate.

The movement, dispersal and mating of *Leptinotarsa decemlineata* was studied by Alyokhin and Ferro (1999) using a mark–recapture technique. Adult dispersal starts within the first 24 hours after eclosion from the pupae and mating produces viable offspring after 34 days. Females do not start laying eggs until a minimum of 51 days after eclosion. Mating decreases female flight activity, but increases it among males.

Bienkowski (1999a) carried out laboratory studies on the mating behaviour of 14 species of *Donacia*. It seems that the male, once on the female elytra, does not attempt to copulate for a long time. Pre-copulatory courtship is observed during the preliminary stages in several species: the male rubs the female's antennal or vertex calli with his fore tarsi. The male can quit first after mating but also the female can pull the male off with her hind tarsi. Females generally feed during amplexus.

The partially sympatric *Leptinotarsa decemlineata* (Say) and *L. juncta* (Germar) are shown to possess a series of reproductive barriers that protect species integrity (Boiteau, 1998). A strongly stereotyped sequence of courtship behaviour ensures that more than 85% of encounters between these congeneric species are terminated before copulation. The courtship behaviour of both species is similar, but differs slightly. Both species use antennal tapping initially and palp tapping for final sex recognition. If courtship behaviour fails to break up, there is no mechanical barrier to mating and sperm transfer takes place. However, there is gametic mortality. Contrary to general thinking, *L. juncta* is more fecund than *L. decemlineata* and less dependent on multiple matings to reach its full reproductive potential. Hybrids between species of Chrysomelidae are extremely rare and only a few reports are known (*Chrysolina*, *Diabrotica*). Generally, the hybrids are sterile. Sympatric *Timarcha* species never copulate in the field and hybrids are unknown in nature. Krysan and Guss (1978)

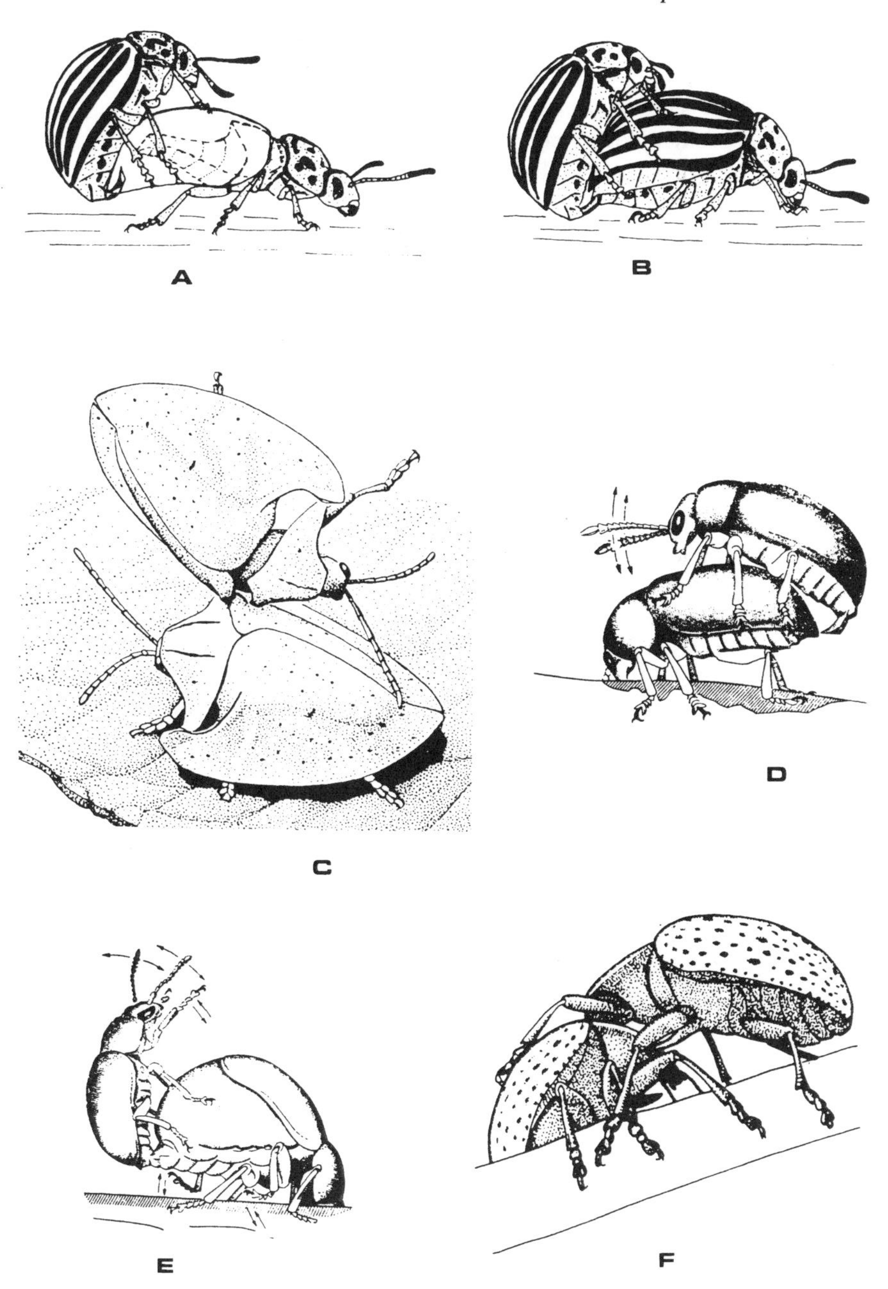

Figure 10.1. Mating habits of various chrysomelids. (A) Mating habits of the Colorado Potato Beetle *Leptinotarsa decemlineata* (Say) (Chrysomelinae). A starving female retracting her abdomen under the elytra, refusing the male penis (after Thibout, 1982); (B) The same. Mating of the male with a well-fed female accepting the male (after Thibout, 1982); (C) Two males of *Acromis sparsa* Boheman (Cassidinae). One male has been lifted from the surface during the fight (after Windsor, 1987); (D) Start of the mating process in *Gastrophysa polygoni* (L.) (Chrysomelinae) (after Medvedev and Pavlov, 1988); (E) Maximum activity of *Gastrophysa polygoni* during mating (after Medvedev and Pavlov, 1988); (F) Two males of *Doryphora* sp. battling on the stem of *Prestonia isthmica* Woodson (Apocynaceae) in Colombia. They use the mesosternal horn during the fight (after Eberhard, 1981).

studied the barriers to hybridization between *Diabrotica virgifera* LeConte and *D. longicornis barberi* Smith and Lawrence in the field, and they noted that hybrids are non-existent. Hybridization between two species of *Gastrophysa* is practically impossible, even under laboratory conditions (Jolivet, 1999b).

Nahrung and Merritt (1999) have shown that single-mated females of the Kenyan alticine *Hemchloda barkeri* (Jacoby) live longer than multiple-mated and unmated females. The beetle has been imported into Australia to fight the prickly *Acacia*, *Acacia nilotica* (L.) Willdenow ex Delile.

Chrysomelid mating happens well exposed on the host plant, but sometimes the male follows the female for a certain distance (Medvedev and Pavlov, 1988). In certain cases (strong wind, strong luminosity), mating can occur in the shade, under a trunk for instance. A certain temperature seems favourable (20–23°C in temperate countries). Adult beetles of *Atrachya menetriesi* Feldermann, a galerucine, starts oviposition in a shorter period in the field than under constant laboratory conditions (Yamashiro *et al.*, 1998). It is very difficult to predict seasonal life cycles in insects. *Chrysolina cavigera* (J. Sahlberg) in Western Siberia, Wrangel Island, is active and mates during the local early 'summer' at around 6°C (Khruleva, 1996). With all the species observed by Medvedev and Pavlov (1988) (*Chrysolina graminis* (L.), *Gastrophysa polygoni* (L.), *Prasocuris phellandrii* (L.), *Galerucella nymphaeae* (L.) and various species of *Donacia*), mating is preceded by a precise ritual and the male position differs, mostly for mechanical reasons, according to the species. Medvedev and Pavlov (1988) also observed mating in *Labidostomis pallidipennis* Gebler, several *Lema*, *Clytra*, *Pachybrachis* spp., *Leptinotarsa decemlineata* (Say), *Plagiodera versicolora* (Laicharting), and several Vietnamese species which do not behave differently from Palaearctic species.

Often, mating is preceded by some kind of male dance and the males usually tap the female's eyes, pronotum and antennae with their own antennae. The shape of the aedeagus determines the position and the way it is inserted into the female and there certainly exists, at least for most of the species, a rather complex 'lock and key' system. This system is also a mechanical barrier to interspecific hybridization. Characteristic mating foreplay, chromosomic differences in number and shape and molecular differences also constitute obstacles to such mating and hybridization. Generally, gametes are sterile between species.

Phillips (1979) described the mating of *Altica lythri* Aubé. When a male meets a female it taps the anterior part of the elytra of the female with its own antennae and then tries to mount rapidly, hitting its mate's thorax and elytra with its maxillary palps. The aedeagus comes out rapidly when the male seizes the dorsal surface of the elytra with its middle and anterior tarsi, the hind tarsi seizing the lateral margin of the abdomen. The aedeagus is bent and is inserted with care into the female vagina. During the copulation, the male antennae and palpi slightly touch the female head and antennae. A paroxysm of antennal oscillations among males is observed before and after aedeagus insertion. Non-receptive females will violently extend the ovipositor, so preventing the insertion of the aedeagus.

Some species react differently when disturbed before mating. Some, like the *Donacia*, part and fly away, to meet again later on. Others, such as *Cryptocephalus* spp., part and drop to the ground, or maintain the pair. Some pairs, like *Donacia*, try to escape by diving into water or below the leaves.

The *Figure 10.1 B, D* and *E* show the mating positions. Anterior male tarsi, sometimes much larger, make the grasping of the smooth female elytra (*Timarcha*) easier. *Labidostomis pallidipennis* Gebler drops its legs onto the ground to adjust the female.

Fights between males for a mate are frequent, for example among Clytrinae, like *Coptocephala* spp. and *Labidostomis longimana* (L.) and Megalopodinae (Schulze, 1996). However, it is a common behaviour among all the genera and species, even if it is not always readily observable in the field. Among clytrines and megalopodines, males attack by raising the anterior part of their bodies and trying to catch or to push their rival away with their anterior legs. They also try to bite their adversary with their mandibles. Parri *et al.* (1998) described frequent fights among the males of *Galerucella nymphaeae* (L.) for the possession of females, but they did not mention the role of female choice.

Certain galerucine or alticine males stimulate the females before mating using the secretions of special glands associated with complex sexual structures on the head or the antennae (*Fleutiauxia, Phyllobrotica, Agetocera*). The male rubs its anterior tarsi on the glands and smears its antennae with this secretion.

Eberhard (1991), Eberhard *et al.* (1993) and Eberhard and Marin (1996) described the sexual behaviour of two species of *Macrohaltica* and *Megalopus armatus* Lacordaire. *Macrohaltica jamaicensis* (F.), in Costa Rica, palpates the female abdomen with its hind tarsi. These movements, together with antennal palpation, are basic during copulation. The female can always prevent the copulation if she is not receptive, by contracting her vaginal muscles so preventing the male from eversion of the internal sac. The giant hind legs of the male of *Megalopus armatus* function as weapons in battles between males over sites where feeding and mating occur. These places (on members of the Solanaceae or Asteraceae) would not normally be used for courtship but males do court females actively during copulation. Eberhard (1981) found similar habits in *Doryphora* sp. in Columbia. However, the adults of these beetles display little preliminary courtship behaviour. Mating lasts approximately one hour and the females mate several times. It seems that the aggressive behaviour of the males among themselves is linked not only with competition for a mate but also with host plant and territory defence.

According to Medvedev and Pavlov (1988), mating between chrysomelids is a complex process which can be divided into several phases: activation of the female (ritual dances), amplexus, copulation and end of mating. While the male is more active during the three first periods, the female becomes the most active partner at the final stage. The complexity of this behaviour is associated with the presence of secondary specialized structures and with glands which liberate attractants and excitants (pheromones). Differences in behaviour exist between genera and species of the same subfamily.

Rodriguez (1994a) observed the sexual behaviour of certain Cassidinae in Central America: *Omaspides convexicollis* Spaeth, *O. bistriata* Boheman, *Charidotella* nr. *sexpunctata* (F.), etc. Among those cassidines there is not much foreplay, but methods are used to stimulate the females.

Windsor (1987) described fights between males in the cassidine *Acromis sparsa* Boheman. These fights take place, and for this the males are provided with much enlarged humeral angles in the elytra (*Figure 10.1 C*). Holes in these elytra, which were made during previous fights, help to deter a rival. These male fights happen before the females arrive on the host plant, the climbing *Merremia*. It is possible that the males eclode before the females. Courtship, amplexus, copulation and oviposition make up a normal process for these cassidines.

In a paper published in 1992, Noël Magis summarized nuptial parades among several beetles. There is enormous variation in this act among families. Even the males of peaceful beetles like *Chrysolina aurichalcea* (Mannerheim) fight over the females (Shimizu and Fujiyama, 1986), and probably all species of *Chrysolina* do the same. This is especially obvious when there is a dense population. According to calculations made by these Japanese authors, in over 269 observations on an isolated male and a mated one, the intruder had a 10.4% chance of winning, a 36.5% chance of losing, and a 39.1% chance of being obliged to retire quickly. In certain cases, there was no conflict between the two males (14.1%). Frontal attacks had 5 times more chance of securing a win than an attack from behind. A bigger solitary male was also more likely to win.

We have already seen (Medvedev and Pavlov, 1988) that secretions of specialized glands are used by the males to seduce the females. Several entomologists (Bartlet *et al.*, 1994) have restudied the antennal morphology of *Psylliodes chrysocephala* (L.), an alticine which normally feeds on turnip. There are glands which are common to both males and females, and there are also those which are specific to males, located under a glabrous area on segments 6–10. These antennae are active during pairing and probably secrete a sexual pheromone.

An additional effect of copulation in *Diabrotica virgifera* LeConte is to stimulate development of the ovaries. Females which remain virgin have a limited ovarian development and rarely lay eggs (Sherwood and Levine, 1993). Among *Plagiodera versicolora* (Laicharting) multiple matings are frequent, up to 96% (McCauley and O'Donnell, 1984). By contrast, female *Diabrotica* generally mate only once, but the males mate with several females. The females usually live longer than the males.

Homosexual behaviour between males (*Meloe*) (Bologna and Marangoni, 1986) and females (*Otiorrhynchus*) (Pardi, 1987) is well known. It is more common in the laboratory, due to promiscuity, than in the field. Nothing similar has yet been observed among leaf beetles, but interspecific, intergeneric, interstage (larva and adult) matings are well known in the laboratory. Only interspecific matings have been observed in the field. I have, in laboratory experiments, mated *Gastrophysa viridula* DeGeer from Belgium and *G. cyanea* Melsheimer from California without any clear result. Never has any interspecific mating been observed in the field between two species of *Timarcha*, related or not.

A statistical equality between the respective sizes of the two partners has been observed with *Trirhabda canadensis* Kirby (Brown, 1993). This balanced mating seems due mainly to the ease with which assorted couples can achieve the insertion of the aedeagus and not to a selection of the female. However, the latter hypothesis should not be rejected completely.

During experiments on the attractiveness of the females of *Longitarsus jacobaeae* (Waterhouse) for the males, Zhang and McEvoy (1994) placed males on the leaves of *Senecio jacobaea* L., which had been exposed during the previous 24 hours to males and females of the alticine. The males chose the leaves which had been exposed to the females and rejected the control leaves. The females were not affected by the leaves, whether exposed or not to males. It seems evident that the females of *L. jacobaeae* emit a sexual pheromone which is strongly attractive to the males. It is very likely that this is a general phenomenon. The male of the alticine genus *Gabonia* in Africa, which selects the wilted leaves of *Helotropium* spp., can evoke a group behaviour of the males towards the females (lek behaviour).

A more detailed account of the sexual behaviour of the leaf beetles is given in Jolivet (1999b) (*Figure 10.1*).

Viviparity

Viviparity was described a long time ago among mountain *Oreina* and even among certain species of low altitude, like some *Chrysolina*. Perroud's paper (1855) was probably the first one on this topic about leaf beetles. Most of the literature on viviparity is gathered together in Bontems work (1988). Many papers (Champion and Chapman, 1901; Rethfeld, 1924; Maneval, 1938) describe the phenomenon, which is not uncommon in mountains, arctic areas and the tropics. The main reason for this behaviour seems to be the need to shorten the life of the larval stage either because of the cold environment (arctics and mountains) or because of the abundance of egg predators and parasitoids (tropics).

Some people have separated ovoviviparity and viviparity. Really, the difference is subtle and Bontems unites all the known cases under the name of viviparity. With certain arctic *Chrysolina*, large eggs are laid and these hatch immediately. We do not know anything about the biology of the leaf beetles of Tierra del Fuego, except that they lay very large eggs (*Brachyhelops*) (Brendel *et al.*, 1993). Probably here also, hatching is immediate but can be stopped by a winter egg diapause, as in the case of *Timarcha tenebricosa* F. (Abeloos, 1935; Chevin, 1994; Jolivet, 1994b). *T. tenebricosa* also lays large eggs in limited number (K-strategy) and this behaviour can be considered as an adaptation of one species of a relatively thermophilic genus to a cold climate. Basically, *Timarcha* is a steppic genus which survives in northern and mountain areas through different types of diapause.

Viviparity was for a long time considered as an adaptation to the period of glaciation and to a short summer (*Oreina*, several *Chrysolina*, *Gonioctena*). Almost all arctic *Chrysolina* are viviparous, but viviparity exists also in the tropics at low altitude. The now familiar case of *Platyphora* is astonishing. Many species of this genus and other related species are viviparous in Brazil. As mentioned earlier, this is probably to protect the eggs against Mymaridae. In some cases, the larvae remain isolated and in others they immediately group themselves in cycloalexy, which is another way of protecting against predation.

Viviparity was mentioned recently (Schroder *et al.*, 1994) among *Platyphora quadrisignata* (Germar) in the Sao Paulo region, but this had been known for a while for this species through work on dissections (Bontems, 1985, 1988) or from research in the field (Vasconcellos-Neto and Jolivet, 1988, 1994; Jolivet *et al.*, 1990). In fact, the first observations on *Platyphora* viviparity were reported in Viçosa, Minas Gerais, Brazil by a young Brazilian in a non-published thesis in 1987 (Picanço, 1987, pers. comm.; Picanço *et al.*, 1999). The females of this large species drop the larvae one by one on the leaves of arbustive *Solanum*. Often, these larvae, sometimes solitary, aggregate themselves together in a ring. Not all *Platyphora* are viviparous, but many are. There are four larval stages and they develop under a temperature of 25°C. Schroder *et al.* (1994) also observed viviparity in *Platyphora fasciatomaculata* Stal. As already mentioned by Bontems (1988), the spermatheca is missing in both the species. The eggs are fertilized when they pass through the nutritional chamber where the spermatozoa are accumulated. It seems that, in this case, as the spermatheca has become useless, it has disappeared. The embryo grows in the ovarioles.

Bontems (1988) quotes 50 viviparous species of Chrysomelinae but the phenomenon probably also exists among many other subfamilies and genera where it has never been studied. The phenomenon has been reported in *Chrysolina*, *Gonioctena*, *Oreina*, *Paropsides*, *Platyphora* and *Pyrgoides*. Most of these species do not have a spermatheca, but this is not an absolute rule. If the spermatheca is present, it is always in regression. Certain non-viviparous species also lack a spermatheca, but this is exceptional. Some species can be viviparous in a part of their distribution, and oviparous elsewhere. The phenomenon is linked, as we have said before, to cold temperatures, but it also exists in the tropics and even, very rarely, in lowland temperate areas.

Details, including the timing of the penetration of the spermatozoa into the spermatheca, when it exists, and into the ovarioles, are given by Bontems (1988). For females without a spermatheca, the spermatozoa progress directly into the oviducts and then into the pedicels. According to Rethfeld (1924), the eggs of the viviparous *Chrysolina* are directly fertilized in the follicles by the spermatozoa crossing the oviducts.

Finally, spermatozoa migration is not much modified by the presence or absence of a spermatheca. Among the 50 viviparous species mentioned by Bontems, some of them show a facultative, geographical, viviparity. Gerber *et al.* (1978) assume that all chrysomelids, except viviparous species, have a highly sclerotized spermatheca. However, it is not possible to generalize, since some viviparous species have a spermatheca, and oviparous ones have none at all.

Parthenogenesis

A complete review of parthenogenesis among Chrysomelidae was published by Cox (1996c). We refer the reader to this important paper for the complete bibliography on the topic. Parthenogenesis, or reproduction from an unfertilized ovum, has mostly been studied among beetles in the Curculionidae, among which triploid species survive without an apparent male (Vandel, 1932). A recent paper by Furth (1994b) has attracted our attention to the alticine genus *Longitarsus*, some geographic isolates of which are parthenogenetic. So far, parthenogenesis has been reported among Eumolpinae, Chrysomelinae, Alticinae and Cassidinae. It may exist elsewhere and be more common than thought so far.

There have been some rather confusing observations made in research on chrysomelids. *Gastrophysa viridula* (Degeer), for instance, seems to present some cases of accidental parthenogenesis in the laboratory. Many females of chrysomelids lay eggs even if they are virgins, but normally these eggs do not develop. Parthenogenesis among *Gastrophysa* rarely produces female imagines and these imagines are not viable (Osborne, 1880; Jolivet, 1951b). Phillips (1977a, 1979) and Eberhard *et al.* (1993) mention parthenogenesis as being frequent among Alticinae (*Altica*,

Macrohaltica). *Altica lazulina* LeConte from the USA is said to produce by triploid thelytokous parthenogenesis *in vitro* (Cox, 1996c).

Several years ago, I (Jolivet, 1979) collected many specimens of *Chaetocnema confinis* Crotch on *Ipomoea aquatica* Forsk in La Réunion under the name of *C. etiennei* Jolivet, and they were all females. This American species is actually invading Madagascar, Mauritius, Eastern Africa and most of the Far East (Vietnam, Thailand, Taiwan, Southern Japan, and probably India and continental China). Females predominate everywhere but males are known in America. The beetles already exist in the Galapagos Islands. This colonization is probably by winds and typhoons and its success is due to the availability of the food plants (*Ipomoea aquatica* Forsk and *I. batatas* L.), to the light weight of the insect, and to the alternative of parthogenesis. A solitary imported individual can rapidly start a population (Jolivet, 1998c). It seems that there are parthenogenetic strains among many other Alticinae, such as *Longitarsus melanurus* (Melsheimer) in North America (Furth, 1994b). Many other species of *Longitarsus*, among the 700 described, may be at least partially parthenogenetic. Unfortunately, the cytology has not been studied as it has been with Curculionidae, and we do not know if these forms are triploid or polyploid.

Orfila (1927) described a case of experimental parthenogenesis with incomplete development of the embryos in a cassidine, *Poecilapsis bonariensis* Dejean. The laying of eggs occurs with many virgin females of Chrysomelidae. These eggs divide, but very few produce larvae or adults.

Bromius obscurus (Linné) (Eumolpinae) is geographically parthenogenetic. North American populations are diploid (bisexual), whilst their European counterparts are apomictic triploids (Cox, 1996c). However, the beetle originated in Europe, where it was probably originally bisexual. Most of the older literature is in Vandel (1931, 1932). *B. obscurus* is actually rare in France due to insecticide treatment of vines (*Vitis vinifera* L.) and to its replacement with an American species. Normally, this species, or at least the type, feeds on *Epilobium hirsutum* L. (Onagraceae) in humid places (Bergeal and Doguet, 1992), but the food plant also seems to be rare now.

Lokki *et al.* (1976) studied the genetics of this species. They compared by electrophoresis the degree of polymorphism of the allozymes between a bisexual Canadian population and several triploid Scandinavian parthenogenetic populations. In contrast to some polyploid parthenogenetic populations of curculionids, the parthenogenetic populations of *Bromius obscurus* show very little genetic variation, probably because of migration and the mixing of populations. Apterous curculionids, on the other hand, are restricted to small areas.

In central Japan, several species of *Demotina* and the species *Hyperaxis fasciata* (Baly) (Eumolpinae) have no males and must be thelytokous (Isono, 1988). Geographic parthenogenesis occurs in these species (Cox, 1996c).

Among *Calligrapha* spp. (Chrysomelinae), males are rare in several species (Brown, 1945). Thelytoky is compulsory for certain species of *Calligrapha*, like *C. vicina* Schaeffer, *C. virginea* Brown, *C. alnicola* Brown, *C. apicalis* Notman, *C. ostryae* Brown and *C. scalaris* LeConte. Parthenogenesis is often tetraploid with 48 chromosomes. According to Robertson (1964, 1966), the evolution was as follows: bisexual diploid, diploid facultative thelytoky and, finally, obligatory tetraploid.

Parthenogenesis is still a somewhat unexplored area in the Chrysomelidae. We do not know if it is an exceptional or common phenomenon, whether certain subfamilies are more prone to it than others and, in particular, we know little about the cytology of the known cases.

Regeneration

Several arthropods, including many insects, practice autotomy, or the spontaneous amputation of the legs. This phenomenon seems non-existent among most beetles, and certainly among chrysomelids. Normally, this loss of legs is followed by regeneration.

Abeloos (1933c) wrote a paper on this problem in the *Timarcha*. Bourdon (1937) made further attempts to research what happened in one case, *Timarcha goettingensis* (L.), then a common French species. The results were not clear.

According to Bourdon, the larva of *T. goettingensis* seem not to regenerate; the organs studied, paired appendages and ocelli, do not regenerate after amputation but only heal up.

The removal of a larval organ always provokes a modification of the homologous organ on the other side in the imago. There seems to be a kind of equilibrium between both organs and a reciprocal influence, a kind of regulating development. A recent paper by Emlen (2000) shows that among scarabeid beetles (and probably among all beetles, including horned chrysomelids) there is a trade-off between horns, eyes and their ommatidia, mouth parts, wings and antennae. Horned males, for instance, have proportionally smaller wings or eyes than hornless males or females. McIntyre and Caveney (1998) predicted that large eyes would be more important in nocturnal beetles than in diurnal ones and this was confirmed by experimentation on dung beetles. However, purely nocturnal leaf beetles like *Timarcha cerdo* Stal and *T. intricata* Haldeman seem to have eyes which are proportionally as big as those of diurnal species like *T. tenebricosa* (F.) or *T. goettingensis* (L.). The correlation does not always work. It could also be possible to compare the number of ommatidia between certain diurnal and nocturnal species to see if they are reduced in the nocturnal ones.

These parts are indeterminate in young larvae and so they can regularize the loss. A small loss of the embryonic tissue of the leg can produce a reduced, but complete, appendage. There is also no regeneration in the pupa, everything being fixed at the larval stage. Abeloos (1933c) thought that a certain regeneration was possible at the pupal stage, but his results are rather confusing. Patay (1937a, 1939) and Poisson and Patay (1938), studying the Colorado beetle, had contradictory results. A larva at the last stage could, according to them, regenerate an imaginal leg but wings could not regenerate. Research on the regeneration of the elytra was not conclusive due to the mortality, probably caused by infection, which followed mutilation.

Chrysomelidae are very homogeneous and it seems that these results can be extended to all subfamilies. However, some beetles show rather important regenerative capacities (Balazuc, 1948), as shown with *Tenebrio molitor* larvae or with *Lampyris* spp. (Crowson, 1981). Regeneration among leaf beetles should be thoroughly reinvestigated.

11

Association with other Organisms

Chrysomelids are associated with many useful symbionts and several kinds of external and internal parasites. Trichomycetes, intestinal fungi which are more of commensals than parasites, have never been found in the gut of Chrysomelidae. Although it seems that they do not occur among them, surprising things can sometimes happen and aquatic species, like Donaciinae, especially the larvae, should be carefully investigated.

Several topics will be treated in this chapter: social commensalism (myrmecophily and termitophily), symbiosis, pathogens, parasites, commensals, phoretics. The chapter will also cover biological control, either the use of chrysomelids to destroy invading plants or the use of pathogens and parasites to reduce pest numbers among the chrysomelids themselves.

Social commensalism

The association of chrysomelids with ants is common (myrmecophily), but association with termites is extremely rare and, until now, no real case of termitophily has been described among chrysomelids. It might exist, however, but so far only termitoxenic (visitors) have been mentioned (in Sri Lanka and India).

MYRMECOPHILY. TERMITOPHILY

Only three chrysomelid subfamilies include myrmecophilic species: Clytrinae (all myrmecophilic and submyrmecophilic at the larval stage, some adult myrmecobionts), Cryptocephalinae (very few of them with submyrmecophilic larvae, some adult myrmecobionts), and Eumolpinae (some adults associated with ants). All Clytrinae so far known are more or less closely associated with ants. It should be noted that myrmecobiont adult Chrysomelidae have been found only in East Africa within the foliar stipules of the myrmecophilic *Acacia*. Others will probably be found somewhere one day, but American myrmecophilic *Acacia* do not seem to harbour adult leaf beetles. Actually, Keroplatinae inside ant plants were discovered in Panama recently for the first time (Aiello and Jolivet, 1997). They have now been found in Africa and elsewhere, too (Sri Lanka). It will probably be the same with myrmecophilic adult leaf beetles.

Only one case of termitophily, or at least a case of casual visitors, is known in Sri Lanka and India among the galerucine *Aulacophora*. Species of this genus have been found in abundance in nests of *Termes* sp. in India and of *Termes redemanni* Wasmann in Ceylon (Sri Lanka) (Abdulali, 1948). These beetles are only synoecetes,

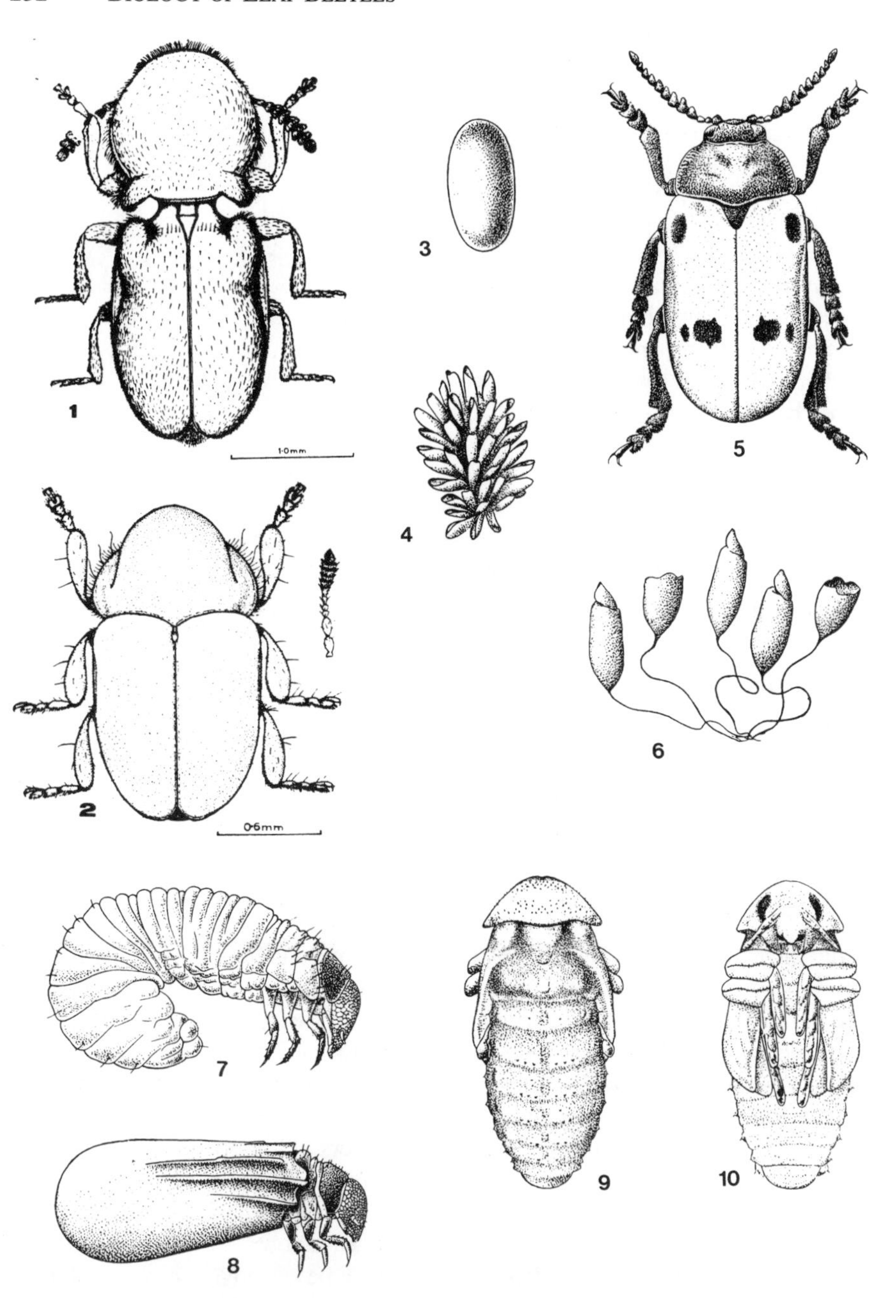

Figure 11.1. Myrmecophily among Clytrinae and Cryptocephalinae. (1) *Hockingia curiosa* Selman (Clytrinae). Symphylic with secreting trichomes. Lives inside the stipular thorns of East African *Acacia* in symbiosis with ants (after Selman, 1962); (2) *Isnus petasus* Selman (Cryptocephalinae). Also symphilic with ants inside stipular thorns of *Acacia* in East Africa; (3) *Clytra quadripunctata* L. (Clytrinae), naked egg (after Jolivet, 1952); (4) Ibid. egg covered with the scatoshell (after Jolivet, 1952); (5) Ibid. imago (after Jolivet, 1952); (6) *Labidostomis tridentata* L. (Clytrinae) eggs together (after Jolivet, 1952); (7) *Clytra quadripunctata* L. (Clytrinae) naked larva (after Jolivet, 1952); (8) Ibid. larva going into its scatoshell (after Jolivet, 1952); (9) Ibid. pupa, dorsal view (after Jolivet, 1952); (10) Ibid. pupa, ventral view (after Jolivet, 1952).

i.e. accidental associates. Unfortunately, nothing is known of their biology. Until now, no real termitobiont or termitophile is known among leaf beetles. It seems it would be difficult for a leaf-eating insect to adapt to underground life with termites, but clytrines could be perfectly able to succeed and too little is known of their life history in the tropics to contradict this notion.

The interrelationship between ants and clytrines was studied by Donisthorpe (1902), Jolivet (1952), Selman (1988b) and Erber (1988). Medvedev (1962) listed the clytrines in ant nests in Russia. Everywhere in the world, ants harbour clytrines. Biological data are scarce or non-existent in India, Australia, Latin America, Madagascar and continental Africa. Clytrinae are rare in Australia, New Guinea and Madagascar. This is not due to a shortage of ants but to the clytrine ability to disperse. Clytrines are almost always absent in volcanic islands, possibly because they are too heavy for aerial transport. However, the small volcanic island of Pantellaria, between Sicilia and Tunisia, harbours *Lachnaea paradoxa* (Olivier), which probably originated from neighbouring Sicily. The species is a good flyer. It feeds on *Pistacia* or *Quercus* trees and is potentially polyphagous. There is no adaptation problem for this species in an island where ants are abundant.

PROTECTION AGAINST PREDATION

Other chrysomelids apart from the clytrines are tolerated by ants. On *Cecropia* trees, the parasol trees in South America (Cecropiaceae), the adults and larvae of a variety of galerucines feed openly upon the leaves: *Syphaxia*, *Dircema*, *Monocesta*, *Coelomera*, and many more. They are well protected from the ants by reflex bleeding, blood toxicity, cycloalexy (*Coelomera*), oral or anal regurgitation, toxic secretions, etc. Some of them, such as *Coelomera*, protect their eggs in an ootheca or lay them inside the internodes of the *Cecropia* trees where they are well protected from predators. The *Azteca* queen does the same, and there is competition between the ant and the beetle to decide which one will occupy a given internode. *Coelomera* eggs laid inside the stem are also aggregated into an ootheca, and for this the females penetrate into a less resistant area called the prostoma. The opening is then closed and sealed by the female immediately after egg-laying. After hatching, the larvae will reopen the prostoma. Similar behaviour is seen in *Azteca* and *Pachycondyla* queens. They lay their eggs inside the internodes of the tree and then competition arises between the ants and the *Coelomera* females for stem occupancy. We never find both insects in the same internode.

This method is much more efficient than egg-laying at the end of the leaflets as parasitism and predation are practically non-existent inside the internode. When the larvae hatch, they go to the leaves and immediately aggregate themselves into a cycloalexic ring. That is a very effective defence system against ant and hemipteran predators, and even against parasitoids, though some have been known to succeed in breaking it down. Several other leaf beetles (*Platyphora*, some Cassidinae, Galerucinae and Criocerinae) do the same. In south-east Asia, *Hoplasoma*, *Haplosomoides* (Galerucinae), *Phyllocharis* (Chrysomelinae) and some others live happily on *Clerodendrum* species (Verbenaceae), plants which are always associated with ants. Mites, ants, beetles and other insects frequent their extrafloral nectaries in perfect peace. The immunity of the adults and larvae of these chrysomelids is probably due to the repugnatorial glands of the larvae and the toxicity of the adults. Reflex bleeding also occurs, and the ants learn to avoid any unpleasant encounters.

All Camptosomata, and also the Lamprosomatinae, cover their eggs with excreta and secretions and the larvae live inside a scatoshell. Sometimes, mostly among clytrines, this scatoshell has more or less the shape and colour of a seed. In the Clytrinae, like many Phasmids, the eggs are carried by the ants into their nests. In the case of the Phasmids, there is a reward, the capitulum, which resembles a seed elaiosome. For the clytrine eggs there must be (as for the seeds in tropical ant gardens) an attractant, a kind of pheromone or pheromone mimic, since no other camptosome eggs are taken by the ants. In both cases (clytrines and Phasmids), the eggs have a striking resemblance to seeds (Compton and Ware, 1991; Hughes and Westoby, 1992; Windsor *et al.*, 1999), but clytrine eggs do not carry a capitulum. It is possible that other chrysomelid eggs which drop to the ground are also taken by ants to their nests, but so far this seems to occur only for some rare cryptocephaline larvae. No one knows how eumolpine, clytrine and cryptocephaline adults have penetrated into the foliar stipules of the East Asian *Acacia* trees. Clytrine larvae may also be taken by ants into their nests, suggesting some special attraction in the chrysomelid. In the nest or just outside it, the ants are sometimes likely to attack the larvae or adults of Clytrinae. The adults will retract their legs and head appendages and the larvae will withdraw their heads into the scatoshell in the event of such an attack.

As we can see, chrysomelids live with the ants without any major problem and very rarely do they fall victim to them. If, for any reason, the cycloalexic ring is broken, some larvae may be captured by ants or pentatomids.

Mining, boring or root-feeding larvae generally escape the ants. There are, however, ants which live underground and feed on the roots. It is probably the reason why *Diabrotica* larvae demonstrate double reflex bleeding in the front and in the rear when they are inside the gallery (Wallace and Blum, 1971; Blum, 1994). Reflex bleeding, and oral or anal regurgitation, are very efficient means of defence for free larvae, as well as the nine pairs of dorso-lateral glands of certain species, or the movements of the anal plate of others. Salicylic aldehyde, or a cyanhydric compound, are sometimes produced in addition to other toxins. The hair density of certain species can also help in protection against predators (Paropsini, Entomoscelina). The dorsal shield of the cassidine larvae or the faecal cover of the criocerine larvae are also useful against aggressive ants. While cassidine larvae move their shields against predators, *Coelomera* larvae use the supra-anal plate to repel them, and *Platyphora* larvae tend to bite their attackers. As we can see, living with the ants is not always without danger. That is why different means of defence have been devised.

According to Selman (1988b), adult *Trachymela* in Australia are active on leaves during the night and their wax-secreting glands on the pronotum and elytra provide good protection against ants. Crowson (1981) mentioned gin-trap armory in certain species. The faecal shield of cassidine larvae seems very efficient against ants. The larva of the tortoise beetle, *Hemisphaerota cyanea* (Say), constructs a thatch from the long filamentous faecal strands beneath which it is concealed (Eisner and Eisner, 2000). Deterrency compounds from the food plant are also present in the shield. The shield rests on the fork that projects from the abdominal tip. This larva has an exceptionally long hind-gut which probably relates to the production of the faecal strands.

SYMPHILES AND SYNOEKETES

While most (but not all) clytrine larvae live inside or around ant nests, a few clytrine, cryptocephaline and eumolpine adults, which were probably taken into the nest as eggs or larvae, do the same. Clytrines and cryptocephalines are symphilic (myrmecobies) with the ants, but eumolpines seem to be only synoeketes, i.e. they are not morphologically modified for life in the nest. They must have ways to escape ant aggressivity.

Clytrine larvae are mostly obligatory parasites inside ant nests. Very few cryptocephaline are known to frequent ant nests and most of the larvae of this subfamily are free-living at the base of the plants or on the bark. Clytrine larvae, as well as other Camptosomata and Lamprosomatinae, are shield-bearing. In the case of Clytrinae and Cryptocephalinae, these scatoshells are made of excremental matter, sometimes with vegetable debris. The shields vary greatly from genus to genus, even from species to species. They offer good protection from ants and, in case of danger, the larva retracts itself inside the shell, its vertex completely closing the aperture. Nymphosis takes place inside the scatoshell. During each moult, the scatoshell is attached to twigs inside the nest. The ants bring clytrine eggs and even larvae into the nest. This suggests they are attractive to the ants. The adults are found on the low branches of small trees and bushes and the females drop their eggs, which are covered with excreta and look like seeds, to the ground. *Syneta*, *Iscadida*, other leaf beetles which also drop their eggs to the ground, do not seem to have any special relationships with the ants, and the eggs are not specially protected.

Coptocephala and *Labidostomis* females attach their eggs, which are covered with excreta, to the leaves, either with peduncles or directly (*Ischiopachys*). Ants cut the peduncle and take the eggs to the nest, or nearby. It may be noted that, if the egg is accidentally lacking its excremental cover, it is immediately eaten by the ants.

Labidostomis larvae, like Cetoniidae or Tenebrionidae, live normally around the nests and in indirect contact with the ants.

The association (synoecy) between ants and clytrines may be less significant than people believe. Some details of the biology remain obscure. *Clytra* larvae hatch after around 20 days. They feed on debris, ant eggs, excreta, twigs and probably ant cadavers. The larvae retract inside the shell in case of danger, but do not provide the ants with sugary secretions. Their hosts generally ignore them. Why are they, and not other camptosome eggs and larvae, brought inside by the ants? Normally, only the head, the legs and part of the pronotum remain outside the scatoshell. Moulting and pupation occur inside the shell.

In Africa, ponerine columns (*Megaponera* spp.), when moving in procession, may carry some clytrine larvae with their scatoshells. More research is needed in tropical America, India, Africa, SE Asia on these associations. It seems that, according to what we have learnt from a variety of sources, clytrine larvae have the same general biology, but with some divergence in detail. People are not generally willing to dig into ant nests.

There seems to be an association between adult *Syagrus* spp. (Eumolpinae) and ants (*Crematogaster*) in the stipular thorns of the East African *Acacia* (Hocking, 1970, 1975; Jolivet, 1986a, 1996a). A more thorough study of the behaviour and life history of the beetles is needed. Where do their larvae live when most of the eumolpine larvae are root-feeders? How do they get to the stipules?

Hockingia curiosa Selman (Clytrinae) and *Isnus petasus* Selman (Cryptocephalinae) are true symphiles with yellow setae, very probably secretory, on the elytra and pronotum (*Hockingia*) or only on the pronotum (*Isnus*). Wassmanian mimetism seems clear between the ants and their guests. Both myrmecophiles live like *Syagrus* in the stipular thorns (pseudogalls) of Eastern African *Acacia* with many other insects, generally not modified. The stipules are inhabited by the *Crematogaster* ants. These beetles must feed inside the stipules on vegetable and animal debris, probably also on ant eggs. How they penetrate into the foliar stipules remains a mystery. At the adult stage they are too big to be able to leave through ant holes. When the stipules decay or dry up, then they could leave, but most are condemned to stay there until their death. They are probably brought in at the egg stage, or they reproduce inside the stipules.

Such cases are not isolated. The stipular thorns of the American *Acacia* do not seem to harbour any chrysomelid, but surprising things can happen, such as in Paraguay where *Acacia* harbour ants without the Beltian bodies.

Another case of a loose association between chrysomelids and ants is *Lilioceris nigripes* (Fabricius) and *Iridomyrmex purpureus* (F. Smith) in Australia. *L. nigripes* feeds on *Cycas ophialitica* K. Hill fronds, a local cycad, at the larval and adult stage. The beetle eats the epidermic tissue of the pinnae, starting at the extremity and going on to the rachis of the frond. These larvae are protected by their excreta and the ants also protect them and feed on their dry excreta (Wilson, 1993).

Microbial symbiosis

Symbionts have been well studied among chrysomelids. They help in the digestion of cellulose and in the synthesis of vital substances like vitamins. They are very irregularly spread among genera and species and are situated in a variety of places. They can be in the Malpighian tubules, the intestinal pouches and connected or not with the gut. Larvae too may have symbionts, which sometimes can survive in the adults. These are either bacteria or fungi and are located in mycetomes.

Nothing has yet been discovered in chrysomelids which can compare with the information gathered about the elytra of the *Gymnopholus*, weevils of the New Guinean mountains. Epizoic symbiosis (Gressitt, 1966) seems to occur only on weevils and a few other beetles, and only in New Guinea. With this elytra phenomenon, in pre-existing cavities, mucus secreted by certain glands allows the development of a varied flora (lichens, mosses, liverworts, algae, fern prothalli, etc.), and among this miniature forest an invertebrate population flourishes (mites, nematodes, insects). In these moss forests, humidity is constant and favours the development of this complex structure. Algae, but only algae, on weevil elytra are known also in the Panamean (*Geobyrsa nodifera* Pasc.) or Bornean mountains and probably exist elsewhere. Nothing comparable with the New Guinea phenomenon has been found anywhere else. A few other New Guinea beetle families carry the algae and fungi. It is probably the long life span of these beetles which allows the development of this elytral flora. So far, no chrysomelid has been found to carry the plants, probably because many tropical chrysomelids (alticines, galerucines, hispines) are small and, even the large ones (Sagrinae, *Promechus*), short-lived and often frequent lowlands. Besides, the elytra of chrysomelids are smooth even among the mountainous species, such as the neotropical *Elytrosphaera*. Moss forests in Borneo,

New Guinea, South America seem to be potential natural laboratories and may give rise to similar associations.

The endosymbionts among the chrysomelids are bacteria and yeasts. They are generally found in mycetomes associated with the gut, mostly among larvae, but often persisting among the adults (Crowson, 1981). Mycetomes are common among Donaciinae, Eumolpinae and Cassidinae, but they exist also among other subfamilies. The synthesis of vitamins and steroids and the digestion of cellulose seem to be their main functions.

Stammer (1935–1936) described the mycetomes of the mid-gut of *Donacia semicuprea* Panzer, *Bromius obscurus* (L.), and *Cassida viridis* Linné. All these structures are transmitted by larvae to the adults, except among *Donacia*, where the symbionts move to the Malpighian tubules. Among *Bromius* and *Cassida*, symbiont pouches open into the common oviduct of the females (Crowson, 1981). Similar mycetomes are found among Lamprosomatinae, Hispinae and Cassidinae, and the larvae become infected by eating the eggshell. Recent work by Ferronato (1988) and Becker (1994) deals with the transmission of symbionts among neotropical eumolpines. Another mode of transmission of symbionts is the transovarian transmission, when symbionts are incorporated into the oocytes. This is frequent among the Curculionidae, but it seems rather rare among the Chrysomelidae.

Recent researches (Kasap and Crowson, 1980; Mann and Crowson, 1983a) on symbiont pouches show that they open in the common oviduct (Sagrinae) or in the vaginal region. This type of symbiont lodging exists only among Aulacoscelinae, Eumolpinae, Cassidinae, and Hispinae. There is no vaginal pouch among Donaciinae.

As these pouches are present among primitive subfamilies, Sagrinae and Aulacoscelinae, Mann and Crowson (1983c) think that the structures existed in the ancestors of Chrysomelidae and were later lost from entire subfamilies and some genera and species.

In *Percolaspis ornata* (Germar) from Brazil, Becker (1994) describes the symbionts as microorganisms, rod-shaped and of bacterial type. They are all grouped into globoid bodies. To be transmitted, the symbiont, having been sprayed over the egg, must survive on the chorion for some time. Among *Donacia*, symbionts are immersed into a gelatinous matter surrounding the egg. Among *Cassida* and many others, symbionts are packed inside impermeable spheres at the pole of the egg. While many eumolpines have symbionts, there are also rare exceptions without these microorganisms.

Pathogens

Among the pathogens are viruses, mycoplasmas, rickettsiae (*Wolbachia*), microsporidia and fungi. This inventory remains very incomplete, but it gives a preliminary idea of the diseases affecting the Chrysomelidae. Protozoa present in the gut of many other beetles are certainly sometimes present in chrysomelids, but they should be classified as inoffensive commensals.

VIRUSES

The latest revision of chrysomelid viruses was carried out by Selman (1988a). Selman mentions that the first iridescent virus to kill chrysomelids was found in 1975 among

Chrysomela vigintipunctata (Scopoli) in Japan (Ohba, 1975). Iridescent viruses were mainly studied among mosquitoes, but no one succeeded in manipulating them for insect control. Viruses can be divided in two groups: those pathogenic in chrysomelids which cause disease, and those transmitted by leaf beetles to their host plants. Sometimes, a transmitted virus can also be pathogenic to the insects.

Entomopathogenic viruses

Some viruses can occasionally give rise to epizooties, but often the virus remains latent in the insect tissues. Most of those viruses are baculoviruses (nuclear polyhedrosis and granulosis viruses) or iridoviruses. These viruses can sometimes control large populations and the beetle becomes infected through its mouth or through wounds, or even through transovarian transmission. In his paper, Selman quotes viruses from *Cerotoma trifurcata* Förster, *Diabrotica undecimpunctata* Jacoby, *Coelaenomenodera minuta* Uhmann, *Leptinotarsa decemlineata* (Say), etc. All leaf beetles can get virus diseases. Some of these viruses have been used experimentally in biological control trials but with debatable results.

Plant pathogenic viruses

Most plant viruses are transmitted by sucking insects like Homoptera, but some can be transmitted by phyllophagous insects like the leaf beetles. We are rather ignorant in that field, but we can quote Criocerinae, Chrysomelinae, Alticinae and Galerucinae among the most common potential carriers for the crops. Among the affected plants, Fabaceae, Cucurbitaceae, Poaceae, Solanaceae are often quoted, but wild plants are also affected by viruses and diseases carried by insects. Leaf beetles, like some Homoptera, probably also carry mycoplasma and bacterial diseases to plants. Chrysomelids can remain infected and are potential virus carriers for about three weeks. It seems that the larvae do not carry plant viruses to the adults and viruses do not appear to multiply inside the chrysomelid vector. *Oulema melanopus* (L.) can be infected simultaneously with two different viruses. Adult chrysomelids infected with plant pathogenic viruses can sometimes show pathological symptoms when the infection is high (*Oulema melanopus*). So, plant pathogenic viruses can cause significant mortality in beetle vectors (Selman, 1988a). It is therefore difficult to decide whether a particular virus is a plant or an insect pathogen, or both.

MYCOPLASMAS (MOLLICUTES)

Lipa discovered spiroplasmosis among Colorado Potato Beetles in Russia and in Bielorussia in 1991, with three populations affected. Over 36% of the specimens were infected.

The Colorado Potato Beetle spiroplasma (CPBS) has spiral coils and has been isolated repeatedly from the gut of the insect. It attaches itself to mid-gut microvilli. It has also been isolated from the gut of *Leptinotarsa texana* Schaeffer in Texas (Hackett and Lipa, 1996). Although this spiroplasma does not seem to be pathogenic to the beetle, Hackett proposes to use it for biological control by incorporating insecticidal genes into its genome.

This spiroplasma grows well in insect tissue cultures. Hackett and Henegar (1992) did not find spiroplasmas among *Platyphora quadrisignata* (Germar) in Brazil, another doryphorini.

Spiroplasmas were found among *Diabrotica undecimpunctata howardi* Barber on alfalfa. They are frequent in *Diabrotica* haemolymph but not in the gut. Interspecies infection is common among *Diabrotica* spp. Probably a great number of chrysomelids contain intestinal or coelomic spiroplasmas. Research has concentrated only on agricultural pests. Actually, mollicutes have been found among species of *Diabrotica, Leptinotarsa, Donacia, Galerucella, Chrysomela* and *Calligrapha* (Hackett and Lipa, 1996).

RICKETTSIACEAE (WOLBACHIAE)

Bacteria of the genus *Wolbachia* are related to *Rickettsia* and are reproductive parasites of arthropods and worms. They are cytoplasmically inherited, like mitochondria, and their phenotypic effects on their hosts range from the induction of parthenogenesis in certain hymenopteran groups, to the feminization of genetic males in isopod crustaceans, and to the induction of cytoplasmic incompatibility and sterility in many insects. *Wolbachia* do not seem to play a very important role in insect speciation by generating reproductive isolation (Rokas, 2000).

Wolbachia are not restricted to the gonads, however, and are distributed throughout the somatic tissues, including the gut, salivary glands, and haemolymph. They could interfere with the ability to transmit disease agents (Wilkinson, 1998). There is a long way to go before *Wolbachia* could be of use in insect control.

Some of these bacteria have been found in the testicles and spermatheca of *Diabrotica virgifera* LeConte, an American species recently introduced into Western Europe. These insects showed a high incidence of abnormal sperms and reduced fertility (Degrugillier, 1996). A virus (picornavirus) was also abundant in the spermathecal cells of *D. virgifera*. The *Wolbachia* were surrounded by a double unit membrane and colonies were enclosed within a cellular, vacuolar membrane.

Wolbachia infections in chrysomelids have been recently reviewed by Giordano and Jackson (1999) who, strangely enough, do not quote Degrugillier's review (1996). *Wolbachia* are a commonly-occurring infection in arthropods, but the only studied cases are those involving *Diabrotica virgifera* LeConte and related species. The possible use of the bacteria in the biological control of chrysomelid pests is being considered but, as for microsporidia or mycoplasma, manipulation of the bacteria remains largely untested.

BACTERIA

Peterson and Schalk (1994) reviewed the bacteria known among chrysomelids. It is evident that our knowledge is very limited and it is mostly *Diabrotica* and *Leptinotarsa* spp. which have been investigated. The authors report the symbiotic bacteria already mentioned here together with free pathogenic bacteria. We know that various strains of *Bacillus thuringiensis* have been used against the CPB.

Intestinal bacteria of *Diabrotica balteata* LeConte and *D. undecimpunctata howardi* Barber are numerous. Some are bacteria from plants and even from soil and are not really pathogenic. A very rich microflora has been isolated. For instance, *Bacillus leptinotarsae* is extremely virulent against the larvae and adults of the CPB. It is not really understood how the beetle becomes infected and how the transmission occurs. The bacterium reproduces easily in the gut, the haemolymph, and even in other beetle organs. It seems that wounds are necessary to start and maintain septicaemy.

Only three species of Chrysomelidae have been studied for pathogenic bacteria. Thousands of species of bacteria must exist among all the chrysomelids.

MICROSPORIDIA

Toguebaye *et al.* (1988) reviewed the microsporidia known among the chrysomelids. Since then, few data have been available. Microsporidia are protozoa and intercellular parasites. Their characteristics and morphology are rather peculiar. Microsporidia mainly parasitize invertebrates and fishes, but some are known among vertebrates and even man (Jolivet, 1999e).

Seventeen species of microsporidia belonging to the genera *Nosema*, *Unikaryon*, *Pleistophora*, and *Microsporidia* are known from chrysomelids. They parasitize the eggs, larvae, pupae and imagines of their hosts. The infection can be very strong and pathogenic and the transmission can be by the mouth or transovarian. The disease persists for many years among some populations but, so far, no one has succeeded in manipulating the pathogen for controlling an insect. One cannot mass-produce the spores or infest insects either in the field or the laboratory.

FUNGI

Fungi which are parasitic on Coleoptera are not all pathogenic. Some of them, such as Laboulbeniales and Trichomycetes, are respectively ectoparasites and endoparasites and seem completely harmless to the insects. Trichomycetes feed mainly on the gut contents, although they may stick themselves tightly with roots to the intestine wall. No Trichomycete has yet been found among chrysomelids. They are pure commensals, like Gregarines, and are mostly abundant among coprophagous and saprophagous insects, such as scarabeids. Very few Laboulbeniales penetrate deeply the integument of insects.

Many fungi are saprophagous, but many are really pathogenic and kill the insect host. These include *Entomophtora*, *Beauveria*, *Metarhizium*, etc. Many cover the insects with a white film. Some of them are used in biological control.

Fungal pathogens known to affect chrysomelid beetles were listed by Humber (1996). He also proposed the use of these fungi for the biological control of chrysomelid pests, as well as for a way of protecting leaf beetles from fungal attacks, especially in laboratory cultures. We know how difficult it is to rear in the laboratory the larvae of Orsodacninae, Megascelinae, Aulacoscelinae and others because of early fungal infection of the larvae, especially in humid tropics.

Humber listed the fungal species which affect chrysomelids: *Beauveria bassiana*, *Metarhizium anisopliae*, *Poecilomyces* sp., *Verticillium lecanii* and a few other Entomophtorales. *B. bassiana* seem to be the most promising species for use against *Leptinotarsa decemlineata*. Early last century, the last named fungus was used against the cockchafer in Europe, but no proper evaluation was done.

Parasites: Helminths

Several helminths have been identified in the gut or the haemocoel of chrysomelids. These are nematodes, cestodes, trematodes and nematomorpha. Mermithidae and Nematomorpha (Gordioidea) are parasites, at the larval stage, in the haemocoel of the host (protelian parasites). Many worms are permanent parasites of the insect. When *Howardula* spp. (Alletonematidae) of *Diabrotica* and *Phyllotreta*, for instance, invade the general cavity and the oviducts, they reduce the fecundity of the insect and eventually kill it, but after the completion of their cycle. Gordians also sterilize their hosts. Protelian parasites probably escape from their hosts during heavy rains or in morning dew. However, in temperate areas and in the tropics, many hosts are attracted to water pools or phytotelmata by some kind of hydrotropism generated by hormones (Jolivet, 1998e). For butterflies parasitized with Mermithidae it is a complete mystery, but they probably release the worms during heavy rains.

Many attempts have been made to use parasitic nematodes as biological control agents against the Colorado Potato Beetle and against *Diabrotica*, but so far without any convincing success. A revision of chrysomelid nematodes was carried out by Poinar (1988) and the author mentions the following groups of Mermithidae: Allantonematinae, Steinernematinae and Heterorhabitinae. The biology differs according to the genera. With *Filipjevinermis leisandra* (Mermithidae), parasite of *Diabrotica* and *Systena*, the young worm penetrates through the cuticle and invades the haemocoele, then the ganglia, to avoid encystment. The nematode quits the host to become an adult, which penetrates into the humid soil. Generally, the parasite kills its host, but only after escaping.

Specificity of all these worms appears relative and we know little about it. *Howardula* parasitizes its host at the adult stage. A strong infestation sterilizes the beetle host and kills the larvae. In this case, specificity is important and encapsulation is rare.

Trials of biological control have been done with *Neoplectana* and *Heterorhabditis* because of their non-specificity, their ability to kill the host in 48 hours, the fact that they are easily cultured on artificial media, and because they are safe.

However, it was found recently that *Leptinotarsa decemlineata* was less sensitive to nematodes than other chrysomelids (Thurston *et al.*, 1994). Colorado beetle haemocoele can encapsulate up to 21 nematodes. The small susceptibility of the CPB to parasites is part of the reason of its success as a crop pest.

Commensals

These organisms seem harmless, except that they may share some of the host's nourishment. So far, no Ciliate and Flagellate Protozoa are known from the gut of chrysomelids. *Nyctotherus* (Ciliates) are known from xylophagous tenebrionids and others, but phyllophagy in chrysomelids does not seem to be favourable to their development. Omnivorous insects like cockroaches and many beetles are their normal hosts. The ciliates help in the digestion of woody cellulose. Donaciinae have not been seriously investigated for intestinal fauna and the larvae would be the most likely candidates for some unusual hosts. Only Gregarines (Apicomplexa) and Laboulbeniales (Fungi) may be included here.

GREGARINES (APICOMPLEXA)

Enzyme distribution in the various parts of the gut of chrysomelids was studied by Prasad and Singh (1991) in the eumolpine *Platycorinus peregrinus* Herbst., which feeds on *Calotropis procera* and other Asclepiadaceae. The pH of the fore-gut is slightly acidic (5.6–6.0) and it becomes slightly alkaline towards the end of the middle gut (6.8–7.4). The hind intestine is 6.7–7.0. This acido-alkalinity of the intestine has a direct effect on the distribution of gregarines, and sometimes various species or genera are present in the same beetle species but in different parts of the intestinal tract. Eumolpines rarely harbour gregarines and, strangely enough, the neotropical subtribe Doryphorina has never been found with any. That is probably due to a peculiar chemistry of the gut or the toxicity of the solanines. However, gregarines often adapt themselves to very toxic plant juice, but perhaps Solanaceae, Asclepiadaceae, Apocynaceae have some repellent effects on these Protozoa.

Only the families Gregarinidae, Hirmocystidae and Actinocephalidae have been found among chrysomelids. The host specificity is not very rigid and a polyinfestation is always possible. Actinocephalidae parasitize mostly tenebrionids and carabids and chrysomelids only exceptionally. Until now, gregarines have been found only among Criocerinae, Clytrinae, Chrysomelinae, Galerucinae, Alticinae, Cassidinae and Hispinae. They probably exist in other chrysomelids too, but have not been specifically investigated in other subfamilies, including the most primitive.

The genus *Gregarina* is the most common and often present in a mixed infection: for instance, *G. crenata* and *G. munieri*. The latter also parasitizes many weevils. As reported by Théodoridès (1988), trophozoites are generally in the mid-gut and gamonts and cysts in the rear gut, near to the point of evacuation. No coelomic gregarines have been found among chrysomelids.

Both larvae and adults can be infested. Infestation occurs with the ingestion of leaves soiled with excreta containing the cysts. Many subfamilies have still to be examined for gregarines: Chlamisinae, Zeugophorinae, Orsodacninae, Aulacoscelinae, Donaciinae, etc. Some tribes, such as Paropsini in Australia, have so far never been dissected for gregarines. This important group seems to have descended from a small group of ancestors and is over diversified.

LABOULBENIALES (FUNGI)

Chrysomelid Laboulbeniales were revised by Balazuc (1988) and later on by Weir and Beakes (1996). So far, Laboulbeniales are known as external parasites among Alticinae, Galerucinae, Criocerinae, Chrysomelinae, Eumolpinae, Cryptocephalinae, Cassidinae and Hispinae. Donaciinae, being aquatic, do not carry the fungi, and it is probable that Clytrinae are protected by their life with ants. Other groups should be more thoroughly investigated. According to Weir, 30% of currently recognized beetle families are known as hosts for laboulbenialean fungi. In fact, many species act as commensals, whereas others appear to be truly parasitic, leading to the mortality of their hosts. We do not know much about the transmission of the ascospores, which appear to be transmitted most frequently during direct contact between host individuals (copulation). Normally, Laboulbeniales are external parasites on the insect diplopod and mite cuticle. They generally parasitize the adults, and infection takes place by contact with the spores. The haustorium penetrates the host integument. Among the chrysomelids, no species send rhizoids into the haemocoel. Scarabeidae and Curculionidae show similar patterns in their association with Laboulbeniales. Staphylinidae are parasitized by 47 genera of the fungi, while chrysomelids support only three genera. The presence or absence of the fungi seems to be due to a combination of rather poorly researched biological or ecological factors. As far as we can tell so far, primitive subfamilies such as Sagrinae do not harbour the fungi.

So, only three genera of Laboulbeniales are known among the chrysomelids out of the 137 described genera: the monoecious genus *Laboulbenia* and the dioecious genus *Dimeromyces* and the genus *Rickia*, the most diverse among these fungi. Host specificity among the Laboulbeniales is not very rigid.

In the chrysomelids, these fungi are generally found on the rear part of the elytra, but they can also appear on the pronotum and the sternal side, or on the antennae or the legs. Most of the Laboulbeniales of the chrysomelids have been found in the tropics.

A list of the species was drawn up by Balazuc (1988) and completed by Weir and Beakes (1996). They list 54 species. There are probably many more and they will be discovered only accidentally.

Phoretic (and parasitic) Acari

Mites parasitic on chrysomelids are generally free, a few being anchored to the cuticle. Canestriniids are often free under the elytra and are carried without causing any damage. They feed on skin exudates, dry blood and probably, eventually, on fungi. There is an abundant flora and fauna on the back of *Gymnopholus* weevils in New Guinea and the mites there can be either predaceous or phytophagous.

Santiago-Blay and Fain (1994) studied and listed the mites known in 1994 to parasitize chrysomelids. New genera and new species of Canestriniids are being discovered frequently on Hispinae and Chrysomelinae, and about one hundred are actually being studied (Don Windsor, pers. comm.). Canestriniids were well known among Chrysomelinae and Cassidinae, but they were first discovered on large neotropical Hispinae (*Alurnus*, *Coraliomela*, *Mecistomela*) by Haitlinger (1989).

The most common families on Chrysomelidae are Heterocoptidae, Canestriniidae, Podapolipidae and Hemisarcoptidae. Heterocoptids and Canestriniids are mostly commensals on Cassidinae and Hispinae. Canestriniids are also frequent on Chrysomelinae and, except those found on *Timarcha*, the species are relatively specific. None of these mites have any effect on population control. They are all phoretics.

Santiago-Blay and Fain (1994) mentioned that heterocoptids feed on exudates collected in the subelytral space and they can also feed on fungi, including Laboulbeniales growing on the host. This situation represents, though on a smaller scale, the epizoic symbiosis of New Guinea, but chlorophyllian cryptogams are missing.

In fact, these phoretic mites can be considered as pure commensals. So far, most of the new species are from the neotropics.

Two mite families are really parasitic on chrysomelids, the Podapolipidae and the Hemisarcoptidae. True predation by mites on chrysomelids seems not to exist, perhaps because of a lack of suitable prey. No biological control is possible using mites since they do not kill the host.

Parasitoids

The definition of a parasitoid is as follows: an internal or external parasite, like Hymenoptera or Diptera, that slowly kills the host at the end of the parasite's larval development. Parasitoids are protelian parasites, since their adult stage is free.

Many Hymenoptera and Diptera parasitize the chrysomelids. A complete revision of the group was carried out by Cox (1994b) and an incredible number of potential parasites of the egg, the larvae and the adult are now known to exist throughout the world. What is known is only the tip of the iceberg. Loss of populations by parasitoids is enormous. Even if certain genera are never vulnerable to predators (*Timarcha*), they can be the victims of several parasitic Hymenoptera and Diptera (Thomas *et al.*, 1999b). However, the braconid (*Perilitus sicheli* Giard) in *Timarcha maritima* Perris, *T. tenebricosa* F. and others does not seem to interfere significantly with the mating process, but it does finally kill the beetle.

Viviparity among certain neotropical Chrysomelinae (*Platyphora*) is probably due to the necessity to shorten the cycle and to avoid mainly oophagous parasitoids. In arctic regions and in mountains, viviparity is a means of shortening the cycle in difficult climatic conditions and a short summer. *Chrysolina* species, imported in Australia to fight St John wort, quickly became the prey of paropsine parasitoids. Since parasitism is not a huge problem in this case, the beetles have survived quite well.

Cox's review (1994b) is extremely detailed and lists the families and subfamilies which parasitize the different stages of chrysomelids. The terminology is extensively discussed and observations are complex since hyperparasites are common among Hymenoptera.

Chalcididae, Pteromalidae, Ichneumonidae and many others are frequent parasitoids among chrysomelids. Mymarids, for instance, are frequently mentioned egg parasites. Practically all apocrite families of Hymenoptera attack chrysomelids, except Evanioidea (Cox, 1994b). Among Diptera, it is mostly the Tachinidae which are parasitoids (endoparasitoids) of chrysomelids. Many attempts have been made to use them as a biological control agent. Also, three other families of Diptera can parasitize chrysomelids: Sarcophagidae, Phoridae, and Rhizophoridae.

As mentioned by Cox (1994b), nothing is known about parasitoids among Aulacoscelinae, Palophaginae, Orsodacninae, Megalopodinae, Megascelinae, Lamprosomatinae and Synetinae. That is because these beetles have only been reared very rarely and, for some of them (3 subfamilies), we have no idea of the larval habits. Megasceline larvae are very probably root-feeders, orsodacnine larvae bud-miners, and aulacosceline larvae stem-feeders, but this may not be the case. Little is known, for the same reason, of Sagrinae (mostly the Gondwanian genera), Zeugophorinae, Clytrinae, Donaciinae and Eumolpinae. The hidden life of these insects (gallicolous, borers, miners, root-feeders, myrmecophilic, aquatic forms) makes observations very difficult. Curiously, Cassidinae which use sophisticated means of protection and possess a complex ootheca are, it seems, the most heavily parasitized.

Many insects in the tropics attach their eggs to peduncles to avoid predators. Some clytrines do the same with their eggs, which are also covered with excreta. Ants cut the peduncles when collecting these eggs. Clytrinae, Cryptocephalinae, Chlamisinae eggs are attacked by various Hymenoptera, which seems to prove that the faecal material and the ants which also protect the eggs are not absolutely effective. The same is true for the scatoshell which surrounds the larvae and pupae. Hymenopterous parasitoids must have a specialized way of penetrating the cover. When sawfly larvae in Australia are in cycloalexy, parasitoids avoid the group by laying eggs on the leaf nearby. The eggs are then ingested by the larvae, which become parasitized. We need to learn more about such complex behaviour. Tachinid flies gain access to the Colorado Potato Beetle's vulnerable abdominal dorsum at the very second it lifts its elytra to initiate flight. Tachinid females do not miss an opportunity to larviposit on their larval or adult hosts (Lopez *et al.*, 1997).

Mining larvae seem also more frequently parasitized, certainly more than root-feeders, gallicolous, and borers. This seems difficult to explain, but perhaps the reason is that miners are easy to spot through the leaf. Moreover, species living on trees and bushes seem to be more parasitized than the ones living on the ground plants and grasses. Flat larvae of hispines living, for instance, among the appressed leaves or water-containing bracts of *Heliconia* spp. seem rather well protected (*Cephaloleia*, *Chelobasis*, etc.). According to Hawkins and Lawton (1987), parasitoids would have more difficulty in locating their potential hosts on annual plants dispersed in a given area than on trees. This hypothesis still remains to be proven, and finding a host on a grass does not necessarily seem more difficult than finding one mining a leaf of a tree.

The recent observations of Lopez *et al.* (1997) on the behaviour of the Colorado Potato Beetle and its Tachinid parasitoids (*Myiopharus doryphorae* (Riley) and *M. aberrans* (Townsend)) are especially noteworthy. A sequence of five defensive behaviours in the different larval stages of the *Leptinotarsa decemlineata* (Say) prevent larviposition in 49% of attempts by the Tachinid fly. In the adult, flies, as we have seen earlier, gain access to the beetle's abdominal dorsum the very second it initiates flight. Females of both *Myiopharus* species actively guard recently parasitized hosts from other *Myiopharus* females for a period of several minutes after larviposition.

Predators

A recent revision of the insect predators of Chrysomelidae by Cox (1999) is very thorough. In his review, Cox even mentions the various means of evaluating the prey, including serology and radioactive tracers. These techniques are mainly used in medical entomology. Many insects and arthropods, and a variety of vertebrates, feed on leaf beetles, despite the high toxicity of many of them. When chickens feed on *Aulacoscelis* in Costa Rica, they rapidly die, the beetle having fed on cycads (*Zamia* spp.), plants which are high in cycasin and other toxic compounds. Toxicity is a protection, but not a complete one, against many potential enemies.

Predators are essentially carnivorous arthropods. Among the insects are the Dermaptera, Odonata, Mantidae, Hemiptera Pentatomidae, Reduviidae, Nabiidae and Piesmidae, Diptera Asilidae and Syrphidae, Hymenoptera Vespidae and Formicidae, Coleoptera Carabidae, Staphylinidae, Histeridae and Coccinellidae, and many others. Among the Arachnids are the Lycosidae, Thomisidae and Phalangidae. Many vertebrates can also be included in the list of predators: lizards, birds, amphibians, bats and many others, including rodents. Carnivorous plants like *Sarracenia*, *Heliamphora*, *Darlingtonia*, *Nepenthes*, and even *Drosera*, often capture chrysomelids, and in tropical mountains ants and beetles are their most common prey. *Longitarsus nigerrimus*, which feed on *Utricularia*, have certainly found a way to avoid capture, like the ants living in the stem of *Nepenthes bicalcarata* Hooker in Borneo and the bugs feeding on carnivorous Martyniaceae in Mexico. The same is true for *Roridula* insects.

As mentioned previously, *Aristobrotica angulicollis* (Erichson), a diabroticite, feeds on adult meloids (Mafra-Neto and Jolivet, 1994). It may be possible that, in some exceptional cases, chrysomelids feed on other chrysomelids. Cannibalism is also practised by the chrysomelids on their own eggs and larvae (Mafra-Neto and Jolivet, 1996). In the case of *Aristobrotica angulicollis*, the attractant is cantharidin, but this is a deviant behaviour. Attraction to cantharidin is common among many other beetles (Dettner, 1997; Nardi and Bologna, 2000). Certain Carabidae, Lebiinae and others are specialists in the capture of the larvae and adults of chrysomelids on their host plants. They also copy their prey (Peckhamian mimicry) to capture them easily. Many cases are reported by Cox (1999), Balsbaugh (1967), Jolivet and Van Parys (1977), Bourdouxhe and Jolivet, 1981, Lindroth (1971), etc.

Snyder and Wise (2000) reported recently that *Diabrotica undecimpunctata howardi* (Barber) demonstrates reduced feeding patterns as a strategy for reducing predation risk by dangerous predators such as the wolf spider. Less dangerous predators are ignored.

Biological control

Recent reviews on this topic have been done by White (1996), Cox (1996a) and Jolivet (1997). The literature is spread among numerous papers dealing mostly with agriculture pests, such as the Colorado Potato Beetle, *Diabrotica* species and many others. White (1996) reviewed leaf beetles as biological control agents against injurious pests in North America.

There are really two economic aspects in biological control. First, numerous species are pests of crops, trees, and flowers. Some species are miners or borers into the stems of cultivated plants, but damage to monocultures is generally limited. Predators or parasitoids can be eventually used against them. The second aspect is the use of chrysomelids against accidentally or deliberately imported plants. They can suddenly become a local pest to the environment or to the crops.

PESTS OF CROPS AND FLOWERS

Several chrysomelids among Chrysomelinae, Alticinae, Galerucinae, Criocerinae, Megalopodinae and a few others feed on crops or garden flowers. The Colorado Potato Beetle (CPB) or *Leptinotarsa decemlineata* (Say) which, from wild species of *Solanum* (*S. rostratum*, *S. angustifolium*), adapted itself during the early nineteenth century to cultivated potato plants (*S. tuberosum*), is a good example. Thousands of papers have been written on this beetle since its discovery in 1824 by Say in Mexico. Its distribution covers a great part of the USA (except Oregon and California), the southern parts of Canada and Mexico, its original area. From France, where it was accidentally introduced in 1922 (in the Bordeaux area), it rapidly invaded Europe and Russia, part of middle Siberia, as far as Lake Balkhach in 1979 (Kim and Vasilev, 1986s), Libya (accidental), Iran (migrations), etc. It then moved eastwards, invading

China, but we do not have clear data from there. We are reduced to pronostics with Vlasova (1978) and Sokolov (1981), but it is progressing fast, even though it cannot go north above 58°N (Zhuravlev and Verba, 1989; Zhuravlev, 1993). *L. decemlineata* cannot invade Central America or India because of the humid climate and its capture in Guatemala results from a confusion with *L. undecimlineata.*

The efficient dispersal of the CPB is due to its great fertility, its adaptability to a certain number (not all) of Solanaceae, its aptitude to migratory flights and its adaptation to cold and dry conditions through a system of quiescence (diapause). Adult dispersal starts within the first 24 hours after eclosion from the pupae (Alyokhin and Ferro, 1999). As stated by Ferro and Voss (1985) and Ferro *et al.* (1991), the complex strategy of the CPB combines migration and diapause to disperse its offspring through space and time, thus reducing all risks linked with its reproduction.

No other American *Leptinotarsa*, even the species accepting *Solanum tuberosum*, the common potato plant, possesses such dispersal and adaptive capacities. Species from Central and South America live mainly on scrub *Solanum* with hard and spiny leaves.

Here below is a list of *Solanum* accepted by *Leptinotarsa.* In reality, Hsiao and Fraenkel (1968), who tested 104 plant species belonging to 87 genera and 39 families, have shown that 36 species are accepted experimentally: 11 Solanaceae, 2 Asclepiadaceae (*Asclepias syriaca* and *A. tuberosa*), 1 Brassicaceae (*Capsella bursa-pastoris*) and 1 Asteraceae (*Lactuca sativa*). In the field, the CPB is attracted only by certain Solanaceae. It rejects many of them, including most of the bushy types and some European species. Acceptance of *Asclepias* seems quite strange since the plant is rather toxic and produces latex, but Asclepiadaceae, and also Asteraceae, are part of the normal selection of many Doryphorini. However, acceptance of Brassicaceae is quite surprising.

Table 11.1. Host plants known among species of Leptinotarsa (after Hsiao 1988, modified).

Leptinotarsa spp.	A. Solanaceae
L. decemlineata	Solanum rostratum
L. defecta	Solanum elaeagnifolium
L. haldemani	Physalis douglasii
L. juncta	Solanum carolinense
L. rubiginosa	Physalis pubescens
L. texana	Solanum elaeagnifolium
L. tumamoca	Physalis wrightii
L. undecimlineata	Solanum ochraceoferrugineum
	B. Asteraceae
L. behrensi	Montanoa leucantha
L. cacica	Asteraceae
L. calceata	Asteraceae
L. heydeni	Viguiera dentata
L. flavitarsus	Verbesina gigantea
L. lineolata	Hymenoclea monogyra
L. melanothorax	Tagetes micrantha
L. typographica	Lasianthaea ceanothifolia
	C. Zygophyllaceae
L. peninsularis	Kallstroemia grandiflora
L. tlascalana	Kallstroemia rosei

Table 11.2. Solanaceae quoted as hosts for *Leptinotarsa decemlineata* (Say) in the field (after Hsiao, 1986. Simplified).

Plant species	Origin	Relative importance
North America		
Hyoscyamus niger	Introduced	Occasional
Lycopersicum esculentum	Introduced	Occasional
Solanum angustifolium	Mexico	Important
S. carolinense	USA	Frequent
S. dimidiatum	USA	Minor
S. diversifolium	Mexico	Minor
S. dulcamara	Introduced	Frequent
S. elaeagnifolium	Mexico, USA	Occasional
S. melongena	Introduced	Frequent
S. rostratum	Mexico, USA	Important
S. sacharhoides	Introduced	Occasional
S. sisymbriifolium	USA	Minor
S. triquitrum	USA	Minor
S. tuberosum	Introduced	Important
Continental Europe		
Hyoscyamus niger	Indigenous	Minor
Lycopersicum esculentum	Introduced	Minor
S. dulcamara	Indigenous	Minor
S. melongena	Introduced	Frequent
S. rostratum	Introduced	Minor
S. laciniatum	Introduced	Minor
S.tuberosum	Introduced	Important

The oviposition of *L. decemlineata* is not that selective and it prefers *S. rostratum* and *S. nigrum* to *S. tuberosum*. This is to be expected for the former plant, which is one of the first selected in nature, but *S. nigrum* is a European species and is eaten neither by the adults nor by the larvae (Hsiao and Fraenkel, 1968). According to the specialists (Boiteau, 1988a), the losses caused by the CPB are estimated at 60% in Ukraine and at 30–50% in Canada. Personally, I (P.J.) find the figures a little exaggerated, except perhaps in the Russian area where there have been rather sudden uprisings (Voisin, 2000). During the war years, the CPB was very abundant in France and began to progress quickly eastwards.

Criocerine larvae live mainly on monocots and certain species can attack orchids or cultivated cycads. One species, *Stetopachys formosa* Baly, in Queensland feeds on 27 species and 67 varieties of orchids, mainly the *Dendrobium*. Orchids seem to be the only host of the insect (Hawkeswood, 1991; Gough *et al.*, 1994). Criocerinae feed also on *Lilium* and *Cycas* (*Lilioceris*), cultivated and wild cereals like maize, barley, wheat, rice (*Oulema*). Others attack various Solanaceae (*Lema trilineata* Olivier). Kuwayama wrote monographs on *Lema oryzae* Kuwayama, the pest of rice crops in Japan, which was synonymized later with another species. Clytrinae are harmless, as are several other subfamilies: Lamprosomatinae, Cryptocephalinae, Chlamisinae. However, Megalopodinae are stem-borers and tend to be pests of tomatoes, potatoes, *Capsicum*, tobacco and eggplant in the neotropics. However, they do not seem to be specific and can also attack Asteraceae in Brazil.

Among Eumolpinae, some attack the leaves (adults) and the roots (larvae) of cacao trees (Ferronato, 1988, 1999) around Bahia, Brazil. These species are not specific and

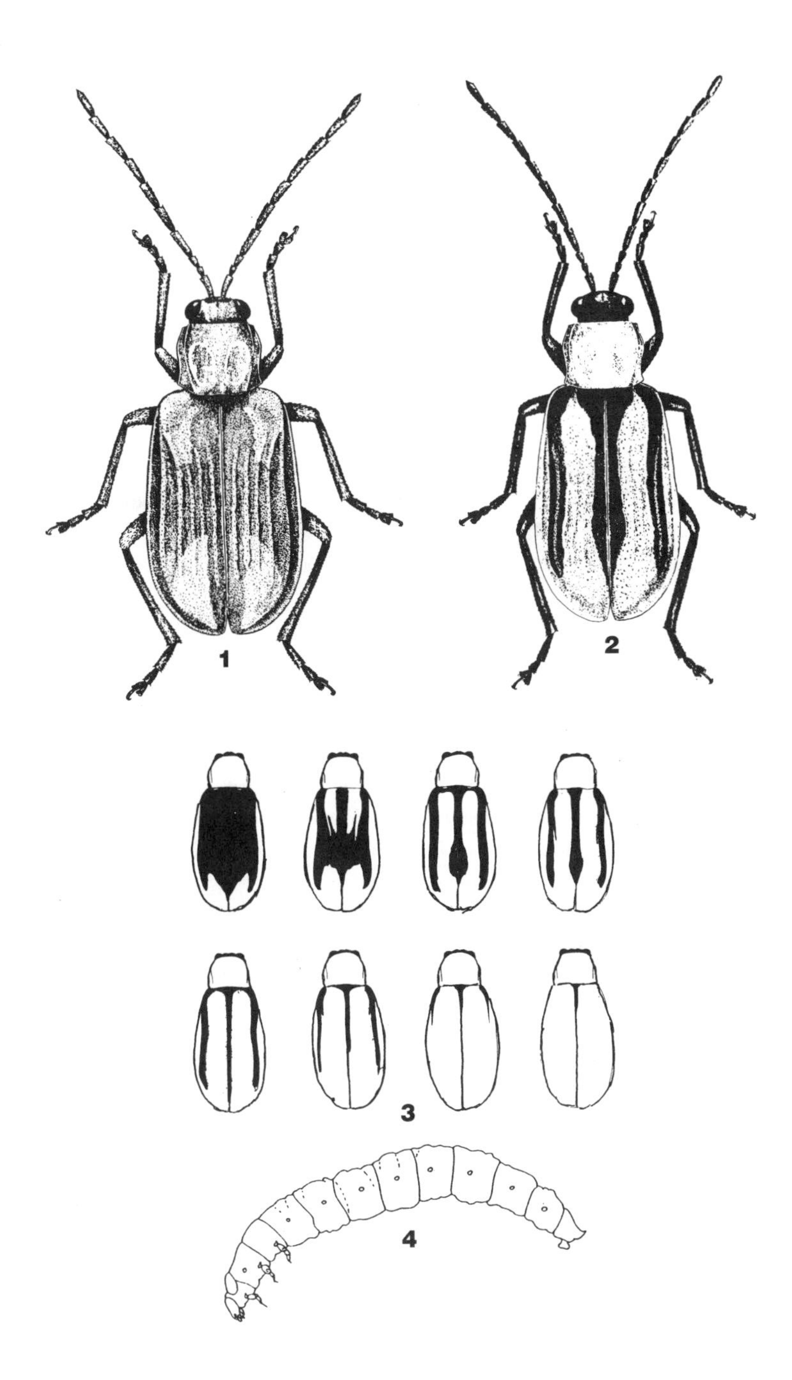

Figure 11.2. Serious crop pests. (1) *Diabrotica virgifera zeae* Krysan and Smith female (after Krysan and Smith, 1987); (2) *Diabrotica virgifera virgifera* LeConte female (after Krysan and Smith, 1987); (3) Variations in the colour of the elytra among specimens of *Diabrotica virgifera* s. str. (after Krysan and Smith, 1987); (4) *Diabrotica virgifera virgifera* LeConte larva, length 11 mm (after Stehr, 1991).

the larvae also feed on the roots of Poaceae. On cocoa trees, the damage is real. *Percolapsis ornata* (Germar), *Rhabdopterus* sp., *Colaspis* sp., *Taimbezinha theobromae* Bryant are present on the leaves among 18 other species. The succulence of the leaves could be an attractive reason for the choice of cocoa, but the plant is native in Brazil and eumolpines often show a polyphagous tendency. Other eumolpines attack the potato plant or the sweet potato (*Typophorus viridicyanus* (Crotch)). Other species feed on cereal roots, banana fruits, etc. Many, like *Nodonota*, are largely polyphagous and feed in America on 40 plant families and 92 genera, some of which have a real economic importance. *Colaspis* spp., which eventually feed on bushy *Solanum* in South America, are also largely polyphagous, even on fruits.

In the Chrysomelinae, there are many more or less important pests. *Gastrophysa viridula* (Degeer) feeds on sorrel; *Phaedon* spp. on *Nasturtium*, *Sinapis*, *Raphanus*; *Prasocuris* spp. on Apiaceae; *Entomoscelis* spp., *Microtheca* spp. and *Colaphellus* spp. on Brassicaceae, etc.

Within the Galerucinae, there are also many agricultural pests, but they also feed on trees and bushes. Many species feed on wild plants and many are polyphagous. *Aulacophora* and *Diabrotica* feed on cucurbits (flowers, leaves, pollen) and the larvae are root-feeders. Those larvae are polyphagous on roots and the adults show oligophagy on cucurbits, while *Diabrotica* is completely polyphagous on many crops. *Diabrotica* feeds on 500 species of plants (Metcalf *et al.*, 1982; Metcalf, 1994; Jolivet and Hawkeswood, 1995; Jolivet, 1999d). The larvae of *Exosoma* feed in a semiaquatic medium inside Amaryllidaceae and Liliaceae bulbs and destroy them. *Galerucella birmanica* is a pest of the water-nut plant (*Trapa* sp.) in the Indian subcontinent.

Many Alticinae feed on crops: cereals, oil plants, fodder plants, cabbage, flowers. Doguet (1994) produced a list for France and western Europe. To that must be added many exotic plants, eaten by *Phyllotreta*, *Podagrica*, *Longitarsus*, which very often belong to the same families eaten in Europe. *Systena*, an American genus, is totally polyphagous and feeds on many crops. However, many alticines feed on wild plants in the tropics. Several genera of alticines feed on *Citrus* in south-east Asia.

Hispines feed on palm trees (Arecaceae) in the tropics and on ornamental garden plants (*Heliconia*, ginger, etc.). These hispines are a flattened, smooth species, with a penny-like larva in the Zingiberaceae. Others feed on Poaceae (rice, wheat, sugar cane, bamboos, etc.). *Coelaenomenodera* damage on oil palms is serious (Mariau, 1988). *Promecotheca* attack palm leaves in Asia (Taylor, 1937), *Hispolepis* in Latin America, *Brontispa* in Indonesia. *Alurnus*, *Coraliomela*, *Mecistomela* are large species and palm pests in South America (Valverde *et al.*, 1994). *Dactylispa* (= *Hispa*) *armigera* Olivier is a pest of paddy in the Orient.

Cassidinae generally live on small plants in temperate areas: Asteraceae, Lamiaceae, Caryophyllaceae, Chenopodiaceae, Convolvulaceae, and on trees, sometimes ornamental in South America. Sporadic attacks occur in Europe, Asia and Africa on beetroot, artichoke, mint, camomilla and sweet potatoes.

PESTS OF TREES AND SHRUBS

Many chrysomelids can fly and can easily reach the trees, even the canopy in the tropics. These include *Zeugophora* on *Salix*, *Populus*, *Juglans*, *Corylus* and *Betula* leaves and several eumolpines which feed on coniferous trees. Aulacoscelinae, as well as several other chrysomelids, certain Sagrines, *Lilioceris*, feed on cycads, including *Zamia*, *Cycas*, *Macrozamia* and others. Certain efficient flyers, such as *Chrysomela*, feed on the young leaves of *Populus* and *Salix*. *Plagiodera versicolora* (Laicharting), imported from Europe around 1940, was a persistent pest of *Salix* and *Populus* (Wade, 1994). Many other species frequent arbustive Salicaceae, such as *Phratora*, *Gonioctena*, in temperate places. *Paropsis* and many Paropsini, in Australia, southern New Guinea and Tasmania, feed on *Eucalyptus* and sometimes on *Acacia*. On the other hand, *Paropsides* in south-east Asia feed on arbustive Rosaceae (*Pyrus*, *Crataegus*) and sometimes also on Myrtaceae. *Calligrapha* in North America attacks many shrubs and trees, but also low plants. *Novocastria* Selman feed on *Nothofagus* in Australia as well as several eumolpines and aulacoscelines in South America (Chile).

Among the Galerucinae, many species feed on trees such as *Sastra* on *Trema* (Ulmaceae) in New Guinea, *Agelastica* on Rosaceae and *Alnus*, *Betula*, *Corylus*, *Salix*, *Populus* and *Carpinus*. *Pyrrhalta luteola* (Müller) feed on *Ulmus*, and many others are tree-feeders in the tropics.

Alticinae we can mention include various genera on *Citrus* (*Clitea*, *Podagricomela*) in China, and in south-east Asia and Africa, *Podagrica* and *Nisotra* on Malvaceae, Tiliaceae, Sterculiaceae, Bombacaceae (Tiliales). Many Hispinae feed on palm trees and Cassidinae mostly on low plants. Certain species fly well and can feed on trees: *Cordia* (Boraginaceae), *Ipomoea arborescens* (Convolvulaceae), *Tabebuia* (Bignoniaceae), etc. In Argentina, *Tabebuia* trees along the sides of roads can be completely defoliated.

Many attempts have been made to use biological control against these pests, with either predators or parasitoids, with mixed success. *Leptinotarsa decemlineata* has a limited number of parasites. Manipulating the predators and the parasitoids is not easy and many experiments were conducted against *Coelaenomenodera* spp. on the Ivory Coast (Mariau, 1988). See also Berti and Mariau (1999). *Bacillus thuringiensis* has been tested several times against CPB with specialized strains. So far, manipulating *Microsporidia*, *Wolbachia* and others seems unrealistic. Parasitoids seem the most promising. Resistance to *B. thuringiensis* should be considered as resistance has developed against most of the insecticides used against the Colorado Potato Beetle. Transgenic plants are a future option for some severe pests.

USE OF CHRYSOMELID BEETLES IN BIOLOGICAL CONTROL AGAINST WEEDS

As certain chrysomelids adapt very well to many indigenous plants, there have been trials to acclimatize them in countries where those plants have been imported and have become noxious. This aspect of the fight against weeds without any toxic herbicides is not new. Many years ago, a cassid (*Physonota alutacea* Boheman) and a galerucid were imported, with mixed results, to Mauritius against imported *Cordia*. One of us (P.J.) saw *Cordia* in Mauritius, still present but under control, 50 years after the first introduction of the beetles. Only the galerucid survived; the cassid died out, probably because of local bird predation. Though biological control of chrysomelid pests using parasitoids has often led to doubtful results, using the insects against plants has been often rather successful. However, the dangers are no less than with the parasitoids. Parasitoids have been known to suppress endemic species, not the target, whether related or not with the insect to be controlled. The case of Guam Island is well documented. There, hundreds of endemic moths were eradicated by introduced parasitoids. In the case of the use of imported phytophagous beetles, the main danger is an adaptation to succulent crops or to ornamentals, and that despite careful screening.

Leaf beetles, like fruit trees, adapt easily from one hemisphere to the other and they automatically invert their cycle and diapause systems. The case of *Microtheca* adapting from the Southern Hemisphere to the north and the case of *Chrysolina* from north to south are well known. A beetle like *Chrysolina varians* adapted from one generation per year during the European summer to many generations in Central Australia.

It is evident that any introduction must be carried out with extreme care so as to avoid catastrophic results. Even the most improbable hosts must be tested in the laboratory. Rigorous tests must be conducted in the country of origin of the plant and complemented with tests in the country of introduction where a safe quarantine laboratory should be available. It is clear that, before introducing an insect feeding on *Rubus*, other possible plants must be tested like *Rosa*, *Fragaria*, raspberries, etc. Because of their lack of specificity, many promising insects must be rejected. Polyphagous and oligophagous insects cannot be chosen. As Balachowski (1951) wrote: "Before loosing and disseminating phytophagous insects in nature, many checkings must be done in laboratory". The USDA has established laboratories at strategic points throughout the world. CSIRO and CABI also have many specialized laboratories, as have ORSTOM (IRD) and CIRAD. Sometimes, the results are excellent (*Chrysolina* on *Hypericum*), sometimes mitigated. *Physonota alutacea* Boheman, a tropical American cassid feeding on *Cordia macrostachya* (Boraginaceae) never established itself in Mauritius, as already mentioned above. Sometimes, the results are poor and the plant is not sufficiently under control. For instance, research is being done in the Merida area (Venezuela) for parasites of a wild *Passiflora* imported to Hawaii. This is a very difficult task since the possibility of adaptation to cultivated species is always a danger. Evidently, complete eradication of a plant is not realistic. Reasonable control is the goal.

It is often possible to multiply an imported phytophagous species in the laboratory. This is always a delicate operation and some genera (*Chlamisus*, eumolpines) tested against *Rubus* are very difficult to multiply *in vitro*. Trials of direct importations of *Chlamisus* in Hawaii were always unsuccessful. The larval biology of the insect is complex, even if not linked with ants, and the damage caused by the adults is insignificant. It can also happen that local parasitoids adapt to the imported species. In Australia, for instance, parasitoids of Paropsini adapted rapidly to the St John's wort *Chrysolina*, but the beetles survived and carried out a good control.

Some cases were a great success, others very much a failure. There were also cases where the introduced beetles after some time were not selective or had modified their trophism (allotrophy). So far, no major catastrophe has been reported.

Hypericum perforatum L., the European St John's wort has been accidentally introduced into Australia and many other countries, together with *Plantago* spp., the latter not considered as a pest. It became very invasive there and grew well above the European norms in size and abundance. Several *Chrysolina* were introduced: *C. hyperici* (Forster), *C. brunvicensis* (Gravenhorst) and *C. varians* (Schaller). It was a rapid success and the *Chrysolina* controlled the plant well, without eliminating it completely. In California and all the Pacific coast of Canada and USA, the beetles were introduced in 1939 and 1946 and then again in 1969 and 1970. Actually, *C. varians* did not survive in British Columbia, but *C. quadrigemina* is still common in the Rockies, south of British Columbia as far as Colorado, Oregon, Washington, Idaho, Montana and California. This species also adapted to local species of *Hypericum* and remains common on *H. perforatum. C. quadrigemina* is better adapted to dry areas, while *C. hyperici* prefers humidity.

Another classical example is the fight against *Cordia macrostachya* Roem in Mauritius. We have seen previously that the cassid *Physonota alutacea* did not survive, but the galerucine *Metrogaleruca obscura* Degeer (= *Schematiza cordiae* Barber) is still common. In Brazil, *Physonota alutacea* is common in forest clearings on *Cordia* leaves and its density is not causing a problem to the plant.

In 1960, one of us (P.J.) proposed, in order to combat *Rubus moluccanus* L. and *R. alceaefolius* Poiret (introduced 200 years ago in La Réunion), the naturalization of *Phaedon fulvescens* Weise, a common leaf beetle of mid-altitude in North Vietnam (1500–2000 m). The insect seems specific (Jolivet, 1984) but the Montpellier team (GERDAT) who study them, under the leadership of Thomas Le Bourgeois, found that it feeds also on a local *Rubus* (*R. apetalus*) in La Réunion. They seem to prefer an eumolpine, *Cleorina modiglianii* Jacoby and a sawfly from Sumatra. The specificity of the *Cleorina* seems questionable and the larva feeding on roots can be polyphagous. Other possible pests against *Rubus*, like eumolpines and chlamisines, are not very efficient or selective. The problem was similar in Chile, New Zealand, and Hawaii, with several species of *Rubus*, and nowhere has a suitable solution been found.

Several introductions are presently being undertaken, like *Agasicles hygrophila* Selman and Vogt in Florida. This is an alticine used to fight *Alternanthera philoxeroides*, an amphibious Amaranthaceae from South America. Pupation takes place inside the stem in or over the water. A flavone has been isolated from the plant and is responsible for the specificity of the insect. The introduction was carried out in Florida with specimens from Mar der Plata in temperate Argentina and it was finally successful (Vogt *et al*., 1979, 1992; Buckingham *et al*., 1981, 1983).

Longitarsus flavicornis (Stephens), an alticine introduced from France into Tasmania in 1979, seems efficient in controlling *Senecio jacobaea* L. (Ireson *et al*., 1991, 2000). However, although a large (more than 90%) reduction in ragwort densities has been recorded, herbicides and flooding restrict the efficacy of the insect in many properties. Whapshere (1982–1983) studied *Longitarsus jacobaeae* Waterhouse for a long time in Montpellier, France, in view of its importation in Australia. He studied also *L. echii* Kutschera, a European species, also for Australia, to fight *Echium plantaginaceum* L. (Boraginaceae). Both species are univoltine, but *L. echii* larvae feed on the central part of the root and those of *L. aeneus* rasp the cortex and the radicles. The introduction of *L. jacobaeae* effectively controlled *Senecio jacobaea* in North America (Windig, 1991, 1993). The insect has also been released in Australia and New Zealand.

To fight *Baccharis salicifolia* (R.1.P.) Pers. in USA and Mexico, Boldt (1989), Boldt *et al*. (1991) proposed various leaf beetles, including *Stolas fuscata* (Klug) from Argentina, which seems to be a good control agent. Boldt and Staines (1983) and Boldt and White (1992) proposed also *Pentispa suturalis* (Baly) (Hispinae) and

Exema elliptica Karren (Chlamisinae). All the above beetles seem to control the weed quite well. In Italy, Fornasari (1993) studied the biology of *Aphthona abdominalis* (Duftschmid) feeding on *Euphorbia esula* L. (Euphorbiaceae) leaves. The larvae feed on the roots of the same plant. *A. abdominalis* seems to be a good candidate for the control of euphorbs in North America.

Another example of the introduction of a beetle is the case of *Zygogramma suturalis* (F.) in Russia. The beetle is used to combat the imported *Ambrosia artemisiaefolia* L. and *A. psilostachya* D.C. *Z. suturalis* was introduced from Canada and USA in 1978 by Kovalev. Many papers have been published in Russia on this insect and its biology. Pantyukhov (1992) made a special study of its ways of hibernation and its survival. See also Kovalev *et al.* (1983), Kovalev and Medvedev (1983) and many other papers on the topic. In India, in Bangalore, it is *Zygogramma bicolorata* Pallister, a Mexican species, more adapted to hot climates, which has been imported (Jayanth and Nagarkatti, 1987; Srikanth and Pushpalatha, 1991; Jayanth and Bali, 1993; Jayanth, *et al.*, 1993). The beetle was supposed to feed on *Parthenium hysterophorus* (Asteraceae), but it feeds also on *Helianthus annuus* L., the sunflower. It seems to be a minor problem linked with the presence of pollen of *Parthenium* on sunflower leaves. While this introduced leaf beetle is well established in 200 000 sq km area at and around Bangalore, causing large-scale defoliation of *Parthenium*, it is not likely to do so well in those parts of the Indian subcontinent which have severe summers, as eggs fail to hatch at 40°C and diapausing adults in soil survive only for less than a day at 45°C (Jayanth and Bali, 1993; pers. comm. from Jayanth, 1995).

Several South African beetles have been tested for introduction to Australia. *Ageniosa electoralis* (Vogel), for instance, was selected to combat *Chrysanthemoides monilifera* (L.), an Asteraceae (Adair and Scott, 1993). This chrysomelid was finally rejected because of non-specificity. It also feeds on various other Asteraceae and Apiaceae, as do many *Ageniosa*, a selection of the type *Chrysochloa* (another Chrysomelinae). In Transvaal, one of us has often seen *Ageniosa nugaria* (Vogel) on *Plectrantus* sp., an ornamental Lamiaceae. Various *Chrysolina* were also tested but were also not used because they were non-selective.

Another chrysomeline, *Calligrapha pantherina* Stal, from Mexico, has been released in Australia (Forno *et al.*, 1992) to control various Malvaceae, such as *Sida acuta* Burm, and the pantropical *Sida rhombifolia* L. All originated from North America.

One Argentinian cassid, *Gratiana spadicea* Klug, was tested in South Africa (Hill and Hulley, 1993; Hill *et al.*, 1993) against imported *Solanum* spp., namely *S. sisymbriifolium*. The cassid also feeds and lays eggs on five indigenous species and on the eggplant. It would probably have been worse with any species of *Leptinotarsa*. *G. spadicea* has been rejected. It is true that in the laboratory, tested insects are less selective than in nature. Moreover, any mutation can occur and a diet change in the field must always be expected.

Many other cassids have been tested against imported *Solanum*, including *S. elaeagnifolium*. For instance, in Uruguay, *Metriona elatior* (Klug) was tested with a certain success, but the lack of selectivity of these cassids remains a potential danger for the crops: tomatoes, potatoes and eggplants (Ponce de Leon *et al.*, 1993).

Lythrum salicaria L. (Lythraceae), an ornamental weed from Europe, became a pest in the USA. Two sympatric galerucines were selected: *Galerucella calmariensis*

L. and *G. pusilla* Duftschmid (Malecki *et al.*, 1993; Blossey *et al.*, 1994). Earlier, several other insects associated in Europe with the purple loosestrife, *Lythrum salicaria,* were listed and eventually tested (Batra *et al.*, 1986). Finally, it was found that the benefits from the introduction did not match the possible risks of attacks against native plants like *Lythrum alatum* L. and *Decodon verticillatum* (L.), also Lythraceae. The introduction has been abandoned. There is currently research being done in North Africa on *Tamarix* spp. (Tamaricinae) introduced into the USA. Galerucines are a possible choice and seem very selective.

Any introduction of species must be carefully planned. *Mesoplatys cincta* Olivier from Africa was proposed against *Sesbania pachycarpa* D.C. in Argentina (Le Bourgeois, 1992). Since this insect also feeds on a certain number of arbustive Fabaceae: *Indigofera, Aeschynomene, Sesbania, Erythrina* (Jolivet and Van Parys, 1977; Bourdourxhe and Jolivet, 1981; Jolivet and Hawkeswood, 1995) and there was obvious risk of oligophagy, the project has been abandoned.

Weiseana barkeri Jacoby feeds on leaves of *Acacia nilotica* (L.) Willd. in Kenya and seems really specific (Marohasy, 1994). It is on the way to being imported to Queensland for use against the African *A. nilotica* naturalized there. The danger remains that monophagy becomes oligophagy on all species of *Mimosa*.

The above examples of biological control involving chrysomelids have been picked out from among hundreds of different projects. Some experiments were fully successful, such as the release of various *Chrysolina* in USA, *Longitarsus* spp. in Australia, etc. For instance, *Longitarsus flavicornis* (Stephens) is well established now on *Senecio jacobaea* L. in Tasmania (Ireson *et al.*, 2000). Other experiments failed, either because the insect was not 100% specific or because it slowly became oligophagic. Prolonged and rigorous experiments in the country of origin and a strict quarantine in the country of importation are necessary before any release. Thankfully, many introduced insects did not survive for many reasons; nature makes up her own rules.

12

Phylogeny of Subfamilies*

Introduction

As noted under 'Introduction' (Chapter 1), Chrysomelidae, with more than 37 000 described species and probably 50 000 species awaiting discovery, is one of the largest families of beetles and, perhaps, of insects. Curculionidae and Staphylinidae are other such big beetle families.

In addition to its size, Chrysomelidae presents a great diversity among its members. It is, therefore, divided into 20 extant subfamilies. This includes a fossil subfamily, Protoscelinae, from the upper Jurassic and the probable ancestors of the living Aulacoscelinae, which show some cerambycid features.

It is difficult to make out the phylogenic relations among the chrysomelid subfamilies. Pioneering efforts have been made by hard-working chrysomelid researchers. These include Terry Seeno of the USA, who edits and publishes a very useful newsletter, *Chrysomela*, D.G. Furth, also of USA, who organizes International Symposia on the Chrysomelidae and P. Jolivet, M.L. Cox and associates, who edit and publish a series of volumes on Chrysomelidae biology. As a consequence, we are now in a much better position to unravel chrysomelid phylogenetic relations. This present chapter is an attempt in this direction.

Chrysomelid subfamilies

In this review, the list of 19 subfamilies (plus the Palophaginae recently described) adopted by Seeno and Wilcox (1982) will be adhered to, with small modifications like the position of Megascelinae. Differences of opinion have been expressed as to the ranking of these groups. It has been suggested that some clusters of the subfamilies should be raised to the family level. This controversy will not be entered into here as there are no non-arbitrary definitions for categories above the species level, and the delimiting of higher categories is particularly difficult up to the family level (Mayr and Ashlock, 1991). Moreover, one has to agree with Schmitt's (1996) remark "... from a phylogenetic-systematic point of view, mere changes of ranks in classification do not solve any phylogenetic problem". Suzuki (1996) presented a general review of the various classifications with a cladogram of his own.

*This review has been included in *Some Aspects of the Insight of Insect Biology* (Editors Sobti and Yadav) published in New Delhi in 1999. In view of the limited circulation and availability of the publication, the review is being reproduced here, with some minor changes, as Chapter 12.

Phylogeny of the subfamilies

As Suzuki (1994) pointed out, five groups of chrysomelid subfamilies are generally accepted as monophyletic groups. These groups and Cerambycidae are shown in *Figure 12.1* (encircled by a solid line). In addition, a number of groups or clades have been drawn up on the basis of significant synapomorphies (in some cases, plesiomorphies) which has been pointed out in recent publications. Such clades are shown in the figure by enclosing them in a broken line. These groups have been numbered, and are referred to in the text by these numbers. This method of showing groups or clades is preferred, as it is less hypothetical than showing lineages.

Given that the generally accepted monophyletic groups are well established, we will discuss the numbered groups (encircled by broken lines in *Figure 12.1*).

GROUP 1

On the basis of similarities in the internal reproductive organs and hindwing venation, Suzuki (1994) suggested a monophyletic origin for the Orsodacninae, Megalopodinae, Zeugophorinae and some Cerambycidae. Schmitt (1992b) finds details of the construction of the mesoscutopronotal stridulatory organ very similar in Zeugophorinae, Megalopodinae, Palophaginae and the majority of Cerambycidae. Another synapomorphy pointed out by this author is an elongated U-like arrangement of digitiform sensilla on the distal-most joint of the maxillary palp in all these groups. A similar arrangement of the sensilla has been pointed out in *Sagra* and Bruchidae by Mann and Crowson (1984). Schmitt, however, holds that in Sagrinae the arrangement of the sensilla is rather irregular, and not quite comparable with that of Donaciinae, Zeugophorinae, etc.

Chen (1985) mentions that the larva of the megalopodine *Temnaspis nankinea* (Pic) has dorsal and ventral locomotor ampullae on its abdominal segments, as in cerambycid larvae.

Cox (1994a) made a comparative study of egg bursters in the first larva of Chrysomelidae, and found that the primitive arrangement of the bursters in the chrysomelid subfamilies Orsodacninae, Zeugophorinae, Megalopodinae and in Cerambycidae is closely comparable. In support of the naturalness of the grouping, he points out another fact, that megalopodine larvae are known to burrow in fairly hard stems, and those of Orsodacninae probably into leaf petioles or terminal buds.

Cerambycid aedeagus in the chrysomelid subfamilies in the cluster (Verma, 1996c) is yet further evidence in support of the naturalness of the group.

GROUP 2

The male efferent genital system in Zeugophorinae and Megalopodinae shows some notable similarity with the corresponding part of anatomy in Cassidinae, Hispinae, Clytrinae, Cryptocephalinae, Chlamisinae and Lamprosomatinae. Verma (1996c) points out the following in this context in the two clusters.

(a) *Vasa deferentia* abbreviated, and origin of accessory glands close below the testis follicles.
(b) Median ejaculatory duct long, slender and winding, with differentiation of a sperm pump or ejaculatory sac.

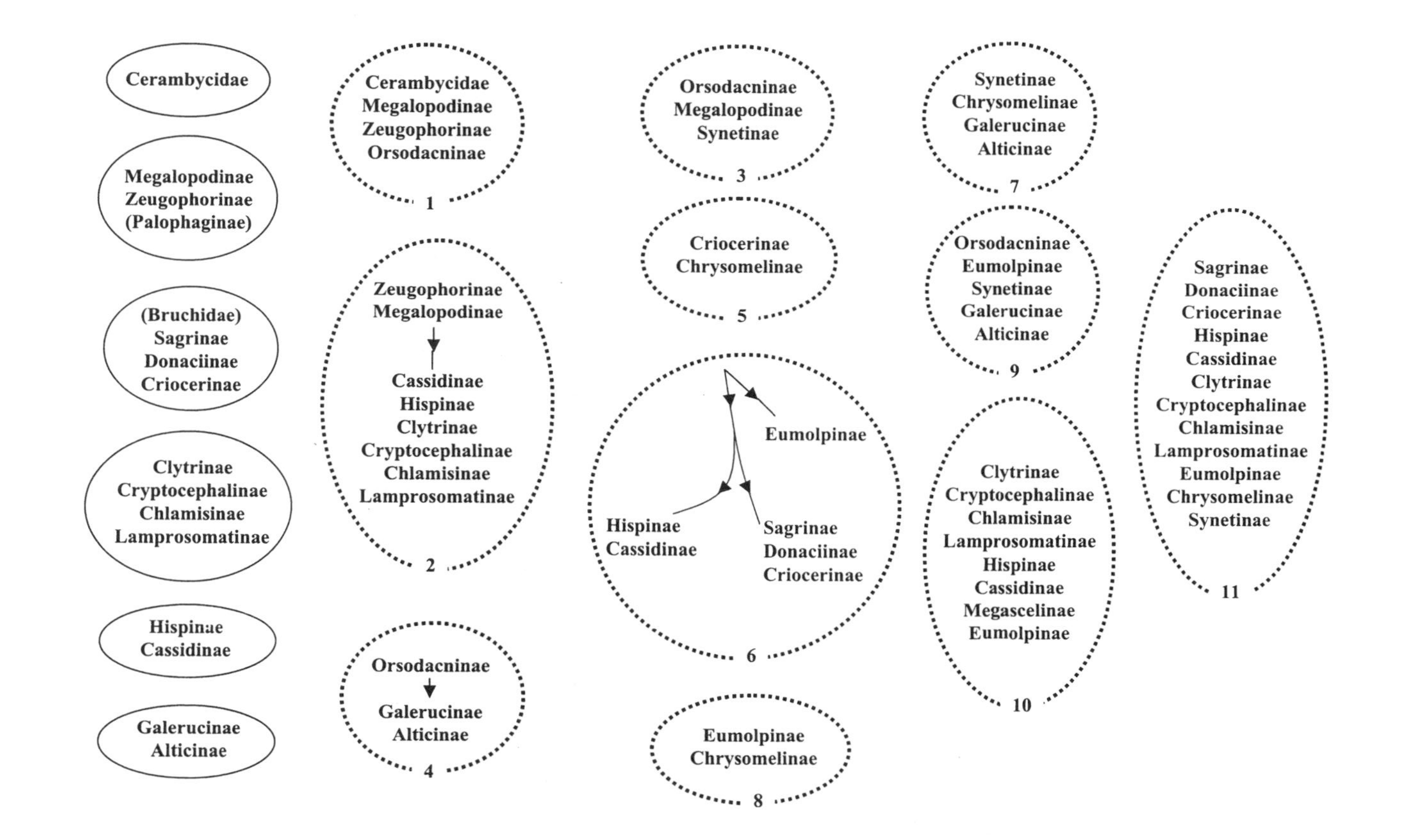

Figure 12.1. Grouping among chrysomelid subfamilies. The subfamily names, encircled by a solid line on the left, are generally accepted monophyletic clades. Those encircled by broken lines, numbered 1–11, are groupings based on significant synapomorphies (in some cases, plesiomorphies) which have been pointed out in recent publications. These groupings (that is those encircled by broken lines) have been referred to by their numbers in the text. (In the figure, Cerambycidae and Bruchidae have also been included.)

GROUP 3

Hsiao (1994a) tried to work out the phylogeny of chrysomelid subfamilies from his study of mitochondrial DNA. One of the clades he produced includes Orsodacninae, Megalopodinae and Synetinae (*Figure 12.1*).

GROUP 4

Crowson and Crowson (1996) comparatively studied the external features and male and female terminalia in Galerucinae and Alticinae and also in seemingly related subfamilies, and inferred that Orsodacninae were forerunners of Galerucinae and Alticinae. The main details of evidence in support of this are:

(a) Larval features of *Orsodacne* are quite similar to those in Galerucinae.
(b) Details of rhabdomere arrangement is remarkably similar in Orsodacninae and Galerucinae–Alticinae.The authors referred to the work of Schmitt *et al.* (1982) on rhabdom structure in Chrysomeloidea.
(c) In adults of *Orsodacne* and in *Exosoma*, and in a few other members of Galerucinae, the tip of the median lobe of the aedeagus is split or bifid. This feature is not known in any other Chrysomelidae.

Synetinae and Chrysomelinae have often been regarded as being related to Galerucinae–Alticinae. Crowson and Crowson (1996), however, believe that any similarity between the two pairs of subfamilies is due to parallelism rather than to common ancestry.

GROUP 5

Criocerinae and Chrysomelinae may be taken as related as they share similar defence glands (Pasteels *et al.*, 1989). Hsiao (1994a), from his study of mitochondrial DNA, has come to a similar conclusion.

Jolivet (1988, 1997), on the basis of feeding habits, wings and male genitalia, regarded Criocerinae and Chrysomelinae, along with Galerucinae and Alticinae, as belonging to one monophyletic group.

Suzuki (1994) found hindwing venation to be similar in Criocerinae and Chrysomelinae.

Verma (1996c) found Criocerinae and Chrysomelinae quite similar in the organization of the male genital system. Points of similarity include:

(a) The two testis follicles of the same side are not closely pressed together. They remain quite separate.
(b) Presence of long and tubular accessory glands.
(c) Aedeagus with basal muscular bulb, but the first spiculum or the terminal apodeme extends posteriorly for some distance beyond the ventroposterior edge of the basal orifice.

GROUP 6

Bruchidae, Sagrinae, Donaciinae, Criocerinae, Hispinae and Cassidinae have bifid tarsal setae which do not occur in any other beetle taxon; hence, they have been taken as a monophyletic cluster (Mann and Crowson, 1981; Schmitt, 1989).

In Criocerinae, Donaciinae, Cassdinae and Hispinae there is a median longitudinal groove on the frons (Schmitt, 1989). In *Sagra tristis* Fabricius, a median longitudinal depression is present, and it seems to correspond to the groove. Under high magnification in SEM, Schmitt noted that the area of the groove and the corresponding area in Sagrinae presented the same detailed microstructure.

Petitpierre (1989) points out a good cytogenetic support for the view that Hispinae–Cassidinae have branched off from the lineage leading to Sagrinae, Donaciinae and Criocerinae. In this study, there is some support for the theory that Eumolpinae are related to the complex.

The study of egg bursters in larvae of Chrysomeloidea by Cox (1988–1994a) also lends support to the view that the subfamilies under discussion are a monophyletic bunch.

Samuelson (1996) found a notable similarity in the arrangement of microtrichia on the under surface of elytra among Eumolpinae, Chrysomelinae, Cassidinae and Hispinae.

Accepting a close relationship between Criocerinae–Donaciinae–Sagrinae and Hispinae–Cassidinae, Chen (1985) pointed to an additional support for the view, viz. that primitive genera within Hispinae had stem-boring habit as larva, a feature also shown by Sagrinae.

Cox (1998b) noted similar pupal features in Sagrinae and Donaciinae.

GROUP 7

Chen (1985) placed Chrysomelinae, Galerucinae, Alticinae and Synetinae under one family because of similar aedeagal structure, similar hindwing venation, subprognathous head, distinct anteclypeus with truncate apex and pronotum with lateral margins.

Jolivet (1957–1959) and Suzuki (1994) have also noted a similarity in hindwing venation of these subfamilies.

Cox (1994a) found that details of egg bursters in the first larva suggest grouping together Chrysomelinae, Alticinae and Galerucinae.

Verma (1996c) noted that the aedeagal structure is comparable in Synetinae, Chrysomelinae, Galerucinae and Alticinae.

Mann and Crowson (1981) noted another point of similarity among these subfamilies, viz. a short and broad median ejaculatory duct. Further, they reported that the testes of the two sides in *Syneta* are close together and enclosed in a loose common membranous sheath, a point of resemblance with Galerucinae–Alticinae.

Cox (1998b) inferred phylogenetic relations between Chrysomelinae on the one hand and Galerucinae–Alticinae on the other on the basis of pupal features.

On the basis of pupal morphology and chaetotaxy, Cox (1996b) kept Chrysomelinae, Galerucinae and Alticinae in the same cluster.

Jolivet (1988) tried to work out chrysomelid phylogeny on the basis of food and feeding habits (also on wing and male genitalia structure). He, too, bunched together Chrysomelinae, Galerucinae and Alticinae.

Reid (1995) attempted a cladistic analysis of subfamilial relationships, and he finds Chrysomelinae, Galerucinae and Alticinae in the same clade.

GROUP 8

From their study of internal reproductive organs, Pajni *et al.* (1987a) inferred "... Cassidinae and Hispinae are closely related, while Chrysomelinae and Eumolpinae represent another similar combination.... The subfamily Eumolpinae represents a connecting link between these two combinations".

Chen (1985), on the basis of head features, regarded Eumolpinae and Chrysomelinae as having evolved from a common ancestor.

GROUP 9

From their comparative study of the aedeagal internal sac in Chrysomelidae, Mann and Crowson (1996) suggested the following phyletic line: Orsodacninae → Eumolpinae → Synetinae → Galerucinae → Alticinae.

Cox (1996b), on the basis of his study of pupal morphology and chaetotaxy in Chrysomeloidea, grouped together Synetinae and Eumolpinae.

As also noted earlier, Verma (1996c) found the aedeagal structure comparable in Synetinae, Galerucinae and Alticinae.

GROUP 10

From his studies on internal reproductive organs and hindwing venation, Suzuki (1988 and 1994) inferred that the eight subfamilies included here form a monophyletic group.

Cox (1994a) regarded five of these subfamilies, viz. Clytrinae, Cryptocephalinae, Chlamisinae, Lamprosomatinae and Eumolpinae, as closely related, as they show very similar egg bursting apparatus.

Chen (1985) also took these five subfamilies as having close kinship in view of features of the head, larval case and aedeagal structure.

GROUP 11

Verma (1996c) took the 12 subfamilies included here as a monophyletic group, keeping in view the septate organization of testes, a basal muscular bulb for the aedeagus, and a corresponding reduction of the first spiculum or the tegminal apodeme.

Let us separately discuss the phyletic position of Aulacoscelinae among Chrysomelidae. This small subfamily, confined to Central and South America, is obviously very primitive. Generally, it has been placed close to Chrysomelinae (Jolivet, 1957 and 1988; Mann and Crowson, 1981; Suzuki, 1988), but Suzuki (1992), having himself examined aulacosceline specimens, is of the opinion that the subfamily should be placed between Megalopodinae and Sagrinae. The fact that *Timarcha* should be placed in its own subfamily, Timarchinae, and the position of this subfamily have been discussed at some length by Sharp and Muir (1912), Chen (1934a), Powell (1941), Verma (1996c) and Jolivet (1999a).

Are Chrysomelidae polyphyletic?

As Chrysomelidae present great diversity, and among its subfamilies there are some well-defined clusters (see groups encircled by solid lines), they have been often described as polyphyletic. Daccordi (1985, 1996), in his study of the phylogeny of Chrysomelidae, reviewed the various systems and, from the spermatological point of view, Baccetti and Daccordi (1988) believe in the close affinity between Bruchidae, Chrysomelidae and Curculionoidea. The study does not solve the problem of mono- or polyphyletism of the group. However, in the previous section, we should have realized that the gaps between the solidly encircled groups are not clear. They seem to be crossed by several phyletic lines. How can we explain this situation? Perhaps, the only reasonable explanation is that the family Chrysomelidae is closely polyphyletic or parallelophyletic in origin, to use a term from Mayr and Ashlock (1991).

In close polyphyly, the groups, evolving independently from a common ancestral stock, have much in common in their genetic potentials. Hence, they may separately evolve very similar features. Mayr and Ashlock (1991) say in this context "A . . . serious problem is posed by propensity of genotypes to produce a certain phenotype, such as stalked eyes in certain acalyptrate dipterans, which is manifested in only some of the possessors of a genotype. Parallelism in this case can be defined as homologous similarity since the common ancestor evidently had the genetic propensity, even if it was not expressed phenotypically".

Now, when we realize that the Chrysomelidae are a product of parallelophyly among Phytophaga, should this group be retained as a family, or should we break it up into some separate families? Mayr and Ashlock (1991), on pages 255–258 have convincingly pleaded for taking a case of parallelophyly as one of monophyly for all practical purposes, when attempting an evolutionary classification. Hence, there appears to be no reason to break up the family Chrysomelidae. Moreover, we have to remember that an important function of a classification is information storage and retrieval, which would be impaired on de-recognizing the well-established family, Chrysomelidae.

Perhaps it would not be out of place to point out that Chrysomeloidea are also closely polyphyletic or monophyletic in origin. Besides other common features, *'retournement'* of the aedeagus seem to be universal among Chrysomeloidea (Kumar and Verma, 1980; Verma, 1994), and sclerotic and muscular components of the aedeagal apparatus in this superfamily, though much varied, are derivable from a common basic plan (Verma, 1996c). Among the conventionally recognized three families of Chrysomeloidea, Chrysomelidae, Cerambycidae and Bruchidae are adapted to their particular ecological zones, and show morphological and developmental adaptations for this. As Chrysomeloidea are closely polyphyletic, it is not surprising that some (Sagrinae) among Chrysomelidae show a stem-boring habit, which comes close to the seed-boring in Bruchidae, and some others (Megalopodinae) as larvae bore into fairly hard stems, a habit not far removed from the wood-boring habit of cerambycid larvae. Though some authors (e.g. Lingafelter and Pakaluk, 1997) believe that ecological considerations are not important in classification, we do feel the need for criteria, other than only morphological, when attempting a natural classification. According to Mayr and Ashlock (1991), "Like the genus, but perhaps to an even greater degree, the family tends to be distinguished by certain adaptative characters that fit it for a particular adaptive zone, e.g. woodpeckers or the family Picidae, the leaf beetles or the family Chrysomelidae". Adaptive shifts in the evolution of Cerambycidae and Bruchidae have been pointed out by Verma and Saxena (1996). After separation of these two families, all that remains of Chrysomeloidea is the family Chrysomelidae. Thus in a way, Chrysomelidae are a *'port-manteau'* category.

Concluding remarks

As inferred from this review of recent publications, the subfamilies of Chrysomelidae, though they present great diversity, are a product of parallelophyly. Though at present most specialists regard the family as polyphyletic in origin, we have pleaded for close polyphyly or parallelophyly.

Postface

We have seen that certain leaf beetles can be serious crop pests. Several different control techniques have been tested for use against them. These include chemical and biological insecticides, pheromones, GMOs (genetically modified organisms), parasites and predators, and cultural practices (integrated control). Parasitoids can adapt to introduced species (Paropsini parasitoids feeding on *Chrysolina* in Australia) but introduced parasitoids and predators can also be as disastrous as insecticides in devastating the native fauna (as in Guam Island, for instance). Chrysomelids are endangered everywhere, not only because of chemicals, nitrates, urbanization, highway construction and fragmentation of the habitat, but also because many of them are wingless or unable to fly and cannot recolonize an area after destruction. The construction of a new road in Amazonia leads to the destruction of whole forests for miles around. One of us (P.J.) has entitled one of his papers: Apterism, the sadness of not being able to fly (Jolivet, 1994a). *Timarcha*, *Elytrosphaera* (Vasconcellos-Neto and Jolivet, 1998), *Cyrtonastes*, *Cecchiniola* (probably extinct), *Cyrtonus* and many other non-flying species are now almost certainly heading for extinction. Beenen (1999) was probably the first one to plead for leaf beetles and the necessity of special management for them.

Whether it is for the management of a pest or for the conservation of a fast dwindling species, our work will be more effective if we are familiar with the biology of the species concerned. We hope that, through this book, we have introduced the reader to the biology of a very interesting group of insects. Our work will perhaps provide a basis on which a particular chrysomelid may be studied in greater detail and a suitable pest management or insect conservation strategy could be drawn up. While protection of our food and food crops is important, the conservation of biodiversity is also vital in order to maintain the fragile equilibrium of nature and to protect against the disappearance of a species. Fluttering of the proverbial butterfly in Beijing may induce a tempest in America. The loss of these living jewels, a buprestid or the colourful and metallic chrysomelids, would certainly make the world a duller and less attractive place.

References

ABDULALI, H. (1948). Bat migration in India and other notes on bats. *J. Bombay Nat. Hist. Soc.* **47** (3), 52.

ABELOOS, M. (1933a). Anomalies des pattes de *Timarcha violaceo-nigra* de Geer. *Bull. Soc. Linn. Norm.* **8** (6), 33.

ABELOOS, M. (1933b). Sur le développement et la métamorphose de *Timarcha violaceo-nigra* de Geer. *Bull. Soc. Linn. Norm.* **8** (7), 20–22.

ABELOOS, M. (1933c). Sur la régénération des pattes chez le Coléoptère *Timarcha violaceo-nigra* de Geer. *C.R. Séances Soc. Biol.* **113**, 17–18.

ABELOOS, M. (1935). Diapause larvaire et éclosion chez le Coléoptère *Timarcha tenebricosa* Fab. *C.R. Ac. Sc. Paris* **200**, 2112–2114.

ABELOOS, M. (1937a). Mécanisme de l'éclosion des œufs des Timarchas. *Bull. Soc. Sc. Bretagne* **14** (1–2), 70–74.

ABELOOS, M. (1937b). Sur la biologie et l'hibernation des Timarchas. *C. R. Soc. Biol.* **124**, 511–513.

ABELOOS, M. (1938a). Les problèmes de l'hibernation des insectes. *Bull. Soc. Sc. Bretagne* **15** (3–4), 125–131.

ABELOOS, M. (1938b). Etude comparative de la croissance dans deux espèces de *Timarcha* (Col. Chrys.). *Travaux Station Zool. Wimereux* **13**, 1–16.

ABELOOS, M. (1939). Etude biométrique des caractères sexuels secondaires (tarses) dans deux espèces de Timarchas (Col.). *C. R. Soc. Biol.* **2** (131), 563–565.

ABELOOS, M. (1941). Diapause embryonnaire inconstante chez le coléoptère *Timarcha violaceo-nigra* de Geer. *C. R. Acad. Sc. Paris* **212** (17), 722–724.

ADAIR, R.J. AND SCOTT, J.K. (1993). Biology and host specificity of *Ageniosa electoralis* (Col. Chrys.), a prospective biological control agent for *Chrysanthemoides monilifera* (Asteraceae). *Biological Control* **3** (3), 191–198.

AEC (1989). Abstracts volume. International Congress of Coleopterology, pp 1–156. Proc. Int. Congr. Coleop., Barcelona, Spain.

AGARWALA, B.K. (1991). Why do ladybirds (Col. Cocc.) cannibalize? *Journal of Biosciences* **162** (3), 103–109.

AGARWALA, B.K. AND DIXON, A.F.G. (1993). Kin recognition: egg and larval cannibalism in *Adalia bipunctata* (Col. Cocc.). *European Journal of Entomology* **90** (1), 45–50.

AIELLO, A. AND JOLIVET, P. (1997). Myrmecophily in Keroplatidae (Dipt. Siaroidea). *J. New York Entomol. Soc.* **104** (3–4), 226–230.

ALYOKHIN, A.V. AND FERRO, D.N. (1999). Reproduction and dispersal of summer-generation of Colorado Potato Beetle (Col. Chrys.). *Environ. Entomol.* **28** (3), 425–430.

ANATHAKRISHNAN, T.N. AND VISHWANATHAN, T.R. (1976). *General animal ecology*. Pp 1–324. Delhi: Macmillan Co. of India Ltd.

ANDERSEN, J.F., PLATTNER, R.D. AND WEISLEDER, D. (1988). Metabolic transformation of cucurbitacins by *Diabrotica virgifera virgifera* LeConte and *D. undecimpunctata howardi* Barber. *Ins. Biochem.* **98**, 71–77.

ANDERSON, J.M. (1950). A cytological and histological study of testicular cyst cells in the Japanese beetle. *Physiol. Zool.* **23**, 308–316.

ANDRADE, J.C. DE (1981). *Biologia de* Cecropia lyratiloba *Miq. var.* nana *Andr. & Car. (Moraceae) na restinga do recreio dos Bandeirantes*. Pp 1–71. Thesis, Rio de Janeiro.

ANGIOSPERM PHYLOGENY GROUP (1998). An ordinal classification for the families of flowering plants. *Ann. Miss. Bot. Garden* **85**, 531–553.

ANGUS, R.B. (1965). A note on the swimming of *Bagous limosus* (Gyll.) (Coleoptera Curculionidae). *Entom. Month. Mag.* **101**, 102.

ARNETT, R.H. (1968a). *The beetles of the United States*. xii+11112 pp. Ann Arbor, MI, USA: American Entomological Institute.

ARNETT, R.H. (1968b). Pollen feeding by Oedemeridae (Col.). *Bull. Entomol. Soc. Am.* **14**, 184.

ARNETT, R.H. (1993). *American insects. A handbook of the insects of America, North of Mexico*. Pp 1–850. Gainesville, FL, USA: The Sandhill Crane Press, Inc.

ARNQUIST, G., EDVARDSSON, M., FRIBERG, U. AND NILSSON, T. Sexual conflict promotes speciation in insects. *PNAS* **97** (19), 10460–10464.

ASKEVOLD, J.S. (1990). Reconstructed phylogeny and reclassification of the genera of Donaciinae (Col. Chrys.). *Questiones Entomologicae* **26**, 601–664.

ATYEO, W.T., WEEKMAN, G.T. AND LAWSON, D.E. (1964). The identification of *Diabrotica* species by chorion structure. *J. Kansas Entomol. Soc.* **37**, 9–11.

BACETTI, H. AND DACCORDI, M. (1988). Sperm structure and phylogeny of the Chrysomelidae. In: *Biology of Chrysomelidae*. Eds. P. Jolivet, E. Petitpierre and T.H. Hsiao, pp 357–378. Dordrecht, The Netherlands: Kluwer Academic Publishers.

BALACHOWSKI, A.S. (1951). *La lutte contre les insectes*. Pp 1–380. Paris: Payot Publishers.

BALACHOWSKI, A.S. (1963). *Entomologie appliquée à l'agriculture. 1 (2): Coléoptères*. Pp 568–1391. Paris: Masson and Cie. Publishers.

BALAZUC, J. (1948). La tératologie des Coléoptères et expériences de transplantation sur *Tenebrio molitor* L. *Mem. Mus. Nat. Hist. Nat.* **25**, 1–293.

BALAZUC, J. (1988). Laboulbéniales (Ascomycetes) parasitic on Chrysomelidae. In: *Biology of Chrysomelidae*. Eds. P. Jolivet, E. Petitpierre and T.H. Hsiao, pp 389–398. Dordrecht, The Netherlands: Kluwer Academic Publishers.

BALCELLS, E. (1946). Estudio bioclimatologico de *Plagiodera versicolora* Laich. (Col. Chrys.). *P. Inst. Biol. Apl.* **3**, 57–77.

BALCELLS, E. (1953). Estudio biologico de *Haltica lythri* subespecie *ampelophaga* Guérin-Méneville (Col. Halt.). *P. Inst. Biol. Apl.* **14**, 5–54.

BALCELLS, E. (1954). Estudio ecologico de *Haltica lythri* subespecie *ampelophaga* Guérin-Méneville (Col. Halt.). *P. Inst. Biol. Apl.* **17**, 5–37.

BALCELLS, E. (1955). Estudio ecologico de los crisomelididos del aliso. *P. Inst. Biol. Apl.* **20**, 47–61.

BALSBAUGH, E.U. (1967). Possible mimicry between certain Carabidae and Chrysomelidae. *Col. Bull.* **21**, 139–140.

BALSBAUGH, E.U. (1988). Mimicry and the Chrysomelidae. In: *Biology of Chrysomelidae*. Eds. P. Jolivet, E. Petitpierre and T.H. Hsiao, pp 261–284. Dordrecht, The Netherlands: Kluwer Academic Publishers.

BALSBAUGH, E.U. AND FAUSKE, G. (1991). Possible Muellerian mimicry of Galerucinae with Criocerinae (both Col. Chrys.) and with *Maepha opulenta* (Lep. Arctiidae). *Col. Bull.* **45** (3), 227–231.

BAMEUL, F. (1999). Observations sur l'Altise *Chaetocnema aerosa* (Letzner): distribution, habitat, plantes associées et adaptation au milieu aquatique. *Nov. Rev. Ent. (N.S.)* **16** (3), 199–209.

BAO, S.N. (1991). Morphogenesis of the flagellum in the spermatids of *Coelomera lanio* (Col. Chrys.). Ultrastructural and cytochemical studies. *Cytobios* **66**, 157–167.

BAO, S.N. (1996). Spermiogenesis in *Coelomera lanio* (Chrys., Galerucinae): Ultrastructural and cytochemical studies. In: *Chrysomelidae biology 3. General studies*. Eds. P. Jolivet and M.L. Cox, pp 119–132. Amsterdam, The Netherlands: SPB Academic Publishers.

BARBOSA, M.G.V., FONSECA, C.R.V. AND GUERRA, J.A.O. (1999). Occurrence of *Spaethiella coccinea* Boheman (Col. Chrys. Hispinae) on *Theobroma grandiflorum* (Wildenow ex Sprengel) Schumman (Sterculiaceae) in Manaus, Amazonas, Brazil. *Acta Amazonica* **29** (2), 313–317.

BARTH, R. (1954). O aparelho saltatoria do halticineo *Homophoeta sexnotata* Har. (Col.). *Mem. Inst. O. Cruz.* **52**, 365–376.

BARTLET, E., ISIDORO, N. AND WILLIAMS, I.H. (1994). Antennal glands in *Psylliodes chrysocephala* and their possible role in reproductive behaviour. *Phys. Entomol.* **19**, 241–250.

BASILEWSKY, P. (1972). La faune terrestre de l'île de Sainte-Hélène. 33. Chrysomelidae. *Ann. Mus. R. Afr. Centr. Tervueren, Zool.* **192**, 238–247.

BASSET, Y. (1992). Host specificity of arboreal and free-living insect herbivores in rain forests. *Biol. J. Linn. Soc.* **47**, 115–133.

BASSET, Y. (1994). Palatability of tree foliage to chewing insects: a comparison between temperate and a tropical site. *Acta Oecologica* **15**, 181–191.

BASSET, Y. (1996). Local communities of arboreal herbivores in Papua New Guinea: Predictors of insect variables. *Ecology* **77**, 1906–1919.

BASSET, Y. AND SAMUELSON, G.A. (1996). Ecological characteristics of an arboreal community of Chrysomelidae in Papua New Guinea. In: *Chrysomelidae biology 2. Ecological studies.* Eds. P. Jolivet and M.L. Cox, pp 243–262. Amsterdam, The Netherlands: SPB Academic Publishers.

BATRA, S.W.T., SCHROEDER, D., BOLDT, P.E. AND MENDL, W. (1986). Insects associated with Purple Loosestrife (*Lythrum salicaria* L.) in Europe. *Proc. Entomol. Soc. Wash.* **88** (4), 748–759.

BAUER, F., REDTENBACHER, J. AND GANGLBAUER, L. (1889). Fossile Insekten aus der Juraformation Ost-Sibiriens. *Mem. Acad. Sc. St Petersburg* **36** (15), 1–22.

BECERRA, J.X. (1994a). Chrysomelid behavioral counterploys to secretive canals in plants. In: *Novel aspects of biology of Chrysomelidae.* Eds. P. Jolivet, M.L. Cox and E. Petitpierre, pp 327–330. Dordrecht, The Netherlands: Kluwer Academic Publishers.

BECERRA, J.X. (1994b). Squirt-gun defense in *Bursera* and the Chrysomelid counterploy. *Ecology* **75** (7), 1991–1996.

BECERRA, J.X. (1997). Insects on plants: Macroevolutionary chemical trends in host use. *Science* **276**, 253–256.

BECERRA, J.X. AND VENABLE, D.L. (1999). Macroevolution of insect–plant associations: the relevance of host biogeography to host affiliation. *PNAS* **96** (22), 12626–12631.

BECHYNE, J. (1952). Los insectos de las Islas Juan Fernandez. II. Alticinae. *Rev. Chilena Entom.* **2**, 117–118.

BECHYNE, J. (1959). Beitrage zur Kenntnis des Alticidenfauna Boliviens (Col. Phyt.). *Neotrop. Fauna* **1** (4), 269–381.

BECHYNE, J. (1997). Evaluacion de las datos sobre los phytophaga dañinos en Venezuela (Col.) I and II. Ed. V. Savini. *Bull. Entomol. Venezolana. Serie Monografias* **N°1**, 1–459.

BECHYNE, J. AND BECHYNE, B. (1966). *Lista preliminar de los phytophaga de importancia agricola en Venezuela.* Maracaibo, 5 pp.

BECHYNE, J. AND BECHYNE, B. (1969). La posicion sistematica de *Megascelis* Chevrolat (Col. Phyt.). *Revta. Fac. Agron. Univ. Centr. Venez.* **5** (3), 65–76.

BECHYNE, J. AND BECHYNE, B. (1971a). Elementos faunisticos en el Peru en base de los Coleoptera phytophaga. *Rev. Per. Entom.* **14** (1), 192–194.

BECHYNE, J. AND BECHYNE, B. (1971b). Reconocimiento faunistico de los Coleoptera Phytophaga en el Peru. *Rev. Per. Entom.* **14** (1), 195–196.

BECHYNE, J. AND BECHYNE, B. (1973). Notas sobre algunos phytophaga de origen paleantartico. *Rev. Chil. Ent.* **7**, 25–30.

BECHYNE, J. AND BECHYNE, B. (1977). Zur Phylogenesis einiger neotropischen Alticiden (Col. Phytophaga). *Studies Neotrop. Fauna Environ.* **12** (2), 81–145.

BECKER, M. (1994). The female organs of symbiont transmission in the Eumolpinae. In: *Novel aspects of the biology of Chrysomelidae.* Eds. P. Jolivet, M.L. Cox and E. Petitpierre, pp 363–370. Dordrecht, The Netherlands: Kluwer Academic Publishers.

BECKER, M. AND FRIERO-COSTA, F.A. (1988). Natality and mortality in the egg stage in *Gratiana spadicea* (Klug, 1829) (Col. Chrys. Cass.), a monophagous cassidine beetle of an early successional Solanaceae. *Rev. Bras. Biol.* **48**, 467–475.

BECKER, M. AND PIRES FREIRE, A.J. (1996). Population ecology of *Gratiana spadicea* (Klug), a monophagous Cassidine on an early successional Solanaceae in Southern Brazil. In: *Chrysomelidae biology 2. Ecological studies.* Eds. P. Jolivet and M.L. Cox, pp 271–287. Amsterdam, The Netherlands: SPB Academic Publishers.

BEENEN, R. (1998). Patterns in the distribution of Galerucinae in the Netherlands. In: *Proceedings XX ICE, Firenze, Italy, 1996. Mus. Reg. Sc. Nat. Torino.* Eds. M. Biondi, M. Daccordi and D.C. Furth, pp 7–11.

BEENEN, R. (1999). Possibilities for conservation and rehabilitation of populations of

Chrysomelidae in a cultivated environment. In: *Advances in Chrysomelidae biology. I.* Ed. M.L. Cox, pp 307–319. Leiden, The Netherlands: Backhuys Publishers.

BEGOSSI, A. AND BENSON, W.W. (1988). Host-plants and defense mechanisms in Oedionychina (Alticinae). In: *Biology of Chrysomelidae*. Eds. P. Jolivet, E. Petitpierre and T.H. Hsiao, pp 57–71. Dordrecht, The Netherlands: Kluwer Academic Publishers.

BERGEAL, M. AND DOGUET, S. (1992). Catalogue des Coléoptères de l'Ile de France. *ACOREP, Paris* **15**, 1–78.

BERNAYS, E.A. AND CHAPMAN, R.F. (1994). *Host-plant selection by phytophagous insects*. xiii+312 pp. New York, USA: Chapman and Hall.

BERNAYS, E. AND GRAHAM, M. (1988). On the evolution of host specificity in phytophagous arthropods. *Ecology* **69** (4), 886–892.

BERTI, N. AND MARIAU, D. (1999). *Coelaenomenodera lameensis* n. sp. ravageur des palmiers à huile (Col. Chrys.). *Nouv. Rev. Ent. (N.S.)* **16** (3), 253–267.

BERTRAND, H. (1924). Eclosion de l'œuf chez quelques Chrysomelides. *Bull. Soc. Ent. Fr.*, 54–57.

BHATTACHARYA, S. AND VERMA, K.K. (1982). Ovariole structure and adult diapause in the tortoise beetle, *Aspidomorpha miliaris* F. (Col. Chrys.). *Entomologist's Monthly Mag.* **118**, 101–109.

BIENKOWSKI, A.O. (1996). Life cycles of Donaciinae (Col. Chrys.). In: *Chrysomelidae biology. 3. General studies*. Eds. P. Jolivet and M.L. Cox, pp 155–171. Amsterdam, The Netherlands: SPB Academic Publishers.

BIENKOWSKI, A.O. (1999a). Mating behaviour in Donaciinae (Col. Chrys.). In: *Advances in Chrysomelidae biology*. Ed. M.L. Cox, pp 411–420. Leiden, The Netherlands: Backhuys Publishers.

BIENKOWSKI, A.O. (1999b). Guide to the identification of leaf-beetles (Col. Chrys.) of Eastern Europe. *Moscow*, 1–204.

BIONDI, M. (1994). Contribution à l'histoire naturelle de l'île de Chypre (Col. Alt.). *Biocosme Mésogéen, Nice* **11** (1), 9–25.

BIONDI, M. (1999). The Black *Longitarsus* species associated with Boraginaceae in South Africa (Col. Chrys. Alt.). In: *Advances in Chrysomelidae biology. 1. The classification, phylogeny and genetics*. Ed. M.L. Cox, pp 515–531. Leiden, The Netherlands: Backhuys Publishers.

BIONDI, M., DACCORDI, M. AND FURTH, D.G. (Eds.) (1998). *Proceedings of the Fourth International Symposium on the Chrysomelidae*. Pp 1–327. Mus. Reg. Sci. Nat. Torino.

BLOSSEY, B., SCHROEDER, D., HIGHT, S.D. AND MALECKI, R.A. (1994). Host specificity and environmental impact of two leaf beetles (*Galerucella calmariensis* and *G. pusilla*) for biological control of purple loosestrife (*Lythrum salicaria*). *Weed Science* **42** (1), 134–140.

BLUM, M.S. (1987). Biosynthesis of arthropod exocrine compounds. *Ann. Rev. Entomol.* **32**, 381–413.

BLUM, M.S. (1994). Antipredator devices in larvae of the Chrysomelidae, a unified synthesis for defensive eclectism. In: *Novel aspects of the biology of Chrysomelidae*. Eds. P. Jolivet, M.L. Cox and E. Petitpierre, pp 277–288. Dordrecht, The Netherlands: Kluwer Academic Publishers.

BLUM, M.S., BRAND, J.M., WALLACE, J.B. AND FOLES, H.M. (1972). Chemical characterization of the defensive secretion of a chrysomelid larva (*Chrysomela interrupta*). *Life Sci.* **2**, 525–531.

BLUM, M.S., WALLACE, J.B., DUFFIELD, R.M., BRAND, J.M., FALES, H.M. AND SOKOLOSKI, E.A. (1978). Chrysomelidial in the defensive secretion of the leaf beetle *Gastrophysa cyanea* Meslheimer. *J. Chem. Ecol.* **4** (1), 47–53.

BÖCHER, J. (1988). The Coleoptera of Greenland. *Meddel. Grönl. Biosc.* **26**, 1–100.

BÖCHER, J. (1989). Boreal insects in northernmost Greenland: palaeoentomological evidence from the Kap Köbenhavn formation (Plio-Pleistocene), Peary Land. *Fauna Norv.* Ser. B **36**, 37–43.

BOITEAU, G. (1988a). Control of the Colorado potato beetle, *L. decemlineata* (Say): learning from the Soviet experience. *Bull. Entomol. Soc. Can.* **20** (1), 9–14.

BOITEAU, G. (1988b). Sperm utilization and post-copulatory female guarding in the Colorado potato beetle, *Leptinotarsa decemlineata*. *Entom. Experiment. & Appl.* **47**, 183–187.

BOITEAU, G. (1998). Reproductive barriers between the partially sympatric Colorado and false potato beetles. *Entom. Experiment. & Appl.* **89** (2), 147–153.

BOLDT, P.E. (1989). Host specificity studies of *Stolas fuscata* (Klug) (Col. Chrys.) for the biological control of *Baccharis salicifolia* (R. and P.) Pers. (Asteraceae). *Proc. Entomol. Soc. Washington* **91** (4), 502–508.

BOLDT, P.E. AND STAINES, C.L. (1993). Biology and description of immature stages of *Pentispa suturalis* (Baly) (Col. Chrys.) on *Baccharis bigelovii* (Asteraceae). *Col. Bull.* **47** (2), 215–220.

BOLDT, P.E. AND WHITE, R.E. (1992). Life history and larval description of *Exema elliptica* Karren (Col. Chrys.) on *Baccharis halimifolia* L. (Asteraceae) in Texas. *Proc. Entomol. Soc. Washington* **94** (1), 83–90.

BOLDT, P.E., CORDO, H.A. AND GANDOLFO, D. (1991). Life history of *Stolas* (*Anacassis*) *fuscata* (Klug) (Col. Chrys.) on seepwillow, *Baccharis salicifolia* (R. and P.) Pers. (Asteraceae). *Proc. Entomol. Soc. Washington* **93** (4), 839–844.

BOLOGNA, M.A. AND MARANGONI, C. (1986). Sexual behaviour in some palaearctic species of *Meloe* (Col. Mel.). *Bol. Soc. Ent. Ital., Genova* **118** (4–7), 65–82.

BOLSER, R.C. AND HAY, M.E. (1998). A field test of inducible resistance to specialist and generalist herbivores using the water lily *Nuphar luteum*. *Oecologia* **116** (1–2), 143–153.

BONDAR, G. (1931). Notas biologicas sobre alguns hispineos observados na Bahia. *O Campo* **2** (6), 74–75.

BONDAR, G. (1938). Hispineos (Col. Chrys.) da Bahia e ssuas plantas hospedeiras. *Rev. Entom.* **8** (1–2), 17–20.

BONTEMS, C. (1985). La viviparité chez les Chrysomelinae. *Bull. Soc. Ent. Fr.* **89**, 973–981.

BONTEMS, C. (1988). Localization of spermatozoa inside viviparous and oviparous females of Chrysomelinae. In: *Biology of Chrysomelidae*. Eds. P. Jolivet, E. Petitpierre and T.H. Hsiao, pp 299–316. Dordrecht, The Netherlands: Kluwer Academic Publishers.

BOPPRÉ, M. AND SCHERER, G. (1981). A new species of Flea Beetle (Alticinae) showing male-biased feeding at withered *Heliotropium* plants. *Syst. Entomol.* **6** (4), 347–354.

BORDY, B. (2000). Coléoptères Chrysomelidae. IV. Hispinae and Cassidinae + S. Doguet Addenda and Corrigenda. II. Alticinae. *Faune de France, Paris* **85**, xii + 250.

BORGES, P.A.V. (1990). A checklist of the Coleoptera from the Azorea with some systematic and biogeographic comments. *Bol. Mus. Mun. Funchal* **42**, 87–136.

BORGES, P.A.V. (1992). Biogeography of the Azorean Coleoptera. *Bol. Mus. Mun. Funchal* **44**, 5–76.

BOROWIEC, L. (1999). *A world catalogue of the Cassidinae (Col. Chrys.)*. 470 pp. Polish Taxonomical Society, Wroclaw.

BORROR, D.J., DELONG, D.M. AND TRIPLEHORN, C.A. (1976). *An introduction to the study of insects*. 852 pp. New York, USA: Holt, Rinehart and Winston Publishers.

BOURDON, J. (1937). Recherches expérimentales sur la régénération chez un Coléoptère *Timarcha goettingensis* L. *Bull. Biol. France Belg.* **71** (4), 466–499.

BOURDOUXHE, L. AND JOLIVET, P. (1981). Nouvelles observations sur le complexe mimétique de *Mesoplatys cincta* Olivier au Sénégal. *Bull. Soc. Linn. Lyon* **50** (2), 46–48.

BOUSQUET, Y. (1991). Checklist of beetles of Canada and Alaska. *Agriculture Canada, Ottawa* **1861/E**, 300–323.

BÖVING, A.G. AND CRAIGHEAD, F.C. (1930). An illustrated synopsis of the principal larval forms of the order Coleoptera. *Entom. Americana* **11** (1), 1–351.

BRANDSTETTER, C.M. AND KAPP, A. (1996). Die Blatt- und Samenkäfer von Voralberg und Liechtenstein. II. *Bürs*, 1–845.

BRAUER, F., REDTENBACHER, J. AND GANGLBAUER, L. (1889). Fossil Insekten aus der Juraformation Ost-Siberien. *Mém. Acad. Imp. Sc. St Petersbourg* (séries 7) **36** (15), 1–22.

BREDEN, F. AND WADE, M.J. (1985). The effect of group size and cannibalism rate on larval growth and survivorship in *Plagiodera versicolora*. *Entomography* **3**, 455–463.

BREDEN, F. AND WADE, M.J. (1987). An experimental study of the effect of group size on larval growth and survivorship in the imported willow leaf beetle *Plagiodera versicolora* (Col. Chrys.). *Environmental Entomology* **16** (5), 1082–1086.

BREDEN, F. AND WADE, M.J. (1989). Selection within and between kin groups of the imported willow leaf beetle: *Plagiodera versicolora*. *American Naturalist* **134**, 35–50.

BRENDELL, M.J., DACCORDI, M. AND SHUTE, S.L. (1993). On the systematic position of the genus *Brachyhelops* Fairmaire (Col. Chrys.). *Boll. Mus. Civ. Sc. Nat. Verona* **17**, 265–276.

BROWN, K.S. (1999). Deep green rewrites evolutionary history of plants. *Science* **285**, 990–991.

BROWN, L. AND SIEGFRIED, B.D. (1983). Effects of male horn size on courtship activity in the forked fungus beetle, *Bolitotherus cornutus* (Col. Ten.). *Ann. Entomol. Soc. Amer.* **76** (2), 253–255.

BROWN, W.D. (1993). The cause of size-assortative mating in the leaf beetle, *Trirhabda canadensis* (Col. Chrys.). *Behav. Ecol. Sociobiol.* **33** (3), 151–157.

BROWN, W.J. (1940). Notes on the American distribution of some species of Coleoptera common to the European and North American continents. *Canad. Entomol.* **72** (4), 65–78.

BROWN, W.J. (1943). The Canadian species of *Exema* and *Arthrochlamys*. *Canad. Entomol.* **75**, 119–131.

BROWN, W.J. (1944). The dimorphism in the male copulatory organ of the chrysomelid *Arthrochlamys bebbianae* Brown. *Canad. Entomol.* **76**, 70–72.

BROWN, W.J. (1945). Food plants and distribution of the species of *Calligrapha* in Canada with descriptions of new species (Col. Chrys.). *Canad. Entomol.* **77**, 117–133.

BROWN, W.J. (1950). The extralimital distribution of some species of Coleoptera. *Canad. Entomol.* **82** (10), 197–205.

BROWN, W.J. (1952). New species of Phytophaga (Col.). *Canad. Entomol.* **84**, 335–342.

BROWN, W.J. (1956). The New World species of *Chrysomela* L. *Canad. Entomol.* **88** Suppl. 3, 1–54.

BROWN, W.J. (1959). Taxonomic problems with closely related species. *Ann. Rev. Ent.* **4**, 77–98.

BROWN, W.J. (1962). The American species of *Chrysolina* Motsch. (Col. Chrys.). *Canad. Entomol.* **94**, 58–74.

BRUCK, T. (1978). *The structure of the male genital system of* Alagoasa bicolor *(L.) (Col. Chrys.) with special regard to sperm transportation*. 114 pp. M.Sc. Thesis, Univ. Puerto Rico, Rio Piedras, Puerto Rico.

BUCKINGHAM, G.R. AND BUCKINGHAM, M. (1981). A laboratory biology of *Pseudolampsis guttata* (Leconte) (Col. Chrys.) on waterfern, *Azolla caroliniana* Willd. (Pterid. Azollaceae). *Col. Bull.* **35** (2), 181–188.

BUCKINGHAM, G.R., BOUCIAS, D. AND THERIOT, R.F. (1983). Reintroduction of the Alligatorweed Flea Beetle (*Agasicles hygrophila* Selman and Vogt) into the United States from Argentina. *J. Aquat. Plant. Management* **21**, 101–102.

BUCKINGHAM, G.R., HAAG, K.H. AND HABECK, D.H. (1986). Native insect enemies of aquatic macrophytes. Beetles. *Aquatics* **8** (2), 28–34.

BUDHRAJA, K., RAWAT, R.R. AND SINGH, O.P. (1979). Feeding behavior of *Dicladispa armigera*. *International Rice Newsletter* **4** (6), 15–16.

BURGES, W.C. (1981). Why are there so many kinds of flowering plants? *BioScience* **31** (8), 572–581.

BUZZI, Z.J. (1988). Biology of neotropical Cassidinae. In: *Biology of Chrysomelidae*. Eds. P. Jolivet, E. Petitpierre and T.J. Hsiao, pp 559–580. Dordrecht, The Netherlands: Kluwer Academic Publishers.

BUZZI, Z.J. AND DJUNKO MIYAZAKF, R.D. (1999). Description of immature stages and life cycle of *Stolas lacordairei* (Boheman) (Col. Chrys. Cass.). In: *Advances in Chrysomelidae biology* . I. Ed. M.L. Cox, pp 581–597. Leiden, The Netherlands: Backhuys Publishers.

CARNE, P.B. (1966). Ecological characteristics of the Eucalypt-defoliating chrysomelid *Paropsis atomaria* Ol. *Austr. J. Zool.* **14**, 647–672.

CARPENTER, F.M. (1992). Arthropoda 4- Superclass Hexapoda. In: *Treatise on invertebrate paleontology*. Pp 1–655. Boulder, CO, USA: The Geol. Soc. America.

CHABROL, L. (1995). *Chaetocnema confinis* Crotch, altise nouvelle pour les Iles de la Société (Col. Chrys.). *Nouv. Rev. Entom. (N.S.)* **11** (4), 380.

CHAMPION, G.C. AND CHAPMAN, T.A. (1901). Observations on some species of *Orina*, a genus of viviparous and ovo-viviparous beetles. *Trans. Ent. Soc. London* 1–17.

CHAPUIS, F. (1874). *Genera des Coléoptères*. Pp 1–452. Paris.
CHEN, S.H. (1934a). Sur la position systématique du genre *Timarcha* Latr. *Bull. Soc. Ent. Fr.* **39** (2), 35–39.
CHEN, S.H. (1934b). *Recherches sur les Chrysomelidae de la Chine et du Tonkin*. Thesis, Fac. Des Sciences, Paris Univ. and Ann. Soc. Ent. Fr. **104**, 127–145; **105**, 145–176; **106**, 283–323.
CHEN, S.H. (1940). Attempt at a new classificationof the leaf beetles. *Sinensia* **11** (5–6), 451–481.
CHEN, S.H. (1943). The relation of elytra to flight in the cucumber beetles. *Sinensia* **14**, 33–35.
CHEN, S.H. (1946). Evolution of the insect larva. *Trans. R. Ent. Soc. London* **97** (15), 381–404.
CHEN, S.H. (1964). Evolution and classification of the chrysomelid beetles. *Acta Entom. Sinica* **13** (4), 469–483.
CHEN, S.H. (1973). The classification of leaf-beetles. *Acta Ent. Sinica* **16** (1), 47–56.
CHEN, S.H. (1985). Phylogeny and classification of the Chrysomeloidea. *Entomography* **3**, 465–475.
CHEN, S.H. AND WANG, S.Y. (1962). On the distribution and desert adaptations of the Chrysomelid beetles of Sinkiang. *Acta Zool. Sinica* **14** (3), 337–354.
CHEN, S.H. AND WANG, S.Y. (1984a). Flea-beetles from Hengduan Mountains, Yunnan. Genera *Hespera* and *Yunohespera* (Col. Chrys.). *Acta Ent. Sinica* **27** (3), 308–322.
CHEN, S.H. AND WANG, S.Y. (1984b). New chrysomeline beetles from Hengduan mountains, Yunnan. *Acta Zootaxon. Sinica* **9** (2), 170–175.
CHEN, S.H. AND WANG, S.Y. (1987). Two alpine genera of the flea beetles from Yunnan (Col. Chrys.). *Acta Zootaxon. Sinica* **9** (2), 170–175.
CHEN, S.H. AND ZIA, Y. (1966). On the Chinese citrus fleabeetles. *Acta Zootaxon. Sinica* **3**, 67–75.
CHEN, S.H., JIANG, S.Q. AND WANG, S.Y. (1987). New alpine Galerucinae from Hengduan mountains of Yunnan and Sichuan (Col. Chrys.). *Sinozoologia* **5**, 61–71.
CHERNOV, Y.I. (1973). Brief review of trophic groups of invertebrates of the subzone of typical tundras of western Taimyr island. *Bioneotsenozyu Taimyrska Tundry Iikh Produktionost, Nauka, Leningrad*, 166–179.
CHERNOV, Y.I. (1978). Adaptive peculiarities of life cycles of insects in the tundra zone. *Zh. Obsch. Biol.* **39** (3), 394–402.
CHERNOV, Y.I., MEDVEDEV, L.N. AND KHRULEVA, O.A. (1994). Leaf beetles (Col. Chrys.) in the Arctic. *Entom. Rev.* **73** (2), 152–167.
CHEVIN, H. (1965). Caractères biologiques et écologiques de *Timarcha normanna* Reiche. *Bull. Soc. Ent. Fr.* **70** (11–12), 250–258.
CHEVIN, H. (1985–1987). Contribution à la biologie des *Timarcha*. *Cah. Liaison OPIE* **56**, 7–10; **57**, 7–14; **60**, 17–21; **64**, 21–25.
CHEVIN, H. (1994). Food selection and life-cycle of the Old World *Timarcha* Latreille 1829 (Col. Chrys.). In: *Novel aspects of the biology of Chrysomelidae*. Eds. P. Jolivet, M.L. Cox and E. Petitpierre, pp 533–539. Dordrecht, The Netherlands: Kluwer Academic Publishers.
CHEVIN, H. AND TIBERGHIEN, G. (1968). Existence d'un régime alimentaire mixte chez *Timarcha maritima* Perris (Col. Chrys.). *Bull. Soc. Ent. Fr.* **73** (11–12), 203–206.
CHOE, J.C. (1989). Maternal care in *Labidomera suturella* Chevrolat (Col. Chrys.) from Costa Rica. *Psyche* **96**, 63–67.
CHOWN, S.L. AND SCHOLTZ, C.H. (1989). Curculionidae (Col.) from the sub-antarctic Prince Edward islands. *Col. Bull.* **43** (2), 165–169.
CHU, H.F. (1949). *How to know the immature insects?* 234 pp. Dubuque, Iowa: Brown Co.
CHÛJÔ, M. (1953, 1954). A taxonomic study on the Chrysomelidae with special reference to the fauna of Formosa. *Techn. Bull. Kagawa Agric. College* **5** (1), 19–36; **5** (2), 121–136.
CHÛJÔ, M. (1959). Chrysomelid beetles of Japan (2): subfamily Synetidae. *Ent. Rev. Japan* **10** (1), 18–23.
CLARK, A.P. (1930). *Paropsis dilatata* Er. in New Zealand. Preliminary account. *New Zealand J. Sci. Techn.* **12**, 114–123.
CLAVAREAU, H. (1913). Chrysomelidae. Sagrinae. Donaciinae. Orsodacninae. Criocerinae. In: *Coleopterorum Catalogus*. Pp 1–103. Berlin: Junk Publs.

CLOUDSLEY-THOMPSON, J.L. (1996). *Biotic interactions in arid lands*. Pp 1–208. Berlin: Springer-Verlag.
CLOUDSLEY-THOMPSON, J.L. (1999). Multiple factors in the evolution of animal coloration. *Naturwissenchaften* **86**, 123–132.
CLOUDSLEY-THOMPSON, J.L. (2000). *Thermal and water relations of desert beetles*. 39 pp.
COMPTON, S.G. AND WARE, A.B. (1991). Ants disperse the elaiosome bearing eggs of an African stick insect. *Psyche* **198**, 207–213.
CONNIN, R.V., JANTZ, O.K. AND BOWERS, W.S. (1967). Termination of diapause in the cereal leaf bettle by hormones. *J. Econ. Entomol.* **60**, 1752–1753.
CORNET, B. (1994). Dicot-like leaf and flowers from the late Triassic Tropical Newark supergroup rift zone, USA. *Modern Geology* **19**, 81–89.
CORSET, J. (1931). Les coaptations chez les insectes. Thèse, Paris, 235 pp, and Suppl. *Bull. Biol. France et Belgique* **13**, 1–337.
COSTA, C., VANIN, S.A. AND CASARI-CHEN, S.A. (1988). *Larvas de Coleoptera do Brasil*. viii+282 pp. Museum of Zoology, University of Sao Paulo.
COSTA, G. (1995). *Behavioural adaptations of desert animals*. Pp 1–198. Berlin: Springer-Verlag.
COSTA-LIMA, A. DA (1953). Sobre especies sudamericanas de *Phaedon* (Col. Chrys.). *Dusensia* **4** (5–6), 429–432.
COSTA-LIMA, A. DA (1955). Insetos do Brasil. 9 tomo Coleopteros. 3 parte. *Escola Nacional de Agronomia, Seria didactica* **II**, 1–289.
COULSON, J.R. (1977). Biological control of Alligator Weed, 1959–1972. A review and evaluation. *USDA Techn. Bull.* 1547, 98 pp.
COX, M.L. (1976). *The taxonomy and biology of the British Chrysomelidae*. Ph.D. Thesis. Univ. of Newcastle-upon-Tyne.
COX, M.L. (1981). Notes on the biology of *Orsodacne* Latreille with a subfamily key to the larvae of the British Chrysomelidae (Coleoptera). *Entomologist's Gazette* **32** (2), 123–135.
COX, M.L. (1982). Larvae of the British genera of Chrysomeline beetles (Col. Chrys.). *Syst. Entomol.* **7**, 297–310.
COX, M.L. (1988). Egg bursters in the Chrysomelidae, with a review of their occurrence in the Chrysomeloidea and Curculionoidea (Col.). *Syst. Entomol.* **13**, 393–432.
COX, M.L. (1991). The larvae of British *Phaedon* (Col. Chrys.). *Entomologist's Gazette* **42** (4), 267–280.
COX, M.L. (1994a). Egg bursters in the Chrysomelidae with a review of their occurrence in the Chrysomeloidea (Col.). In: *Novel aspects of the biology of Chrysomelidea*. Eds. P. Jolivet, M.L. Cox and E. Petitpierre, pp 75–110. Dordrecht, The Netherlands: Kluwer Academic Publishers.
COX, M.L. (1994b). The Hymenoptera and Diptera parasitoids of Chrysomelidae. In: *Novel aspects of the biology of Chrysomelidae*. Eds. P. Jolivet, M.L. Cox and E. Petitpierre, pp 419–467. Dordrecht, The Netherlands: Kluwer Academic Publishers.
COX, M.L. (1994c). Diapause in the Chrysomelidae. In: *Novel aspects of the biology of Chrysomelidae*. Eds. P. Jolivet, M.L. Cox and E. Petitpierre, pp 469–502. Dordrecht, The Netherlands: Kluwer Academic Publishers.
COX, M.L. (1996a). Insect predators of Chrysomelidae. In: *Chrysomelidae biology. 2. Ecological studies*. Eds. P. Jolivet and M.L. Cox, pp 23–91. Amsterdam, The Netherlands: SPB Academic Publishers.
COX, M.L. (1996b). The pupae of Chrysomeloidea. In: *Chrysomelidae biology. 1. The classification, phylogeny and genetics*. Eds. P. Jolivet and M.L. Cox, pp 119–265. Amsterdam, The Netherlands: SPB Academic Publishers.
COX, M.L. (1996c). Parthenogenesis in the Chrysomeloidea. In: *Chrysomelidae biology. 3. General studies*. Eds. P. Jolivet and M.L. Cox, pp 133–151. Amsterdam, The Netherlands: SPB Academic Publishers.
COX, M.L. (1997). The larva of the flea beetle, *Mniophila muscorum* (Koch, 1803) (Col. Chrys. Alt.), not a leaf miner. *Entomologist's Gazette* **48**, 275–283.
COX, M.L. (1998a). The first instar larva of *Megascelis puella* Lacordaire (Col. Chrys. Megascelinae) and its value on the placement of the Megascelinae. *J. Nat. Hist.* **32**, 387–403.

COX, M.L. (1998b). The pupae of Chrysomeloidea and their use in phylogeny (Coleoptera). In: *Proceedings of the Fourth International Symposium on the Chrysomelidae, Firenze, 1996*. Eds. M. Biondi, M. Daccordi and D.G. Furth, pp 73–90. Museo Reg. Sci. Nat. Torino.

COX, M.L. (Ed.) (1999). *Advances in Chrysomelidae biology. 1*. Pp 1–674 . Leiden, The Netherlands: Backhuys Publishers.

COX. M.L. AND WINDSOR, D.M. (1999a). The first instar of *Aulacoscelis* sp. and *Megascelis puella* Lacordaire (Col. Chrys. Aulacoscelinae, Megascelinae) and their value in the placement of the Aulacoscelinae and Megascelinae. In: *Advances in Chrysomelidae biology. 1*. Ed. M.L. Cox, pp 5–70. Leiden, The Netherlands: Backhuys Publishers.

COX, M.L. AND WINDSOR, D.M. (1999b). The first instar larva of *Aulacoscelis appendiculata* n. sp. (Col. Chrys. Aulacoscelinae) and its value in the placement of the Aulacoscelinae. *J. Nat. Hist.* **33**, 1049–1087.

CRAWFORD, C.S. (1981). *Biology of desert invertebrates*. Pp 1–314. Berlin: Springer-Verlag.

CREPET, W.L. AND FRIS, E.M. (1987). The evolution of insect pollination. In: *The origins of angiosperms and their biological consequences*. Eds. E.M. Fris, W.G. Chaloner and P.R. Crane, 1821 pp. Cambridge University.

CROIZAT, L. (1961). *Principia Botanica*. 1821 pp. Caracas.

CRONQUIST, A. (1981). *An integrated system of classification of flowering plants*. 1262 pp. New York, USA: Columbia University Press.

CRONQUIST, A. (1988). *The evolution and classification of flowering plants*. 555 pp. Bronx, NY: New York Botanical Garden.

CROWSON, R.A. (1946). A revision of the genera of the chrysomelid group Sagrinae. *Trans. R. Entomol. Soc., London* **97**, 75–115.

CROWSON, R.A. (1953). The classification of the families of British Coleoptera. *Ent. Month. Mag.* **89**, 181–198.

CROWSON, R.A. (1960). The phylogeny of Coleoptera. *Ann. Rev. Entomol.* **5**, 111–134.

CROWSON, R.A. (1967). *The natural classification of the families of Coleoptera*. 214 pp. Middlesex, UK: E.W. Classey Publishers.

CROWSON, R.A. (1975). The evolutionary history of Coleoptera as documented by fossil and comparative evidence. *Att. × Congr. Naz. Ital. Entom., Sassari*, 47–90.

CROWSON, R.A. (1980). The amphipolar distribution patterns in some cool climate groups of Coleoptera. *Entomologia Generalis* **6** (2–4), 281–292.

CROWSON, R.A. (1981). *The biology of Coleoptera*. 802 pp. London: Academic Press.

CROWSON, R.A. (1989). *Relations of beetles to cycads*. Eds. Zunino *et al.*, pp 13–28. European Ass. Coleopt., Int. Congress Coleopterol., Barcelona, Spain.

CROWSON, R.A. (1991). *Relations of beetles to cycads*. Pp 13–28. International Congress of Coleopterology. Abstracts Volume. European Association of Coleopterology.

CROWSON, R.A. (1994). A long perspective on Chrysomelid evolution. In: *Novel aspects of the biology of Chrysomelidae*. Eds. P. Jolivet, M.L. Cox and E. Petitpierre, pp xix–xxiii. Dordrecht, The Netherlands: Kluwer Academic Publishers.

CROWSON, R.A. (1998). Green algae and chrysomelid evolution? *Chrysomela* **36,** 4.

CROWSON, R.A. AND CROWSON, E.A. (1996). The phylogenetic relations of Galerucinae-Alticinae. In: *Chrysomelidae biology. 1. The classification, phylogeny and genetics*. Eds. P. Jolivet and M.L. Cox, pp 97–118. Amsterdam, The Netherlands: SPB Academic Publishers.

CUÉNOT, L. (1894). Le rejet de sang comme moyen de défense chez quelques Coléoptères. *C. R. Acad. Sc., Paris* **118**, 875–877.

CUÉNOT, L. (1896a). La saignée réflexe et moyens de défense de quelques insectes. *Arch. Zool. Exp. Gen.* **24**, 655–680.

CUÉNOT, L. (1896b). La saignée réflexe chez les insectes. *Mém. Soc. Scinet. Antonio Alzate* **10**, 39–48.

DACCORDI, M. (1985). *Contributo ad uno studio filogenetico delle Crissomeline ed analisi della loro evoluzione nella regione australiana*. 280 pp. Thesis, Univ. Padova.

DACCORDI, M. (1994). *Notes for phylogenetic study of Chrysomelinae with descriptions of new taxa and a list of all known genera (Col. Chrys.)*. Proc. 3rd. Int. Symp. Chrysomelidae, Beijing, China. Ed. D.G. Furth, pp 60–84. Leiden, The Netherlands: Backhuys Publishers.

DACCORDI, M. (1996). Notes on the distribution of the Chrysomelinae and their possible origin. In: *Chrysomelidae biology. 1. The classification, phylogeny and genetics*. Eds. P. Jolivet and M.L. Cox, pp 399–412. Amsterdam, The Netherlands: SPB Academic Publishers.

DACCORDI, M. AND LESAGE, L. (1999). Revision of the genus *Labidomera* Dejean with a description of two new species (Col. Chrys.). In: *Advances in Chrysomelidae biology. 1.* Ed. M.L. Cox, pp 437–461. Leiden, The Netherlands: Backhuys Publishers.

DALLAI, R., AFZELIUS, B.A., LUZETTI, P. AND OSELLA, G. (1998). Sperm structure of some Curculionidea and their relationship with Chrysomeloidea. In: *Taxonomy, ecology and distribution of Curculionidea*. Pp 27–50. XXICE, Firenze, Italy. Mus. Reg. Sci. Nat. Torino.

DALY, H.V., DOYEN, J.T. AND EHRLICH, P.R. (1978). *Introduction to insect biology and diversity*. Pp 1–564. NY: McGraw Hill.

DAMMAN, H. AND CAPPUCCINO, N. (1991). Two forms of egg defence in a chrysomelid beetle: Egg clumping and excrement cover. *Ecol. Entomol.* **16** (2), 163–167.

DA SILVA, W. (1997). On the trail of the lonesome pine. *New Scientist*, 36–39.

DEGRUGILLIER, M.E. (1996). Ultrastucture and distribution of *Rickettsia*-like organisms in reproductive tissues of the Western Corn Rootworm, *Diabrotica virgifera virgifera*. In: *Chrysomelidae biology. 2. Ecological studies*. Eds. P. Jolivet and M.L. Cox, pp 135–138. Amsterdam, The Netherlands: SPB Academic Publishers.

DEGRUGILLIER, M.E., DEGRUGILLIER, S.S. AND JACKSON, J.J. (1991). Nonoccluded, cytoplasmic virus particles and *Rickettsia*-like organisms in testes and spermathecae of *Diabrotica virgifera*. *J. Invertebr. Pathol.* **57** (1), 50–58.

DEMILO, A.B., LEE, C.J., SCHRODER, R.F.W., SCHMIDT, W.F. AND HARRISON, D.J. (1998). Spectral characterization of cucurbitacins in a bitter mutant of Hawkesbury watermelon (*Citrullus vulgaris* Schrad) that elicit a feeding response to diabroticite beetles. (Col. Chrys.). *J. Entomol. Sci.* **33** (4), 343–354.

DEROE, C. AND PASTEELS, J.M. (1977). Defensive mechanisms against predation in the Colorado beetle (*Leptinotarsa decemlineata* (Say)). *Arch. Biol. (Bruxelles)* **88**, 289–304.

DETTNER, K. (1997). Inter- and intraspecific transfer of toxic insect compound cantharidin. In: *Vertical food web interactions. Ecological studies*. Pp 115–145. Berlin, Germany: Springer-Verlag.

DE WILDE, J. AND DE LOOF, A. (1973). In: *Physiology of insects .1.* Ed. M. Rockstein, 128 pp. New York, USA: Academic Press.

DE WILDE, J., BONGERS, W. AND SCHOONEVELD, H. (1969). Effects of host-plant age on phytophagous insects. *Entom. Experimental. & Applic.* **12**, 714–720.

DHILLON, N.P.S. (1993). The lack of a relationship between bitterness and resistance of cucurbits to red pumpkin beetle (*Aulacophora foveicollis*). *Plant Breeding* **110** (1), 73–76.

DICENTA, A. AND BALCELLS, E. (1963). Notas ecologicas: *Chrysolina* (= *Chrysomela*) *banksi* F. (Col. Chrys.). *Graellsia, Rev. de Entom. Esp.* **20**, 11–117.

DICKINSON, J.L. (1988). Determinants of paternity in the milkweed leaf-beetle. *Behav. Ecol. Sociobiol.* **23**, 9–19.

DICKINSON, J.L. (1992a). Egg cannibalism by larvae and adults of the milkweed leaf beetle (*Labidomera clivicollis*, Col. Chrys.). *Ecol. Entomol.* **17** (3), 209–218.

DICKINSON, J.L. (1992b). Scramble competition polygyny in the milkweed leaf beetle: Combat, mobility and the importance of being there. *Behav. Ecol.* **3** (1), 32–41.

DICKINSON, J.L. (1994). Trade-offs between postcopulatory riding and mate location in the blue milkweed beetle. *Behavioral Ecology* **6** (3), 280–286.

DIZER, Y.B. (1955). On the physiological role of the elytra and the subelytral cavity of steppe and desert Tenebrionidae. *Zool. Journ.* **34** (2), 319–322.

DOBLER, S. AND ROWELL-RAHIER, M. (1994). Production of cardenolides versus sesquestration of pyrrolizidine alkaloids in larvae of *Oreina* species (Col. Chrys.). *J. Chem. Ecol.* **20** (3), 555–558.

DOBLER, S., MARDULYN, P., PASTEELS, J.M. AND ROWELL-RAHIER, M. (1996). Host plant switches and the evolution of viviparity and chemical defense in the leaf beetle genus *Oreina*. *Evolution* **50** (6), 2373–2386.

DODSON, E.O. (1968). *Evolution: process and product* (Indian edition). New Delhi: Affiliated East–West Press.

DOGUET, S. (1984). Contribution à l'étude des espèces d'Afrique du Nord du genre *Phyllotreta* (Col. Chrys.). *Nouv. Rev. Entom. (N.S.)* **1**, 243–265.

DOGUET, S. (1994). Coléoptères Chrysomelidae. 2. Alticinae. *Faune de France* **80**, 1–693.

DOLCH, R. AND TSCHARNTKE, T. (2000). Defoliation of alders (*Alnus glutinosa*) affects herbivory by leaf beetles on undamaged neighbours. *Oecologia* **125**, 504–511.

DONISTHROPE, H.J.K. (1902). The life-history of *Clythra quadripunctata* L. *Trans. Entom. Soc. London* **50**, 11–24.

DRANEY, M.L. (1993). The subelytral cavity of desert tenebrionids. *Florida Entomologist* **76** (4), 539–549.

DUFFEY, S.S. AND PASTEELS, J.M. (1993). Transient uptake of hypericin by chrysomelids is regulated by feeding behaviour. *Physiol. Entomol.* **18**, 119–129.

DURCHON, M. (1946). Les adultes du doryphore peuvent s'attaquer à leurs propres œufs. *C.R. Séances Acad. Sc., Paris* **222**, 340–342.

DUSSOURD, D.E. AND DENNO, R.F. (1991). Deactivation of plant defense: correspondence between insect behavior and secretory canal architecture. *Ecology* **72**, 1383–1396.

DUSSOURD, D.E. AND EISNER, T. (1987). Vein cutting behavior: insect counterploy to the latex defense of plants. *Science* **237**, 898–901.

DUTRILLAUX, B. AND CHEVIN, H. (1969). Etude cytogénétique de *Timarcha goettingensis* L. et de *T. normanna* Reiche. *Bull. Soc. Ent. France* **74** (11–12), 219–224.

EBEN, A. (1999). Host plant breadth and importance of cucurbitacins for the larvae of *Diabrotica* (Galerucinae, Luperini). In: *Advances in Chrysomelidea biology. 1.* Ed. M.L. Cox, pp 361–374. Leiden, The Netherlands: Backhuys Publishers.

EBEN, A. (2000). New host plant record for *Amphelasma cavum* Barber (Chrys. Gal., Luperini). *Coleopt. Bull.* **54** (3), 408.

EBEN, A. AND ESPINOSA DE LOS MONTEROS, A. (2000). The evolution of host plant breadth in Diabroticites (Col. Chrys.). XXI Int. Congress of Entomology, Iguaçu, Brazil, a poster.

EBERHARD, W.G. (1981). The natural history of *Doryphora* sp. (Col. Chrys.) and the function of its sternal horn. *Ann. Entomol. Soc. Am.* **74**, 445–448.

EBERHARD, W.G. (1982). Beetle horn dimorphism: making the best of bad lot. *Amer. Nat.* **119** (3), 420–426.

EBERHARD, W.G. (1991). Copulatory courtship and cryptic female choice in insects. *Biol. Rev.* **66**, 1–31.

EBERHARD, W.G. (1994). Evidence for widespread courtship during copulation in 131 species of insects and spiders and implications for cryptic female choice. *Evolution* **48**, 711–7773.

EBERHARD, W.G. (1997). Sexual selection by cryptic female choice in insects and arachnids. In: *The evolution of mating systems in insects and arachnids.* Eds. J.C. Choe and J. Crespi, J., pp 32–57. Cambridge, UK: Cambridge University Press.

EBERHARD, W.G. AND KARIBO, S.J. (1996). Copulation behavior inside and outside the beetle *Macrohaltica jamaicensis* (Col. Chrys.). *J. Ethol.* **14**, 59–72.

EBERHARD, W.G. AND MARIN, M.C. (1996). Sexual behavior and the enlarged hind legs of male *Megalopus armatus* (Col. Chrys. Megal.). *J. Kansas Entomol. Soc.* **69** (1), 1–8.

EBERHARD, W.G., ACHOY, H., MARIN, M.C. AND UGALDE, J. (1993). Natural history and behavior of two species of *Macrohaltica* (Col. Chrys.). *Psyche* **100**, 93–119.

EDWARDS, J.G. (1953). Species of the genus *Syneta* of the world (Col. Chrys.). *Wasmann J. Biol.* **11** (1), 23–82.

EDWARDS, J.S. (1961). On the reproduction of *Prionoplus reticularis* (Col. Cerambycidae), with general remarks on reproduction in the Cerambycidae. *Q. J. Microsc. Sci.* **102**, 519–529.

EDWARDS, M.A. AND SEABROOK, W.D. (1997). Evidence for an airborne sex pheromone in the Colorado potato beetle, *Leptinotarsa decemlineata. Canad. Entomol.* **129**, 667–672.

EHRLICH, P.R. AND RAVEN, P. (1964). Butterflies and plants: A study of coevolution. *Evolution* **18**, 586–608.

EHRLICH, P.R. AND RAVEN, P. (1967). Butterflies and plants. *Sci. Am.* **216**, 104–113.

EICKWORT, K.R. (1971). The Ecology of *Labidomera clivicollis*, a relatively rare milkweed beetle. Ph.D. dissertation, 123 pp. Ann Arbor, Michigan: Cornell University, University microfilms.

EICKWORT, K.R. (1973). Cannibalism and kin selection in *Labidomera clivicollis* (Col. Chrys.). *Amer. Naturalist* **107**, 452–453.

EICKWORT, K.R. (1977). Population dynamics of a relatively rare species of milkweed beetle (*Labidomera*). *Ecology* **58**, 527–538.

EISNER, T. (1971). Chemical ecology: on arthropods and how they live as chemists. *Verhandlungen dt. Zool. Ges.* **65**, 123–137.

EISNER, T. (1980). Chemistry, defense, and survival: case studies and selected topics. In: *Insect biology in the future*. Eds. M. Locke and D.S. Smith, pp 847–878. London, UK: Academic Press.

EISNER, T. AND ANESHANSLEY, D.J. (2000). Defense by foot adhesion in a beetle (*Hemisphaerota cyanea*). *PNAS* **97** (12), 6568–6573.

EISNER, T. AND EISNER, M. (2000). Defensive use of a fecal thatch by a beetle larva (*Hemisphaerota cyanea*). *PNAS* **97** (6), 2632–2636.

EISNER, T. AND KAFATOS, F.C. (1962). Defense mechanisms of arthropods. X. A pheromone promoting aggregation in an aposematic distateful insect. *Psyche* **69** (2), 53–61.

EISNER, T. AND WILSON, E.O. (Eds.) (1977). *The insects. Readings from the Scientific American*. Pp i–iv+1–334. San Francisco: W.H. Freeman Publishers.

EISNER, T., VAN TASSEL, E.E. AND CARREL, J. (1967). Defensive use of a 'fecal shield' by a beetle larva. *Science, N.Y.* **158**, 1471–1473.

EISNER, T., HICKS, K., EISNER, M. AND ROBSON, D.S. (1978). Wolf in sheep's-clothing' strategy of a predaceous insect larva. *Science, N.Y.* **4330**, 790–794.

ELGAR, M.A. AND CRESPI, B.J. (Eds.) (1992). *Cannibalism, ecology and evolution among diverse taxa*. 361 pp. Oxford: Oxford Univ. Press.

ELIAS, S.A. (1994). *Quaternary insects and their environments*. xiii+284 pp. Washington, USA: Smithsonian Institution Press.

ELIAS, S.A. (2000). Late Pleistocene climates of Beringia, based on analysis of fossil beetles. *Quarternary Research* **53**, 229–235.

ELTON, C. (1927). *Animal ecology*. New York: Macmillan Publishers.

EMLEN, D.J. (2000). Integrating development with evolution. A case study with beetle horns. *BioScience* **50** (5), 403–441.

ERBER, D. (1969). Beitrag zur Entwicklungsbiologie mitteleuropaischer Clytrinen und Cryptocephalinen (Col. Chrys.). *Zool. Jb. Syst.* **96**, 453–477.

ERBER, D. (1984). The Chrysomelidae of Madeira. *Bol. Mus. Mun. Funchal* **38** (174), 43–69.

ERBER, D. (1988). Biology of Camptosomata: Clytrinae, Cryptocephalinae, Chlamisinae, Lamprosomatinae. In: *Biology of Chrysomelidae*. Eds. P. Jolivet, E. Petitpierre and T.H. Hsiao, pp 513–552. Dordrecht, The Netherlands: Kluwer Academic Publishers.

ERBER, D. AND MEDVEDEV, L.N. (1993). The larva of *Chrysolina fragariae* (Wollaston, 1864). *Bocagiana, Funchal* **166**, 1–5.

ERWIN, D.H. (1998). The end and the beginning recoveries from mass extinctions. *TREE* **13** (9), 344–349.

ERWIN, T.L. (1982). Tropical forests: their richness in Coleoptera and other arthropod species. *Col. Bull.* **36** (1), 74–75.

ERWIN, T.L. (1983a). Tropical forest canopies: the last biotic frontier. *Bull. American Entomol. Soc.* **29** (1), 14–19.

ERWIN, T.L. (1983b). Beetles and other insects of tropical forest canopies at Manaus, Brazil, sampled by insecticide fogging. In: *Tropical rain forest: ecology and management*. Special Publ. 2 of British Ecological Society. Eds. S.L. Sutton, T.C. Whitmore and A. Chadwick. Boston, Oxford: Blackwell Scientific Publications.

EVANS, A. AND BELLAMY, C.L. (1996). *An inordinate fondness for beetles*. Pp 1–208. New York: Henry Holt Publishers.

FARRELL, B.D. (1998). 'Inordinate fondness' explained: Why are there so many beetles? *Science, N.Y.* **281**, 555–559.

FARRELL, B.D. AND ERWIN, T.L. (1988). Leaf-beetle community structure in an Amazonian rainforest canopy. In: *Biology of Chrysomelidae*. Eds. P. Jolivet, E. Petitpierre and T.H. Hsiao, pp 73–90. Dordrecht, The Netherlands: Kluwer Academic Publishers.

FARRELL, B.D. AND MITTER, C. (1998). The timing of insect/plant diversification: might

Tetraopes (Col. Cerambycidae) and *Asclepias* (Asclepiadaceae) have co-evolved? *Biological Journal of The Linnean Society* **63** (4), 553–577.

FARRELL, B.D., MITTER, C. AND FUTUYMA, D.J. (1992). Diversification of the insect–plant interface. *BioScience* **42** (1), 34–42.

FAUVEL, A. (1907). Faune analytique des Coléoptères de la Nouvelle–Calédonie. *Revue d'Entomologie* **26** (5–6), 149–152.

FERGUSON, J.E. AND METCALF, R.L. (1985). Cucurbitacins. Plant-derived defense compounds for Diabroticites (Col. Chrys.). *J. Chem. Ecol.* **11**, 311–317.

FERRO, D.N. AND VOSS, R.H. (1985). Proceedings of the Symposium on the Colorado Potato Beetle. XVII Int. Congress of Entom., Hamburg. *Mass. Agric. Exp. Station Research Bull.* **704**, 1–144.

FERRO, D.N., TUTTLE, A.F. AND WEBER, D.C. (1991). Ovipositional and flight behavior of overwintered Colorado Potato Beetle (Col. Chrys.). *Env. Entomol.* **20** (5), 1309–1314.

FERRONATO, E.M.O. (1986). Observaçoes preliminares sobre as plantas hospedeiras das formas imaturas dos principais crisomelideos (Col. Chrys.) que ocorrem em cacauais. *Revista Theobroma* **16** (2), 107–110.

FERRONATO, E.M.O. (1988). Eumolpinae associated with cacao trees (*Theobroma cacao* L.) in Southeast Bahia. In: *Biology of Chrysomelidae*. Eds. P. Jolivet, E. Petitpierre and T.H. Hsiao, pp 553–558. Dordrecht, The Netherlands: Kluwer Academic Publishers.

FERRONATO, E.M.O. (1999). Fluctuation in the abundance of immature forms and images of Eumolpinae (Col. Chrys.) in *Theobroma cacao* (Sterculiaceae) in the South of Bahia, Brazil. In: *Advances in Chrysomelidae biology. I.* Pp 271–297. Leiden, The Netherlands: Backhuys Publishers.

FIEBRIG, K. (1910). Cassiden and Cryptocephaliden Paraguays. *Zool. Jahbr. Suppl.* **12**, 161–264.

FITZGERALD, G.F. (1992). Filial cannibalism in Fishes: Why do parents eat their offspring? *TREE* **7** (1), 7–10.

FLOATE, K.D. AND WHITHAM, T.G. (1994). Aphid–ant interaction reduces chrysomelid herbivory in a cottonwood hybrid zone. *Oecologia* **97**, 215–221.

FLOWERS, R. (1991). Aggregations of Cassidinae (Chrys.) in Santa Rosa and Guanacaste National Parks, Costa Rica. *Biotropica* **23** (3), 308–310.

FLOWERS, R.W. (1999). Internal structure and phylogenetic importance of male genitalia in the Eumolpinae. In: *Advances in Chrysomelidae biology. 1.* Ed. M.L. Cox, pp 71–93. Leiden, The Netherlands: Backhuys Publishers.

FLOWERS, R.W. AND JANZEN, D.H. (1997). Feeding records of Costa Rican leaf beetles (Col. Chrys.). *Florida Entomol.* **80** (3), 334–359.

FLOWERS, R.W. AND TIFFER, R. (1992). Comportamiento gregario de la vaquita *Hypolampsis* sp. (Col. Alticinae) en Guanacaste, Costa Rica. *Brenesia* **37**, 135–138.

FLOWERS, R.W., FURTH, D.G. AND THOMAS, M.C. (1994). Notes on the distribution and biology of some Florida leaf beetles (Col. Chrys.). *Col. Bull.* **48** (1), 79–89.

FORNASARI, L. (1993). Life history of the flea beetle, *Aphthona abdominalis* Duftschmid, on *Euphorbia esula* L. (leafy spurge) in Italy. *Biological Control* **3** (3), 161–175.

FORNASARI, L. (1996). Biology and ethology of *Aphthona* spp. (Col. Chrys. Alticinae) associated with *Euphorbia* spp. (Euphorbiaceae). In: *Chrysomelidae biology. I. General studies*. Eds. P. Jolivet and M.L. Cox, pp 293–313. Amsterdam, The Netherlands: SPB Academic Publishers.

FORNO, I.W., KASSULKE, R.C. AND HARLEY, K.L.S. (1992). Host specificity and aspects of the biology of *Calligrapha pantherina* (Col. Chrys.) a biological control agent of *Sida acuta* (Malvaceae) and *S. rhombifolia* in Australia. *Entomophaga* **37** (3), 409–417.

FOX, L.R. (1975). Cannibalism in natural populations. *Ann. Rev. Ecol. System.* **6**, 87–106.

FRANK, L. (1994). When hyenas kill their own. *New Scientist*, 5 March 1994. 1915, 38–41.

FRIIS, E.M., CHALONER, W.G. AND CRANE, P.R. (1987). *The origins of angiosperms and their biological consequences*. x + 358 pp. Cambridge, UK: Cambridge University Press.

FUNK, D.J., FUTUYMA, D.J., ORTI, G. AND MEYER, A. (1995). Mitochondrial DNA sequences and multiple data sets: A phylogenetic study of Phytophagous beetles. (Chrysomelidae: *Ophraella*). *Mol. Biol. Evol.* **12** (4), 627–640.

FURTH, D.G. (1979a). Zoogeography and host plant ecology of the Alticinae of Israel, especially *Phyllotreta* with descriptions of three new species. *Israel J. Zool.* **28** (1), 1–37.

FURTH, D.G. (1979b). Zoogeography and host plants of *Longitarsus* in Israel, with descriptions of six new species (Col. Chrys.). *Israel J. Entomol.* **13**, 79–124.

FURTH, D.G. (1979c). Wing polymorphism, host plant ecology and biogeography of *Longitarsus* in Israel (Col. Chrys.). *Israel J. Entom.* **13**, 125–148.

FURTH, D.G. (1980). Inter-generic differences in the metafemoral apodeme of flea beetles (Chrys. Alt.). *Syst. Entomol.* **5**, 263–271.

FURTH, D.G. (1982a). The metafemoral spring of flea beetles (Chrys. Alt.). *Spixiana Suppl.* **7**, 11–27.

FURTH, D.G. (1982b). *Blepharida* biology, as demonstrated by the sacred sumac flea beetle (*B. sacra* Weise) (Col. Chrys. Alt.). *Spixiana Suppl.* **7**, 43–52.

FURTH, D.G. (1983). Alticinae of Israel: Psylliodes (Col. Chrys.). *Israel J. Entomology* **17**, 37–58.

FURTH, D.G. (1985a). The natural history of a sumac tree, with an emphasis on the entomofauna. *Trans. Conn. Acad. Arts and Sc.* **46**, 137–234.

FURTH, D.G. (1985b). Some flea beetles and their food plants from Kenya (Chrys. Alt.). *Col. Bull.* **39** (3), 259–263.

FURTH, D.G. (1988). The jumping apparatus of flea beetles (Alticinae). The metafemoral spring. In: *Biology of Chrysomelidae*. Eds. P. Jolivet, E. Petitpierre and T.H. Hsiao, pp 285–297. Dordrecht, The Netherlands: Kluwer Academic Publishers.

FURTH, D.G. (1992). The new world *Blepharida* group, with a key to genera and description of a new species (Col. Chrys.). *J. New York Entomol. Soc.* **100** (2), 399–414.

FURTH, D.G. (Ed.) (1994a). *Proceedings of the Third International Symposium on the Chrysomelidae. Beijing, 1992.* 150 pp. Leiden, The Netherlands: Backhuys Publishers.

FURTH, D.G. (1994b). A new case of parthenogenesis in beetles: *Longitarsus melanurus* (Melsheimer) (Col. Chrys.). *J. New York Entomol. Soc.* **102** (3), 310–317.

FURTH, D.G. AND SEENO, T.N. (Eds.) (1985). First International Symposium on the Chrysomelidae. XVII Int. Congress of Entomology, Hamburg. *Entomography* **3**, 371–503.

FURTH, D.G. AND SEENO, T.N. (Eds.) (1989). Second International Symposium on the Chrysomelidae. Vancouver. *Entomography* **6**, 343–552.

FURTH, D.G. AND SUZUKI, K. (1990a). Comparative morphology of the tibial flexor and extensor tendons in insects. *Syst. Entomol.* **15**, 433–441.

FURTH, D.G. AND SUZUKI, K. (1990b). The metatibial extensor and flexor tendons in Coleoptera. *Syst. Entomol.* **15**, 443–448.

FURTH, D.G. AND SUZUKI, K. (1992). The independent evolution of the metafemoral spring in Coleoptera. *Syst. Entomol.* **17**, 341–349.

FURTH, D.G. AND SUZUKI, K. (1994). Character correlation studies of problematic genera of Alticinae in relation to Galerucinae (Col. Chrys.). *Proc. Third Int. Symp. Chrysomelidae, Beijing, 1992.* Ed. D.G. Furth, pp 116–135. Leiden, The Netherlands: Backhuys Publishers.

FURTH, D.G. AND SUZUKI, K. (1998). Studies of Oriental and Australian Alticinae genera based on the comparative morphology of the metafemoral spring, genitalia and hind wing venation. *Proc. Fourth Int. Symp. Chrysomelidae. XX ICE, Firenze, 1996.* Eds. M. Biondi, M. Daccordi and D.G. Furth, pp 91–124. Mus. Reg. Sci. Nat. Torino, Italy.

FURTH, D.G. AND YOUNG, D.A. (1988). Relationships of herbivore feeding and plant flavonoids (Coleoptera Chrysomelidae and Anacardiaceae: *Rhus*). *Oecologia* **74** (4), 496–500.

FURTH, D.G., TRAUB, W. AND HARPAZ, I. (1983). What makes *Blepharida* jump? A structural study of the metafemoral spring of a flea beetle. *J. Exp. Zool.* **227**, 43–47.

FUTUYMA , D.J. (1992). Genetics and the phylogeny of insect-plant interactions. In: *Proc. 8th Int. Symp. Insect–Plant Relationships.* Eds. S.B. Menken, J.H. Visser and P. Harrewijn, pp 191–200. Dordrecht, The Netherlands: Kluwer Academic Publishers.

FUTUYMA, D.J. (2000). Some current approaches to the evolution of plant-herbivore interactions. *Plant Species Biology* **15**, 1–9.

FUTUYMA, D.J. AND KEASE, M.C. (1992). *Evolution and coevolution of plants and phytophagous arthropods in herbivores: their interactions with secondary plant metabolites.* Ed. D.J. Futuyma, pp 439–475. New York, USA: Academic Press.

FUTUYMA, D.J. AND MCCAFFERTY, S.S. (1990). Phylogeny and the evolution of host plant associations in the leaf beetle genus *Ophraella* (Col. Chrys.). *Evolution* **44** (8), 1885–1913.

FUTUYMA, D.J. AND MITTER, C. (1996). Insect–plant interactions: the evolution of component communities. *Phil. Trans. R. Soc. Lond. B* **351**, 1361–1366.

FUTUYMA, D.J., HERRMANN, C., MILSTEIN, S. AND KEESE, M.C. (1992). Apparent transgenerational effects of host plant in the leaf beetle, *Ophraella notulata* (Col. Chrys.). *Chemoecology* **3** (3–4), 365–372.

FUTUYMA, D.J., KEESE, M.C. AND FUNK, D.J. (1995). Genetic constraints on macroevolution: the evolution of host affiliation in the leaf beetle genus *Ophraella*. *Evolution* **49** (5), 797–809.

GADEAU DE KERVILLE, H. (1900). L'accouplememnt des Coléoptères. *Bull. Soc. Ent. Fr.* 101–107.

GAGE, M.J.C. (1992). Removal of rival sperm during copulation in a beetle, *Tenebrio molitor*. *Anim. Behav.* **44**, 587–589.

GAGE, M.J.C. AND BAKER, R. (1991). Ejaculate size varies with socio-sexual situation in an insect. *Ecol. Entomol.* **16**, 331–337.

GAHAN, C.J. (1891). Mimetic resemblances between species of the Coleopterous genera *Lema* and *Diabrotica*. *Trans. R. Ent. Soc. London*, 367–374.

GAHAN, C.J. (1903). Coleoptera. In: *The natural history of Socotra and Abd-el-Kuri*. Ed. H.O. Forbes, pp 263–291. Liverpool.

GAHAN, C.J. (1912–1913). Mimicry in Coleoptera. *Proc. S. Lond. Ent. Nat. Hist. Soc.*, 28–38.

GARIN, C.F., JUAN, C. AND PETITPIERRE, E. (1999). Mitochondrial DNA phylogeny and the evolution of host plant use in Palearctic *Chrysolina* (Col. Chrys.) leaf beetles. *J. Molecular Evolution* **48**, 435–444.

GASTON, K.J. (1991). The magnitude of global insect richness. *Conservation Biology* **5** (3), 283–296.

GEISTHARDT, M. AND VAN HARTEN, A. (1992). Noxious beetles of the Cape Verde Islands. Pp 1–242. Wiesbaden, Germany: Verlag Christa Hemmen.

GERBER, G.H., NEILL, G.B. AND WESTRAL, P.H. (1978). The anatomy and histology of the internal reproductive organs of the sunflower beetle, *Zygogramma exclamationis* (Col. Chrys.). *Canad. J. Zool.* **56**, 2542–2553.

GHILAROV, M.S. (1964). The main directions in insect adaptations to the life in the desert. *Zool. Zhurn. Moscow* **43**, 443–454.

GIORDANO, R. AND JACKSON, J.J. (1999). *Wolbachia* infections in Chrysomelids. In: *Advances in Chrysomelidae biology. 1*. Ed. M.L. Cox, pp 185–196. Leiden, The Netherlands: Backhuys Publishers.

GOMEZ-ZURITA, J., GARIN, C.F., JUAN, C. AND PETITPIERRE, E. (1999). Mitochondrial 16SrDNA sequences and their use as phylogenetic markers in leaf beetles with special reference to the subfamily Chrysomelidae. In: *Advances in Chrysomelidae biology. 1*. Ed. M.L. Cox, pp 25–38. Leiden, The Netherlands: Backhuys Publishers.

GOMEZ-ZURITA, J., JUAN, C. AND PETITPIERRE, E. (2000a). The evolutionary history of the genus *Timarcha* (Col. Chrys.) inferred from mitochondrial COII gene or partial 16SrDNA sequences. *Molecular Phylogeny and Evolution* **14** (2), 304–317.

GOMEZ-ZURITA, J., PETITPIERRE, E. AND JUAN, C. (2000b). Nested cladistic analysis, phylogeography and speciation in the *Timarcha goettingensis* complex (Col. Chrys.). *Molecular Ecology* **9**, 557–570.

GOUGH, N., BARTRAEAU, T. AND MONTGOMERY, B.L. (1994). Distribution, hosts and pest status of the orchid beetle, *Stethopachys formosa* Baly (Col. Chrys.). *Austr. Entomol.* **21** (2), 49–54.

GRÉGOIRE, J.C. (1988). Larval gregariousness in the Chrysomelidae. In: *Biology of Chrysomelidae*. Eds. P. Jolivet, E. Petitpierre and T.H. Hsiao, pp 253–260. Dordrecht, The Netherlands: Kluwer Academic Publishers.

GRESSITT, J.L. (1966). Epizoic symbiosis: the Papuan weevil genus *Gymnopholus* (Leptopiinae) symbiotic with cryptogamic plants, oribatid mites, rotifers and nematodes. *Pacific Insects* **8** (1), 211–280.

GRESSITT, J.L. AND HART, A.D. (1974). Chrysomelid beetles from the Papuan Subregion. 8. (Chrysomelinae 1). *Pacific Insects* **16** (2–3), 261–306.

GRESSITT, J.L. AND KIMOTO, S. (1961–1963). The Chrysomelidae (Col.) of China and Korea. 1 and 2. *Pacific Insects Monogr.*, pp 1–1026. Honolulu, Hawaii: Bernice Bishop Museum.

GRIMALDI, D.A. (1990). Insects from the Santana formation, Lower Cretaceous of Brazil. *Bull. Amer. Mus. Nat. Hist., New York* **195**, 1–191.

GROBBELAAR, E. AND JOLIVET, P. (1996). Preliminary biological observations on the Southern African chrysomeline genus *Iscadida* Chevrolat, 1843. *Chrysomela* **31**, 8–9.

GUILLET, G., PODESZFINSKI, C., REGNAULT-ROGER, C., ARNASON, J.T. AND PHILOGENE, B.J.R. (2000). Behavioral and biochemical adaptations of generalist and specialist herbivorous insects feeding on *Hypericum perforatum* (Guttiferae). *Environ. Entomol.* **29** (2), 135–139.

HACKETT, K.J. AND HENEGAR, H.R. (1992). Distribution of Colorado Potato Beetle spiroplasmas in North America. *Biol. Control* **2**, 218–225.

HACKETT, K.J. AND LIPA, J.J. (1996). Mollicutes associated with the Chrysomelidae. In: *Chrysomelidae biology. 2. Ecological studies.* Eds. P. Jolivet and M.L. Cox, pp 139–146. Amsterdam, The Netherlands: SPB Academic Publishers.

HAITLINGER, R. (1989). New Canestrinid mites (Acari, Astigmata, Canestriniidae) associated with beetles of the subfamily Hispinae (Ins. Col. Chrys.). *Polskie Pismo entomol.* **59**, 281–292.

HAMMOND, P.M. (1979). Wing folding mechanisms of beetles with special reference to investigations of adephagan phylogeny (Col.). In: *Carabid beetles their evolution, natural history and classification.* Eds. T. Erwin, G.E. Ball and D.R. Whitehead, pp 113–180. The Hague, The Netherlands: Junk Publishers.

HARNISCH, W. (1915). Uber den männlichen Begattungsapparat einiger Chrysomeliden. *Z. Wiss. Zool.* **114**, 1–94.

HARTMANN, F.A. (1977). The ecology and evolution of common milkweed (*Asclepias syriaca*, Ascl.) and milkweed beetles (*Tetraopes tretrophtalmus*, Cerambycidae). Dissert. University Michigan, Ann Arbor, Mich., USA (quoted by J. Becerra).

HARTMANN, T., WITTE, L., EHMKE, A., THEURING, C., ROWELL-RAHIER, M. AND PASTEELS, J.M. (1997). Selective sequestration and metabolism of plant derived pyrrolizidine alkaloids by chrysomelid leaf beetles. *Phytochemistry* **45**, 489–497.

HARTMANN, T., THEURING, C., SCHMIDT, J., ROWELL-RAHIER, M. AND PASTEELS, J.M. (1999). Biochemical strategy of sequestration of pyrrolizidine alkaloids by adults and larvae of chrysomelid leaf-beetles. *J. Insect Physiol.* **45**, 1085–1095.

HAWKESWOOD, T.J. (1977). Those mighty plant defoliators: the Chrysomelidae. *Wildlife in Australia* **14** (1), 11–14.

HAWKESWOOD, T.J. (1985). Notes on some beetles (Col.) associated with *Xanthorrhoea johnsonii* (Xanthorrhoeaceae) in the Brisbane area, S.E. Queensland. *Victorian Nat.* **102**, 162–166.

HAWKESWOOD, T.J. (1987). *Beetles of Australia.* viii+248 pp. Sydney, Australia: Angus and Robertson Publishers.

HAWKESWOOD, T.J. (1991). Review of the biology and control of the orchid beetle, *Stethopachys formosa* Baly (Col. Chrys.). *Victorian Entomologist* **21**, 129–131.

HAWKESWOOD, T.J. (1992). Notes on the biology and host plants of the Australian leaf beetle, *Lilioceris* (*Crioceris*) *nigripes* (F.) (Col. Chrys.). *Entomol.* **111**, 210–212.

HAWKESWOOD, T.J. AND FURTH, D.G. (1994). New host plant records for some Australian Alticinae (Col. Chrys.). *Spixiana* **17** (1), 43–49.

HAWKESWOOD, T.J. AND JOLIVET, P. (1988). Notes on the biology and behavior of the Australian leaf beetles, *Cleptor inermis* Lefèvre. *Bull. Ann. Soc. Ent. Belg.* **124**, 189–194.

HAWKINS, B.A. AND LAWTON, J.H. (1987). Species richness for parasitoids of British phytophagous insects. *Nature* **326**, 788–790.

HAYASHI, Y., FUJIYAMA, S. AND SUEKUNI, J. (1994). Life-cycle synchronization in *Chrysolina aurichalcea* (Col. Chrys.) to its host *Artemisia princeps*: Effects of host leaf age on survival. *Appl. Entomol. Zool.* **29** (2), 149–155.

HEBERDEY, R.F. (1928). Ein Beitrag zur Entwicklungsgeschichte der männlichen Geschlechtsapparates des Coleopteren. *Z. f. Morphol. u. Okol. D. Tiere* **10**, 533–575.

HEIKERTINGER, F. (1950). Bestimmungstabellen europaïscher Käfer. 82. Fam. Chrysomelidae.

5. Subfam. Halticinae. 2. Gatt. Crepidodera. *Verwandschaft weitesten Sinmes Koleopt. Rundshau.* **32**, 1–84.

HEMP, CL., HEMP, A. AND DETTNER, K. (1999). Attraction of the colour beetle species *Pallenothriocera rufimentris* by cantharidin (Cleridae, Col.). *Entomol. Generalis* **24** (2), 115–123.

HENDRIX, S.D. (1980). An evolutionary and ecological perspective of the insect fauna of ferns. *Amer. Naturalist* **115**, 171–196.

HERING, M. (1950). Monophagie und Xenophagie. Die Nahrungswahl phytophager Insekten und die geographische Herkunft ihrer wirtspflanzen. *Naturwissenchaften* **37** (23), 531–536.

HERING, M. (1951). *Biology of leaf miners.* Pp 1–420. Den Haag, The Netherlands: Junk Publishers.

HERON, H.D.C. (1992). Cycloalexy in two South African Tortoise Beetles (Chrys. Cass.). *Chrysomela* **27**, 3.

HERON, H.D.C. (1999). The biology of *Conchyloctenia punctata* (Fabricius). A cycloalexic cassid (Chrys. Cassidinae). In: *Advances in biology of Chrysomelidae. 1.* Ed. M.L. Cox, pp 565–580. Leiden, The Netherlands: Backhuys Publishers.

HESPENHEIDE, H.A. (1976). Reversed sex-limited mimicry in a beetle. *Evolution* **29** (4), 780–783.

HESPENHEIDE, H.A. (1996). Chrysomelidae of the subfamily Clytrinae as models for mimicy complexes. In: *Chrysomelidae biology. 2. Ecological studies.* Eds. P. Jolivet and M.L. Cox, pp 227–239. Amsterdam, The Netherlands: SPB Academic Publishers.

HESPENHEIDE, H.A. AND DANG, V. (1999). Biology and ecology of leaf mining Hispinae (Col. Chrys.) of the La Selva Biological Station, Costa Rica. In: *Advances in Chrysomelidae biology. 1.* Ed. M.L. Cox, pp 375–389. Leiden, The Netherlands: Backhuys Publishers.

HEYMONS, R. AND LUHMANN, M. (1933). Die vasa Malpighi von *Galerucella viburni* Payk (Col.). *Zool. Anz.* **102**, 78–86.

HILKER, M. (1994). Egg deposition and protection of eggs in Chrysomelidae. In: *Novel aspects of the biology of Chrysomelidae.* Eds. P. Jolivet, M.L. Cox and E. Petitpierre, pp 263–276. Dordrecht, The Netherlands: Kluwer Academic Publishers.

HILKER, M. AND MEINERS, T. (1999). Chemical cues mediating interactions between chrysomelids and parasitoids. In: *Advances in Chrysomelidae biology. I.* Ed. M.L. Cox, pp 197–216. Leiden, The Netherlands: Backhuys Publishers.

HILL, M.P. AND HULLEY, P.E. (1993). Biology and host-range of *Gratiana spadicea* (Chrys. Cass.), a potential biological control agent for the weed *Solanum sisymbriifolium* (Col.) in South Africa. Proc. Ninth Ent. Congr. Ent. Soc. S. Africa, 177 pp.

HILL, M.P., HULLEY, P.E. AND OLCKERS, T. (1993). Insect hervivores on the exotic weeds *Solanum elaeagnifolium* Cavanilles and *S. sisymbrifolium* Lamarck (Solanaceae) in South Africa. *African Entomology* **1** (2), 175–182.

HINTON, H.E. (1946). The 'gin-traps' of some beetle pupae: a protective device which appears to be unknown. *Trans. R. Ent. Soc. London* **97**, 473–491.

HINTON, H.E. (1951). On a little known protective device of some chrysomelid pupae (Col.). *Proc. R. Ent. Soc. London* **26**, 67–73.

HINTON, H.E. (1973). Some recent work on the colours of insects and their likely significance. *Proc. Trans. Brit. Ent. Soc.* **6**, 43–54.

HINTON, H.E. (1981). *Biology of insect eggs. 1, 2 and 3.* 1135 pp. Oxford, UK: Pergamon Press.

HOCKING, B. (1970). Insect associations with the swollen thorn acacias. *Trans. R. Ent. Soc. London* **122** (7), 211–255.

HOCKING, B. (1975). Ant–plant mutualism: evolution and energy. In: *Coevolution of animals and plants.* Eds. L. Gilbert and P.H. Raven, pp 78–90. Austin, TX: University of Texas Press.

HOFFMAN, C.E. (1940). The relation of *Donacia* larvae (Chrys. Col.) to dissolved oxygen. *Ecology* **21** (2), 176–183.

HOLLANDE, A.CH. (1911a). Etude histologique comparée du sang des insectes à hémorrhée et des insectes sans hémorrhée. *Arch. Zool. Exp. Gén.* **6** (9), 283–323.

HOLLANDE, A.CH. (1911b). *L'autohémorrhée ou le rejet de sang chez les insectes (toxicologie du sang).* 148 pp. Paris: Thesis.

HOLLANDE, A.CH. (1926). La signification de l'autohémorrhée des insectes. *Arch. Anat. Micr. Morph. Exp.* **22**, 374–412.

HOLLISTER, B. AND MULLIN, C.A. (1999). Isolation and identification of primary metabolite feeding stimulants for adult western corn rootworm, *Diabrotica virgifera virgifera* LeConte from host pollen. *J. Chemical Ecology* **25** (6), 1263–1280.

HONOMICHL, L.K. (1980). Die digitiformen sensillen auf dem maxillarpalpus von Coleoptera. I. Vergleichend-topographische Untersuchung des Kutikulären Apparates. *Zool. Ang.* **20**, 1–12.

HOULIHAN, D.F. (1969). Respiratory physiology of the larva of *Donacia simplex*, a root-piercing beetle. *J. Insect Physiol.* **15**, 1517–1536.

HOULIHAN, D.F. (1970). Respiration in low oxygen partial pressures: the adults of *Donacia simplex* that respire from the roots of aquatic plants. *J. Insect Physiol.* **16**, 1607–1622.

HSIAO, T.H. (1974). Chemical influence on feeding behavior of *Leptinotarsa* beetles. In: *Experimental analysis of insect behaviour*. Pp 237–248. Heidelberg & New York: Springer-Verlag Publishers.

HSIAO, T.H. (1981). Ecophysiological adaptations among geographic populations of the Colorado potato beetle in North America. In: *Advances in potato pest management*. Eds. J.H. Lashomb and R. Casagrande, pp 69–85. Stroudsburg, PA: Hutchinson Ross Publishing Co.

HSIAO, T.H. (1986). Specificity of certain chrysomelid beetles for Solanaceae. In: *Solanaceae*. Ed. W.G. D'Arcy, pp 345–363. New York, USA: Columbia University Press.

HSIAO, T.H. (1988). Host specificity, seasonality and bionomics of *Leptinotarsa* beetles. In: *Biology of Chrysomelidae*. Eds. P. Jolivet, E. Petitpierre and T.H. Hsiao, pp 581–599. Dordrecht, The Netherlands: Kluwer Academic Publishers.

HSIAO, T.H. (1994a). *Molecular phylogeny of chrysomelid beetles inferred from mitochondrial DNA data*. Proc. 3rd Int. Symp. Chrysomelidae. Beijing 1992. Ed. D.G. Furth, pp 9–17. Leiden, The Netherlands: Backhuys Publishers.

HSIAO, T.H. (1994b). Molecular techniques for studying systematics and phylogeny of Chrysomelidae. In: *Novel aspects of the biology of Chrysomelidae*. Eds. P. Jolivet, M.L. Cox and E. Petitpierre, pp 237–248. Dordrecht, The Netherlands: Kluwer Academic Publishers.

HSIAO, T.H. AND FRAENKEL, G. (1968). Selection and specificity of the Colorado potato beetle for solanaceous and non-solanaceous plants. *Ann. Entomol. Soc. Amer.* **61**, 493–502.

HSIAO, T.H. AND HSIAO, C. (1983). Chromosomal analysis of *Leptinotarsa* and *Labidomera* species (Col. Chrys.). *Genetica* **60**, 139–150.

HSIAO, T.H. AND PASTEELS, J.M. (1999). Evolution of host plant affiliation and chemical defense of *Chrysolina-Oreina* leaf beetles as revealed by mt DNA phylogenies. In: *Advances in Chrysomelidae biology. 1*. Ed. M.L. Cox, pp 321–342. Leiden, The Netherlands: Backhuys Publishers.

HSIAO, T.H. AND WINDSOR, D.M. (1999). Historical and biological relationships among Hispinae inferred from 12S MTDNA sequence data. In: *Advances in Chrysomelidae biology. 1*. Ed. M.L. Cox, pp 39–50. Leiden, The Netherland: Backhuys Publishers.

HUGHES, L. AND WESTOBY, M. (1992). Capitula on stick insect eggs and elaiosomes on seeds: convergent adaptations for burial by ants. *Functional Ecology* **6**, 642–648.

HUGHES, N.F. (1976). *Palaeobiology of angiosperm origins*. vii+242 pp. New York, USA: Cambridge University Press.

HUMBER, R.A. (1996). Fungal pathogens of the Chrysomelidae and prospects for their use in biological control. In: *Chrysomelidae biology. 2. Ecological studies*. Eds. P. Jolivet and M.L. Cox, pp 93–115. Amsterdam, The Netherlands: SPB Academic Publishers.

HUTCHINSON, G.E. (1957). Concluding remarks. *Cold Spring Harbor Symp. Quart. Biol.* 22, pp 415–427. NY, USA: Cold Spring Harbor.

IHSSEN, G. (1936). Über die Lebenweise von *Longitarsus nigerrimus* Gyll. *Koleopter. Rundshau* **22**, 253–255.

IRESON, J.E., FRIEND, D.A., HOLLOWAY, R.J. AND PATERSON, S.C. (1991). Biology of *Longitarsus flavicornis* (Stephens) (Col. Chrys.) and its effectiveness in controlling ragwort (*Senecio jacobaea* L.) in Tasmania. *J. Austr. Entomol. Soc.* **30**, 129–141.

IRESON, J.E., LEIGHTON, S.M., HOLLOWAY, R.J. AND CHATTERTON, W.S. (2000). Establishment and redistribution of *Longitarsus flavicornis* (Stephens) (Col. Chrys.) for the biological control of ragwort (*Senecio jacobaea* L.) in Tasmania. *Austr. J. Entomol.* **39**, 42–46.

ISONO, M. (1988). Differentiation in life history pattern and oviposition behavior and thelytoky in *Demotina* and *Hyperaxis* beetles (Col. Chrys.) in western Japan. *Kontyu* **56**, 402–409.

IUGA, V.G. AND KONNERTH, A.K. (1963). La structure de l'apex abdominal chez les Halticinae comparée à celle des Cérambycidae. *Trav. Mus. Hist. Nat. Gr. Antipa, Bucarest (Ent.)* **4**, 201–216.

IYENGAR, V.K. AND EISNER, T. (1999). Female choice increases offspring fitness in the arctiid moth (*Utetheisa ornatrix*). *PNAS* **96** (26), 15013–15016.

JACOBSON, G. (1910). Uber die Chrysomelidae, Elateridae and Curculionidae der Samlung der Russischen Nordpolar Expedition. *Lap. Imper. Akad. Nauk, Sér.* **8**, 53–66.

JACOBSON, G. (1925). The geographical distribution of the species of the genus *Timarcha*. *C.R. Acad. Sc. Russia*, 45–46.

JACOBY, M. (1908). *The fauna of British India, including Ceyland and Burma. Col. Chrysomelidae. 1.* xx+534 pp. London: Taylor and Francis Publishers.

JACQUES, R.L. (1988). Potato beetles: the genus *Leptinotarsa* in North America (Col. Chrys.). In: *Flora and fauna handbook 3.* 144 pp. Leiden, The Netherlands: Brill Publishers.

JAYANTH, K.P. AND BALI, G. (1993). Diapause behaviour of *Zygogramma bicolorata* (Col. Chrys.) a biological control agent for *Parthenium hysterophorus* (Asteraceae) in Bangalore, India. *Bull. Entomol. Research* **83**, 383–388.

JAYANTH, K.P. AND NAGARKATTI, S. (1987). Investigations on the host- specificity and damage potential of *Zygogramma bicolorata* Pallister introduced into India for the biological control of *Parthenium hysterophorus*. *Entomon* **12** (2), 141–145.

JAYANTH, K.P., MOHANDAS, S., ASOKAN, R. AND VISALAKSHY, P.N. (1993). *Parthenium* pollen feeding by *Zygogramma bicolorata* (Col. Chrys.) on sunflower (*Helianthus annuus*) (Comp.). *Bull. Entomol. Res.* **83**, 595–598.

JEANNEL, R. (1942). *La genèse des faunes terrestres.* 514 pp. Paris, France: Presses Univ.

JEANNEL, R. (1955). *L'édéage.* 155 pp. Paris, France: Publ. Museum National d'Histoire Naturelle 16.

JEANNEL, R. AND PAULIAN, R. (1944). Morphologie abdominale des Coléoptères. *Rev. Française Entomol.* **11** (2), 66–110.

JEANNEL, R. AND PAULIAN, R. (1949). Ordre des Coléoptères. In: *Traité de zoologie*. Ed. P.P. Grassé, pp 771–1077. Paris, France: Masson & Co Publishers.

JEREZ, V. (1995). *Stenomela pallida* Erichson, 1847. Redescripcion, ontogenia y afinidad con el genero *Hornius* (Chrys. Eumolpinae). *Gayana Zool.* **59** (1), 1–12.

JEREZ, V. (1996). Biology and phylogenetic remarks on the subantarctic genera *Hornius*, *Stenomela* and *Dictyneis* (Chrys. Eumol.). In: *Chrysomelidae biology. 3. General studies.* Eds. P. Jolivet and M.L. Cox, pp 239–258. Amsterdam, The Netherlands: SPB Academic Publishers.

JEREZ, V. (1999). Biology and ecology of the genus *Procalus* Clark, 1865, endemic to the Andino-patagonian region (Alt.). In: *Advances in Chrysomelidae biology. 1.* Ed. M.L. Cox, pp 545–555. Leiden, The Netherlands: Backhuys Publishers.

JEREZ, V. AND IBARRA-VIDAL, H. (1992). Morfologia y bionomia de *Hornius grandis* (Phil. and Phil., 1864) (Chrys. Eumol.). *Bol. Soc. Biol. Concepcion, Chile* **63**, 93–100.

JERMY, T., SZENTESI, A. AND HORVATH, J. (1988). Host plant finding in phytophagous insects: the case of the Colorado potato beetle. *Entomol. Exp. Appl.* **49** (1–2), 83–98.

JOFFREE, C.E. AND JOFFREE, E.P. (1996). Redistribution of the potential geographical ranges of Mistletoe and Colorado beetle in Europe in response to the temperature component of climate change. *Functional Ecology* **10**, 562–577.

JOLEAUD, L. (1939). *Atlas de paléogéographie.* 99 pls. Paris, France: Paul Lechevalier Publishers.

JOLIVET, P. (1948). Introduction à la biologie des *Timarcha. Miscell. Ent.* **45** (1), 1–32.

JOLIVET, P. (1950a). Les parasites, prédateurs et phorétiques des Chrysomeloidea (Col.) de la faune franco-belge. *Bull. Inst. Roy. Sc. Nat. Belgique* **26**, 1–39.

JOLIVET, P. (1950b). Contribution à l'étude des *Microtheca* Stal (Col. Chrys.). *Bull. Inst. R. Sc. Nat. Belgique* **26** (48), 1–27.

JOLIVET, P. (1951a). Contribution à l'étude du genre *Gastrophysa* Chevrolat (Col. Chrys.). *Bull. Inst. R. Sc. Nat. Belgique* **27** (9), 1–11.

JOLIVET, P. (1951b). Contribution à l'étude du genre *Microtheca* Stal (Col. Chrys.) (note 2). *Bull. Inst. R. Sc. Nat. Belgique* **27** (38), 1–7.

JOLIVET, P. (1952a). Quelques données sur la myrmécophilie des Clytrides (Col. Chrys.). *Bull. Inst. Roy. Sc. Nat. Belgique* **28** (8), 1–12.

JOLIVET, P. (1952b). Notes biométriques sur quelques Chrysomeloidea (Col.). *Bull. Inst. R. Sc. Nat. Belgique* **28** (44), 1–8.

JOLIVET, P. (1953). Les Chrysomeloidea des Iles Baléares. *Mem. Inst. Roy. Sc. Nat. Belgique* **2** (50), 1–88.

JOLIVET, P. (1954a). L'aile des Chrysomeloidea (Col.). *Miscell. Ent.* **48** (61–62), 85–101.

JOLIVET, P. (1954b). Notes écologiques préliminaires sur les Chrysomeloidea de la Sierra Nevada. *Bull. Ann. Soc. Entom. Belgique* **90** (3–4), 69–72.

JOLIVET, P. (1957–1959). Recherches sur l'aile des Chrysomeloidea. *Mem. Inst. Roy. Sc. Nat. Belgique* 2 sér. **51**, 1–180; **58**, 1–152, 40 pls.

JOLIVET, P. (1965). Notes sur l'écologie des *Timarcha* marocaines (Col. Chrys.). *Bull. Soc. Sc. Nat. Phys. Maroc* **45**, 159–190.

JOLIVET, P. (1967a). Les Alticides vénéneux de l'Afrique du Sud. *L'Entomologiste* **23** (4), 100–111.

JOLIVET, P. (1967b). Notes écologiques sur les *Timarcha* tuniso-lybiens (Col. Chrys.). *Bull. Soc. Ent. Fr.* **72** (5), 224–239.

JOLIVET, P. (1971). Sélection trophique et adaptation écologique chez le genre *Timarcha* Latreille (Col.). *Proc. XIII Int. Congr. Entomol., Moscow* **1**, 505–506.

JOLIVET, P. (1972). An interpretation of the host plants selected by species of the genus *Timarcha* (Col. Chrys.). *Korean J. Entomol., Seoul* **2** (1), 21–26.

JOLIVET, P. (1975). Une excursion entomologique à l'île de Quelpart (Cheju-Dô) et découverte d'une espèce alpine nouvelle pour la Corée de *Chrysolina* Motschulsky (Col. Chrys.). Réflexions sur la plantagophagie. *Bull. Soc. Linn. Lyon* **44** (2), 57–64; **44** (3), 72–81.

JOLIVET, P. (1976). Notes préliminaires sur la biologie des *Timarcha* du Pacifique Nord Occidental Américain (Col. Chrys.). *Cahiers Pacifique* **19**, 153–165.

JOLIVET, P. (1977). Sélection trophique chez les Eupoda (Col. Chrys.). *Bull. Soc. Linn. Lyon* **46** (9), 321–336.

JOLIVET, P. (1978). Sélection trophique chez les Clytrinae, Cryptocephalinae et Chlamisinae (Cryptosomata) et les Lamprosomatinae (Cyclica) (Col. Chrys.). *Acta Zool. Pathol. Antwerp* **70**, 167–200.

JOLIVET, P. (1979). Réflexions sur l'écologie, l'origine et la distribution des Chrysomelidae (Col.) des Iles Mascareignes, Océan Indien, avec la description de deux espèces nouvelles. *Bull. Soc. Linn. Lyon* **48** (9–10), 524–528; 606–607; 641–649.

JOLIVET, P. (1982). Les Eumolpinae (Col. Chrys.) des Apocynaceae et des Asclepiadaceae (Gentianales). *Bull. Soc. Linn. Lyon* **51** (7), 214–222.

JOLIVET, P. (1984). *Phaedon fulvescens* Weise (Col. Chrys.), un auxiliaire possible pour le contrôle des *Rubus* aux tropiques. *Bull. Soc. Linn. Lyon* **53** (7), 235–246.

JOLIVET, P. (1985). Comments on the Chrysomelidae of the Cape Verde Islands. *Entomography* **3**, 502.

JOLIVET, P. (1986a). *Les fourmis et les plantes. Un exemple de coévolution.* 254 pp. Paris, France: Boubée Publishers.

JOLIVET, P. (1986b). *Insects and plants.* 197 pp. New York, USA: Brill Publishers.

JOLIVET, P. (1987a). Remarques sur la biocénose des *Cecropia* (Cecropiaceae). Biologie des *Coelomera* Chevrolat avec la description d'une nouvelle espèce du Brésil (Col. Chrys.). *Bull. Soc. Linn. Lyon* **56** (8), 255–276.

JOLIVET, P. (1987b). Premières données sur la biologie *d'Elytrosphaera xanthopyga* Stal (Col. Chrys.). *Bull. Soc. Ent. Fr.* **91** (5–6), 163–170.

JOLIVET, P. (1987c). Sélection trophique chez les Megascelinae et les Eumolpinae (Cyclica) (Col. Chrys.). *Bull. Soc. Linn. Lyon* **56** (6), 199–208 and **56** (7), 217–240.

JOLIVET, P. (1988a). Food habits and food selection of Chrysomelidae. Bionomic and evolutionary perspectives. In: *Biology of Chrysomelidae*. Eds. P. Jolivet, E. Petitpierre and T.H. Hsiao, pp 1–24. Dordrecht, The Netherlands: Kluwer Academic Publishers.

JOLIVET, P. (1988b). Les soins parentaux chez les Chrysomélides (Col.). *L'Entomologiste, Paris* **44** (2), 93–101.

JOLIVET, P. (1989a). A propos des *Timarcha* nord-américains. *L'Entomologiste, Paris* **45** (1), 27–34.

JOLIVET, P. (1989b). Un genre en danger de mort: *Timarcha. L'Entomologiste, Paris* **45** (6), 301–310.

JOLIVET, P. (1989c). The Chrysomelidae of *Cecropia* (Cecropiaceae). A strange cohabitation. *Entomography* **6**, 391–395.

JOLIVET, P. (1989d). Sélection trophique chez les Hispinae (Col. Chrys. Cryptostoma). *Bull. Soc. Linn. Lyon* **58** (9), 297–317.

JOLIVET, P. (1991). Le Doryphore menace l'Asie, *Leptinotarsa decemlineata* (Say) (Col. Chrys.). *L'Entomologiste, Paris* **47** (1), 29–47.

JOLIVET, P. (1992). Contribution à la taxonomie, la distribution et la biologie des *Chrysolina* nord-américains. *L'Entomologiste, Paris* **48** (1), 29–51.

JOLIVET, P. (1993). Mimétisme comportemental sous les tropiques. *Bull. ACOREP* **18**, 29–36.

JOLIVET, P. (1994a). Dernières nouvelles de la progression du Doryphore: *Leptinotarsa decemlineata* (Say, 1824) (Col. Chrys.). *L'Entomologiste, Paris* **50** (2), 105–111.

JOLIVET, P. (1994b). *Remarks on the biology and biogeography of* Timarcha *(Chrys.).* Proc. 3rd Int. Symp. Chrysomelidae, Beijing 1992, pp 85–97. Leiden, The Netherlands: Backhuys Publishers.

JOLIVET, P. (1994c). Physiological colour changes in tortoise beetles. In: *Novel aspects of the biology of Chrysomelidae*. Pp 331–335. Dordrecht, The Netherlands: Kluwer Academic Publishers.

JOLIVET, P. (1995). Preliminary observations on the genus *Iscadida* Chevrolat, 1843 in South Africa (Col. Chrys.). *Nouv. Rev. Ent. (N.S.)* **12** (4), 305–308.

JOLIVET, P. (1996a). *Ants and plants. An example of coevolution*. Pp 1–303. Leiden, The Netherlands: Backhuys Publishers.

JOLIVET, P. (1996b). A la poursuite du *Timarcha* perdu, *Timarcha melittensis* Weise (Col. Chrys.). *L'Entomologiste, Paris* **52** (6), 203–211.

JOLIVET, P. (1997). *Biologie des Coléoptères Chrysomélides*. Pp 1–279. Paris, France: Boubée Publishers.

JOLIVET, P. (1998a). Jurassic Park ou les Coléoptères des Cycadales. *Le Coléoptériste, Paris* **33**, 77–85.

JOLIVET, P. (1998b). *Interrelationship between insects and plants*. 309 pp. Boca Raton, FL, USA: CRC Press.

JOLIVET, P. (1998c). Les nouveaux envahisseurs ou les Chrysomélides voyageurs. *L'Entomologiste, Paris* **54** (1), 33–44.

JOLIVET, P. (1998d). Host plants of the Entomoscelina. Proc. *XXth Int. Congress Entomology, Firenze, Italy, 1996*. Eds. M. Biondi, M. Daccordi and D.G. Furth, pp 125–136. Torino, Italy: Museo Regionale Sc. Nat.

JOLIVET, P. (1998e). Manipulation du comportement chez les fourmis et les Coléoptères sous l'influence de leurs parasites. *L'Entomologiste, Paris* **54** (5), 211–222.

JOLIVET, P. (1999a). Timarchophilia or Timarchomania. Reflexions on the genus *Timarcha. Nouv. Rev. Ent. (N S.)* **16** (1), 11–18.

JOLIVET, P. (1999b). Sexual behaviour among Chrysomelidae. In: *Advances in Chrysomelidae biology. 1*. Ed. M.L. Cox, pp 391–409. Leiden, The Netherlands: Backhuys Publishers.

JOLIVET, P. (1999c). Du nouveau sur les *Metallotimarcha* Motschulsky (Col. Chrys.). *Le Coléoptériste, Paris* **37**, 197–201.

JOLIVET, P. (1999d). Les espèces du genre *Aulacophora* Chevrolat sont-elles polyphages comme celles des *Diabrotica* Chevrolat (Col. Chrys. Gal.)? *L'Entomologiste, Paris* **55** (6), 251–258.

JOLIVET, P. (1999e). La menace des Insectes. Un nouveau casse-tête pour les entomologistes: les bactéries des Insectes attaquent-elles aussi l'Homme? *L'Entomologiste, Paris* **55** (2), 73–78.

JOLIVET, P. (2001). What is a Chrysomelid? *Nouv. Revue Ent. (N.S.)* **18** (2), 135–146.

JOLIVET, P. (2002). *Subaquatic Chrysomelidae*. XXI Intern. Congress of Entomology, Iguaçu, Brazil, 2000. Ed. D.G. Furth. FISCB proceedings (in print).

JOLIVET, P. AND COX, M.L. (Eds.) (1996). *Chrysomelidae biology. 1. The classification, phylogeny and genetics. 2. Ecological studies. 3. General studies*. 1: 1–444; 2: 1–465; 3: 1–365. Amsterdam, The Netherlands: SPB Academic Publishers.

JOLIVET, P. AND HAWKESWOOD, T.J. (1995). *Host-plants of the Chrysomelidae of the world*. 281 pp. Leiden, The Netherlands: Backhuys Publishers.

JOLIVET, P. AND MAES, J.M. (1996). Un cas de cycloalexie chez un Curculionide: *Phelypera distigma* (Boheman) (Hyperinae) au Nicaragua. *L'Entomologiste, Paris* **52** (3), 97–100.

JOLIVET, P. AND PETITPIERRE, E. (1973). Plantes-hôtes connues des *Timarcha* Latreille. Quelques considérations sur les raisons possibles du trophisme sélectif. *Bull. Soc. Ent. Fr.* **78** (1–2), 9–25.

JOLIVET, P. AND PETITPIERRE, E. (1976a). Les plantes-hôtes connues des *Chrysolina*. Essai sur les types de sélection trophique. *Ann. Soc. Ent. Fr. (N.S.)* **12** (1), 123–149.

JOLIVET, P. AND PETITPIERRE, E. (1976b). Sélection trophique et évolution chromosomique chez les Chrysomelinae (Col. Chrys.). *Acta Zool. Pathol. Antwerp* **66**, 59–90.

JOLIVET, P. AND PETITPIERRE, E. (1981). Biology of Chrysomelidae (Col.). *Bull. Inst. Cat. Hist. Nat.* **47** (Sec. Zool. 4), 105–138.

JOLIVET, P. AND THÉODORIDÈS, J. (1951–1953). Les parasites, phorétiques et prédateurs des Chrysomeloidea (Col.). *Bull. Inst. Roy. Sc. Nat. Belgique* **27**, 1–55; **28**, 1–19; **29**, 1–15.

JOLIVET, P. AND VAN PARYS, E. (1977). Un cas inédit de mimétisme agressif entre un Chrysomélide (*Mesoplatys cincta* Olivier) et un Carabique (*Cyaneodinodes ammon* Fabricius) (Col.). *Bull. Soc. Linn. Lyon* **46** (6), 168–180.

JOLIVET, P. AND VASCONCELLOS-NETO, J. (1993). Un genre aptère de Coléoptères Chrysomélides: *Elytrosphaera* en voie d'extinction et sa distribution dans le SE Brésilien (Col. Chrys.). *Nouv. Rev. Ent. (N.S.)* **10** (4), 321–325.

JOLIVET, P., PETITPIERRE, E. AND DACCORDI, M. (1986). Les plantes-hotes des Chrysomelinae. Quelques nouvelles précisions et additions. *Nouv. Rev. Ent. (N.S.)* **3** (3), 341–357.

JOLIVET, P., PETITPIERRE, E. AND HSIAO, T.H. (Eds.) (1988). *Biology of Chrysomelidae*. xxvii+615 pp. Dordrecht, The Netherlands: Kluwer Academic Publishers.

JOLIVET, P., VASCONCELLOS-NETO, J. AND WEINSTEIN, P. (1990). Cycloalexy: a new concept in the larval defense of insects. *Insecta Mundi* **4** (1–4), 133–142.

JOLIVET, P., COX, M.L. AND PETITPIERRE, E. (Eds.) (1994). *Novel aspects of the biology of Chrysomelidae*. xxiii+582 pp. Dordrecht, The Netherlands: Kluwer Academic Publishers.

JONES, D.L. (1993). *Cycads of the world*. Pp 1–312. Washington, DC: Smithsonian International Press.

JUANJIE, T. AND WANG, S.Y. (1984). New species of Eumolpinae (Col.) from the Hengduan Mountains, Yunnan. *Acta Zootaxonomica Sinica* **9** (1), 55–58.

KALAICHELVAN, T. AND VERMA, K.K. (2000). Fecal cover for eggs of Indian Cassidines (Col. Chrys.). *Insect Environment* **6** (1), 41–42.

KALAICHELVAN, T., VERMA, K.K. AND SHARMA, B.N. Experimental, morphological and ecological approach to the taxonomy of some Oriental species of *Lema* (Col., Chrys.) (in print).

KALSHOVEN, L.G.E. (1951). Coleoptera Chrysomelidae Hispidae. In: *De Plagen vand de Culturgewassen in Indonesie, Jakarta* 2. Pp 724–762.

KALSHOVEN, L.G.E. (1957). An analysis of ethological, biological and taxonomic data on Oriental Hispinae (Col. Chrys.). *Tijdschr. Entomol.* **100** (1), 5–24.

KALSHOVEN, L.G.E. (1981). *Pests of crops in Indonesia*. Pp 438–458. Jakarta: Ichtiar Baru-Van Hoeve.

KANGAS, E. (1967). Identification of the Coleoptera collected by the Finnish Spitzbergen expedition. *Ann. Entomol. Fenn.* **33** (1), 41–43.

KARREN, J.B. (1964). Protective coloration and form in the North American genus *Exema* (Chrys. Col.). *Proc. North Central Branch E.S.A.* **19**, 77–79.

KASAP, H. AND CROWSON, R.A. (1976). On systematic relations of *Oomorphus concolor* (Sturm) (Col. Chrys.), with description of its larva and of an aberrant cryptocephaline larva from Australia. *J. Nat. Hist.* **10** (1), 99–112.

KASAP, H. AND CROWSON, R.A. (1979). The male reproductive organs of Bruchidae and Chrysomelidae (Col.). *Türkiye Bitki Koruma Dergisi* **3** (4), 199–216.
KASAP, H. AND CROWSON, R.A. (1980). The female reproductive organs of the Bruchidae and Chrysomelidae (Col.). *Türkiye Bitki Koruma Dergisi* **4**, 85–102.
KASTON, B.J. (1936). The morphology of the elm bark beetle (*Hylurgopinus rufipes*). *Conn. Agric. Exp. Sta. Bull.* **387**, 613–650.
KASUYA, E. (1985). Size-disassortive mating in the chrysomelid beetle *Chrysolina aurichalcea* (Col. Chrys.). *Evolution* **39** (3), 705–707.
KATO, M. (1991). Leaf-mining chrysomelids reared from pteridophytes. *Jap. J. Entomol.* **59** (3), 671–674.
KAUFMANN, T. (1970). Studies on biology and ecology of *Pyrrhalta nymphaeae* (Col. Chrys.) in Alaska with special reference to population dynamics. *American Midland Nat.* **83** (2), 496–509.
KHATIB, S.M.H. (1946a). The external morphology of *Galerucella birmanica* Jacoby (Col. Phytophaga Chrys. Gal.). *Proc. Ind. Acad. Sci.* **23**, 1–38.
KHATIB, S.M.H. (1946b). The internal anatomy of *Galerucella birmanica* Jac. *Proc. Ind. Acad. Sci.* **24**, 35–54.
KHRULEVA, O.A. (1994). Life cycle of the leaf cutting beetle *Chrysolina subsulcata* (Col. Chrys.) on Wrangel Island. *Entomol. Rev.* **73** (7), 117–125.
KHRULEVA, O.A. (1996). Biology of Arctic leaf beetle *Chrysolina cavigera* on Wrangel Island. In: *Chrysomelidae biology. 3. General studies*. Eds. P. Jolivet and M.L. Cox, pp 259–270.
KHRULEVA, O.A. AND KOROTYAEV, B.A. (1999). Weevils (Col. Apionidae. Curculionidae) of Wrangel Island. *Entomol. Rev.* **79** (9), 119–137.
KIM, P.I. AND VASIL'EV, YU.E. (1986). Focus of Colorado potato beetle liquidated. *Zashchita Rastenii* **10**, 43.
KIMOTO, S. (1964–1966). The Chrysomelidae of Japan and the Ryukyu islands. *J. Fac. Agr. Kyushu Univ.* **13** (1), 99–164; **13** (2), 235–308; **13** (3), 369–459; **13** (4), 602–671.
KIMOTO, S. (1984). Chrysomelidae. In: *The Coleoptera of Japan in color*. Eds. Hayashi *et al.*, pp 147–224. Osaka, Japan 4: Hoikusha Publishing Company.
KIMOTO, S. AND TAKIZAWA, H. (1994). *Leaf beetles (Chrysomelidae) of Japan*. 539 pp. Tokyo, Japan: Tokai University Press.
KIMOTO, S. AND TAKIZAWA, H. (1997). *Leaf beetles (Chrysomelidae) of Taiwan*. 581 pp. Tokyo, Japan: Tokai University Press.
KINCAID, T. (1900). The metamorphosis of some Alaska Coleoptera. *Proc. Wash. Acad. Sci.* **2**, 367–388.
KINGSOLVER, J.M. (1995). On the family Bruchidae. *Chrysomela* **30**, 3.
KINGSOLVER, J.M. AND PFAFFENBERGER, G.S. (1980). Systematic relationship of the genus *Rhaebus* (Col. Bruchidae). *Proc. Entomol. Soc. Washington* **82** (2), 293–311.
KIRK, H.M. (1988). Cannibalism in a chrysomelid beetle, *Gastrophysa viridula*. Doctoral Dissertation, Liverpool University, pp 1–289.
KIRKPATRICK, T.W. (1957). *Insect life in the tropics*. Pp xiv + 311. London: Longmans, Green & Co. Publishers.
KLAUSNITZER, B. AND FÖRSTER, G. (1971). Zur Eimorphologie einiger mitteleuropaïscher Chrysomelidae (Col.). *Polskie Pismo Ent.* **41**, 429–437.
KNAB, F. (1909). Nuptial colors in the Chrysomelidae. *Proc. Entomol. Soc. Washington* **11**, 151–153.
KNOWLES, L.L., LEVY, A., MCNELLIS, J.M., GREENE, K.P. AND FUTUYMA, D.J. (1999a). Tests of inbreeding effects on host-shift potential in the phytophagous beetle *Ophraella communa*. *Evolution* **53** (2), 561–567.
KNOWLES, L.L., FUTUYMA, D.J., EANES, W.F. AND RANNALA, B. (1999b). Insight into speciation from historical demography in the phytophagous beetle genus *Ophraella*. *Evolution* **53** (6), 1846–1856.
KOGAN, M. AND GOEDEN, R.D. (1970). The biology of *Lema trilineata daturaphila* (Col. Chrys.) with notes on efficiency of food utilization by larvae. *Ann. Entomol. Soc. America* **63** (2), 537–546.
KONTKANEN, P. (1959). Uber einige *Chrysolina*- Arten der Sectio *Pleurosticha* Motsch. Sensu Jacobson 1910 (Col. Chrys.). *Ann. Entomol. Fenn.* **25**, 27–35.

KOVALEV, O.V. AND MEDVEDEV, L.N. (1983). Theoretical principles for the introduction of Ambrosia Leaf Beetles of the genus *Zygogramma* Chevr. (Col. Chrys.) into the USSR for the biological control of *Ambrosia*. *Entomol. Rev., Washington* **62** (1), 1–19.

KOVALEV, O.V., REZNIK, S.YA. AND CHERKASHIN, V.N. (1983). Features of the method for using leaf beetles of the genus *Zygogramma* Chevr. (Col. Chrys.) in the biological control of ragweeds (*Ambrosia artemisiifolia* L., *A. psilostachya* D.C.). *Entomol. Rev., Washington* **62** (2), 169–175.

KRAFSUR, E.S. (1995). Gene flow between univoltine and semivoltine Northern Corn Rootworm (Col. Chrys.) populations. *Ann. Entomol. Soc. Am.* **88** (5), 699–704.

KRAFSUR, E.S. (1999). Allozyme gene diversity in some leaf beetles (Col. Chrys.). *Biochem. Genetics* **37** (7–8), 215–226.

KRAFSUR, E.S. AND NARIBOLI, P. (1995). Elm leaf beetles have greatly reduced levels of gene diversity. *Biochem. Genetics* **33** (3–4), 91–95.

KRAFSUR, E.S., NARIBOLI, P. AND TOLLEFSON, J.J. (1993). Gene diversity in natural *Diabrotica barberi* Smith and Lawrence population (Col. Chrys.). *Ann. Entomol. Soc. Amer.* **86** (4), 490–496.

KRESLAVSKII, A.G., SOLOMATIN, V.M., MIKHEEV, A.V. AND GRITSENKO, V.V. (1976). Intrapopulation ecological differentiation in the chrysomelid beetle *Chrysochloa cacaliae*. *Zool. Zhurnal* **55** (8), 1163–1171.

KRYSAN, J.L. (1999). Selected topics in the biology of *Diabrotica*. In: *Advances in Chrysomelidae biology. 1*. Ed. M.L.Cox, pp 479–513. Leiden, The Netherlands: Backhuys Publishers.

KRYSAN, J.L. AND GUSS, P.L. (1978). Barriers to hybridization between *Diabrotica virgifera* and *D. longicornis barberi* (Col. Chrys.). *Ann. Entomol. Soc. Am.* **71**, 931–934.

KRYSAN, J.L. AND SMITH, R.F. (1987). Systematic of the *virgifera* group of *Diabrotica* (Col. Chrys.). *Entomography* **5**, 375–484.

KRZEMINSKA, E., KRZEMINSKI, W., HAENNI, J.P. AND DUFOUR, C. (1992). *Les Fantômes de l'Ambre*. 142 pp. Switzerland: Musée d'Histoire Naturelle Neuchâtel.

KRZYMANSKA, J., WALIGORA, D. AND WODA-LESNIEWSKA, M. (1993). The development of the larvae of Colorado potato beetle (*Leptinotarsa decemlineata* Say) on an artificial diet with the addition of hexane extract from potato leaves. *Roczniki Nauk Rolniczych. Seria E., Ochrona Roslin* **22** (1–2), 79–82.

KUDO, S. AND ISHIBASHI, E. (1995). Notes on maternal care in the ovoviviparous leaf beetle *Gonioctena japonica* (Col. Chrys.). *Canadian Entomol.* **127**, 275–276.

KUDO, S., ISHIBASHI, E. AND MAKINO, S. (1995). Reproductive and subsocial behavior in the ovoviviparous leaf beetle *Gonioctena sibirica* (Col. Chrys.). *Ecol. Entomol.* **20**, 367–373.

KULIK, A.V. (1991). What does the Colorado beetle feed on? *Zashchita Rastenii* **7**, 38.

KULIK, A.V. AND TIMOSHIN, A.A. (1986). Unusual tastes of the Colorado Beetle. *Proroda Mask.* **10**, 98.

KUMAR, D. AND VERMA, K.K. (1971). 'Retournement' of the aedeagus in Chrysomelidae (Col. Phytophaga). *J. Nat. Hist.* **5**, 635–642.

KUMAR, D. AND VERMA, K.K. (1978). An endocrine control of 'retournement' of the aedeagus in the tortoise beetle, *Aspidomorpha miliaris* F. *Indian J. Exp. Biol.* **16**, 639–641.

KUMAR, D. AND VERMA, K.K. (1980). Aedeagal musculature in Phytophaga (Col.). *J. Nat. Hist.* **14**, 237–270.

KÜNTZEN, H. (1919). Skizze zur Verbreitung einiger flugunfähiger Blattkäfer (*Metallotimarcha*). Sizungb. der Gesell. Naturf. *Freunde zu Berlin*, 228–250.

KURCHEVA, G.F. (1958). Soil dwelling larvae of Eumolpinae (Col. Chrys.) the commonest subfamily of leaf-beetles in the European part of the USSSR. *Acta Soc. Entomol. Czeckoslov*. **55**, 383–393.

KURCHEVA, G.F. (1967). Leaf-beetle larvae of the subfamily Eumolpinae and of *Syneta betulae* F. (Col. Chrys.). *Entomol. Rev., Washington* **46**, 132–137.

KURCHEVA, G.F. (1975). Larvae of *Basilepta fulvipes* Motsch. (Col. Chrys.) from Southern Primorge. *Trud. Biol. Pech. Inst. Vlad.* **28**, 107–113.

KURSAR, T.A. AND COLEY, P.D. (1992). Delayed greening in tropical leaves: an antiherbivore defense? *Biotropica* **24** (2b), 256–262.

KUSCHEL, G. AND MAY, B.M. (1990). Palophaginae, a new subfamily for leaf-beetles feeding as adult and larva on Araucarian pollen in Australia (Col. Megalop.). *Invertebrate Taxonomy* **3**, 697–719.

KUSCHEL, G. AND MAY, B.M. (1996a). Palophaginae, their systematic position and biology. In: *Chrysomelidae biology. 3. General studies*. Eds. P. Jolivet and M.L. Cox, pp 173–185. Amsterdam, The Netherlands: SPB Academic Publishers.

KUSCHEL, G. AND MAY, B.M. (1996b). Discovery of Palophaginae (Col. Megal.) on *Araucaria araucana* in Chile and Argentina. *New Zealand Entomol.* **19**, 1–13.

LABANDEIRA, C.C. (1998a). The role of insects in Late Jurassic to Middle Cretaceous ecosystem. *New Mexico Museum of Nat. Hist. and Science Bull.* **14**, 105–124.

LABANDEIRA, C.C. (1998b). How old is the flower and the fly? *Science* **280**, 57–59.

LABANDEIRA, C.C. (1999). Insects and other Hexapods. In: *Encyclopedia of paleontology. 1 (A–L)*. Ed. R. Singer, pp 603–624.

LABANDEIRA, C.C. AND SEPKOSKI, J.J. (1993). Insect diversity in the Fossil Record. *Science* **261**, 310–315.

LABOISSIÈRE, V. (1917). Diagnoses de Galerucini nouveaux d'Afrique. *Bull. Soc. Ent. Fr.* **18**, 327–329.

LABOISSIÈRE, V. (1921–1922). Etude des Galerucini de la collection du Musée du Congo Belge. *Rev. Zool. Afr.* **9** (1), 33–130; **10** (1), 87–130; **10** (3), 131–183.

LABOISSIÈRE, V. (1924). Description de trois espèces nouvelles de Galerucini recueillies en Afrique méridionale par R. Ellenberger (1914–1915). *C.R. Congr. Soc. Savantes. Sciences*, 232–239.

LABOISSIÈRE, V. (1929–1933–1936). Observations sur les Galerucini asiatiques principalement du Tonkin et du Yunnan et descriptions de nouveaux genres et espèces. *Ann. Soc. Entom. Fr.* **98**, 251–288; **102**, 51–72; **105**, 239–262.

LABOISSIÈRE, V. (1932). *Synopsis des genres de Galerucini de Madagascar*. Pp 575–592. Paris: Livre Centenaire Soc. Ent. Fr.

LABOISSIÈRE, V. (1934). Galerucinae de la Faune Française (Col.). *Ann. Soc. Ent. Fr.* **103**, 1–108.

LABOISSIÈRE, V. (1939). Notes sur les Halticinae de la collection du Musée du Congo. *Rev. Zool. Bot. Afr.* **32** (3–4), 394–407.

LACORDAIRE, TH. (1845–1848). *Monographie des Coléoptères Subpentamères de la famille des Phytophages*. 1: liii+740 pp; 2: vi+890 pp. Liège.

LADLE, R.J. AND FOSTER, E. (1992). Are giant sperm copulatory plugs? *Acta Oecologica* **13** (5), 635–638.

LANDA, V. (1960). Development and function of imaginal reproductive organs of the cockchafer, *Melolontha melolontha* L. In: *Ontogeny of insects*. Ed. D. Hardy. London and New York: Academic Press.

LARSSON, S.G. (1959). Zoology of Iceland. Coleoptera 2. General remarks. *E. Munksgaard Publs., Copenhagen* **3** (466), 1–85.

LARSSON, S.G. (1978). *Baltic amber. A paleobiological study*. 192 pp. Klampenborg, Denmark: Scandinavian Science Press Ltd.

LARSSON, S.G. AND GIGJA, G. (1959). Zoology of Iceland. Coleoptera 1. Synopsis. *E. Munksgaard, Copenhagen* **3** (46a), 1–218.

LAWRENCE, J.F. (1990). Order Coleoptera. In: *Immature insects. 2*. Ed. F.W. Stehr, pp 144–658. Dubuque, Iowa, USA: Kendall/Hunt.

LAWRENCE, J.F. (1992). Coleoptera. In: *Synopsis and classification of living organisms*. Ed. S.P. Parker, pp 482–553.

LAWRENCE, J.F. AND BRITTON, E.B. (1994). *Australian beetles*. 192 pp. Australia: Melbourne University Press.

LAWRENCE, J.F. AND NEWTON, A.F. (1995). Families and subfamilies of Coleoptera. In: *Biology, phylogeny and classification of Coleoptera. Papers celebrating the 80th birthday of Roy A. Crowson*. Eds. J. Pakaluk and S.A. Slipinsky, pp 779–1092. Warsaw, Poland: Muzeum i Instytut Zoologii.

LAWRENCE, J.F., HASTINGS, A.M., DALLWITZ, M.J., PAINE, T.A. AND ZURCHER, E.J. (1999a). *Beetles of the world. A key and information system for families and subfamilies*. CD-ROM, Version I. Melbourne, Australia: CSIRO Publishing.

LAWRENCE, J.F., HASTINGS, A.M., DALLWITZ, M.J., PAINE, T.A. AND ZURCHER, E.J. (1999b). *Beetle larvae of the world*. CD-ROM. Melbourne, Australia: CSIRO Publishing.

LE BOURGEOIS, T. (1992). Exemple de destruction d'une adventice tropicale *Sesbania pachycarpa* (Fabaceae) par un ravageur naturel *Mesoplatys cincta* (Chrys.) au Nord Cameroun. *Entomophaga* **37** (4), 609–611.

LECAILLON, A. (1898a). Sur les enveloppes ovulaires de quelques Chrysomélides. *Arch. Anat. Microsc.* **2**, 89–117.

LECAILLON, A. (1898b). *Recherches sur l'œuf et sur le développement embryonnaire de quelques Chrysomélides*. 219 pp. Thése, Fac. Sc. Paris.

LEE, J.E. (1990). Description of first instar larva of *Syneta adamsi* Baly from Japan, with notes on the systematic position of Synetinae(Col. Chrys.). *Esakia* **29**, 77–81.

LEE, J.E. (1993). Phylogenetic studies on the larvae of the Chrysomelidae (Col.) from Japan. *Jpn. J.Ent.* **61** (3), 409–424.

LESAGE, L. (1982). The immature stage of *Exema canadensis* Pierce (Col. Chrys.). *Coleopt. Bull.* **36** (2), 318–327.

LESAGE, L. (1984). Egg, larva and pupa of *Lexiphanes saponatus* (Col. Chrys. Crypt.). *Canad. Entomol.* **116** (4), 537–548.

LESAGE, L. (1988a). Notes on European *Longitarsus* species introduced in North America (Col. Chrys. Alt.). *Canad. Entomol.* **120**, 1133–1145.

LESAGE, L. (1988b). *Longitarsus huberi* n. sp. and the related species *L. mancus* LeConte (Col. Alt.). *Col. Bull.* **42** (2), 167–172.

LESAGE, L. (1997). Description d'une nouvelle espèce de *Clavicornaltica* en provenance de Taiwan (Col. Chrys.). *Nouv. Rev. Ent. (N.S.)* **14** (3), 239–247.

LESAGE, L. AND DENIS, J. (1999). The flea-beetle *Altica corni* Woods in North America (Col. Chrys. Alt.). In: *Advances in Chrysomelidae biology*. Ed. M.L. Cox, pp 533–544. Leiden, The Netherlands: Backhuys Publishers.

LESAGE, L. AND STIEFEL, V.L. (1996). Biology and immature stages of the North American Clytrines *Anomoea laticlavia* (Forster) and *A. flavokansiensis* Moldenke (Col. Chrys. Clytrinae). In: *Chrysomelidae biology. 3. General studies*. Eds. P. Jolivet and M.L. Cox, pp 217–238. Amsterdam, The Netherlands: SPB Academic Publishers.

LEVER, R. (1930). A new endosternal organ in the hind leg of Halticinae. *Zool. Anz.* **92**, 287–288.

LEWIS, S.E. AND CAROLL, M.A. (1991). Coleopterous eggs deposit on alder leaves from the Klondike Mountain (middle Eocene) northeastern Washington. *J. Paleontol.* **65** (2), 334–335.

LI, S.W. (1989). Allozyme variation in six species of *Leptinotarsa* (Col. Chrys.). *Acta Entomol. Sinica* **32** (3), 263–270.

LINDROTH, C.H. (1971). Disappearance as a protective factor. A supposed case of Batesian mimicry among beetles (Col. Carabidae and Chrysomelidae). *Entomol. Scand.* **2**, 41–48.

LINGAFELTER, S.W. AND KONSTANTINOV, A.S. (2000). The monophyly and relative rank of alticine and galerucine leaf beetles. A cladistic analysis using adult morphological characters (Col. Chrys.). *Ent. Scand.* **30**, 397–416.

LINGAFELTER, S.W. AND PAKALUK, J. (1997). Comments on the bruchine Chrysomelidae. *Chrysomela* **33**, 3.

LOKKI, J., SAURA, A., LANKINEN, P. AND SUOMALAINEN, E. (1976). Genetic polymorphism and evolution in parthenogenetic animals. V. Triploid *Adoxus obscurus* (Col. Chrys.). *Genet. Res.* **28**, 27–38.

LOPATIN, I.K. (1984). *Leaf beetles (Chrys.) of Central Asia and Kazakhstan*. 416 pp. New Delhi: Oxonian Press.

LOPATIN, L.K. (1996). High altitude fauna of the Chrysomelidae of Central Asia: Biology and Biogeography. In: *Chrysomelidae biology. 3. General studies*. Eds. P. Jolivet and M.L. Cox, pp 3–12. Amsterdam, The Netherlands: SPB Academic Publishers.

LOPATIN, L.K. (1999). Biology and biogeography of desert leaf-beetles of Central Asia. In: *Advances in Chrysomelidae biology. 1*. Ed. M.L. Cox, pp 159–168. Leiden, The Netherlands: Backhuys Publishers.

LOPEZ, E.R., ROTH, L.C., FERRO, D.N., HOSMER, D. AND MAFRA-NETO, A. (1997). Behavioral ecology of *Myiopharus doryphorae* (Riley) and *M. aberrans* (Townsend), tachinid parasitoids of the Colorado Potato Beetle. *J. Insect. Behav.* **10** (1), 49–78.

MACARTHUR, R.H. AND WILSON, E.O. (1967). *The theory of island biogeography*. 203 pp. Princeton: Princeton University Press.

MACEDO, M.V., VASCONCELLOS-NETO, J. AND JOLIVET, P. (1998). New biological data on the apterous beetle *Elytrosphaera lahtivirtai* Bechyne (Chrys.) and remarks on the biology and distribution of the genus. *Proc. XX ICE, Florence, Italy*. Eds. M. Biondi, M. Daccordi and D.C. Furth, pp 271–279. Torino, Italy: Museo Regionale di Scienze Naturali.

MAES, J.M. (1999). *Insectos de Nicaragua*. 3 vols. 1898 pp. Leon, Nicaragua.

MAES, J.M. AND STAINES, C.L. (1991). Catalogo de los Chrysomelidae (Col.) en Nicaragua. *Rev. Nicarag. Entomol.* **18**, 1–53.

MAFRA-NETO, A. AND JOLIVET, P. (1994). Entomophagy in Chrysomelidae; adult *Aristobrotica angulicollis* (Erichson) feeding on adult Meloids. In: *Novel aspects of the biology of Chrysomelidae*. Eds. P. Jolivet, M.L. Cox and E. Petitpierre, pp 171–178. Dordrecht, The Netherlands: Kluwer Academic Publishers.

MAFRA-NETO, A. AND JOLIVET, P. (1996). Cannibalism in leaf-beetles. In: *Chrysomelidae biology. 2. Ecological studies*. Eds. P. Jolivet and M.L. Cox, pp 195–211. Amsterdam, The Netherlands: SPB Academic Publishers.

MAGIS, N. (1992). Les parades nuptiales chez les Coléoptères. *Revue Verviétoise d'Hist. Nat.* **1**, 3–9.

MALECKI, R.A., BLOSSEY, B., HIGHT, S.D.,SCHROEDER, D., KOK, L.T. AND COULSON, J.R. (1993). Biological control of Purple Loosestrife. *BioScience* **43** (10), 680–686.

MANEVAL, H. (1938). La ponte ovovivipare de *Chrysochloa viridis*. *Misc. Entom.* **39**, 99–101.

MANGUIN, S., WHITE, R., BLOSSEY, B. AND HIGHT, S.D. (1993). Genetics, taxonomy, and ecology of certain species of *Galerucella* (Col. Chrys.). *Ann. Entomol. Soc. America* **86** (4), 397–410.

MANI, M.S. (1974). *Fundamentals of high altitude biology*. Bombay: Oxford and IBM Publishing Co.

MANN, J.S. (1985). Sensory response and feeding behaviour of chrysomelids (Col. Chrys.) associated with Ber plant. *J. Res. Punjab Agric. Univ.* **22** (2), 417–420.

MANN, J.S. AND CROWSON, R.A. (1981). The systematic positions of *Orsodacne* Latr. and *Syneta* Lacord. (Col. Chrys.) in relation to characters of larvae, internal anatomy and tarsal vestiture. *J. Nat. Hist.* **15**, 727–749.

MANN, J.S. AND CROWSON, R.A. (1983a). Phylogenetic significances of the ventral nerve cord in the Chrysomeloidea. *Syst. Entomol.* **8**, 103–119.

MANN, J.S. AND CROWSON, R.A. (1983b). On the internal male reproductive organs and their taxonomic significance in the leaf beetle (Col. Chrys.). *Entomol. Gener.* **9** (1–2), 75–99.

MANN, J.S. AND CROWSON, R.A. (1983c). On the occurrence of mid-gut caeca and organs of symbiont transmission in leaf beetles (Col.). *Col. Bull.* **37** (1), 1–15.

MANN, J.S. AND CROWSON, R.A. (1983d). Observations on the internal anatomy and classification of Donaciinae (Col. Chrys.). *Entomol. Month. Mag.* **119**, 17–27.

MANN, J.S. AND CROWSON, R.A. (1984). On the digitiform sensilla of adult leaf beetles (Col. Chrys.). *Entomol. Gener.* **9** (3), 121–133.

MANN, J.S. AND CROWSON, R.A. (1991). Some observations on the genitalia of Sagrinae (Col. Chrys.). In: *Advances in coleopterology*. Eds. Zunino *et al.*, pp 35–60. Barcelona: AEC.

MANN, J.S. AND CROWSON, R.A. (1996). Internal sac structure and phylogeny of Chrysomelidae. In: *Chrysomelidae biology. I. The classification, phylogeny and genetics*. Eds. P. Jolivet and M.L. Cox, pp 291–316. Amsterdam, The Netherlands: SPB Academic Publishers.

MANN, J.S. AND SINGH, J.P. (1979). Ovariole number in the family Chrysomelidae (Col. Phyt.) from Northern India. *J. Ent. Res.* **3** (2), 217–222.

MANSINGH, A. (1971). Physiological classification of dormancies in insects. *Canad. Entomol.* **103**, 983–1009.

MARDULYN, P., MILINKOVITCH, M.C. AND PASTEELS, J.M. (1997). Phylogenetic analyses of DNA and allozyme data suggest that *Gonioctema* leaf beetles (Col. Chrys.) experienced convergent evolution in their history of host-plant family shifts. *Syst. Biol.* **46** (4), 722–747.

MARIAU, D. (1988). The parasitoids of Hispinae. In: *Biology of Chrysomelidae*. Eds. P. Jolivet,

E. Petitpierre and T.H. Hsiao, pp 449–461. Dordrecht, The Netherlands: Kluwer Academic Publishers.

MARIAU, D. (1999). Les Coléoptères Chrysomélides inféodés au Palmier à huile et au Cocotier et leurs parasitoides. *Ann. Soc. Ent. Fr. (N.S.)* (Suppl.), 230–237.

MAROHASY, J. (1994). Biology and host specificity of *Weisena barkeri* (Col. Chrys.), a biological control agent for *Acacia nilotica* (Mimosaceae). *Entomophaga* **39** (3–4), 335–340.

MARSHALL, J.E. (1979). The larvae of the British species of *Chrysolina* (Chrys.). *Syst. Entomol.* **4**, 409–417.

MARTYNOV, A.V. (1935). Note on the fossil Mesozoic deposits in Cheliabinsk district. *Trudy Paleozoologicheskago Inst., Akad. Nauk SSSR* **4**, 37–48.

MATHEWS, S. AND DONOGHUE, M.J. (1999). The pedigree of the Angiosperm phylogeny inferred from duplicate phytochrome genes. *Science* **286**, 947–950.

MATSUDA, K. (1982). Reflex bleeding in *Gallerucida nigromaculata* Baly (Col. Chrys.). *Appl. Entomol. Zool.* **17** (2), 277–278.

MATSUDA, K. (1988). Feeding stimulants of leaf beetles. In: *Biology of Chrysomelidae*. Eds. P. Jolivet, E. Petitpierre and T.H. Hsiao, pp 41–56. Dordrecht, The Netherlands: Kluwer Academic Publishers.

MATSUDA, R. (1976). *Morphology and evolution of the insect abdomen*. Oxford, UK and New York, USA: Pergamon Press.

MATTHEWS, R.W. AND MATTHEWS, J.R. (1978). *Insect behavior*. xiii + 507 pp. New York, NY, USA: Wiley and Sons Publishers.

MAULIK, S. (1919). Coleoptera Chrysomelidae: Hispinae and Cassidinae. In: *Fauna of British India*. 439 pp. London, UK: Taylor and Francis Publishers.

MAULIK, S. (1926). Coleoptera Chrysomelidae: Chrysomelinae and Halticinae. In: *Fauna of British India*. 442 pp. London, UK: Taylor and Francis Publishers.

MAULIK, S. (1929a). New injurious Hispinae. *Bull. Entomol. Res.* **20** (1), 81–94.

MAULIK, S. (1929b). Injurious Hispinae from the Solomon Islands. *Bull. Entomol. Res.* **20** (2), 233–239.

MAULIK, S. (1932). On a structure in the antennae of beetles of the chrysomelid genus *Agetocera*. *Proc. Zool. Soc. London* **4**, 943–956.

MAULIK, S. (1936). Coleoptera Chrysomelidae: Galerucinae. In: *Fauna of British India*. 645 pp. London, U K: Taylor and Francis Publishers.

MAULIK, S. (1937). Distributional correlation between hispine beetles and their host-plants. *Proc. Zool. Soc. London A* **2**, 129–159.

MAULIK, S. (1939). The geographic distribution of European hispine beetles (Chrys. Col.). *Proc. Zool. Soc. London B* **109**, (2), 131–152.

MAULIK, S. (1941). Biology and morphology of the Sagrinae (Col. Chrys.). *Ann. Mag. Nat. Hist.* **2** (7), 235–254.

MAY, R.M. (1988). How many species are there on Earth? *Science* **241**, 1441–1449.

MAYR, E. AND ASHLOCK, P.D. (1991). *Principles of systematic zoology (2nd ed.)*. 475 pp. New York, USA: McGraw Hill Inc.

MCBRIDE, J.A., BACH, C.E. AND WALKER, G.K. (2000). Developmental changes in the caudal and lateral processes of larvae of *Aspidomorpha deusta* (Fabricius) (Col. Chrys. Cass.). *Australian J. Entomol.* **39**, 167–170.

MCCAULEY, D.E. AND O'DONNELL, R. (1984). The effect of multiple mating and generic relatedness in larval aggregations of the imported willow leaf beetle (*Plagiodera versicolora*). *Behav. Ecol. Sociobiol.* **15**, 287–291.

MCINTYRE, P. AND CAVENEY, S. (1998). Superposition optics and the tune of flight in Onitine dung beetles. *Comparative Physiology A* **183**, 45–60.

MEDVEDEV, L.N. (1962). On the functional importance of secondary sexual characters in chrysomelid beetles (Col. Chrys.). *Zool. Zhurnal.* **41** (1), 77–84.

MEDVEDEV, L.N. (1968). Leaf-beetles of the Kara Tau Jurassic. In: *Symposium of Jurassic Insects of Kara Tau*. Pp 155–165. Moscow, Russia: Nauka Publishing. House.

MEDVEDEV, L.N. (1971). The ways of evolution and phylogeny of Chrysomelidae (Col.). *Proc. XIII Int. Congress of Entomology, Moscow 1968*, 271–272.

MEDVEDEV, L.N. (1996a). Leaf beetles in the Arctic. In: *Chrysomelidae biology. 3. General studies*. Eds. P. Jolivet and M.L. Cox, pp 57–62. Amsterdam, The Netherlands: SPB Academic Publishers.

MEDVEDEV, L.N. (1996b). The Chrysomelidae of Arabia. In: *Fauna of Saudi Arabia*. Pp 211–263. Basle, Switzerland 15: Natural History Museum.

MEDVEDEV, L.N. (1998a). The first record of Chrysomelidae (Col.) from Severnaya Zemlya Archipelago. *Russian Entomol.* **7** (1–2), 41–42.

MEDVEDEV, L.N. (1998b). To the knowledge of the North American larvae of Clytrinae (Col.). *Latvijas Entomologs* **36**, 36–43.

MEDVEDEV, L.N. AND DANG, THI DAP (1982). Relations trophiques des Chrysomélides du Vietnam. *Animal Kingdom, Vietnam Moscow*, 84–97.

MEDVEDEV, L.N. AND KHRULEVA, S.A. (1986). *Chrysomelidae (Col.) of Wrangel Island*. Pp 135–145. Vladivostock, Russia: Akad. Nauk.

MEDVEDEV, L.N. AND OKHRIMENKO, N.V. (1991). Contribution to knowledge of leaf beetles of the genus *Chrysolina* Motsch. (Col. Chrys.) in the Caucasus. *Entomol. Obozrenie* **70** (4), 866–874.

MEDVEDEV, L.N. AND PAVLOV, S.I. (1988). Mating behavior of the Chrysomelidae (Col.). *Entomol. Rev., Washington* **67** (3), 100–108.

MEIDELL, E.M. (1983). Diapause, aerobic and anaerobic metabolism in alpine adult *Melasoma collaris* (Col.). *Acta Oecologica Scand.* **41**, 239–244.

METCALF, R.L. (1994). Chemical ecology of Diabroticites. In: *Novel aspects of the biology of Chrysomelidae*. Eds. P. Jolivet, M.L. Cox and E. Petitpierre, pp 153–169. Dordrecht, The Netherlands: Kluwer Academic Publishers.

METCALF, R.L., RHODES, A.M., METCALF, R.A., FERGUSON, J., METCALF, E.R. AND LU, P.Y. (1982). Cucurbitacin contents and diabroticite (Col. Chrys.) feeding upon *Cucurbita* spp. *Environmental Entomology* **11** (4), 931–937.

METCALFE, M.E. (1932). The structure and development of the reproductive system in the Coleoptera with notes on its homologies. *Quartely J. Microscop. Sci.* **75**, 49–129, pls. 7–10.

MITCHELL, B.K., ROLSETH, B.M. AND MCCASHIN, B.G. (1990). Differential responses of galeal gustatory sensilla of the adult Colorado Potato Beetle, *Leptinotarsa decemlineata* (Say), to leaf saps from host and non-host plants. *Physiol. Entomol.* **15** (1), 61–72.

MITTER, C. AND FARRELL, B. (1991). Macroevolutionary aspects if insect plant relationships. In: *Insect plant interactions*. Ed. E. Bernays, pp 35–78. Boca Raton, FL, USA: CRC Press.

MOHAN, J. AND VERMA, K.K. (1981). Histology of testis and proximal part of the male efferent genital system in *Aspidomorpha miliaris* F. (Col. Phyt. Chrys.). *Folia Morph. (Czech.)* **29**, 384–403.

MOHR, K.H. (1962). Bestimmungstabelle und Faunistik der Mitteleuropaischen *Longitarsus*-Arten. *Ent. Blätt.* **58** (2), 55–118.

MOLDENKE, A.R. (1971). Host-plant relations of phytophagous beetles in Mexico (Col. Bruch., Chrys. Curcul.). *Pan-Pacific Entomol.* **47** (2), 105–116.

MOLDENKE, A.F. AND BERRY, R.E. (1999). Biological control of Colorado Potato Beetle, *Leptinotarsa decemlineata* (Say) (Col. Chrys.). In: *Advances in Chrysomelidae biology. 1*. Ed. M.L. Cox, pp 169–184. Leiden, The Netherlands: Backhuys Publishers.

MONROS, F. (1949). Descripcion de los metamorfosis de *Lamprosoma chorisiae* Monros y consideraciones taxonomicas, sobre Laprosominae (Col. Chrys.). *Acta Zool. Lillioana. Inst. Miguel Lillo* **7**, 449–466.

MONROS, F. (1952). Notas sobre algunas Eumolpinae neotropicales (Col. Chrys.). *Rev. Chil. Entomol.* **2**, 187–196.

MONROS, F. (1953). Aulacoscelinae, eine neue Chrysomeliden-Unterfamilie, mit Beschreiburg einer neuen bolivianischen Gattung. *Ent. Arb. Mus. G. Frey* **4**, 19–25.

MONROS, F. (1954a). *Megalopus jacobyi*, nueva plaga de Solanaceas en le Noroeste Argentino, con notas sobre biologia y taxonomis de Megalopinae (Col. Chrys.). *Rev. Agronomica Nor. Arg.* **1** (2), 167–179.

MONROS, F. (1954b). Revision of the Chrysomelid subfamily Aulacoscelinae. *Bull. Mus. Comp. Zool.* **112** (4), 321–359.

MONROS, F. (1955a). Biologia y descripcion de la larva de *Atalasis sagroides* (Col. Chrys.). *Rev. Agronomica Nor. Arg.* **1** (3), 275–281.

MONROS, F. (1955b). Remarques sur les affinités des familles de Cerambycoidea (Col.). *Bull. Inst. R. Sc. Nat. Belgique* **31** (31), 1–7.

MONROS, F. (1956). Sur le genre *Megamerus* MacLeay. *Rev. Fr. Ent.* **23** (2), 104–115.

MONROS, F. (1958). Considerationes sobre la fauna del sur de Chile y revision de la tribus Stenomelini (Col. Chrys.). *Acta Zool. Lill.* **15**, 143–153.

MONROS, F. (1959). Los generos de Chrysomelidae (Col.). *Opera Lilloana* **3**, 1–337.

MOREIRA, C. (1913). Métamorphoses de quelques coléoptères du Brésil. *Ann. Soc. Ent. Fr.* **86**, 743–751.

MOZNETTE, G.F. (1916). The fruit tree leaf beetle *Syneta*. *J. Econ. Entomol.* **9** (5), 458–461.

MUIR, F. AND SHARP, D. (1904). On the egg cases and early stages of some Cassidinae. *Trans. R. Ent. Soc. London* **1**, 1–23.

MÜLLER, H.J. (1966). Probleme der Insektendiapause. *Verhandlungen der Deutschen Zoologischen Gessellschaft 1965 (Zool. Anzeig. Supplementband 29)*, 192–222.

MURRAY, F. AND TIEGS, O.W. (1935). The metamorphosis of *Calandra oryzae*. *Quart. J. of Microsc. Sci.* **77**, 405–495, pls.

NAHRUNG, H. AND MERRITT, D. (1999). Effect of mate availability on female longevity, fecundity and egg development of *Hemichloda barkeri* (Jacoby) (Col. Chrys.). *Col. Bull.* **53** (4), 329–332.

NAKAMURA, K. AND ABBAS, J. (1987). Preliminary life table of the spotted tortoise beetle, *Aspidomorpha miliaris* (Col. Chrys.) in Sumatra. *Res. Popul. Ecol.* **29**, 229–236.

NAKAMURA, K. AND ABBAS, J. (1989). Seasonal change in abundance and egg mortality of two tortoise beetles under a humid-equatorial climate in Sumatra. *Entomography* **6**, 487–495.

NARDI, G. AND BOLOGNA, M.A. (2000). Cantharidin attraction in *Pyrochroa* (Col. Pyrochr.). *Entomol. News* **111** (1), 74–76.

NICHOLS, S. W. (Ed.) (1989). *The Torre Bueno glossary of entomology*. 840 pp. American Museum Natural History, New York, NY, USA: New York Entomol. Soc.

NIIELSEN, J.K. (1988). Crucifer-feeding Chrysomelidae. Mechanisms of host plant finding and acceptance. In: *Biology of Chrysomelidae*. Eds. P. Jolivet, E. Petitpierre and T.H. Hsiao, pp 25–40. Dordrecht, The Netherlands: Kluwer Academic Publishers.

NIKLAS, K.J., CREPET, W.L. AND NIXON, K.C. (1999). Early plant history: something borrowed, something new? *Science* **285**, 1673.

NISHIDA, R. AND FUKAMI, H. (1990). Sequestration of distasteful compounds by some pharmacophagous insects. *J. Chem. Ecology* **16** (1), 151–164.

NISHIDA, R., YOKOYAMA, M. AND FUKAMI, H. (1992). Sequestration of cucurbitacin analogs by New and Old World chrysomelid leaf beetles in the tribe Luperini. *Chemoecology* **3** (1), 19–24.

NOKKALA, C. AND NOKKALA, S. (1998). Species and habitat races in the chrysomelid *Galerucella nymphaeae* species complex in Northern Europe. *Entomol. Exp. Appl.* **89** (1), 1–13.

NOKKALA, C., NOKKALA, S. AND NORDELL-PAAVOLA, A. (1998). European and North American populations of *Galerucella nymphaeae* (Col. Chrys.): two separate species revealed by chorion polypeptide analysis. *European J. Entomol.* **95** (2), 269–274.

NORDELL-PAAVOLA, A., NOKKALA, S., KOPONEN, S. AND NOKKALA, C. (1999). The utilization of chorion ultrastructure and chorion polypeptide analysis in recognizing taxonomic units in North European Galerucini (Col. Chrys.). In: *Advances in Chrysomelidae Biology. 1*. Ed. M.L. Cox, pp 95–104. Leiden, The Netherlands: Backhuys Publishers.

NORSTOG, K.J. AND NICHOLLS, T.J. (1997). *The biology of cycads*. Pp 1–364. Ithaca, NY, USA: Cornell University Press.

NOVOTNY, V., BASSET, Y., SAMUELSON, G. A. AND MILLER, S.E. (1999). Host use by chrysomelid beetles feeding on Moraceae and Euphorbiaceae in New Guinea. In: *Advances in Chrysomelidae biology. I.* Ed. M.L. Cox, pp 343–360. Leiden, The Netherlands: Backhuys Publishers.

NYFFELER, R. (1999). A new ordinal classification of the flowering plants. *TREE* **14** (5), 168–170.

OBERMAIER, E. AND ZWÖLFER, H. (1999). Plant quality or quantity? Host exploitation strategies in three Chrysomelidae species associated with Asteraceae host plants. *Entom. Exper. Applic.* **92**, 165–177.

OHBA, M. (1975). Studies on the pathogenesis of Chilo iridescent virus. 3. Multiplication of CIV in the silkworm *Bombyx mori* L. and field insects. *Sci. Bull. Fac. Agron. Kyushu Univ.* **30**, 71–81.

OLIVIER, A.G. (1791). Encyclopédie méthodique. Histoire Naturelle. *Insectes* **5** (2), 369–793, **6** (1), 1–368.

OLMSTEAD, K.L. (1994). Waste products as chrysomelid defenses. In: *Novel aspects of the biology of Chrysomelidae*. Eds. P. Jolivet, M.L. Cox and E. Petitpierre, pp 311–318. Dordrecht, The Netherlands: Kluwer Academic Publishers.

OLMSTEAD, K.L. (1996). Cassidine defenses and natural enemies. In: *Chrysomelidae biology. 2. Ecological studies*. Eds. P. Jolivet and M.L. Cox, pp 3–21. Amsterdam: SPB Academic Publishers.

OLMSTEAD, K.L. AND DENNO, R.F. (1993). Effectiveness of tortoise beetle larval shields against different predator species. *Ecology* **74**, 1394–1405.

ORFILA, R.N. (1927). Observaciones sobre partenogenesis. *Rev. Soc. Ent. Argentina* **1** (4), 71.

OSBORNE, J.A. (1880). Some facts in the life-history of *Gastrophysa raphani*. *Entom. Month. Mag.* **17**, 49–57.

OWEN, J.A. (1997). Some notes on the life-history of *Cryptocephalus 6-punctatus* Linnaeus (Col. Chrys.). *Entomologist's Rec. J. Var.* **109**, 43–48.

OWEN, J.A. (1999). Notes on the biology of *Cryptocephalus coryli* (Linnaeus) (Col. Chrys.). *Entomologist's Gazette* **50**, 199–204.

PAJNI, H.R. (1968). Development of the female genital ducts and the associated structures in *Callosobruchus maculatus* (F.) (Bruchidae, Col.). *Research Bull. Punjab Univ., N.S.* **19**, 341–348.

PAJNI, H.R., DEVI, N. AND TEWARI, P.K. (1985a). A comparative account of the internal reproductive organs in family Chrysomelidae (Col.). 1. Subfamilies Alticinae and Galerucinae. *Ann. Entomol.* **31** (1), 1–18.

PAJNI, H.R., DEVI, N. AND TEWARI, P.K. (1985b). A comparative account of the internal reproductive organs in family Chrysomelidae (Col.). 11. Subfamilies Clytrinae, Cryptocephalinae and Criocerinae. *Uttar Pradesh J. Zool.* **5** (2), 130–138.

PAJNI, H.R., DEVI, N. AND TEWARI, P.K. (1987a). A comparative account of the internal reproductive organs in the family Chrysomelidae (Col.). 11. Subfamilies Cassidinae, Hispinae, Chrysomelinae and Eumolpinae. *Ann. Entomol.* **5** (2), 25–45.

PAJNI, H.R., GANDHI, S.S. AND SINGLA, S.R. (1987b). Studies on the oviposition of some Indian Chrysomelidae (Col.). *Res. Bull. Punjab Univ.* **38** (1–2), 87–96.

PALMER, J.O. (1985). Phenology and dormancy in the Milkweed Leaf Beetle, *Labidomera clivicollis* (Kirby). *American Midland Nat.* **114** (1), 13–18.

PALMER, M., PONS, G., ALONSO-ZARAZAYA, M.A., BELLES, X., DE FERRER, J., FERRER, J., OUTERELO, R., PETITPIERRE, E., PLATA, P., RUIZ, J.L., SANCHEZ-RUIZ, M., VASQUEZ, X.A., VIVES, E. AND VIVES, J. (1999). Coleopteros de las Islas Chafarinas (N. Africa): catalogo faunistico e implicaciones biogeographicas. *Boll. Soc. Hist. Nat. Balears* **42**, 147–166.

PANTYUKHOV, G.A. (1992). Conditions of hibernation and survival of adult Ragweed Striped Leaf-beetle *Zygogramma suturalis* F. (Col. Chrys.) in Stavropol territory. *Entom. Rev.*, 25–31.

PARDI, L. (1987). La 'pseudocopula' delle femmine di *Otiorrhynchus pupillatus cyclophtalmus* (Sol.) (Col. Curc.). *Boll. Ist. Entomol. Guido Grandi, Bologna* **41**, 355–363.

PARRI, S., ALATALO, R.V. AND MOPPES, J. (1998). Do female leaf beetles *Galerucella nymphaeae* choose their mates and does it matter? *Oecologia* **114** (1), 127–132.

PASTEELS, J.M. (1993). The value of defensive compounds as taxonomic characters in the classification of leaf-beetles. *Biochem. Syst. Ecology* **21** (1), 135–142.

PASTEELS, J.M. AND ROWELL-RAHIER, M. (1989). Defensive glands and secretions as taxonomical tools in the Chrysomelidae. *Entomography* **6**, 423–432.

PASTEELS, J.M. AND ROWELL-RAHIER, M. (1991). Proximate and ultimate causes for host-plant influence on chemical defense of leaf beetles (Col. Chrys.). *Entomol. Gener.* **15** (4), 227–235.

PASTEELS, J.M. GRÉGOIRE, J.C. AND ROWELL-RAHIER, M. (1983). The chemical ecology of defense in arthropods. *Ann. Rev. Entomol.* **28**, 263–289.

PASTEELS, J.M., BRAEKMAN, J.C. AND DALOZE, D. (1988a). Chemical defence in the Chrysomelidae. In: *Biology of Chrysomelidae*. Eds. P. Jolivet, E. Petitpierre and T.H. Hsiao, pp 233–252. Dordrecht, The Netherlands: Kluwer Academic Publishers.

PASTEELS, J.M., ROWELL-RAHIER, M. AND RAUPP, M.J. (1988b). Plant derived defense in Chrysomelid beetles. In: *Novel aspects of insect–plant interactions*. Eds. P. Barbosa and D. Letourneau, pp 235–272. New York, USA: John Wiley and Sons Inc. Publishers.

PASTEELS, J.M., ROWELL-RAHIER, M., BRAEKMAN, J.C., DALOZE, D. AND DUFFEY, S. (1989). Evolution of exocrine chemical defense in leaf-beetles (Col. Chrys.). *Experientia* **45**, 295–300.

PASTEELS, J.M., DUFFEY, S. AND ROWELL-RAHIER, M. (1990). Toxins in chrysomelid beetles. *Journal Chemical Ecology* **16**, 211–222.

PASTEELS, J.M., EGGENBERGER, F., ROWELL-RAHIER, M., EHMKE, A. AND HARTMANN, T. (1992). Chemical defense in chrysomelid leaf beetles. *Naturwissenschaften* **79**, 521–523.

PASTEELS, J.M., ROWELL-RAHIER, M., BRAEKMAN, J.C. AND DALOZE, D. (1994). Chemical defence of adult leaf beetles updated. In: *Novel aspects of the biology of Chrysomelidae*. Eds. P. Jolivet, M.L. Cox and E. Petitpierre, pp 289–301. Dordrecht, The Netherlands: Kluwer Academic Publishers.

PASTEELS, J.M., ROWELL-RAHIER, M., ELMKE, A. AND HARTMANN, T. (1996). Hostplant derived pyrrolizzidine alkaloids in *Oreina* leaf beetles; physiological, ecological and evolutionary aspects. In: *Chrysomelidae biology. 2. Ecological studies*. Eds. P. Jolivet and M.L. Cox, pp 213–225. Amsterdam, The Netherlands: SPB Academic Publishers.

PATAY, R. (1937a). Sur la régénération des pattes chez le Coléoptère Chrysomélide *Leptinotarsa decemlineata* Say. *C.R. Soc. Biol.* **126**, 283–285.

PATAY, R. (1937b). Quelques observations anatomiques et physiologiques sur le Doryphore. *Bull. Soc. Sc. Bretagne* **14** (1–2), 93–102.

PATAY, R. (1939). *Contribution à l'étude d'un Coléoptère* (Leptinotarsa decemlineata (*Say*)). *Evolution des organes au cours du développement*. Pp 1–145. Thesis, Rennes Univ.

PATERSON, N.F. (1930). The bionomics and morphology of the early stages of *Paraphaedon tumidulus* Germ. (Col. Phyt. Chrys.). *Proc. Zool. Soc. London* **3**, 627–676, pls.

PATERSON, N.F. (1931). The bionomics and comparative morphology of the early stages of certain Chrysomelidae (Col. Phyt.). *Proc. Zool. Soc. London* **4**, 879–949.

PATERSON, N.F. (1941). The early stages of some South African Chrysomelidae (Col.). *J. Entomol. Soc. South Africa* **4**, 1–15.

PATHAK, S.C., KULSHRESTHA, V., CHOUBEY, A.K. AND PARULEKAR, A.H. (2001). Coleoptera among insects of terrestrial origin over Indian Ocean. *Advances in Biosciences* **20** (11), 77–82.

PAULIAN, R. (1942). L'endosquelette femoral chez les Sagrides. *Bull. Soc. Zool. Fr.* **67**, 184–186.

PAULIAN, R. (1943). *Les Coléoptères*. 396 pp. Paris: Payot Publishers.

PAULIAN, R. (1949). Ordre des Coléoptères. Partie Systématique. In: *Traité de Zoologie*. Ed. P.P. Grassé, pp 892–1077.

PAULIAN, R. (1950). La vie larvaire des insectes. *Mém. Mus. Hist. Nat., Nlle. Sér., Paris* **30** (1), 1–205.

PAULIAN, R. (1961). *La zoogéographie de Madagascar et des îles voisines*. 485 pp. Tananarive: Institut de Recherche Scientifique.

PAULIAN, R. (1988). *Biologie des Coléoptères*. 719 pp. Paris: Paul Lechevalier-Masson Publishers.

PAULIAN, R. (1993). *Les Coléoptères à la conquête de la Terre*. 245 pp. Paris: Boubée Publishers.

PAVAN, M. (1953). Studi sugli antibiotici e insetticidi di origine animale I. Gui principio attivo della larva di *Melasoma populi* L. (Col. Chrys.). *Arch. Zool. Ital.* **38**, 157–184.

PECK, S.B. AND THOMAS, M.C. (1998). A distributional checklist of the beetles (Col.) of Florida. *Arthropods of Florida, DPI, Gainesville, FL, USA* **16**, 1–180.

PENG, C.W. AND WEISS, M.J. (1992). Evidence of an aggregation pheromone in the flea beetle *Phyllotreta cruciferae* (Goeze) (Col. Chrys.). *J. Chemical Ecology* **18** (6), 875–884.
PENG, C.W., BARTELT, R.J. AND WEISS, M.J. (1999). Male crucifer flea beetles produce an aggregation pheromone. *Physiol. Entomol.* **24**, 98–99.
PERROUD, B.P. (1855). Notice sur la viviparité et l'ovoviviparité des *Oreina speciosa* et *superba*. *Ann. Soc. Linn. Lyon* **2**, 402–408.
PETERSON, A. (1960). *Larvae of insects. II.* 415 pp. Ann Arbor, MI, USA: Edwards Brothers, Inc.
PETERSON, J.K. AND SCHALK, J.M. (1994). Internal bacteria in the Chrysomelidae. In: *Novel aspects of the biology of Chrysomelidae*. Eds. P. Jolivet, M.L. Cox and E. Petitpierre, pp 393–405. Dordrecht, The Netherlands: Kluwer Academic Publishers.
PETITPIERRE, E. (1968a). Sistematica i constitucio cromosomica de *Timarcha*. El cariotypus d'algunes formes catalanes. *Treballs Soc. Cat. Biol.* **25**, 13–17.
PETITPIERRE, E. (1968b). Sistematica, citologia y genetica del genero *Timarcha* (Col. Chrys.). *Genet. Iberica* **20**, 13–21.
PETITPIERRE, E. (1970). Cytotaxonomy and evolution of *Timarcha* Latr. (Col. Chrys.). *Genet. Iberica* **22**, 67–120.
PETITPIERRE, E. (1973). *Estudios sistematicos, citogeneticos y evolutivos sobre el genero Timarcha (Col. Chrys.).* 261 pp. Tesis Doctoral, Univ. Barcelona, Spain.
PETITPIERRE, E. (1985). Notas faunisticas y ecologicas sobre Chrysomelidae de Mallorca y Catalunya. *Bull. Soc. Hist. Nat. Balears* **29**, 31–36.
PETITPIERRE, E. (1988). Cytogenetics, cytotaxonomy and genetics of Chrysomelidae. In: *Biology of Chrysomelidae*. Eds. P. Jolivet, E. Petitpierre and T.H. Hsiao, pp 131–159. Dordrecht, The Netherlands: Kluwer Academic Publishers.
PETITPIERRE, E. (1989). Recent advances in the evolutionary cytogenetics of the leaf beetles (Col. Chrys.). *Entomography* **6**, 433–442.
PETITPIERRE, E. (1990). Karyological evolution and cytotaxonomy of the leaf-beetles. *The Nucleus* **33** (1–2), 30–40.
PETITPIERRE, E. (1999). The cytogenetics and cytotaxonomy of *Chrysolina* Mots. and *Oreina* Chevr. (Col. Chrys.). *Hereditas* **131**, 55–62.
PETITPIERRE, E. AND JOLIVET, P. (1976). Phylogenetic position of the American *Timarcha* Latr. (Col. Chrys.) based on chromosomal data. *Experientia* **32**, 157–158.
PETITPIERRE, E. AND JUAN, C. (1994). Genome size, chromososmes and egg chorion ultra-structure in the evolution of Chrysomelinae. In: *Novel aspects of the biology of Chrysomelidae*. Eds. P. Jolivet, M.L. Cox and E. Petitpierre, pp 213–225. Dordrecht, The Netherlands: Kluwer Academic Publishers.
PETITPIERRE, E., JUAN, C. AND ALVAREZ-FUSTER, A. (1991). Evolution of chromosomes and genome size in Chrysomelidae and Tenebrionidae. In: *Advances in Coleopterology*. Eds. Zunino *et al.*, pp 129–144. Barcelona, Spain: AEC.
PEYERIMHOFF, P. DE (1911–1915). Notes sur la biologie de quelques Coléoptères Phytophages du Nord Africain (1 and 2 sér.). *Ann. Soc. Ent. Fr.* **80**, 283–314; **84**, 19–61.
PEYERIMHOFF, P. DE (1934). Les Coléoptères remontent-ils au Permien? *Bull. Soc. Ent. Fr.* **39** (2), 39–44.
PHILLIPS, W.M. (1976). Effect of leaf age on feeding preference and egg laying in the chrysomelid beetle, *Haltica lythri*. *Physiol. Entomol.* **1**, 223–226.
PHILLIPS, W.M. (1977a). Observations on the biology and ecology of the chrysomelid genus *Haltica* Geoffr. in Britain. *Ecol. Entomol.* **2**, 205–216.
PHILLIPS, W.M. (1977b). Some aspects of the host plant relations of the chrysomelid genus *Haltica* with special reference to *Haltica lythri*. *Entomol. Exper. Appl.* **21**, 261–274.
PHILLIPS, W.M. (1978). Sensilla types from the ovipositor of the flea beetle *Altica lythri* (Col. Chrys.). *Entomol. Exp. Appl.* **24**, 399–400.
PHILLIPS, W.M. (1979). A contribution to the study of species relations within the chrysomelid genus *Altica* Müller in Britain. *Zool. J. Linn. Soc.* **66**, 289–308.
PIC, M. (1896). Coléoptères d'Asie Mineure et de Syrie. *Miscell. Entom.* **4** (2), 35–36.
PICANÇO, M., LEITE, G.L.D., BASTOS, C.S., SUINAGA, F.A. AND CASALE, V.W. (1999). Coleopteros associados ao jiloeiro (*Solanum gilo* Raddi). *Rev. Bras. Entomol.* **43** (1–2), 131–137.

PIETRYKOWSKA, E. (2000). Morphology of the egg and first instar larva of *Coptocephala rubiconda* (Laicharting, 1781) and notes on its biology. *Genus* **11** (1), 37–44.

PILSON, D. (1999). Plant hybrid zones and insect host range expansion. *Ecology* **80** (2), 407–415.

POINAR, G.O. JR. (1988). Nematode parasites of Chrysomelidae. In: *Biology of Chrysomelidae*. Eds. P. Jolivet, E. Petitpierre and T.H. Hsiao, pp 433–448. Dordrecht, The Netherlands: Kluwer Academic Publishers.

POINAR, G.O. JR. (1992). *Life in Amber*. 350 pp. Stanford, Ca., USA: Stanford University Press.

POINAR, G.O. (1999). Chrysomelidae in fossilized resin: behavioural inferences. In: *Advances in Chrysomelidae biology*. Ed. M.L. Cox, pp 1–16. Leiden, The Netherlands: Backhuys Publishers.

POINAR, G.O. AND POINAR, R. (1994). *The quest for life in Amber*. xiii+217 pp. Reading, Mass., USA: Addison-Wesley Publishing Co.

POINAR, G.O. AND POINAR, R. (1999). *The Amber forest. A reconstruction of a vanished world*. xviii+239 pp. Princeton, NJ, USA: Princeton University Press.

POINAR, G., JOLIVET, P. AND GRAFTEAUX, A. (2001). New food-plants provide clues for the origin and distribution of *Timarcha* (Col. Chrysomelidae Chrysomelinae). *Lambilionea, Brussels* **101** (2), 1–5.

POISSON, R. AND PATAY, R. (1938). Sur quelques modalités de la régénération des pattes et des ailes chez la larve du Doryphore, *Leptinotarsa decemlineata* Say (Col. Chrys.). *C.R. Soc. Biol.* **129**, 126–128.

POLAK, M., STARMER, W.T. AND BARKER, J.S. (1998). A mating plug and male mate choice in *Drosophila hibisci* Beck. *Animal Behaviour* **56** (4), 919–926.

POLL, M. (1932). Contribution à l'étude des tubes de Malpighi des Coléoptères. Leur utilité en phylogénie. *Recueil Inst. Zool. Torley-Rousseau* **4**, 47–80.

PONCE DE LEON, R., MORELLI, E. AND GONZALEZ VAINER, P. (1993). Observaciones de campo sobre la biologia de *Metriona elatior* (Col. Chrys.) en *Solanum elaeagnifolium* (Solanaceae) del Uruguay. *Entomophaga* **38** (4), 461–464.

PONOMARENKO, A.G. (1969). Historical development of Archostemata beetles. *Tr. Paleontol. Inst. Akad. Nauk. SSSR* **125**, 1–240.

POWELL, E.F. (1941). Relationships within the family Chrysomelidae (Col.) as indicated by the male genitalia of certain species. *Amer. Mid. Nat.* **25** (1), 148–195.

PRASAD, P. AND SINGH, B.K. (1991). Physiology of digestion of adult insect *Platycorinus peregrinus* (Col. Chrys.). *Environment and Ecology* **9** (3), 664–666.

RABAUD, E. (1915). Notes sommaires sur la biologie des Cassides. 1. Mode de ponte et alimentation. 2. Le cycle évolutif; les mues et le paquet d'excréments. *Bull. Soc. Ent. Fr.* **196–198**, 209–212.

RABAUD, E. (1919). L'immobilisation réflexe et l'activité normale des Arthropodes. *Bull. Biol. Fr. Belgique* **53**, 1–149.

RANK, N.E. (1991). Effects of plant chemical variation on a specialist herbivore: willow leaf beetles in the eastern Sierra Nevada. In: *Natural history of Eastern California and high altitude research*. Eds. Hall, Doyle-Jones and Widawski, pp 161–181. Berkeley, USA: University of California Press.

RATTI, E. (1986). Recerche faunistiche del Museo Civico di Storia Naturale di Venezia nell' isola di Pantellaria. VI. Col. Cerambycidae and Chrysomelidae. *Bull. Mus. Civ. St. Nat. Venezia* **37**, 47–55.

RAUPP, M.J. (1985). Effects of leaf toughness on mandibular wear of the leaf beetle, *Plagiodera versicolora*. *Ecol. Entomol.* **10**, 73–79.

RAUPP, M.J. AND SADOF, C.S. (1991). Responses of leaf beetles to injury-related changes in their salicaceous hosts. In: *Phytochemical induction by herbivores*. Eds. D.W. Tallamy and M.J. Raupp, pp 183–204. New York, USA: John Wiley and Sons Publishers.

REID, C.A. (1995). A cladistic analysis of subfamilial relationships in the Chrysomelidae sensu lato (Chrysomeloidea). In: *Biology, phylogeny and classification of Coleoptera: Papers celebrating the 80th birthday of Roy A. Crowson*. Pp 560–631. Warszawa, Poland.

REID, C.A. (2000). Spilopyrinae Chapuis: a new subfamily in the Chrysomelidae and its systematic placement (Coleoptera). *Invertebrate Taxonomy* **14**, 837–862.

RETHFELD, CHR. (1924). Die viviparität bei *Chrysomela varians* Schall. *Zool. Jahrb. Anatomie* **46** 245–302.

RICKELMANN, K.M. AND BACH, C.E. (1991). Effects of soil moisture on the pupation behavior of *Altica subplicata. The Great Lakes Entomol.* **24** (4), 231–237.

RICKSON, F.R., CRESTI, M. AND BEACH, J.H. (1990). Plant cells which aid in pollen digestion within a beetle's gut. *Oecologia* **82** (3), 424–426.

ROBERTSON, J.G. (1964). The chromosomal cytology of bisexual and parthenogenetic *Calligrapha* spp. (Col. Chrys.). *Canad. Entomol.* **96**, 144.

ROBERTSON, J.G. (1966). The chromosomes of bisexual and parthenogenetic species of *Calligrapha* (Col. Chrys.) with notes on sex ratio, abundance and egg number. *Canad. J. Genet. Cytol.* **8** (4), 695–732.

ROBINSON, G.S. (1984). *Insects of the Falkland Islands*. 38 pp. London, UK: British Museum.

ROCKWOOD, L.L. (1974). Seasonal changes in the susceptibility of *Crescentia alata* leaves to the flea beetle, *Oedionychus* sp. *Ecology* **55**, 142–148.

RODRIGUEZ, V. (1994a). Sexual behavior in *Omaspides convexicollis* Spaeth and *O. bistriata* Boheman with notes on maternal care of eggs and young. *Coleopt. Bull.* **48** (2), 140–144.

RODRIGUEZ, V. (1994b). Function of the spermathecal muscles in *Chelymorpha alternans* Boheman (Col. Chrys. Cass.). *Physiol. Entomol.* **19**, 198–202.

RODRIGUEZ, V. (1995a). Relation of flagellum length to reproductive success in male *Chelymorpha alternans* Boheman (Col. Cass.). *Coleopt. Bull.* **49** (3), 201–205.

RODRIGUEZ, V. (1995b). Copulatory courtship in *Chelymorpha alternans* Boheman (Col. Chrys. Cass.). *Coleopt. Bull.* **49** (4), 327–331.

ROHDENDORF, B.B. (1957). Paleoentomological research in the USSR. *Moskva Academii Nauk SSSR, Paleontologicheskii Institut, Trudy* **66**, 1–100.

ROHDENDORF, B.B. AND RAZNITSIN, A. (Eds.) (1980). The historical development of the class Insecta. *Trudy Paleontol. Inst. Moscow* **175**, 1–268.

ROKAS, A. (2000). *Wolbachia*, as a speciation agent. *TREE* **15** (2), 44–45.

ROOT, R.B. (1967). The niche exploitation pattern of the blue-grey gnatcatcher. *Ecol. Monogr.* **37**, 317–350.

ROUBIK, D.W. AND SKELLEY, P.E. (2001). *Stenotarsus subtilis* Arrow, the aggregating fungus beetle of Barro Colorado Island Nature Management, Panama (Col. Endomychidae). *Col. Bull.* **55** (3), 249–263.

ROUDIER, A. (1957). Localités nouvelles françaises ou espagnoles de Curculionides. Description d'une sous-espèce nouvelle. *L'Entomologiste, Paris* **13** (2–3), 24–36.

ROWELL-RAHIER, M. AND PASTEELS, J.M. (1986). Economics of chemical defense in Chrysomelinae. *Journ. Chem. Ecol.* **12** (5), 1189–1203.

ROWELL-RAHIER, M. AND PASTEELS, J.M. (1990). Phenolglucosides and interactions at three trophic levels: Salicaceae, herbivores, predators. In: *Insect–plant interactions*. Ed. E.A. Bernays, pp 75–94. Boca Raton, FL, USA: CRC Press.

ROWELL-RAHIER, M. AND PASTEELS, J.M. (1992). *Genetic relationships between* Oreina *species with different defensive strategies*. Proc. 8th Int. Symp. Insect–Plant Relationships. Pp 341–342. Dordrecht, The Netherlands: Kluwer Academic Publishers.

ROWELL-RAHIER, M., PASTEELS, J.M., ALONSO-MEJIA, A. AND BROWER, L.P. (1995). Relative unpalatability of leaf beetles with either biosynthetized or sequestered chemical defence. *Anim. Behav.* **49**, 709–714.

ROWLEY, W.A. AND PETERS, D.C. (1972). Scanning electron microscopy of the egg-shell of four species of *Diabrotica* (Col. Chrys.). *Ann. Ent. Soc. Amer.* **65**, 1188–1191.

RUNGIUS, H. (1911). Der Darmkanal (der Imago und Larve) von *Dytiscus marginalis. Zeitschr. wiss. Zool.* **96**, 179.

RUSCHKAMP, F. (1927). Der Flugapparat der Käfer. *Zool. Stuttgart* **28** (75), 1–88.

SACARES, A. AND PETITPIERRE, E. (1999). Noves cites de Chrysomelidae (Col.) d'Eivissa i Formentera (Ille Pitiuses). *Boll. Soc. Hist. Nat. Balears* **42**, 33–37.

SACCHI, R. AND BUSARDO, G. (1935). Variabilita e correlazione in specie molto affini. Noti Appunti sper. *Ent. agraria* **3**, 1–24.

SACHET, M.H. (1962). Monographie physique et biologique de l'île Clipperton. *Ann. Institut Océanographique Monaco, Nlle. Série* **40** (1), 1–107.

SAINI, R.S. (1964). Histology and physiology of the cryptonephridial system of insects. *Trans. R. Ent. Soc. Lond.* **116** (14), 347–392.

SAITOH, S. AND KATAKURA, H. (1996). Strictly parapatric distribution of flightless leaf beetles of the *Chrysolina angusticollis* species complex (Col. Chrys.) in the vicinity of Sapporo, northern Japan. *Biol. J. Linnean Soc.* **57** (4), 371–384.

SAMSINAK, K. (1965). Pripad phoresie u larve rodu *Clytra* Laich. *Vestn. Cesko-sl. Zool. Spolecnosti* **20** (4), 375–376.

SAMUELSON, G.A. (1967). Alticinae of the Solomon Islands (Col. Chrys.). *Pacific Insects* **9** (1), 139–174.

SAMUELSON, G.A. (1973). Alticinae of Oceania (Col. Chrys.). *Pacific Insects Monogr.* **30**, 1–165.

SAMUELSON, G.A. (1979). A new *Argopistes* from Tahiti, marking a range extension for endemic leaf beetles in Oceania (Col. Chrys.). *Pacific Insects* **20** (4), 404–409.

SAMUELSON, G.A. (1984). Plant associated Alticinae from the Bismarck Range, Papua New Guinea (Col. Chrys.). *Esakia* **21**, 31–47.

SAMUELSON, G.A. (1988). *Pollen feeding in Alticinae (Col. Chrys.).* Proc. Xviii Int. Congress Entom.,Vancouver, 37.

SAMUELSON, G.A. (1994). Pollen consumption and digestion by leaf beetles. In: *Novel aspects of the biology of Chrysomelidae*. Eds. P. Jolivet, M.L. Cox and E. Petitpierre, pp 179–183. Dordrecht, The Netherlands: Kluwer Academic Publishers.

SAMUELSON, G.A. (1996). Binding sites: elytron-to-body meshing structures of possible significance in the higher classification of Chrysomeloidea. In: *Chrysomelidae biology. 1. The classification, phylogeny and genetics*. Eds. P. Jolivet and M.L. Cox, pp 267–290. Amsterdam, The Netherlands: SPB Academic Publishers.

SANTIAGO-BLAY, J.A. (1994). Paleontology of leaf beetles. In: *Novel aspects of the biology of Chrysomelidae*. Eds. P. Jolivet, M.L. Cox and E. Petitpierre, pp 1–68. Dordrecht, The Netherlands: Kluwer Academic Publishers.

SANTIAGO-BLAY, J.A. AND CRAIG, P.R. (1999). Preliminary analyses of chrysomelid paleodiversity with a new record and a new species from Dominican amber (early to middle Miocene). In: *Advances in Chrysomelidae biology. 1*. Ed. M.L. Cox, pp 17–24. Leiden, The Netherlands: Backhuys Publishers.

SANTIAGO-BLAY, J.A. AND FAIN, A. (1994). Phoretic and ectoparasitic mites (Acari) of the Chrysomelidae. In: *Novel aspects of the biology of Chrysomelidae*. Eds. P. Jolivet, M.L. Cox and E. Petitpierre, pp 407–417. Dordrecht, The Netherlands: Kluwer Academic Publishers.

SANTIAGO-BLAY, J.A., POINAR, JR., G.O. AND CRAIG, P.R. (1996). Dominican and Mexican Amber Chrysomelids with descriptions of two new species. In: *Chrysomelidae biology. 1. The classification, phylogeny and genetics*. Eds. P. Jolivet, M.L. Cox and E. Petitpierre, pp 413–424. Amsterdam, The Netherlands: SPB Academic Publishers.

SANTOS, H.R. DOS (1980). Distribuçao e danos do *Agathomerus sellatus* (Germar, 1824) (Col. Chrys.) em cultura de tomateiro. *Rev. Setor Ciencias Agrarias* **2**, 157–162.

SANTOS, H.R. DOS (1981a). Redescriçao de *Agathomerus sellatus* (Germar, 1824) (Col. Chrys.), praga de tomateiro (*Lycopersicum esculentum*). *Rev. Setor Ciencias Agrarias* **3**, 107–116.

SANTOS, H.R. DOS (1981b). Biologia de *Agathomerus sellatus* (Germar, 1824), broca de tomateiro. *Rev. Bras. Entomol.* **25** (2), 165–170.

SCHAEFFER, P.W. (1987). Trench-feeding behavior of *Aulacophora foveicollis* Motschulsky (Col. Chrys.) on a cucurbit in Central China. *Coleopt. Bull.* **41** (2), 136.

SCHERER, G. (1969). The Alticinae of the Indian subcontinent (Col. Chrys.). *Pacific Insects Monogr.* **22**, 1–251.

SCHERER, G. (1974). *Clavicornaltica*, a new genus from Ceylon (Sri Lanka) (Col. Chrys.). *Rev. Suisse Zool.* **81** (1), 57–68.

SCHERER, G. (1979). *Clavicornaltica* recorded also from the Philippine Islands (Col. Chrys. Alt.). *Rev. Suisse Zool.* **86** (3), 713–714.

SCHERER, G. (Ed.) (1982). First International Alticinae Symposium, Munich 11–15 August 1980. *Spixiana, München, Suppl.* **7**, 3–65.

SCHERER, G. (1986). Die Halticiden der Kapverden (Col. Chrys. Alt.). *Cour. Forsch. Inst. Senkenberg* **81**, 65–68.

SCHERER, G. (1988). The origins of the Alticinae. In: *Biology of Chrysomelidae*. Ed. P. Jolivet, E. Petitpierre and T.H. Hsiao, pp 115–130. Dordrecht, The Netherlands: Kluwer Academic Publishers.

SCHERER, G. (1989). Ground living flea beetles from the Himalayas (Col. Chrys.). *Spixiana* **12** (1), 31–55.

SCHERER, G. AND BOPPRÉ, M. (1997). Attraction of *Gabonia* and *Nzerekorena* to pyrrolizidine alkaloids with description of 13 new species and notes on male structural peculiarities. *Spixiana* **20** (1), 7–38.

SCHERF, H. (1956). Zum feineren Bau der Eigelege von *Galeruca tanaceti* (Col. Chrys.). *Zool. Anz.* **157**, 124–130.

SCHERF, H. (1966). Beobachtungen aan Ei und Gelege von *Galeruca tanaceti* (Col. Chrys.). *Biol. Zbl.* **85**, 7–17.

SCHLEGTENDAL, A. (1934). Beitrag zum Farbensinn der Arthropoden. *Z. vergl. Physiol.* **20**, 545–583.

SCHMITT, M. (1988). The Criocerinae: Biology, phylogeny and evolution. In: *Biology of Chrysomelidae*. Eds. P. Jolivet, E. Petitpierre and T.H. Hsiao, pp 475–495. Dordrecht, The Netherlands: Kluwer Academic Publishers.

SCHMITT, M. (1989). On the phylogenetic position of the Bruchidae within the Chrysomeloidea (Col.). *Entomography* **6**, 531–537.

SCHMITT, M. (1992a). Stridulatory devices of Leaf Beetles (Chrys.) and other Coleoptera. In: *Advances in Coleopterology*. Eds. M. Zunino, X. Belles and M. Blas, pp 263–280. Barcelona, Spain: Asociacion Europea de Coleopterologia, 1991.

SCHMITT, M. (1992b). The position of Megalopodinae and Zeugophorinae in a phylogenetic system of the Chrysomeloidea (Ins. Coleopt.). In: *Proc. 3rd Int. Symp. Chrysomelidae, Beijing, China, 1992*. Ed. D.G. Furth, pp 38–44. Leiden, The Netherlands: Backhuys Publishers.

SCHMITT, M. (1994). Stridulation in leaf beetles (Col. Chrys.). In: *Novel aspects in the biology of Chrysomelidae*. Eds. P. Jolivet, M.L. Cox and E. Petitpierre, pp 319–325. Dordrecht, The Netherlands: Kluwer Academic Publishers.

SCHMITT, M. (1996). The phylogenetic system of the Chrysomelidae. History of ideas and present state of knowledge. In: *Chrysomelidae biology. 1. The classification, phylogeny and genetics*. Eds. P. Jolivet and M.L. Cox, pp 57–96. Amsterdam, The Netherlands: SPB Academic Publishers.

SCHMITT, M. (1998). Again Bruchid classification. *Chrysomela* **36**, 3–4.

SCHMITT, M., MISCHE, U. AND WACHMANN, E. (1982). Phylogenetic and functional implications of the rhabdom patterns in the eyes of Chrysomeloidea. *Zoologica Scripta* **11**, 31–44.

SCHÖLLER, M. (1995). Petalophagie bei *Cryptocephalus nubigena* Franz 1982 auf La Palma (Kanarische Inseln). *Mitt. Internat. Entomol. Ver.* **20** (3–4), 121–127.

SCHÖLLER, M. (1996a). Zoosaprophagy and phytosaprophagy in chrysomelid beetle larvae, *Macrolenes dentipes* and *Pachybrachis anoguttatus* (Col. Chrys. Clytrinae and Cryptocephalinae). *Proc. XXI ICE, Firenze, Italy*. Eds. M. Biondi, M. Daccordi and D.C. Furth, pp 281–285. Torino.

SCHÖLLER, M. (1996b). Ökologie mitteleuropaïscher Blattkäfer, Samenkäfer und Bratrüssler (Col. Chrysomelidae, einschliessich Bruchinae, Anthribidae). In: *Die Blatt- und Samenkäfer von Voralberg and Liechtensstein*. Eds. M. Clemens, M. Brandstetter and A. Kapp. *Burs.* **2**, 1–65.

SCHÖLLER, M. (1997). Räumliche Verteilung der Larven von *Cryptocephalus moraei* (Linnaeus, 1758) (Col. Chrys. Crypt.). *Novias* **22** (2), 511–514.

SCHÖLLER, M. (1999). Field studies of Cryptocephalinae biology. In: *Advances in Chrysomelidae biology. 1*. Ed. M.L. Cox, pp 421–436. Leiden, The Netherlands: Backhuys Publishers.

SCHÖPS, SYRETT, P. AND EMBERSEN, R.M. (1996). Summer diapause in *Chrysolina hyperici* and *C. quadrigemina* (Col. Chrys.) in relation to biological control of St John's wort, *Hypericum perforatum* (Clusiaceae). *Bull. Entomol. Res.* **86**, 591–597.

SCHRODER, R.F.W., PUTTLER, B., IZHEVSKY, S.S. AND GANDOLFO, D. (1994). Viviparity and larval development of *Platyphora quadrisignata* (Germar) (Col. Chrys.) in Brazil. *Coleopt. Bull.* **48** (3), 237–243.

SCHULZE, L. (1996). Life-history and description of early stages of *Sphondylia tomentosa* (Lacordaire) (Col. Chrys. Megalop.). In: *Chrysomelidae biology. 3. General studies*. Eds. P. Jolivet and M.L. Cox, pp 187–199. Amsterdam, The Netherlands: SPB Publishers.

SEENO, T.N. AND WILCOX, J.A. (1982). Leaf beetle genera (Col. Chrys.). *Entomography* **1**, 1–221.

SÉGUY, E. (1967). *Dictionnaire des termes d'Entomologie*. 465 pp. Paris: Lechevalier Publishers.

SEIFERT, R.P. (1982). Neotropical *Heliconia* insect communities. *Quartely Review Biology* **57**, 1–28.

SEIFERT, R.P. AND SEIFERT, F.H. (1976a). A community matrix analysis of *Heliconia* insect communities. *Amer. Natur.* **110**, 462–483.

SEIFERT, R.P. AND SEIFERT, F.H. (1976b). Natural history of insects living in inflorescences of two species of *Heliconia*. *J. New York Entomol. Soc.* **84**, 233–242.

SEIFERT, R.P. AND SEIFERT, F.H. (1979). Utilization of *Heliconia* (Musaceae) by the beetle *Xenarescus monocerus* (Olivier) (Chrys. Hispinae) in a Venezuelan forest. *Biotropica* **11** (1), 51–59.

SELMAN, B.J. (1962). Remarkable new Chrysomelids found in the nests of arboreal ants in Tanganyika. *Ann. Mag. Nat. Hist.* **5**, 295–299.

SELMAN, B.J. (1972). Eumolpinae (Col. Chrys.). *Expl. Parc. Nat. Garamba, Brussels, Belgium* **55**, 1–95.

SELMAN, B.J. (1985). The use and significance of color in the separation of paropsine chrysomelid species. *Entomography* **3**, 477–479.

SELMAN, B.J. (1988a). Viruses and Chrysomelids. In: *Biology of Chrysomelidae*. Eds. P. Jolivet, E. Petitpierre and T.H. Hsiao, pp 379–387. Dordrecht, The Netherlands: Kluwer Academic Publishers.

SELMAN, B.J. (1988b). Chrysomelids and ants. In: *Biology of Chrysomelidae*. Eds. P. Jolivet, E. Petitpierre and T.H. Hsiao, pp 463–473. Dordrecht, The Netherlands: Kluwer Academic Publishers.

SELMAN, B.J. (1994a). Eggs and oviposition in Chrysomelid beetles. In: *Novel aspects of the biology of Chrysomelidae*. Eds. P. Jolivet, M.L. Cox and E. Petitpierre, pp 69–74. Dordrecht, The Netherlands: Kluwer Academic Publishers.

SELMAN, B.J. (1994b). The biology of the paropsine eucalyptus beetles of Australia. In: *Novel aspects of the biology of Chrysomelidae*. Eds. P. Jolivet and M.L. Cox, pp 555–565. Dordrecht, The Netherlands: Kluwer Academic Publishers.

SERVADEI, A. (1936). Reperrti sulla biologia e morphologia della *Galerucella nymphaeae*. *Redia* **22**, 1–31.

SHARP, D. AND MUIR, F. (1912). The comparative anatomy of the male genital tube in Coleoptera. *Trans. Ent. Soc. London* **3**, 477–642.

SHEAR, W.A. AND KUKALOVA-PECK, J. (1990). The ecology of Paleozoic terrestrial arthropods: the fossil evidence. *Canad. J. Zool.* **68**, 1807–1834.

SHEPARD, W.D. (1997). *Lilioceris* sp. (Col. Chrys.), herbivory on *Cycas siamensis* Miguel. *Pan-Pacific Entomol.* **73** (1), 36–39.

SHERWOOD , D.R. AND LEVINE, E. (1993). Copulation and its duration affects female weight, oviposition, hatching patterns, and ovarian development in the western corn rootworm (Col. Chrys.). *J. Econ. Entomol.* **86** (6), 1664–1671.

SHIMIZU, N. AND FUJIYAMA, S. (1986). Aggressive encounters between males in *Chrysolina aurichalcea*. *J. Ethol.* **4** (1), 11–15.

SHRIVASTAVA, R.K. (1986). *Studies on morphology and physiology of the digestive system of Aspidomorpha miliaris F. (Col. Chrys.)*. Ph.D. Thesis. Ravishankar University, Raipur, India.

SHRIVASTAVA, R.K. AND VERMA, K.K. (1982). Peritrophic membrane cells and their digestive function in the tortoise beetle *Aspidomorpha miliaris* F. (Col. Chrys.). *Indian J. Exper. Biol.* **20**, 595–599.

SHRIVASTAVA, R.K. AND VERMA, K.K. (1983). Rectal strand and its function in the tortoise beetle, *Aspidomorpha miliaris* F. (Col. Chrys.). *Indian J. Exper. Biol.* **21**, 465–467.

SHUTE, S.L. (1980). Wing-polymorphism in British species of *Longitarsus* beetles (Chrys. Alt.). *System. Entomol.* **5**, 437–448.

SHUTE, S.L. (1983). Key to the genera of Galerucine beetles of New Guinea, with a review of *Sastra* and related new taxa (Chrys.). *Bull. British Museum (Nat. Hist.). Entomol.* **46** (3), 205–266.

SIEW, Y.C. (1965). The endocrine control of adult reproductive diapause in the chrysomelid beetle, *Galeruca tanaceti* (L.). II. *J. Insect Physiol.* **11**, 463–479.

SILFVERBERG, H. (1976). Studies on galerucine genitalia. I. (Col. Chrys.). *Notulae Entomologicae* **56**, 1–9.

SILFVERBERG, H. (1994). Chrysomelidae in the Arctic. In: *Novel aspects of the biology of Chrysomelidae*. Eds. P. Jolivet, M.L. Cox and E. Petitpierre. Dordrecht, The Netherlands: Kluwer Academic Publishers.

SILFVERBERG, H. (1998). Towards a zoogeography of Galerucinae. In: *Proc. of the 4th Int. Symp. Chrysomelidae, Firenze, Italy*. Eds. M. Biondi, M. Daccordi and D.G. Furth, pp 155–160. Torino.

SIMMUL, T.L. AND DE LITTLE, D.W. (1999). Biology of the Paropsini (Chrys.). In: *Advances in Chrysomelidae biology. 1*. Ed. M.L. Cox, pp 463–477. Leiden, The Netherlands: Backhuys Publishers.

SIMONSEN, V., ELMEGAARD, N., KJAER, C. AND NIELSEN, B.O. (1999). Gene flow among populations of the leaf beetle species *Gastrophysa polygoni* and *Lochmaea suturalis* in Denmark (Col. Chrys.). *Entomol. Gener.* **23** (4), 271–279.

SIMPSON, G.G. (1965a). *Principles of animal taxonomy (Indian Edition)*. 247 pp. Calcutta, New Delhi: Oxford Book Co.

SIMPSON, G.G. (1965b). *The geography of evolution*. 200 pp. Philadelphia: Chilton Book Co.

SINGH, J.P. (1976). Studies on the excretion in *Aulacophora foveicollis* Lucas (Col. Chrys.): the rate of ejection of excretory material. *Zool. Beitr.* **22**, 9–18.

SLOBODCHIKOFF, C.N. AND WISMANN, K. (1981). A function of the subelytral chamber of Tenebrionid beetles. *J. Exper. Biol.* **90**, 109–114.

SMART, J. AND HUGHES, N.F. (1973). The insect and the plant: progressive palaeoecological integration. In: *Insect/plant relationships*. Ed. H.F. Van Emden, pp 143–155. Symposium R. Entomol. Soc. London.

SMITH, E.H. (1986). Revision of the genus *Phyllotreta Chevrolat* of America, North of Mexico. Part I. The maculate species (Col. Chrys. Alt.). *Fieldiana, Zoology, New Series* **28** (1364), 1–168.

SNEATH, P.H.A. (1962). The construction of taxonomic groups. In: *Microbial classification*. Eds. G.C. Ainsworth and P.H.A. Sneath, pp 289–332. Cambridge, UK: Cambridge University Press.

SNODGRASS, R.E. (1935). *Principles of insect morphology*. 667 pp. New York and London: McGraw Hill Inc.

SNYDER, W.E. AND WISE, D.H. (2000). Antipredator behavior of Spotted Cucumber Beetle (Col. Chrys.) in response to predators that pose varying risks. *Environ. Entomol.* **19** (1), 30–42.

SOETENS, P., PASTEELS, J.M., DALOZE, D. AND KAISIN, M. (1998). Host plant influence on the composition of the defensive secretion of *Chrysomela vigintipunctata* larvae (Col. Chrys.). *Biochemical Systematics and Ecology* **26**, 703–712.

SOKOLOV, V.E. (Ed.). (1981). *Colorado beetle, L. decemlineata (Say). Phylogeny, morphology, physiology, ecology, adaptation and natural enemies*. 375 pp. Moscow: Nauka.

SOUTHWOOD, T.R.E. (1985). Interactions of plants and animals. Patterns and processes. *Oikos* **44**, 5–11.

SPENCER, K.A. (1990). *Host specialization in the world Agromyzidae (Diptera)*. Series Entomology 45, pp 1–444. Dordrecht, The Netherlands: Kluwer Academic Publishers.

SRIKANTH, J. AND PUSHPALATHA, N.A. (1991). Status of biological control of *Parthenium hysterophorus* L. in India. A review. *Insect Sci. Appl.* **12** (4), 347–359.

STAINES, C.L. (1999). Possible mimetic complexes in Central American *Cephaloleia* (Col. Chrys.). In: *Advances in Chrysomelidae biology. 1*. Ed. M.L. Cox, pp 239–246. Leiden, The Netherlands: Backhuys Publishers.

STAMMER, H.J. (1935). Studien an Symbiosen zwischen Käfern und Mikroorganismen. 1. Die symbiose der Donaciinen. *Z. morph. Ökol. Tiere* **29**, 585–608.

STAMMER, H.J. (1936). Studien an Symbiosen zwischen Käfern und Mikroorganismen. 2. Die Symbiose der *Bromius obscurus* L. und der *Cassida*-Arten. *Z. Morph. Ökol. Tiere* **31**, 682–697.

STEHR, F.W. (1991). *Immature insects*. Pp 1–974. Kendall Publishers.

STEINHAUSEN, W.W. (1978). Bestimmungstabelle für die larven der Chrysomelidae (partim). Ordnung Coleopera (larven). In: *Bestimmungsbücher zur Bodenfauna Europas*. Ed. B. Klausnitzer, pp 336–343. Krefeld, Germany: Joecke and Evers.

STEINHAUSEN, W.R. (1985). Die Bedeutung larven-morphologischer Studien für die Systematik der Blattkäfer (Col. Chrys.). *Mitt. Deutschen Gesell. Für Allgemeine und Angewandte Entomologie* **4** (4–6), 204–207.

STEINHAUSEN, W. (1994). Chrysomelidae Larven. In: *Die Larven der Käfer Mitteleuropas*. Ed. B. Klausnitzer, pp 231–314. Krefeld, Germany: Goeke and Evers Publishers.

STEINHAUSEN, W. (2000). Neue palaearktische Blattkäfer-Larven (Col. Chrys.). *Ent. Bl.* **96**, 57–66.

STEWART, J.G., FELDMAN, J. AND LEBLANC, D.A. (1999). Resistance of transgenic potatoes to attack by *Epitrix cucumeris* (Col. Chrys.). *Canad. Entomol.* **131** (4), 423–431.

STEVENS, L. (1992). Cannibalism in beetles. In: *Cannibalism*. Eds. Elgar and Crespi, pp 156–175. Oxford, UK: Oxford University Press.

STEVENS, L. AND MCCAULEY, D.E. (1989). Mating prior to overwintering in the imported willow leaf beetle, *Plagiodera versicolora* (Col. Chrys.). *Ecol. Entomol.* **14**, 219–223.

STIEFEL, V.L. (1993). The larval habitat of *Pachybrachis pectoralis* (Meslheimer) and *Cryptocephalus fulguratus* LeConte (Col. Chrys.). *J. Kans. Entomol. Soc.* **66** (4), 450–453.

STOCKMANN, R. (1966). Etude de la variabilité de quelques espèces françaises du genre *Timarcha* Latreille (Col. Chrys.). *Ann. Soc. Ent. Fr. (N.S.)* **2** (1), 105–126.

STORK, N.E. (1980a). Experimental analysis of adhesion of *Chrysolina polita* (Chrys. Col.) on a variety of surfaces. *Journal Exp. Biol.* **88**, 91–107.

STORK, N.E. (1980b). A scanning electron microscope study of tarsal adhesive setae in the Coleoptera. *Zool. J. Linn. Soc.* **68** (3), 173–306.

STORK, N.E. (1980c). Role of waxblooms in preventing attachment to brassicas by the mustard beetle, *Phaedon cochleariae*. *Entomologia Exp. Appl.* **28** (1), 100–107.

STORK, N.E. (1983). The adherence of beetle tarsal setae to glass. *Journal Nat. Hist.* **17** (4), 583–597.

STORK, N.E. (1987). Arthropod faunal similarity of Bornean rain forest trees. *Ecol. Entomol.* **12**, 219–226.

STORK, N.E. (1988). Insect diversity: facts, fiction and speculation. *Biol. J. Linnean Soc.* **35**, 321–337.

STORK, N.E. AND GASTON, K. (1990). Counting species one by one. *New Scientist*, 43–47.

STRAND, A. (1942). Die Käferfauna von Svalburd. *Norsk. Ent. Tidskr.* **6**, 53–69.

STRONG, D.R. (1977a). Insect species richness: Hispine beetles of *Heliconia latispatha*. *Ecology* **58** (3), 573–582.

STRONG, D.R. (1977b). Rolled-leaf hispine beetles (Chrys.) and their zingiberales host plants in Middle America. *Biotropica* **9** (3), 156–169.

STRONG, D.R. (1981). The possibility of insect communities without competition: Hispine beetles on *Heliconia*. In: *Insect life history patterns: Habitat and geographic variation*. Eds. R.F. Denno and H. Dingle, pp 183–194. New York, Heidelberg, Germany: Springer-Verlag.

STRONG, D.R. (1982a). Harmonious coexistence of Hispine beetles on *Heliconia* in experimental and natural communities. *Ecology* **63** (4), 1039–1049.

STRONG, D.R. (1982b). Potential interspecific competition and host specificity: hispine beetles on *Heliconia*. *Ecological Entomol.* **7** (2), 217–220.

STRONG, D.R. AND WANG, M.D. (1977) (1978). Evolution of insect life histories and host plant chemistry: hispine beetles on *Heliconia*. *Evolution* **31** (4), 854–862.

SUNDMAN, J.A. AND KING, D.R. (1964). Morphological, histological and histochemical studies of the alimentary canal and Malpighian tubes of the adult boll weevil, *Anthonomus grandis* (Col. Curcul.). *Ann. Ent. Soc. Amer.* **57**, 89–95.

SUZUKI, K. (1974). Ovariole number in the family Chrysomelidae (Ins. Col.). *Journal Coll. Liberal Arts, Toyama Univ., Japan* **7**, 53–70.

SUZUKI, K. (1975). Variation of ovariole number in *Pseudodera xanthospila* (Col. Chrys., Alticinae). *Kontyu, Tokyo* **43** (1), 36–39.

SUZUKI, K. (1978). Discovery of flying population in *Chrysolina aurichalcea* (Mannh.) (Col. Chrys.). *Kontyu, Tokyo* **46** (4), 549–551.

SUZUKI, K. (1988). Comparative morphology of the internal reproductive system of the Chrysomelidae (Col.). In: *Biology of Chrysomelidae*. Eds. P. Jolivet, E. Petitpierre and T.H. Hsiao, pp 317–355. Dordrecht, The Netherlands: Kluwer Academic Publishers.

SUZUKI, K. (1992). The systematic position of the subfamily Aulacoscelinae (Col. Chrys.). In: *Proc. 3rd Intern. Symp. on Chrysomelidae, Beijing (1992)*. Eds. D.C. Furth, pp 45–59. Leiden, The Netherlands: Backhuys Publishers.

SUZUKI, K. (1994). Comparative morphology of the hind-wing venation of the Chrysomelidae (Col.). In: *Novel aspects of the biology of Chrysomelidae*. Eds. P. Jolivet, M.L. Cox and E. Petitpierre, pp 337–354. Leiden, The Netherlands: Kluwer Academic Publishers.

SUZUKI, K. (1996). Higher classification of the family Chrysomelidae (Col.). In: *Chrysomelidae biology. 1. The classification, phylogeny and genetics*. Eds. P. Jolivet and M.L. Cox, pp 3–54. Amsterdam: SPB Academic Publishers.

SUZUKI, K. AND FURTH, D.G. (1992). What is a classification? A case study in insect systematics: potential confusion before order. *Zoological Science* **9** (6), 1113–1126.

SUZUKI, K. AND HARA, A. (1975). Supplementary report on the ovariole number in the family Chrysomelidae (Ins. Col.). *Journal Coll. Liberal Arts, Toyama Univ., Japan* **8**, 87–93.

SUZUKI, K. AND NAKAMURA, H. (1999). Rapid invasion of *Ophraella communa* LeSage, 1986 (Col. Chrys. Gal.) to Fukui and Ishikawa prefectures, Central Honshu, Japan. *Entomol. J. Fukui* **25**, 5–6.

SUZUKI, K. AND TANAKA, C. (1998). Distribution pattern of the sensilla of the hindwing veins of the family Chrysomelidae (Col.) and its systematic significance. *Proc. Fourth Int. Symp. Chrysomelidae. Proc. XX ICE , Firenze, 1996*. Pp 161–203. Torino: Mus. Reg. Sci. Nat.

SUZUKI, K. AND WINDSOR, D.M. (1999). The internal reproductive system of Panamian *Aulacoscelis* sp. (Col. Chrys. Aulacoscelinae) and comments on the systematic position of the subfamily. *Entomol. Sci.* **2** (3), 391–398.

SUZUKI, K. AND YAMADA, K. (1976). Intraspecific variation of ovariole number in some chrysomelid species. *Kontyu, Tokyo* **44** (1), 77–84.

SWAIN, T. (Ed.) (1963). *Chemical plant taxonomy*. Pp 1–543. London: Academic Press.

SZENTESI, A. (1985). Behavioural aspects of female guarding and inter male conflict in the Colorado potato beetle. In: *Proc. Symp. Colorado potato beetle. XVIII Int. Congress Entomol.* Eds. D. N. Ferro and R.H. Voss. *Agric. Exp. Station Res. Bull.* **104**, 127–137.

SZENT-IVANY, J.J.H., WOMERSLEY, J.S. AND ARDLEY, J.H. (1956). Some insects of *Cycas* in New Guinea. *Papua New Guinea Agric.* **11** (2), 53–56.

TAKHTAJAN, A.L. (1969). *Flowering plants. Origin and dispersal*. 310 pp. Edinburgh: Oliver and Boyd.

TAKHTAJAN, A.L. (1980). Outline of the classification of flowering plants (Magnoliophyta). *The Botan. Review* **46** (3), 225–359.

TAKHTAJAN, A.L. (1991). *Evolutionary trends in flowering plants*. 241 pp. New York, NY: Columbia University Press.

TAKHTAJAN, A.L. (1997). *Diversity and classification of flowering plants*. 643 pp. New York, NY: Columbia University Press.

TAKIZAWA, H. AND DACCORDI, M. (1998). Description of a new species of the genus *Gonioctena* Chevrolat from Japan (Col. Chrys.). *Entomol. Sci.* **1** (1), 105–108.

TALLAMY, D.W. (1984). Insect parental care. *BioScience* **34** (1), 20–24.

TALLAMY, D.W. (1985). Squash beetle feeding behavior: an adaptation against induced cucurbit defenses. *Ecology* **66** (5), 1574–1579.

TALLAMY, D.W. (1994). Nourishment and the evolution of paternal investment in subsocial arthropods. In: *Nourishment and evolution in insect societies*. Eds. J.H. Hint and C.A. Nalepa, pp 21–25. Boulder, Co., USA: Westview Press.

TALLAMY, D.W. (1999). Child care among the insects. *Scient. Amer.*, 50–55.

TALLAMY, D.W. AND HALAWEISH, F.T. (1993). Effects of age, reproductive activity, sex, and prior exposure on sensivity to cucurbitacins in southern corn rootworm (Col. Chrys.). *Environ. Entomol.* **22** (5), 925–932.

TANNER, V.M. (1927). A preliminary study of genitalia of female Coleoptera. *Trans. Amer. Entomol. Soc.* **53**, 5–50.

TAYADE, D.S. (1978). Bionomics of *Sagra femorata* Drury (Sagrinae, Col., Chrys.). *Mysore J. Agric. Sci.* **12** (4), 582–587.

TAYLOR, F.H.C. (1937). *The biological control of an insect in Fiji.* 239 pp. London, UK: Imperial Intitute of Entomology.

TELLA, R. (1952). Contribuçoes para a conhecimento do *Agathomerus sellatus* Germar (Col. Megal.). *R. Agric. Piracicaba* **27**, 373–376.

TEMPÈRE, G. (1935). Les Phanérogames Centrospermées et l'instinct botanique de quelques Coléoptères. *Proc. Verb. Soc. Linn. Bordeaux*, 1–5.

TEMPÈRE, G. (1946). L'instinct botanique des insectes phytophages. *L'Entomologiste, Paris* **2**, 219–224.

TEMPÈRE, G. (1967). Un critère méconnu des systématiciens phanérogamistes : l'instinct des insectes phytophages. *Trav. Biol. Vég. Prof. P. Dangeard, Le Botaniste* **50**, 473–482.

TEIXERA, C., MACEDO, M. VALVERDE DE AND MONTEIRO, R.F. (1999). Biology and ecology of the leaf-mining Hispinae *Octuroplata octopustulata* (Baly). In: *Advances in Chrysomelidae biology. 1*. Ed. M.L. Cox, pp 557–564. Leiden, The Netherlands: Backhuys Publishers.

THÉODORIDÈS, J. (1988). Grégarines of Chrysomelidae. In: *Biology of Chrysomelidae*. Eds. P. Jolivet, E. Petitpierre and T.H. Hsiao, pp 417–431. Dordrecht, The Netherlands: Kluwer Academic Publishers.

THIBOUT, E. (1982). Le comportement sexuel du Doryphore, *Leptinotarsa decemlineata* (Say) et son possible contrôle par l'hormone juvénile et les corps allates. *Behaviour* **80** (1–2), 199–217.

THIELE, H.U. (1973). Remarks about Mansingh's and Müller's classification of dormancies in insects. *Canad. Entomol.* **105**, 925–928.

THOMAS, F., OGET, E., GENTE, P., DESMOTS, D. AND RENAUD, F. (1999a). Assortive pairing with respect to parasite load in the beetle *Timarcha maritima* (Chrys.). *J. Evol. Biol.* **12**, 385–390.

THOMAS, F., GEATE, P., OGET, E., DESNOTS, D. AND RENAUD, F. (1999b). Parasitoid infection and sexual selection in the beetle *Timarcha maritima* Perris (Col. Chrys.). *Col. Bull.* **53** (3), 253–257.

THORNE, R.F. (1976). A phylogenetic classification of the Angiospermae. *Evol. Biol.* **9**, 35–106.

THORNE, R.F. (1992). An updated phylogenic classification of the flowering plants. *Aliso* **13** (2), 365–389.

THORNHILL, R. AND ALCOCK, J. (1983). *The evolution of insect mating systems.* 547 pp. Cambridge, USA: Harvard University Press.

THURSTON, G.S., YULE, W.N. AND DUNPHY, G.B. (1994). Explanations for the low susceptibility of *Leptinotarsa decemlineata* to *Steinerma carpocapsae*. *Biol. Control* **4** (1), 53–58.

TIWARY, P.N. AND VERMA, K.K. (1989). Studies on polymorphism in *Callosobruchus analis* (Col. Bruchidae). II. Endocrine control of polymorphism. *Entomography* **6**, 291–300.

TOGUEBAYE, B.S., MARCHAND, B. AND BOUIX, G. (1988). Microsporidia of the Chrysomelidae. In: *Biology of Chrysomelidae*. Eds. P. Jolivet, E. Petitpierre and T.H. Hsiao, pp 399–416. Dordrecht, The Netherlands: Kluwer Academic Publishers.

TROUVELOT, B. AND GRISON, P. (1946). L'alimentation et la croissance du Doryphore (*Leptinotarsa decemlineata* Say) sur divers organes de la pomme de terre et aux dépens de ses propres œufs. *C.R. Hebd. Séances Acad. Agric.* **32**, 320–323.

TURNEY, J. (2000). What fossils don't tell you. *New Scientist* **2231**, 46–47.

UHMANN, E. (1957–1958). *Coleopterorum catalogus suppl. Hispinae.* **135** (1–2), 1–398. The Hague, The Netherlands: Junk Publishers.

URIARTE, M. (2000). Interactions between Goldenrod (*Solidago altissima* L.) and its insect herbivore (*Trirhabda virgata*) over the course of succession. *Oecologia* **122**, 521–528.

VALVERDE, M. DE MACEDO, FERREIRA MONTEIRO, R. AND LEWINSOHN, T.M. (1994). Biology and ecology of *Mecistomela marginata* (Thunberg, 1821) (Hisp. Alurnini) in Brazil. In: *Novel aspects of the biology of Chrysomelidae*. Eds. P. Jolivet, M.L. Cox and E. Petitpierre, pp 567–571. Dordrecht, The Netherlands: Kluwer Academic Publishers.

VANDEL, A. (1931). *La parthénogénèse*. Pp 412. Paris: Doin Publishers.

VANDEL, A. (1932). La spanandrie, la parthénogénèse géographique et la polyploidie chez les Curculionides. *Bull. Soc. Ent. Fr.* **17**, 255–256.

VAN DYKE, E.C. (1953). *The Coleoptera of the Galapagos Islands*. 181 pp. San Francisco: Occ. Papers, Cal. Acad. Sci.

VARLEY, G.C. (1939). On the structure and function of the hind spiracles of the larva of the beetle *Donacia* (Col. Chrys.). *Proc. R. Ent. Soc. London (A)* **14** (9–12), 115–123.

VARMA, B.K. (1955a). Taxonomic value of spermathecal capsules as subfamily characters among the Chrysomelidae. *Indian J. Entomol.* **17**, 189–192.

VARMA, B.K. (1955b). Phylogenetic study of the family Chrysomelidae (Col.). *Curr. Sci.* **24** (1), 18–19.

VARMA, B.K. (1963). A study of the development and structure of the female genitalia and reproductive organs of the *Galerucella birmanica* Jac. (Chrys. Col.). *Indian J. Entomol.* **25**, 224–232.

VASCONCELLOS-NETO, J. (1988). Genetics of *Chelymorpha cribraria*, Cassidinae: colour patterns and their ecological meanings. In: *Biology of Chrysomelidae*. Eds. P. Jolivet, E. Petitpierre and T.H. Hsiao, pp 217–232. Dordrecht, The Netherlands: Kluwer Academic Publishers.

VASCONCELLOS-NETO, J. AND JOLIVET, P. (1988). Une nouvelle stratégie de défense: la stratégie de défense annulaire (cycloalexie) chez quelques larves de Chrysomélides brésiliens. *Bull. Soc. Ent. Fr.* **92** (9–10), 291–299.

VASCONCELLOS-NETO, J. AND JOLIVET. P. (1994). Cycloalexy among chrysomelid larvae. In: *Novel aspects of the biology of Chrysomelidae*. Eds. P. Jolivet, M.L. Cox and E. Petitpierre, pp 303–309. Dordrecht, The Netherlands: Kluwer Academic Publishers.

VASCONCELLOS-NETO, J. AND JOLIVET, P. (1998). Are Brazilian species of *Elytrosphaera* (Col. Chrys.), an apterous genus, threatened of extinction. *Proc. 4th Int. Symp. on Chrysomelidae*. Eds. M. Biondi, M. Daccordi and D.G. Furth, pp 299–309. Torino, Italy: Museo Regionale di Scienze Naturali.

VENCL, F.V. AND MORTON, T.C. (1999). Macroevolutionary aspects of larval shield defences. In: *Advances in Chrysomelidae biology. 1*. Ed. M.L. Cox, pp 217–238. Leiden, The Netherlands: Backhuys Publishers.

VERDYCK, P. (1999). Biochemical systematics of the *Phyllotreta cruciferae* complex (Col. Chrys. Alt.). *Ann. Entomol. Soc. Amer.* **92** (1), 30–39.

VERHOEFF, K.W. (1918). Zur vergleichenden Morphologie des Abdomens der Coleopteren und über die phylogenetischen Bedeutung derselben. *Zeitsch. für Wissenchaft. Zool.* **117**, 130–204.

VERMA, K.K. (1958). Torsion of the male genitalia of *Galerucella birmanica* (Col. Chrys.). *Ann. Mag. Nat. Hist.* **13** (I), 793–794.

VERMA, K.K (1969). Functional and developmental anatomy of the reproductive organs in the male of *Galerucella birmanica* Jac. (Col. Phyt. Chrys.). *Annls. Sci. Nat. (Zool.) Ser* **12, 11** (2), 139–234.

VERMA, K.K. (1975). A basilar membrane in the testis of *Aulacophora foveicollis* Luc. (Col. Phyt. Chrys.). **9**, 249–255.

VERMA, K.K. (1985). Male reproductive organs as taxonomic characters for a broad classification of Chrysomelidae. *Entomography* **3**, 485–487.

VERMA, K.K. (1992). Cycloalexy in the tortoise beetle, *Aspidomorpha miliaris* F. (Col. Chrys. Cass.). *Chrysomela* **26**, 6.

VERMA, K.K. (1994). 'Retournement' of the aedeagus in Chrysomelidae (Col.). In: *Novel aspects of the biology of Chrysomelidae*. Eds. P. Jolivet, M.L. Cox and E. Petitpierre, pp 355–362. Dordrecht, The Netherlands: Kluwer Academic Publishers.

VERMA, K.K. (1996a). *Brumus suturalis* Fabr. (Coccinellidae) and *Cryptocephalus ovulum* Suffr. (Chrysomelidae). *Chrysomela* **32**, 5.

VERMA, K.K. (1996b). Cycloalexy in leaf-beetles (Col. Chrys.). *Insect Environment* **2** (3), 82–84.

VERMA, K.K. (1996c). Inter-subfamily relations among Chrysomelidae as suggested by organisation of the male genital system. In: *Chrysomelidae biology. I. The classification, phylogeny and genetics*. Eds. P. Jolivet and M.L. Cox, pp 317–351. Amsterdam, The Netherlands: SPB Academic Publishers.

VERMA, K.K. (1999). Phylogeny of Chrysomelid subfamilies (Col.). A review. In: *Some aspects on the insight of insect biology*. Eds. Sobti and Yadav, pp 55–61. New Delhi, India: Nareadra Publishing House.

VERMA, K.K. AND JOLIVET, P. (2000). Phylogeny of Synetinae. Reconsidered. *Nouv. Rev. Ent. (N.S.), Paris* **17** (1), 35–49..

VERMA, K.K. AND KUMAR, D. (1972). The aedeagus its musculature and 'retournement' in *Aspidomorpha miliaris* F. (Col. Phyt. Chrys.). *J. Nat. Hist.* **6** (6), 699–719.

VERMA, K.K. AND SAXENA, R. (1996). The status of Bruchidae as a family. *Chrysomela* **32**, 3.

VERMA, K.K. AND SHRIVASTAVA, R.K. (1985). Separate niches for two species of *Aspidomorpha* (Col. Chrys.) living on *Ipomoea fistulosa*. *Entomography* **3**, 437–446.

VERMA, K.K. AND SHRIVASTAVA, R.K. (1986). A method for theoretical visualisation of niche, with particular reference to chrysomelid beetles. In: *Recent advances in insect physiology, morphology and ecology*. Eds. S.C. Pathak and Y.N. Sahai, pp 291–295. New Delhi, India: Today and Tomorrow's Printers and Publishers.

VERMA, K.K. AND SHRIVASTAVA, R.K. (1989). Differential distribution of enzyme activity in the mid-gut of *Aspidomorpha miliaris* F. (Col. Chrys.). *Entomography* **6**, 373–379.

VERMA, K.K. AND VYAS, M. (1987). Protective mimicry shown by some chrysomelids living on 'ber' shrubs. *Chrysomela* **16**, 5–6.

VINCENT, C. AND STEWART, R.K. (1986). Influence of trap color on captures of adult crucifer-feeding flea beetles. *J. Agric. Entomol.* **3** (2), 120–124.

VIRKKI, N. (1957). Structure of testis follicle in relation to evolution in the Scarabeidae (Col.). *Canad. J. Zool.* **35**, 265–277.

VIRKKI, N. (1980). Flea beetles, especially Oedionychina, of a Puerto Rican marshland in 1969–72. *Journal. Agric. Univ. P. Rico* **64** (1), 63–92.

VIRKKI, N. (1984). Additional observations on the life history of the Oedionychine flea beetles: *Alagoasa januaria* Bechyne. *J. Agric. Univ. Puerto Rico* **68** (1), 107–109.

VIRKKI, N. AND BRUCK, T. (1994). Unusually large sperm cells in Alticinae: their formation and transportation in the male genital system and their evolution. In: *Novel aspects of the biology of Chrysomelidae*. Eds. P. Jolivet, M.L. Cox and E. Petitpierre, pp 371–381. Dordrecht, The Netherlands: Kluwer Academic Publishers.

VIRKKI, N. AND ZAMBRANA, I. (1980). Demes of a Puerto Rican flea beetle, *Alagoasa bicolor* (L.), differing in mean body size and food plant association. *J. Agric. Univ. Puerto Rico* **64** (3), 264–274.

VIRKKI, N. AND ZAMBRANA, I. (1983). Life history of *Alagoasa bicolor* (L.) in indoor rearing conditions (Col. Chrys. Alt.). *Entomol. Arb. Mus. G. Frey* **31–32**, 131–155.

VLASOVA, V.A. (1978). A prognosis of the distribution area of the Colorado beetle in the Asiatic part. *Zassheita Rast.* **6**, 44–45.

VOGT, G.B., MCGUIRE, J.U. AND CUSHMAN, A.D. (1979). Probable evolution and morphological variation in South American Disonychine flea beetles (Col. Chrys) their Amaranthaceous hosts. *USDA Techn. Bull.* **1593**, 148 pp.

VOGT, G.B., QUIMBY, P.C. AND KAY, S.H. (1992). Effects of weather on the biological control of Alligator weed in the Lower MississipiValley Region. 1973–1983. *USDA Techn. Bull.* **1766**, 143 pp.

VOISIN, J.F. (1980). Notes on insects of Tristan da Cunha and Gough Island. *Ent. Month. Mag.* **116**, 253–255.

VOISIN, J.F. (2000). Essaimage de Doryphores, *Leptinotarsa decemlineata*. *Le Coléoptériste, Paris* **38**, 72.

WADE, M.J. (1994). The biology of the imported willow leaf beetle, *Plagiodera versicolora* (Laicharting). In: *Novel aspects of the biology of Chrysomelidae*. Eds. P. Jolivet, M.L. Cox and E. Petitpierre, pp 541–547. Dordrecht, The Netherlands: Kluwer Academic Publishers.

WADE, M.J. AND BREDEN, F. (1986). Life history of natural populations of the imported willow leaf beetle, *Plagiodera versicolora* (Col. Chrys.). *Ann. Entomol. Soc. Am.* **79** (1), 73–79.

WAGNER, T. (1998). Influence of tree species and forest type on the chrysomelid community in the canopy of an Ugandan tropical forest. *Proceedings of the 4th Int. Symp. on Chrysomelidae, Firenze 1996*. Eds. M. Biondi, M. Daccordi and D.G. Furth, pp 253–269. Torino, Italy: Museo Regionale di Scienze Naturali.

WAGNER, T. (1999). Arboreal chrysomelid community structure and faunal overlap between different types of forests in Central Africa. In: *Advances in Chrysomelidae biology. 1*. Ed. M.L. Cox, pp 247–270. Leiden, The Netherlands: Backhuys Publishers.

WALIGORA, D., WODA-LESNIEWSKA, M. AND KRZYMANSKA, J. (1993). The development of the larvae of Colorado potato beetle (*Leptinotarsa decemlineata* Say) on an artificial diet with the addition of some enzymes. *Roczniki Nauk Rolniczych. Seria E. Ochrona Roslin* **22** (1–2), 75–77.

WALLACE, J.B. (1970). The defensive function of a case on a chrysomelid larva. *J. Ga. Entomol. Soc.* **5**, 19–24.

WALLACE, J.B. AND BLUM, M.S. (1969). Refined defensive mechanisms in *Chrysomela scripta*. *Ann. Entomol. Soc. Amer.* **62**, 503–506.

WALLACE, J.B. AND BLUM, M.S. (1971). Reflex bleeding: a highly refined defensive mechanism in *Diabrotica* larvae. *Ann. Entomol. Soc. Amer.* **64** (5), 1021–1024.

WANG, S.Y. (1990). Primary discussion on the fauna of Hengduan mountains, China. *Acta Entomol. Sinica* **33** (1), 94–101.

WANG, S.Y. AND CHEN, S.H. (1981). Coleoptera Chrysomelidae Chrysomelinae. *Insects of Xizang* **1**, 509–516.

WARCHALOWSKI, A. (1995). *Chrysomelidae V. in Fauna Poloniae 17*. 359 pp. Warzawa: Polska Akad. Nauk. Muzeum i Inst. Zool. **21**, 135–216.

WEBB, L.A. (1986). Some insect pollinators of *Kunzea ambigua* (Sm.) Druce (Myrtaceae) near Sydney, New South Wales. *Victorian Nat.* **103** (1), 12–15.

WEINSTEIN, P. AND MAELZER, D.A. (1997). Leadership behavior in sawfly larvae, *Perga dorsalis*. *Oikos* **79**, 450–455.

WEIR, A. AND BEAKES, G.W. (1996). Biology and identification of species of *Laboulbenia* Mont. and C.P. Robin (Fungi Ascomycetes) parasitic of Alticine Chrysomelidae. In: *Chrysomelidae biology. 2 . Ecological studies*. Eds. P. Jolivet and M.L. Cox, pp 117–134. Amsterdam, The Netherlands: SPB Academic Publishers.

WHALLEY, P. (1987). Insect evolution during the extinction of the Dinosaurs. *Entom. Gener.* **13** (1–2), 119–124.

WHAPSHERE, A.J. (1982). Life histories and host specificities of the *Echium* flea beetles, *Longitarsus echii* and *L. aeneus*. *Entomophaga* **27** (2), 173–181.

WHAPSHERE, A.J. (1983). Discovery and testing of a climatically adapted strain of *Longitarsus jacobaeae* (Col. Chys.) for Australia. *Entomophaga* **28** (1), 27–32.

WHITE, R.E. (1996). Leaf beetles as biological control agents against injurious plants in North America. In: *Chrysomelidae biology. 2. Ecological studies*. Eds. P. Jolivet and M.L. Cox, pp 373–399. Amsterdam, The Netherlands: SPB Academic Publishers.

WICKLER, W. (1968). *Le Mimétisme animal et végétal*. 254 pp. Paris: Hachette Publishers.

WIEMAN, H.L. (1910a). A study of the germ cells of *Leptinotarsa signaticollis*. *J. Morph.* **21**, 135–216.

WIEMAN, H.L. (1910b). The degenerated cells in the testis of *Leptinotarsa signaticollis*. *J. Morph.* **21**, 485–495.

WIGGLESWORTH, V.B. (1972). *The principles of insect physiology*. 434 pp. London, U K: Chapman and Hall.

WILCOX, J.A. (1971–1975). *Chrysomelidae. Galerucinae. Col. Cat. Suppl.* 770 pp. 's-Gravenhage, The Netherlands: Junk Publishers **78**, 1–2–3–4.

WILCOX, J.A. (1979). *Leaf beetle host plants in Northeastern North America*. 30 pp. Kinderhook, USA: World Natural History Publishers.

WILF, P., LABANDEIRA, C.C., KRESS, W.J., STAINES, C.L., WINDSOR, D.M., ALLEN, A.L. AND JOHNSON, K.R. (2000). Timing the radiations of leaf beetles: Hispines on gingers from latest Cretaceous to recent. *Science* **289**, 2091–294.

WILKINSON, T. (1998). *Wolbachia* come of age. *TREE* **13** (6), 213–214.

WILLIAMS, C.E. (1991). Host plant latex and the feeding of *Chrysochus auratus* (Col. Chrys.). *Coleopt. Bull.* **45**, 195–196.

WILSON, E.O. (1992). *The diversity of life*. 424 pp. Cambridge, Mass., USA: Harvard University Press.

WILSON, G.W. (1993). The relationships between *Cycas ophialtica* K. Hill (Cycadaceae), the butterfly *Theclinesthes onycha* (Lycaenidae), the beetle *Lilioceris nigripes* (Col. Chrys.) and the ant *Iridomyrmex purpureus*. *Proc. Postgrad. Student Assoc. Symp. Univ. Centr. Queensland*, 53–57.

WILSON, H.F. AND LOVETT, A.L. (1911–1912). *Miscellaneous insect pests of orchard and garden*. Pp 160–161. Oregon Agric. College Exp. Station, Corvallis. Biennal Crop Pest and Horticultural Report for 1911–1912.

WILSON, H.F. AND MOZNETTE, G.F. (1913–1914). *Miscellaneous insect pests of orchard and garden*. Pp 96–101. Corvallis, OR: Oregon Agricultural College Experimental Station.

WILSON, S.J. (1934). The anatomy of *Chrysochus auratus* Fab. (Col. Chrys.), with an extended discussion of the wing venation. *J. New York Ent. Soc.* **42**, 65–84.

WINDIG, J.J. (1991). Life cycle and abundance of *Longitarsus jacobaeae* (Col. Chrys.) biocontrol agent of *Senecio jacobaea*. *Entomophaga* **36** (4), 605–618.

WINDIG, J.J. (1993). Intensity of *Longitarsus* jacobaeae herbivory and mortality of *Senecio jacobaea*. *J. Appl. Ecol.* **30**, 179–186.

WINDSOR, D.M. (1982). Advanced parental care and mate selection in a tropical tortoise beetle. In: *The biology of social insects*. Eds. M.D. Breed *et al.*, pp 182–183. Boulder, Co., USA: Westview Press.

WINDSOR, D.M. (1987). Natural history of a subsocial tortoise beetle, *Acromis sparsa* Boheman (Chrys. Cass.) in Panama. *Psyche* **94**, 127–150.

WINDSOR, D.M. AND CHOE, J.C. (1994). Origins of parental care in chrysomelid beetles. In: *Novel aspects of the biology of Chrysomelidae*. Eds. P. Jolivet, M.L. Cox and E. Petitpierre, pp111–117. Dordrecht, The Netherlands: Kluwer Academic Publishers.

WINDSOR, D.M., RILEY, E.G. AND STOCKWELL, H.P. (1992). An introduction to the biology and systematics of Panamian tortoise beetles. In: *Insects of Panama and Mesoamerica*. Eds. D. Quintero and A. Aiello, xxii+692pp. Oxford, UK: Oxford University Press.

WINDSOR, D.M., VALVERDE DE MACEDO, M. AND TOLEDO SIQUEIRA-CAMPOS, A. DE (1995). Flower feeding by species of *Echoma* Chevrolat (Col. Chrys. Cass.) on *Mikania* (Asteraceae) in Panama and Brazil. *Coleopt. Bull.* **49** (2), 101–108.

WINDSOR, D.M., TRAPNELL, D.W. AND AMAT, G. (1996). The egg capitulum of a neotropical walkingstick, *Calynda bicuspis*, induces aboveground egg dispersal by the ponerine ant, *Ectatomma ruidum*. *Journal of Insect Behavior* **9** (3), 353–367.

WINDSOR, D.M., NESS, J., GOMEZ, L.D. AND JOLIVET, P. (1999). Species of *Aulacoscelis* Duponchel and Chevrolat (Chrys.) and *Nomotus* Gorham (Languriidae) feed on fronds of Central American cycads. *Coleopt. Bull.* **53** (3), 217–231.

WOODS, W.C. (1916). Malpighian vessels of *Haltica bimarginata* Say (Col.). *Ann. Entomol. Soc. Amer.* **9**, 391–407.

WOODSON, W.D. (1994). Interspecific and intraspecific larval competition between *Diabrotica virgifera virgifera* and *Diabrotica barberi* (Col. Chrys.). *Environ. Entomol.* **23** (3), 612–616.

YAMASHIRO, C., ANDO, Y. AND MASAKI, S. (1998). Thermoperiod reduces the thermal constant required for oviposition in the leaf beetle *Atrachya menetriesi*. *Entomol. Sci.* **1** (3), 299–307.

YANG, K.K. AND YU, P.Y. (1994). Morphological adaptations to high altitude in *Galeruca* species in China. *Proc. 3rd Int. Symp. Chrysomelidae, Beijing, 1992*. Ed. D.G. Furth, pp 102–115. Leiden, The Netherlands: Backhuys Publishers.

YU, PEIYU (1977). On *Temnaspis nankinea* (Pic.) (Megal. Chrys.), a new pest of *Fraxinus chinensis* Roxb. *Acta Entomol. Sinica* **20** (4), 482–484.

YU, PEIYU (1988). A preliminary study of *Syneta adamsi* Baly. *Scientia Silave Sinicae* **24**, 235–238.

YU, PEIYU (1992). Hispidae. In: *Iconography of forest insects in Hunan China*. Eds. J. Peng and Y. Liu, pp 610–615. Hunan, China: Academia Sinica and Hunan Forestry Institute.

YU, PEIYU AND YANG, X. (1994). Biological studies on *Temnaspis nankinea* (Pic.) (Chrys. Megal.). In: *Novel aspects of the biology of Chrysomelidae*. Eds. P. Jolivet, M.L. Cox and E. Petitpierre, pp 527–531. Dordrecht, The Netherlands: Kluwer Academic Publishers.

YU, PEIYU, YANG, X. AND WANG, S. (1996). Biology of *Syneta adamsi* Baly and its phylogenetic implication. In: *Chrysomelidae biology. 3. General studies*. Eds. P. Jolivet and M.L. Cox, pp 201–216. Amsterdam, The Netherlands: SPB Academic Publishers.

ZACHARIASSEN, K.E. (1977). Ecophysiological studies on beetles from arid regions in East Africa. *Norw. J. Entomol.* **24** (2), 167–170.

ZACHARUK, R.Y., ALBERT, P.J. AND BELLAMY, F.W. (1977). Ultrastructure and function of digitiform sensilla on the labial palp of a larval elaterid (Col.). *Canad. J. Zool.* **55**, 569–578.

ZAITZEV, YU. M. (1982). Larva of *Sagra femorata* (Col. Chrys.) from Vietnam. *Zool. Zhurnal* **61** (3), 458–460.

ZAITZEV, YU. M. (1986). Larvae of *Lycaria westermanni* (Col. Chrys.) and taxonomic position of the genus. *Zool. Zhurnal* **65** (9), 11424–1427.

ZHANG, B.Y. AND LU, H.P. (1989). Study on the bionomics and control of *Podagricomela cuprea* Wang. *Acta Phytophylactica Sinica* **16** (3), 169–174.

ZHANG, Z.Q. AND MCEVOY, P.B. (1994). Attraction of *Longitarsus jacobaeae* males to cues associated with conspecific females. *Environ. Entomol.* **23** (3), 732–737.

ZHANG, Z.Q. AND MCEVOY, P.B. (1995). Responses of ragwort flea beetle *Longitarsus jacobaeae* (Col. Chrys.) to signals from host plants. *Bull. Entomol. Res.* **85**, 437–444.

ZHURAVLEV, V.N. (1993). Is the Colorado beetle always dangerous? *Zashchita Rastenii, Moscow* **5**, 6–8.

ZHURAVLEV, V.N. AND VERBA, YU. P. (1989). Forecasting the distribution of the Colorado potato beetle. *Zashchita Rastenii* **6**, 37.

ZIA, Y. (1936). Comparative studies of the male genital tube in Coleoptera Phytophaga. *Sinensia* **7** (3), 319–352.

ZULUETA, A. DE (1925). La herencia ligada al sexo en el Coleoptero *Phytodecta variabilis*. *Eos* **1**, 203–231.

ZULUETA, A. DE (1929). La mutacion 'jaspeado' del Coleoptero *Phytodecta variabilis* (su aparicion y herencia). *Mem. Real. Soc. Esp. Hist. Nat.* **15**, 819–824.

ZVEREVA, E.L., KOZLOO, M.V. AND NIEMELA, P. (1998). Effects of leaf pubescence in *Salix borealis* on host-plant choice and feeding behaviour of the leaf beetle, *Melasoma lapponica*. *Entom. Exper. Appl.* **89**, 297–303.

Subject Index

Taxonomic Index – Animals

Taxonomic Index – Plants